D0917070

SINGLE NEURON COMPUTATION

This is a volume in
NEURAL NETS: FOUNDATIONS TO APPLICATIONS

Edited by Steven F. Zornetzer, Joel Davis, Clifford Lau, and Thomas McKenna

SINGLE NEURON COMPUTATION

Edited by

THOMAS MCKENNA
JOEL DAVIS
STEVEN F. ZORNETZER

Office of Naval Research
Biological Intelligence Program
Arlington, Virginia

ACADEMIC PRESS, INC.
Harcourt Brace Jovanovich, Publishers
Boston   San Diego   New York
London   Sydney   Tokyo   Toronto

COPYRIGHT © 1992 BY ACADEMIC PRESS, INC.
ALL RIGHTS RESERVED.
NO PART OF THIS PUBLICATION MAY BE REPRODUCED OR
TRANSMITTED IN ANY FORM OR BY ANY MEANS, ELECTRONIC
OR MECHANICAL, INCLUDING PHOTOCOPY, RECORDING, OR
ANY INFORMATION STORAGE AND RETRIEVAL SYSTEM, WITHOUT
PERMISSION IN WRITING FROM THE PUBLISHER.

ACADEMIC PRESS, INC.
1250 Sixth Avenue, San Diego, CA 92101

United Kingdom Edition published by
ACADEMIC PRESS LIMITED
24–28 Oval Road, London NW1 7DX

Library of Congress Cataloging-in-Publication Data:

Single neuron computation / edited by Thomas McKenna, Joel
    Davis, Steven F. Zornetzer.
        p.   cm. — (Neural nets.   Foundations to applications)
    Includes bibliographical references and index.
    ISBN 0-12-484815-X
    1. Neural networks (Computer science)   2. Neural com-
    puters.
    I. McKenna, Thomas M.   II. Davis, Joel, 1948–   III.
    Zornetzer, Steven F.   IV. Series.
    QA76.87.S57   1992   91-31528
    006.3—dc20   CIP

PRINTED IN THE UNITED STATES OF AMERICA

92 93 94   9 8 7 6 5 4 3 2 1

Table of Contents

Contributors   ix

Preface   xiii

I.   COMPUTATION IN DENDRITES AND SPINES   1

Chapter 1.   **Electrotonic Models of Neuronal Dendrites and Single Neuron Computation**   7
William R. Holmes and Wilfrid Rall

Chapter 2.   **Canonical Neurons and Their Computational Organization**   27
Gordon M. Shepherd

Chapter 3.   **Computational Models of Hippocampal Neurons**   61
Brenda J. Claiborne, Anthony M. Zador,
Zachary F. Mainen, and Thomas H. Brown

Chapter 4.   **Hebbian Computations in Hippocampal Dendrites and Spines**   81
Thomas H. Brown, Anthony M. Zador,
Zachary F. Mainen, and Brenda J. Claiborne

Chapter 5.   **Synaptic Integration by Electro-Diffusion in Dendritic Spines**   117
Terrence J. Sejnowski and Ning Qian

Chapter 6.   **Dendritic Morphology, Inward Rectification, and the Functional Properties of Neostriatal Neurons**   141
Charles J. Wilson

Chapter 7. **Analog and Digital Processing in Single Nerve Cells:**
            Dendritic Integration and Axonal Propagation   173
            Idan Segev, Moshe Rapp, Yair Manor,
            and Yosef Yarom

Chapter 8. **Functions of Very Distal Dendrites: Experimental**
            and Computational Studies of Layer I Synapses on
            Neocortical Pyramidal Cells   199
            Larry J. Cauller and Barry W. Connors

II.   ION CHANNELS AND PATTERNED DISCHARGE,
      SYNAPSES, AND NEURONAL SELECTIVITY   231

Chapter 9. **Ionic Currents Governing Input–Output Relations**
            of Betz Cells   235
            Peter C. Schwindt

Chapter 10. **Determination of State-Dependent Processing**
            in Thalamus by Single Neuron Properties and
            Neuromodulators   259
            David A. McCormick, John Huguenard, and
            Ben W. Strowbridge

Chapter 11. **Temporal Information Processing in Synapses,**
            Cells, and Circuits   291
            Philip S. Antón, Richard Granger, and Gary Lynch

Chapter 12. **Multiplying with Synapses and Neurons**   315
            Christof Koch and Tomaso Poggio

Chapter 13. **A Model of the Directional Selectivity Circuit in Retina:**
            Transformations by Neurons Singly and in Concert   347
            Lyle J. Borg-Graham and Norberto M. Grzywacz

III.   NEURONS IN THEIR NETWORKS   377

Chapter 14. **Exploring Cortical Microcircuits: A Combined**
            Anatomical, Physiological, and Computational
            Approach   381
            Rodney J. Douglas and Kevan A. C. Martin

Chapter 15. **Evolving Analog VLSI Neurons**   413
            M. A. Mahowald

Chapter 16. **Relations between the Dynamical Properties**
            of Single Cells and Their Networks in Piriform
            (Olfactory) Cortex   437
            James M. Bower

Chapter 17. **Synchronized Multiple Bursts in the Hippocampus:**
            A Neuronal Population Oscillation Uninterpretable
            without Accurate Cellular Membrane Kinetics   463
            Roger D. Traub and Richard Miles

IV. MULTISTATE NEURONS AND STOCHASTIC MODELS
     OF NEURON DYNAMICS   477

Chapter 18. **Signal Processing in Multi-Threshold Neurons**   481
            David C. Tam

Chapter 19. **Cooperative Stochastic Effects in a Model**
            of a Single Neuron   503
            Adi R. Bulsara, William C. Schieve, and Frank E. Moss

Chapter 20. **Critical Coherence and Characteristic Times**
            in Brain Stem Neuronal Discharge Patterns   525
            Karen A. Selz and Arnold J. Mandell

Chapter 21. **A Heuristic Approach to Stochastic Models**
            of Single Neurons   561
            Charles E. Smith

Chapter 22. **Fractal Neuronal Firing Patterns**   589
            Malvin C. Teich

Index   627

Contributors

Numbers in parentheses indicate pages on which the authors' contributions begin.

Philip S. Antón (291), Bonney Center for the Neurobiology of Learning and Memory, University of California, Irvine, California 92717

Lyle J. Borg-Graham (347), Center for Biological Information Processing, Department of Brain and Cognitive Sciences, Massachusetts Institute of Technology, Cambridge, Massachusetts 02139

James M. Bower (437), Computation and Neural Systems Program, Division of Biology 216-76, California Institute of Technology, Pasadena, California 91125

Thomas H. Brown (61, 81), Department of Psychology, Yale University, New Haven, Connecticut 06510

Adi R. Bulsara (503), Naval Ocean Systems Center, Code 633, Materials Research Branch, San Diego, California 92152-5000

Larry J. Cauller (199), Section of Neurobiology, Division of Biology and Medicine, Brown University, Providence, Rhode Island 02912

Brenda J. Claiborne (61, 81), Division of Life Sciences, University of Texas at San Antonio, San Antonio, Texas 78285

Barry W. Connors (199), Section of Neurobiology, Box G-M, Division of Biology and Medicine, Brown University, Providence, Rhode Island 02912

Rodney J. Douglas (381), MRC Anatomical Neuropharmacology Unit, South Parks Road, Oxford, OX1 3QT, England

Richard Granger (291), Bonney Center for the Neurobiology of Learning and Memory, University of California, Irvine, California 92717

Norberto M. Grzywacz (347), Smith-Kettlewell Eye Research Institute, San Francisco, California 94115

William R. Holmes (7), Department of Zoological and Biomedical Sciences, Irvine Hall, Ohio University, Athens, Ohio 45701

John Huguenard (259), Stanford University School of Medicine, Department of Neurology, Stanford, California 94305

Christof Koch (315), Computation and Neural Systems Program, 216-76, Division of Biology, California Institute of Technology, Pasadena, California 91125

Gary Lynch (291), Bonney Center for the Neurobiology of Learning and Memory, University of California, Irvine, California 92717

M. A. Mahowald (413), Department of Computation and Neural Systems, California Institute of Technology, Pasadena, California 91125

Zachary F. Mainen (61, 81), Department of Psychology, Yale University, New Haven, Connecticut 06510

Arnold J. Mandell (525), Laboratory of Experimental and Constructive Mathematics, Departments of Mathematics and Psychology, Florida Atlantic University, Boca Raton, Florida, 33431

Yair Manor (173), Department of Neurobiology, Institute of Life Science Hebrew University, Jerusalem 91904, Israel

Kevan A. C. Martin (381), Neurobiology Research Center, University of Alabama at Birmingham, Birmingham, Alabama 35294

David A. McCormick (259), Yale University School of Medicine, Section of Neurobiology, 333 Cedar Street, New Haven, Connecticut 06510

Richard Miles (463), Institut Pasteur, Laboratoire de Neurobiologie Cellulaire, INSERM U261, 28 Rue du Dr. ROUX, 75724 Paris Cedex 15, France

Frank E. Moss (503), Physics Department, University of Missouri, St. Louis, Missouri 63121

Tomaso Poggio (315), Center for Biological Information Processing, Department of Brain and Cognitive Sciences and Artificial Intelligence Laboratory, Massachusetts Institute of Technology, Cambridge, Massachusetts 02139

Ning Qian (117), Department of Brain and Cognitive Sciences, Massachusetts Institute of Technology, Room E25-236, Cambridge, Massachusetts 02139

Wilfrid Rall (7), Mathematical Research Branch, National Institute of Diabetes and Digestive and Kidney Diseases, National Institutes of Health, Bethesda, Maryland 20892

Moshe Rapp (173), Department of Neurobiology, Institute of Life Science, Hebrew University, Jerusalem 91904, Israel

William C. Schieve (503), Physics Department and Center for Studies in Statistical Mechanics, University of Texas, Austin, Texas, 78712

Peter C. Schwindt (235), Department of Physiology and Biophysics, SJ-40, University of Washington School of Medicine, Seattle, Washington 98195

Idan Segev (173), Department of Neurobiology, Institute of Life Science, Hebrew University, Jerusalem 91904, Israel

Terrence J. Sejnowski (117), Computational Neurobiology Laboratory, The Salk Institute, P.O. Box 85800, San Diego, California 92186

Karen A. Selz (525), Laboratory of Experimimental and Constructive Mathematics, Departments of Mathematics and Psychology, Florida, Atlantic University, Boca Raton, Florida 33431

Gordon M. Shepherd (27), Section of Neurobiology, Yale University School of Medicine, 333 Cedar Street, New Haven, Connecticut 06510

Charles E. Smith (561), Department of Statistics, Biomathematics Program, North Carolina State University, Raleigh, North Carolina 27695-8203

Ben W. Strowbridge (259), Yale University School of Medicine, Section of Neurobiology, 333 Cedar Street, New Haven, Connecticut 06510

David C. Tam (481), Division of Neuroscience, Baylor College of Medicine, Houston, Texas 77030

Malvin C. Teich (589), Department of Electrical Engineering, Columbia University, 500 West 120 Street, New York, New York 10027

Roger D. Traub (463), IBM Research Division, IBM T. J. Watson Research Center, Yorktown Heights, New York 10598

Charles J. Wilson (141), Department of Anatomy and Neurobiology, University of Tennessee, Memphis, 875 Monroe Avenue, Memphis, Tennessee 38163

Yosef Yarom (173), Department of Neurobiology, Institute of Life Science, Hebrew University, Jerusalem 91904, Israel

Anthony M. Zador (61, 81), Department of Psychology, Yale University, New Haven, Connecticut 06510

Preface

The focus of this book reflects the confluence of two streams of research: biological neural networks and the biophysics of whole neurons. Artificial neural networks have been demonstrated to be useful in signal processing and pattern recognition tasks, yet to handle real world problems the nets must be scaled up or modularized, and they must be implemented in hardware. Artificial neural nets are based largely on their connection patterns, but they have very simple processing elements or nodes. There are theoretical limits to the scale and capacity of artificial neural nets. The large number of connections needed to implement such nets in hardware occupy a large fraction of the chip area. Now biological networks, incorporating features of real neurons and the connectivity of real neural nets, can be shown to exhibit theoretical advantages in scaling and time. This raises the following issues: What kind of processing element is a real neuron? What does a neuron compute? Why are there different kinds of neurons? Do they change their computational abilities under control of specific signals? Can we capture the essential computations of real neurons in models that will provide processing elements for the next generation of biologically inspired neural networks? What do real neuronal properties add to an artificial neural network?

The second stream of research is the biophysics of whole neurons. As outlined in several chapters in this book, the mathematical representation of neurons, historically based on the elaboration of cable equations and the Hodgkin–Huxley equations, is best applied to highly simplified neurons, in which the geometry of the neuron is collapsed into cylinders or spheres. As contemporary biophysicists began to address issues relating neuronal function to its morphology, and the interaction of that morphology with the ion channels in dendrites and dendritic spines, they came increasingly to rely on multi-compartmental simulations of neurons. Recent results from this line of research, presented in many chapters of this book, demonstrate that the dynamic interaction of inputs in dendrites containing voltage-sensitive ion channels is capable of realizing logical operations, nonlinear interactions, and local domains of computation. This raises the possibility that the neuron is itself a network. The work presented in this volume represents the first steps in the definition of the neuron as a computational device, or network of devices.

Unlike other books that survey current flow in cable models, ion channels, or general neural modeling, this volume has a unique focus on the computations performed in neurons. *Single Neuron Computation* represents the cutting edge of research from many outstanding investigators who have contributed chapters with both a broad perspective and sufficient technical detail to serve as a guide to contemporary neural modeling techniques.

The scope of contributions extends from derivations of the mathematics used to model synapses, membranes, and ionic channels in dendrites, to simulations of whole neuron computation, the role of neuron biophysics in biological networks, and models inferred from the statistics of neural spike trains. The book is divided into four sections:

  I. *Computation in Dendrites and Spines* introduces electrotonic modeling of realistic neurons and the interaction of dendritic morphology and voltage-dependent membrane properties on the processing of neuronal synaptic input.
 II. *Ion Channels and Patterned Discharge, Synapses, and Neuronal Selectivity* considers the neuronal properties underlying patterned discharge in single neurons, and the ability of neurons to respond selectively to temporal and spatial patterns of synaptic input.
III. *Neurons in Their Networks* examines the interaction of neuron properties and network properties.
 IV. *Multistate Neurons and Stochastic Models of Neuron Dynamics* consists of formal models of neurons based on stochastic approximations, physical system analogies, and nonlinear dynamical systems.

This collection should prove valuable to neuroscientists, biophysicists, and bioengineers, as well as neural network researchers (engineers and computer scientists) looking for new sources of processing elements for neural networks.

Thomas McKenna
Arlington, Virginia

PART I

COMPUTATION IN DENDRITES AND SPINES

This section introduces the topics of electrotonic modeling of realistic neurons and the interaction of dendritic morphology and voltage-dependent membrane properties on the processing of neuronal synaptic input.

The opening chapter by Holmes and Rall outlines the process of estimating the electrotonic structure (electric circuit equivalent) of neurons from their basic morphology and biophysical measurements. Many of the synapses on vertebrate central neurons contact small spines on the dendrites. Such an arrangement suggests the importance of spine–spine interactions in the local computations performed by neurons, pointing to the emergence of the concept of a neuron as a network or set of processors with a variable degree of interdependence. The authors demonstrate why computational models need to include mechanisms for the modification of computations at the neuronal level.

Shepherd (Chapter 2) traces the history of the neuron as a biological entity. He points out that the soma-centric view (that all inputs sum at the soma) led to the representation of neurons as single nodes, and that this oversimplification hindered the appreciation of the importance of dendrites as the basis for complex information processing in the neuron. He develops the concepts of a canonical neuron and the basic circuits built from these neurons. He then proposes the synapse to be the basic building block of the nervous system, with the spine as the smallest compartment capable of a synaptic input/output function. Dendritic subunits (spine clusters) are then seen as the intermediate level, and the entire neuron at the highest level is considered as a network capable of a range of input/output functions, including the possibility of multiple parallel information channels.

Brown, Claiborne, Mainen, and Zador contribute the next two chapters, which first show explicitly how morphometric and neurophysiological properties can be combined and incorporated into compartmental models that enable efficient simulations, and, secondly, consider the ways in which real neurons differ from the processing elements of artificial neural nets. The first of these chapters (Claiborne et. al.) is noteworthy for its history of morphometric techniques and its discussion of the development of efficient algorithms for finite difference approximations when used for simulation of branched, multi-compartmental model neurons. The second chapter (Brown et. al.) considers three categories of differences between processing elements and neurons that could be computationally significant. These are probability (synaptic release being stochastic), time (time course of synaptic potentials and kinetics of membrane conductance), and space (neuron dendrites and electrotonic structure). Simulations of hippocampal CA3 neurons with large numbers of synaptic inputs (NMDA-like Hebbian synapses on spines) reveal that clusters of enhanced synapses can emerge for some spatio-

3

temporal patterns. This important result implies that the electrotonic structure of neurons is a significant constraint in the self-organization of biological networks. Additionally, a possible mechanism for XOR (exclusive or) computations in dendrites is introduced, which involves clusters of potassium channels.

Sejnowski and Qian (Chapter 5) reinvestigate the possibility that shunting inhibition on spines could produce logical AND-NOT operations. They compare cable models of spines with electrodiffusion models based on Nernst–Planck equations. This approach allows them to account for the changes in equilibrium potential of ionic species due to the changing concentrations of ions in the small spaces within spines. They conclude that for K^--mediated inhibition, placing inhibitory synapses on the dendrite shafts at the base of spines produces effective shunting of excitatory inputs arriving at the spine head, but inhibitory synapses on the spine head would not be effective. However, for Cl^--mediation inhibition, a spine head location would cancel weak excitatory inputs.

Wilson (Chapter 6) examines realistic simulations of spiny projection neurons of the neostriatum. These neurons are bimodal, e.g., either totally quiescent or depolarized and bursting. He shows how the presence of a fast anomalous rectifier (a potassium current activated at hyperpolarized potentials) produces a situation in which the cell's membrane resistivity, time constant, and length constant are all functions of the membrane potential. This means that the effective electrotonic size of the neuron is a function of voltage, and this dependence gives rise to cooperative effects among simultaneously activated synapses. The neuron then becomes a bandpass filter for limited groups of coactive synapses.

Segev, Rapp, Manor, and Yarom (Chapter 7) consider models of Purkinje neurons and neocortical axons. After examining the input/output properties of realistic simulations of these neurons, they conclude that due to the partial electrical decoupling of different parts of the dendritic tree, each region performs a specific I/O function. They point out that since the EPSPs due to distal synaptic inputs peak in the soma 10 msec later than those due to proximal inputs, the neuron is well-suited to detect temporal relations between inputs arriving at these two sites, suggesting that the neuron is an adaptive delay line. They raise the possibility that, given the delays between dendritic events and axon events, including branch point failure, a network analogy for neurons may be more appropriate than a unitary processor.

Cauller and Connors (Chapter 8) consider the functions of the very distal dendrites of the large layer V pyramidal neurons in neocortex. Distal synaptic inputs evoked experimentally *in vitro* produce a long-lasting depolarization at the soma. This may reflect the existence of regenerative current in the apical dendrites, since simulations using passive dendrites show only weak effects on the soma. They point out that the electrical isolation of the apical distal dendritic tufts improves the effectiveness of layer I inputs, an argument that parallels recent concepts of the function of dendritic spines. They conclude that the distal apical tuft in layer I operates as an isolated, nonlinear integrator, whose effectiveness on influencing action potential production may be modulated by synaptic inputs along the apical trunk.

Chapter 1   Electrotonic Models of Neuronal Dendrites and Single Neuron Computation

WILLIAM R. HOLMES

Department of Zoological and Biomedical Sciences
Ohio University
Athens, Ohio

WILFRID RALL

Mathematical Research Branch
National Institute of Diabetes and Digestive and Kidney Diseases
National Institutes of Health
Bethesda, Maryland

I. Introduction

To design enhanced processor elements for use in neural nets, we are going to have to find a means to formalize the computations done by real neurons. Unfortunately, real neurons are very complicated, and it is going to be difficult to capture the essential computational properties of real neurons in any simple form.

If we aim to capture the essential computational properties of a real neuron in a model, there are several things we need (Table I). First, we must have knowledge of the basic morphological and electrotonic structure of the neuron. The morphology can be based on serial reconstructions, but the electrotonic

TABLE 1.

To capture the essential computational properties of a *real* neuron we need:
1. Knowledge of the basic *morphological* and *electrotonic structure* of the real neuron
2. To be able to reproduce the dynamic range of *computational possibilities* exhibited by the real neuron
3. A means to include *modification* of the computation done by the real neuron.

structure is more difficult to determine. Second, the model we develop must be able to produce the dynamic range of computational possibilities exhibited by the real neuron. The computational possibilities are determined by the different types and voltage dependencies of synaptic and non-synaptic inputs to the cell and their spatial and temporal patterns of activation. This is discussed in many of the other chapters. Third, the model must have some means to include synaptic modification. The computation done by the neuron may be much different at one time than at another, and there should be some means to include such changes. Hebbian modification will be discussed by Brown *et al.* in Chapter 4. There is a tremendous number of degrees of freedom available for constructing a model with these properties (Rall, 1990). It is a formidable task to reduce the number of degrees of freedom to a manageable level where useful and productive models can be developed.

In this chapter, we discuss work we have done in each of these three areas that might help to reduce the number of degrees of freedom. First, we discuss the importance for modeling studies of having good estimates of the electrotonic structure of a cell; problems with estimating the electrotonic structure of a cell based on the equivalent cylinder model are discussed, and we describe a method for estimating the electrotonic structure of a cell that can be applied to any neuron. Second, we explore the dynamic range of computational possibilities available to a neuron, by merely considering its possible *resting* states, assuming that a real neuron ever can be considered to be at rest. Finally, we discuss variables that may be important for producing modification in dendritic spines. The focus here will be on the electrical and diffusional resistance of the spine neck.

II.   Estimating the Electrotonic Structure of a Cell

In constructing a model of a real neuron, we first must make some appropriate assumptions about the morphological and electrotonic structure of the neuron. For the morphological structure, do we have the lengths and diameters of every dendritic segment? If so, how should these values be corrected for tissue shrinkage? If not, what simplified structure should we use? Are there dendritic spines on the cell? If so, how should these be included? As for the electrotonic structure, what values should we use for membrane resistivity, intracellular resistivity, and membrane capacity? Should these values be uniform or nonuniform throughout the cell? There is a large number of degrees of freedom for constructing a model, but if one is to develop a useful model, this large number must be reduced appropriately. With too many unknowns, the model can become intractable, and results may not be understandable or interpretable.

A.   Importance of Membrane Resistivity Values in Models

We illustrate the importance of reducing the number of degrees of freedom appropriately by considering the effect that the use of different membrane resistivity values (R_m values) has on computed somatic postsynaptic potentials (or EPSPs). The somatic potential changes, due to brief synaptic conductance changes at distal and mid-dendritic locations in a cortical pyramidal cell, were computed when R_m was 5,000, 20,000, or 80,000 Ωcm^2. As shown in Fig. 1a, the peak of the EPSP at the soma was three times larger when R_m was 20,000 Ωcm^2 than when R_m was 5,000 Ωcm^2; this difference grew to sixfold when R_m was 80,000 Ωcm^2. The membrane resistivity value also affected the time course of the soma potential. When R_m was 5,000 Ωcm^2, the potential had decayed back to rest by 30–40 ms. However, when R_m was 20,000 Ωcm^2, the potential had decayed only to 50% of its peak value at 30 ms; when R_m was 80,000 Ωcm^2, the potential was still 70–80% of its peak value at 50 ms. For the mid-dendritic input, there were small differences in the peak soma potential for the three R_m values (Fig. 1b). Of course, these small differences would have been much larger had the conductance change been longer, but for a brief conductance change at a proximal or mid-dendritic synapse, differences in peak potential with the use of different R_m values are small. A more important difference with the mid-dendritic input was that, as with the distal input, the decay time course was much slower when higher R_m values were used. From these examples, it should be obvious that the value chosen for membrane resistivity can make a tremendous difference in the computation done by a cell, particularly with regard to distal synapses.

B.   Limitations of the Equivalent Cylinder Model for Estimating Cell Parameters

Given the importance of the resistivity parameters, how does one go about choosing appropriate values? The equivalent cylinder model can and has been used in conjunction with measurements from experimental transients to provide

A. Distal input

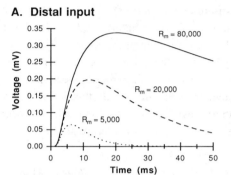

B. Mid-dendritic input

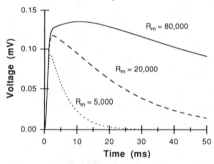

FIGURE 1.   Effect of different membrane resistivity values on EPSPs observed at the soma. A 2 nS conductance change lasting 1 ms was modeled at 10 distal dendritic spines (a) and at one mid-dendritic spine (b). The voltage changes at the soma are shown when membrane resistivity was 5,000, 20,000, and 80,000 Ωcm². The cell modeled is the cell pictured in Fig. 3, and the distal and mid-dendritic input locations are those labeled D and M in Fig. 3a. Values for intracellular resistivity and membrane capacity were 70 Ωcm and 1.0 μF/cm², respectively.

estimates of membrane resistivity R_m and the electrotonic length L of the dendritic tree in many neurons, and these values have been used in many modeling studies, including our own. In the equivalent cylinder model (Rall, 1962), the morphology of the dendritic tree is assumed to be equivalent to a cylinder, and the membrane and intracellular resistivity are assumed to be uniform. In a cylinder with uniform resistivity, the passive voltage response following a long current step can be described as a sum of exponentials (Rall, 1969; Rall, 1977):

$$V(x,t) = V_\infty(x,t) - \sum_{i=0}^{\infty} C_i(x) \exp\left(-\frac{t}{\tau_i}\right). \qquad (1)$$

The first two time constants, τ_0 and τ_1, and their coefficients, C_0 and C_1, can often be estimated from experimental transients. These parameters can be used to compute the electrotonic length of a cell; for example, the formula:

$$L = \frac{\pi}{(\tau_0/\tau_1 - 1)^{1/2}} .$$

(2)

The membrane resistivity R_m can be estimated from $\tau_0 = \tau_m = R_m C_m$, assuming R_m is uniform and C_m, the specific membrane capacitance, is known. (C_m is usually assumed to be 1.0 $\mu F/cm^2$.)

However, the electrotonic length and membrane resistivity estimates obtained with procedures based on the equivalent cylinder model are strictly valid only for neurons that can be approximated as equivalent cylinders with uniform membrane resistivity. Over the last several years, evidence has been accumulated suggesting that many types of neurons cannot be well approximated as equivalent cylinders with uniform membrane. Dendrites do not all end at the same electrotonic distance. Some cells have dendritic spines. Membrane resistivity may not be uniform, either because of a genuine difference in membrane resistivity between the soma and the dendrites, or because of a soma shunt produced by electrode penetration into the cell. Then there is the issue of why membrane resistivity estimates obtained with whole-cell patch clamp recordings are fourfold or more higher than estimates obtained with intracellular recordings. If the whole-cell patch clamp measurements are right, then the intracellular recordings may include a significant soma shunt. Given these facts, it should not be surprising that the *mismatch* of equivalent cylinder formulae to non-equivalent cylinder neurons can produce erroneous estimates of electrotonic length and dendritic membrane resistivity. (See Holmes *et al.* 1991; Holmes and Rall, 1991a,b.)

A problem with many of the equivalent cylinder formulae, including the commonly used Eq. (2), is that the parameters used in the formulae either cannot be estimated accurately or they have an interpretation that is different in cells that are not equivalent to cylinders. For example, if membrane resistivity is not uniform, then τ_0 cannot be equated with τ_m or the product $R_m C_m$; doing so will give an erroneous R_m estimate. Erroneous L estimates can come from use of Eq. (2) in non-equivalent cylinder cells. Equation (2) uses τ_1, but τ_1 has a different interpretation for a cylinder than for a branched non-equivalent cylinder cell (Holmes *et al.*, 1991). Very often, the τ_1 estimated from a voltage transient by exponential peeling (or other methods) bears little or no resemblance to the computed theoretical τ_1 or any other τ_i.

For example, we have peeled time constants from transients generated for cortical pyramidal cells and motoneurons. These transients were generated with the use of Eq. (1) and analytic values for the time constants and coefficients (computed with the *forward computation* described shortly), given the morphology of the neurons and values for R_m, R_i, and C_m. Although the peeled τ_0 and theoretical τ_0 values were similar, the peeled and theoretical τ_1 values were quite different. We should note here that because our peeled estimates come from computed transients, not experimental transients, our estimates were not affected by experimental noise or the activation of voltage-dependent conductances. After τ_0, there were a number of long theoretical time constants, but methods for estimating exponentials from data could not resolve them. For the cortical pyramidal cell, the τ_1 obtained by peeling was somewhere between the actual τ_3 and the actual τ_4. It seemed to be an average of τ_{1-5} (Holmes and Woody, 1989). In the motoneuron, which did not have a dominant long apical dendrite, the peeled τ_1 was approximately equal to the actual τ_{64} (Holmes *et al.*, 1991).

The value of τ_1 estimated from an experimental transient depends on how well the data resembles a perfectly passive noise-free transient and on the method used to estimate time constants from data. Unfortunately, all methods for estimating time constants from data have inherent problems that prevent the actual τ_1 (or a useful τ_1, as seen in Holmes *et al.*, 1991) from being resolved from the data in many cases.

C.   Estimating Electrotonic Structure with Compartmental Models

How can one estimate electrotonic length and resistivity parameters in cells that are not equivalent to cylinders? The approach that we have developed makes use of compartmental models as described by Rall (1964) and in greater detail by Perkel *et al.* (1978, 1981). As has been shown many times before, compartmental models can be used to get the potential in the neuron by solving a system of differential equations of the following form:

$$\frac{dV}{dt} = AV + b, \tag{3}$$

where V is the vector of membrane potentials, A is a matrix of coefficients that includes terms describing coupling between neighboring compartments, and b is a vector describing injected currents.

We apply compartmental models in two different ways, which we call the *forward computation* and the *inverse computation*. In the forward computation, we specify values for all parameters of the model, such as R_m, R_i, and C_m, and we compute a response. The response includes parameters that can be measured or estimated from experimental data, such as time constants and coefficients from voltage transients, τ_0, τ_1, C_0, and C_1, the input resistance R_N, and the first time constant of a current transient following voltage clamp, τ_{vc1}. With the forward computation, we can compute not an infinite number of time constants and coefficients as expressed in Eq. (1), but N time constants and N coefficients, where N is the number of compartments in the model. The time constants come from the eigenvalues of the matrix $\mathbf{A}$ in Eq. (3), and the coefficients come from the eigenfunctions as shown by Perkel *et al.* (1981). In a similar manner, we can compute the time constants in a current transient following voltage clamp. Thus, for a given morphology and given values for R_m, R_i, and C_m (whether they be uniform or nonuniform in the cell), we can compute values for a number of parameters that can be estimated experimentally.

It can become tedious to keep choosing different values for R_m, R_i, and C_m in the forward computation to match computed and experimentally estimated values of τ_0, C_0, τ_{vc1}, and R_N. This is where the inverse computation comes into play. With the inverse computation, estimates of a set of unknown electrotonic parameters are computed from knowledge of a set of parameters that can be estimated experimentally. Given the morphology of a cell, a useful set of computed values to compare with experimentally estimated values can be chosen from τ_0, C_0, τ_{vc1}, and R_N. For example, we may want to estimate R_m, R_i, and C_m (assuming these parameters are uniform throughout the cell), given τ_0, C_0, and R_N. We do this by providing initial guesses for R_m, R_i, and C_m, and computing values for τ_0, C_0, and R_N with the forward computation. The computed values are compared with the experimental values and the guesses for R_m, R_i, and C_m are updated according to a Newton–Raphson algorithm. This procedure proceeds iteratively until there is convergence or no progress toward a solution. We have tried this method on hypothetical neurons and on actual neurons where τ_0 and R_N were known (and sample C_0 and τ_{vc1} were chosen) and have found it to be quite useful for providing estimates for the unknown parameters (Holmes and Rall, 1991b). However, we have yet to use it with experimental data in which four parameters have been estimated experimentally.

We emphasize again here that compartmental models can have thousands of degrees of freedom. We have a great deal of flexibility in how we use the inverse computation. In the inverse computation, we restrict the number of degrees of freedom to just a few and we try to get good estimates of these. In the example in the previous paragraph, we assumed that R_m was uniform. However, in a different inverse computation, we can let R_{ms} be the soma membrane resistivity and R_{md} be the dendritic membrane resistivity. Then we could estimate R_{ms}, R_{md}, R_i, and C_m, given experimental values of τ_0, C_0, τ_{vc1}, and R_N. Different results will be obtained depending on whether one assumes R_m is uniform or nonuniform; however, generally accepted physiological bounds on the parameters to be estimated could lead one to select one set of results over another.

It is important to be specific about the assumptions one makes. One could assume that there is a step increase in membrane resistivity or that membrane resistivity increases linearly (or sigmoidally) away from the soma, for example, as in models studied by Fleshman *et al.* (1988). The inverse computation can be applied to either situation. What it will do is give R_{ms} and R_{md} if one assumes an electrotonic model with a step increase in resistivity and it will give the starting R_m value and the rate of increase if one assumes an electrotonic model with a linear (or sigmoidal) increase in R_m. The inverse computation may provide physiologically acceptable results with either assumption about membrane resistivity; in this case, the choice of which assumption is correct must be based on other knowledge about the cell. The inverse computation gives tremendous flexibility, and can give more realistic values for the electrotonic parameters for a given set of constraints than the equivalent cylinder model, but values will depend on the electrotonic model assumed.

D.   Use of a Simplified Dendritic Geometry with the Inverse Computation

To use the forward or inverse computations, the lengths and diameters of all of the processes obtained from a complete reconstruction can be used if available, but there might be computational limitations preventing one from doing this. Finding the inverse of a 700-by-700 matrix is a simple matter for a supercomputer and many advanced workstations, but it may take much computer time and memory on other machines. For many investigators, working with a reduced or simplified morphology is a necessity. Clements and Redman (1989) reduce motoneuron dendritic trees to what they call *equivalent dendrites*. Similarly, Fleshman *et al.* (1988) plot a normalized $d^{3/2}$ as a function of electrotonic distance for given R_m and R_i values; such a plot gives a condensed representation of motoneuron morphology. Stratford *et al.* (1988) have used a cortical pyramidal cell *cartoon* to provide a simplified morphology of these cells. The advantage of reduced or simplified models with 20–50 compartments is that the inverse computation can be run with these models interactively. Once the reduced models have provided some estimates, these estimates then can be used as starting guesses in the larger inverse computation, which uses the full morphology.

E.   Including Dendritic Spines in a Model

Dendritic spines pose a special difficulty for computational models. If spines are ignored in a model, then the model may be ignoring as much as 50% of the total membrane area. Ignoring this much membrane in the inverse computation will lead to erroneous estimates of electrotonic parameters. Spines pose problems for general models as well. Suppose the dendritic morphology and the estimates for the electrotonic structure of a neuron are both available. If the model uses the dendritic morphology and the correct estimates for the electrotonic parameters, but does not include spines, then the computations done by that model will be erroneous.

For example, we have constructed a hypothetical cortical pyramidal neuron containing a large degree of symmetry, which allowed us to model explicitly almost 14,000 dendritic spines with *only* 714 dendritic segments (Fig. 2a). The voltage response at the soma to a 5 nS conductance change at one dendritic spine lasting 0.5 ms was modeled when the 14,000 spines were included or excluded from the model. As shown in Fig. 2b, the peak voltage response when spines were excluded was more than twice that when spines were included.

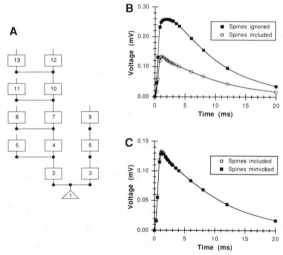

FIGURE 2.   Including dendritic spines in models. (a) A hypothetical cortical pyramidal cell. Regions 5, 8, 11, and 13 represent branches off the apical dendrite (regions 2, 4, 7, 10, and 12), and regions 3, 6, and 9 represent the basilar tree. Regions 5, 8, 11, and 13, as well as regions 3, 6, and 9, were composed of many dendritic branches. All branches within a region were identical. Spines were modeled on dendrites in regions 4–13 at intervals of 2, 4, or 12 μm, and any number of spines could be modeled at each location on a branch within a region. Because all branches and their spines were identical within a region, compartments had to be coded only for one complete branch within a region. This symmetry allowed us to explicitly model almost 14,000 spines with only 714 segments. (b) Voltage transients at the soma due to a 5 nS conductance change lasting 0.5 ms at a spine located near the distal end of region 7 when spines were included or ignored by the model. (c) The voltage transients at the soma when spines were included or mimicked by increasing C_m and reducing R_m according to spine membrane area.

Any spine that receives synaptic activation will have to be explicitly included in a model because of voltage attenuation through the spine neck. However, the tens of thousands of spines that do not receive synaptic activation in a model can be included, without actually coding separate compartments for each spine, by either of two methods. The first method is to reduce R_m and increase C_m on each dendritic segment according to the proportion of membrane area due to spines (Holmes, 1986). For example, if spines comprise half the membrane area of a given dendritic segment, then one can halve R_m and double C_m for that particular segment. A second method for including spines, which might be more useful if one does not want to use modified R_m and C_m values, is a method that has been used by Stratford et al. (1988). In their method, R_m and C_m are unchanged, but the segment length and diameter are increased so that the membrane area of the new segment equals the sum of the spine membrane area and the original segment membrane area. The length and diameter are increased in such a way that length divided by diameter squared is the same as for the original dendritic segment.

The usefulness of the first approximation method is demonstrated in Fig. 2c. Here, we have plotted the somatic voltage response to a brief synaptic conductance change when spines were included explicitly and when the presence of spines was mimicked by reducing R_m and increasing C_m. The two curves almost overlap exactly.

The two methods presented for including spines are equivalent and one of them should be used when one is working with real neurons that have spines, whether one is doing the inverse computation described previously or any other modeling of cells with spines.

To summarize this section, dendritic models concerned with computation must make assumptions about the morphological and electrotonic structure of the neuronal dendrites. It is important to have accurate estimates of parameters that determine the electrotonic structure of a neuron, and we have outlined a method to do this that can be used with a full morphology or a reduced morphology. Finally, we have described simplifications one can use to include dendritic spines in models.

III.   The Dynamic Range of Computational Possibilities Exhibited by Neurons

Once we have an approximate morphological and electrotonic structure for our model neuron we can explore the neuron's dynamic range of computational possibilities. The dynamic range of computational possibilities depends on the spatial distribution and temporal patterns of activation of synaptic and non-synaptic conductances in the neuron. Other authors will discuss some of the computational possibilities that exist when one considers voltage-dependent conductances. In what follows, we will consider only passive conductances. We will further restrict our discussion to some of the possible implications of uniform or nonuniform activation of uniformly or nonuniformly distributed ionic conductances.

A. What is the Resting State of a Neuron?

We begin by asking the question: What is the actual *resting* state of the neuron? This seems like such a simple question, but there really is no simple answer. When we measure a resting potential at the soma, we assume that the neuron is in some resting state. However, the state of a neuron at a given moment, whether or not one considers it to be the resting state (if any state can truly be called the resting state), is determined by the distribution of activated synaptic and non-synaptic conductances in the neuron at that moment.

It is well-known that the distribution of different types of ionic channels is nonuniform in many neurons. For example, in cortical pyramidal cells, we know that inputs to the soma are almost exclusively inhibitory, whereas inputs on dendritic spines are primarily excitatory, and channels mediating inhibition at the soma are permeable to a different set of ions than channels in dendritic spines. Furthermore, inputs to different laminae come from different sources. There is no reason to expect that afferents from different sources, terminating in different laminae, will have similar temporal patterns of activation or will open or close the same type of ionic channels. If a cell is constantly being bombarded by both excitatory and inhibitory inputs from many different sources, then what is the actual resting state of the cell?

Suppose with an electrode in the soma of a cortical pyramidal cell we measure a resting potential of -60 mV and an input resistance of 12 MΩ. This alone tells us little about the resting state of the cell. If excitatory and inhibitory inputs are distributed and activated uniformly, as we usually assume, then the resting potential will be uniform throughout the cell, or -60 mV; but what if these tonically activated synaptic conductances are not distributed uniformly or are not activated uniformly throughout the cell. In this case, we would expect to find a nonuniform resting potential in the cell.

Consider the example of a cortical pyramidal cell as shown in Fig. 3, given the constraints that the soma resting potential is -60 mV and the input resistance is 12 MΩ. If excitatory conductances are located primarily on dendritic spines and inhibitory conductances are located primarily on the soma and proximal

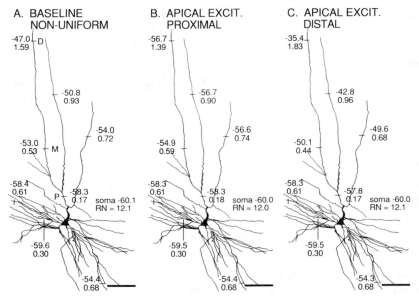

FIGURE 3.   Resting potential and electrotonic distance for different distributions of ton-ically active conductances. At the indicated locations, the upper number is the *resting* potential and the lower number is the electrotonic distance from the soma. (a) Excitatory conductances were activated uniformly on dendritic spine heads, and inhibitory con-ductances were activated uniformly over the soma and proximal dendrites. (b) As in a, except that excitatory conductances on apical dendritic spines were almost exclusively restricted to proximal spines. (c) As in a, except that excitatory conductances on apical dendritic spines were almost exclusively restricted to distal spines. Conductances on basilar dendrites were unchanged. Scale bar is 100 μm. From Holmes and Woody (1989, Fig. 1, p. 15).

dendrites, and if these conductances are activated uniformly, then one might find a potential distribution similar to the one in Fig. 3a. Here, the tips of the basilar dendrites are up to 6 mV depolarized relative to the soma, and the tips of the apical dendrites are up to 15 mV depolarized relative to the soma. We are considering only passive conductances here, but if conductances in the den-drites and spines are voltage-dependent, then such differences might be extremely significant.

On the other hand, what if the resting-state tonic excitation to just the apical tree were restricted primarily to the proximal apical dendrites or primarily to the distal apical dendrites. Then, given the constraints that the resting potential is − 60 mV and the input resistance is 12 MΩ, the resting states, in these cases, would resemble those shown in Figs. 3b and 3c. With apical excitation restricted to proximal regions, the resting potential differences are reduced. However, with apical excitation restricted to distal regions, the difference between soma and distal dendritic potentials was 25 mV or more. Which state is the true resting state of the neuron? If we could get recordings from distal dendrites, we might be able to settle this issue. The few intracellular recordings from dendrites that have been reported seem to indicate that the dendrites are depolarized relative to the soma, but it is difficult to know how much of this is due to shunting (injury) because of the difficulty of penetrating small-diameter processes and how much of it is actual depolarization. However, the fact that inhibitory and excitatory conductances have very different distributions in a cell would seem to indicate that nonuniform resting potentials are to be expected.

B.   *Nonuniform Conductance Activation and Computation*

What consequences do nonuniform distributions of activated conductances have for neuronal computation? We have looked at the somatic response to proximal and distal synaptic inputs, given the three nonuniform distributions of activated conductances shown so far. Responses to proximal inputs were nearly the same in the three cases, but responses to the distal input were quite different as shown in Fig. 4. The differences occurred because of the differences in driving force (because of the different resting potential at the input site) and the differences in electrotonic distances to the soma in the three cases. When tonic apical excitation was primarily proximal, the distal input produced a much larger change in potential at the soma than when tonic apical excitation was primarily distal. Thus, the resting state of the neuron had a large effect on the efficacy of distal inputs.

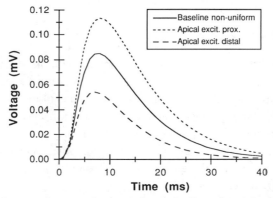

FIGURE 4.   Voltage response at the soma due to 2 nS conductance changes lasting 1.0 ms on 10 dendritic spines at location D in Fig. 3a for three different distributions of activated conductances. From Holmes and Woody (1989, Fig. 2B, p. 16).

What if the resting state can switch among the three distributions of activated conductances described so far? If this were true, then the effectiveness of a distal input will vary over a wide range depending on the activity in the cell at the time of the input. If we are to develop a simple model of neuronal computation that relies on synaptic weighting functions between neurons, then these weighting functions cannot be simple constant functions that might be modified by a synaptic modification rule. These weights, at least for distal synapses, should be allowed to change dynamically with the state of the neuron.

To summarize this section, the dynamic range of computational possibilities for a neuron is immense. Our simple example regarding the range of possibilities for the efficacy of distal inputs barely begins to explore this area. One of the first things we should do, again to restrict our degrees of freedom, is to decide on possible resting states for our model neuron. Given the possible resting states, we can assign variable synaptic weighting functions and then further explore the range of computational possibilities.

IV.   Synaptic Modification in Dendritic Spines

If we have a morphological and electrotonic structure for our neuron, and we know the resting states of the neuron and its range of computational possibilities, then we need some means to incorporate synaptic modification. This brings us to the final subject, synaptic modification in dendritic spines. Dendritic spines are thought to play an important role in synaptic transmission and plasticity. (See also Chapter 4 by Brown *et al.*) What we need to do is to determine a set of parameters that could be used in some rule to describe synaptic modification.

A.   Spine Neck Electrical Resistance

Electrotonic models of dendritic spines, beginning with the studies of Rall and Rinzel some 20 years ago (Rall and Rinzel, 1971a,b), have focused on the large electrical resistance provided by the narrow constriction of the spine neck. These theoretical studies showed that over a certain range of values, changes in the spine neck resistance could produce significant changes in synaptic efficacy. Thus, it might seem that one way to incorporate modification into a model would be to dynamically change the electrotonic structure of spines as a function of synaptic activity, for example.

However, measurements of spine dimensions raise doubts about this being the way one should proceed. For example, if the dimensions of long-thin spines as measured and categorized in dentate granule cells (Desmond and Levy, 1985) are used in computations of spine neck resistance, we find that when R_i in the spine neck is assumed to be 70 Ωcm, spine neck resistance is only about 70 MΩ. This value is too small for increases in spine neck diameter to produce any significant increase in synaptic efficacy. There are reasons to believe that R_i may be larger in spines than dendrites, and hence that spine neck resistance might be larger; for example, the spine apparatus and other organelles may make R_i larger, but such differences remain to be measured. Even if R_i is larger and spine neck resistance is within the range required for changes in spine neck resistance to produce changes in efficacy, it must be demonstrated that changes in spine neck resistance actually take place. Spine neck resistance is most sensitive to changes in spine neck diameter, but there has been much debate as to whether changes in spine neck dimensions actually take place. Unless it is determined that R_i is much higher in spines than in dendrites or that spine necks are much longer and thinner in other types of cells, then dynamically changing the electrotonic structure of spines in a model according to synaptic activity, or another variable, may not be the way one should incorporate modification in a model.

B.   Spine Neck Diffusional Resistance

Besides providing an electrical resistance to current flow, the thin spine neck provides a diffusional resistance to the flow of ions and molecules. Gordon Shepherd mentioned this in the 1979 edition of his book (Shepherd 1979) and Brown *et al.* discuss this in Chapter 4, as do Sejnowski and Qian in Chapter 5.

Because of the role of calcium in long-term potentiation, much attention has focused on the calcium concentration in dendritic spines. Gamble and Koch (1987) showed how large, transient increases in spine head calcium concentration might be attained with repetitive input. Large increases in spine head calcium concentration are restricted to the spine head in long-thin spines because of the diffusional resistance of the spine neck (Holmes, 1990). Perhaps a synaptic modification rule based on spine head calcium concentration would be more appropriate than dynamically changing spine dimensions as a means for incorporating synaptic modification.

C.   Spine Shape and Spine Head Calcium Concentration

What controls spine head calcium concentration? We have found three potentially important variables to be spine shape, calcium buffer concentration, and the strength and duration of the synaptic calcium current (Holmes, 1990). One might expect calcium buffer concentration and the strength and duration of the calcium current to be important, but the degree to which spine shape, and particularly

the diffusional resistance of the spine neck, is critically important may be surprising, and that will be the focus of the present discussion.

The model we have used is a compartmental model with similarities to and
differences from the Gamble and Koch model, and it is described in an earlier
paper (Holmes and Levy, 1990). We have modeled three different spine shapes
corresponding to long-thin, mushroom-shaped, and stubby spines as measured
and categorized from dentate granule cells by Desmond and Levy (1985). Calcium was assumed to enter the outermost spine head compartment, after which
it could diffuse into a neighboring compartment, become bound to a buffer, or
be pumped out of the cell.

For the synaptic calcium current, we first chose a current that was computed
in a model of calcium current through NMDA receptor channels (Holmes and
Levy, 1990). With this calcium current, calcium concentration in the spine head
of the long-thin spine peaked at about 28 μM as shown on the right side of Fig.
5a. (We note that calcium concentration in the spine head was fairly uniform,
but there was a steep calcium gradient down the spine neck and calcium concentration in the dendrite peaked at about 100 nM). When the same current
entered the spine head of a mushroom-shaped spine, calcium concentration in
the spine head peaked at about 1.5 μM. Peak calcium concentration was just
slightly lower in the short stubby spine except that, with no spine neck to speak
of, there was no diffusional barrier keeping calcium elevated in the spine. Because
calcium flow to the dendrite was easier, calcium concentration in the stubby
spine decayed faster than in the previous cases.

The finding that calcium can be concentrated to much higher levels in long-
thin spines than in short, stubby, or mushroom-shaped spines was not dependent
on the shape of the calcium current. Results similar to those shown in Fig. 5a
were found when the calcium current was a half-rectified sine wave or a constant
current step as shown in Figs. 5b and 5c. We have used many other current
wave forms and they all produce results similar to those illustrated, provided
that the strength and duration of the calcium current are comparable to those of
the currents shown.

The large differences in spine head calcium concentration between long-thin
and short, stubby spines did depend on calcium buffer concentration and the
strength and duration of the calcium current. When the magnitude of the current
was halved or calcium buffer concentration was doubled, the differences in spine
head calcium concentration between stubby and long-thin spines was only fourfold instead of the almost 20-fold differences shown in Fig. 5. Changes in rate
parameters associated with binding and unbinding with buffer or the calcium
pump produced only small changes in spine head calcium (Holmes, 1990). Given
these results, the fact that spine shape can control calcium concentration levels

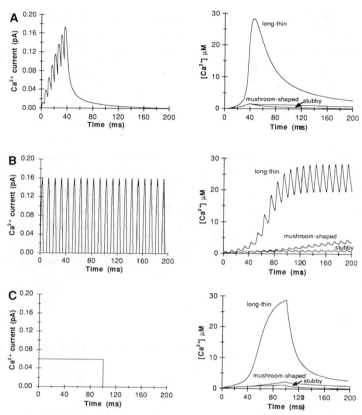

FIGURE 5.   Spine head calcium concentration for different spine shapes given different calcium currents. Calcium currents are shown on the left, and the calcium concentration changes in the spine head due to these calcium currents are shown on the right for three different spine shapes. (a) Calcium current from a model of calcium influx through NMDA-receptor channels (Holmes and Levy 1990). (b) Calcium current is a half-rectified sine wave. (c) Step calcium current. Spine dimensions (taken from Desmond and Levy 1985) for the spine head and neck were 0.55×0.55 and 0.1×0.73 μm (long-thin), 0.77×0.77 and 0.2×0.43 μm (mushroom), and 0.76×0.99 μm (stubby, head, and neck together). Parts of this figure were adapted from Holmes (1990, Figs. 1–3, pp. 339–340).

near a synapse may have important consequences for the relative plasticity and modifiability of long-thin versus short, stubby spines.

To conclude this final section, there are many ways in which synaptic modification can be included in a model. Unless intracellular resistivity is much higher in spines than in dendrites, dynamically changing the electrotonic structure of spines may not be an appropriate procedure. Given the role of calcium in triggering long-term potentiation, it would seem that a rule based at least in part on calcium concentration might be appropriate and the critical variables in such a rule would be spine shape, calcium buffer concentration, and the strength and duration of the calcium signal.

V.   Summary

In summary, if we are to develop a model that captures the essential computational features of a real neuron, we must first know the morphological and electrotonic structure of the real neuron. Second, the model must be able to reproduce the dynamic range of computational possibilities of real neurons. Finally, there must be some mechanism for modification of the computation done by the neuron. The number of degrees of freedom in constructing such a model is tremendous, but we hope that the work described here will help to reduce this number somewhat.

Acknowledgments

We thank Dr. C. D. Woody for providing the morphological data for the cell pictured in Fig. 3.

References

CLEMENTS, J. D., and REDMAN, S. J. (1989). "Cable Properties of Cat Spinal Moto-neurones Measured by Combining Voltage Clamp, Current Clamp and Intracellular Staining," *Journal of Physiology (London)* **409**, 63–87.

DESMOND, N. L., and LEVY, W. B. (1985). "Granule Cell Dendritic Spine Density in the Rat Hippocampus Varies with Spine Shape and Location," *Neuroscience Letters* **54**, 219–224.

FLESHMAN, J. W., SEGEV, I., and BURKE, R.E. (1988). "Electrotonic Architecture of Type-Identified α-Motoneurons in the Cat Spinal Cord," *Journal of Neurophysiology* **60**, 60–85.

GAMBLE, E., and KOCH, C. (1987). "The Dynamics of Free Calcium in Dendritic Spines in Response to Repetitive Synaptic Input," *Science* **236**, 1311–1315.

HOLMES, W. R. (1986). *Cable Theory Modeling of the Effectiveness of Synaptic Inputs in Cortical Pyramidal Cells*. Ph.D. Thesis, University of California, Los Angeles.

HOLMES, W. R. (1990). "Is the Function of Dendritic Spines to Concentrate Calcium?" *Brain Research* **519**, 338–342.

HOLMES, W.R., and LEVY, W. B. (1990). "Insights into Associative Long-Term Poten-tiation from Computational Models of NMDA Receptor-Mediated Calcium Influx and

Intracellular Calcium Concentration Changes," *Journal of Neurophysiology* **63,** 1148–1168.

HOLMES, W. R., and RALL, W. (1991a). Electrotonic Length Estimates in Neurons with Dendritic Tapering or Somatic Shunt (submitted).

HOLMES, W. R., and RALL, W. (1991b). Estimating the Electrotonic Structure of Neurons with Compartmental Models (submitted).

HOLMES, W. R., SEGEV, I., and RALL, W. (1991). Interpretation of Time Constants and Length Estimates in Multi-Cylinder or Branched Neuronal Structures (submitted).

HOLMES, W. R., and WOODY, C. D. (1989). "Effects of Uniform and Non-Uniform Synaptic 'Activation-Distributions' on the Cable Properties of Modeled Cortical Pyramidal Neurons," *Brain Research* **505,** 12–22.

PERKEL, D. H., and MULLONEY, B. (1978). "Calibrating Compartmental Models of Neurons," *American Journal of Physiology* **235,** R93–R98.

PERKEL, D. H., MULLONEY, B., and BUDELLI, R. W. (1981). "Quantitative Methods for Predicting Neuronal Behavior," *Neuroscience* **6,** 823–837.

RALL, W. (1962). "Theory of Physiological Properties of Dendrites," *Annals of the New York Academy of Science* **96,** 1071–1092.

RALL, W. (1964). "Theoretical Significance of Dendritic Trees for Neuronal Input–Output Relations," in R. Reiss (ed.), *Neural Theory and Modeling* (pp. 73–97). Stanford University Press, Stanford, California.

RALL, W. (1969). "Time Constants and Electrotonic Length of Membrane Cylinders and Neurons," *Biophysical Journal* **9,** 1483–1508.

RALL, W. (1977). "Core Conductor Theory and Cable Properties of Neurons," in E. R. Kandel (ed.), *Handbook of Physiology (Sect. 1): The Nervous System I. Cellular Biology of Neurons* (pp. 39–97). American Physiological Society Bethesda, Maryland.

RALL, W. (1990). "Perspectives on Neuron Modeling," in M. D. Binder and L. M. Mendell (eds.), *The Segmental Motor System* (pp. 129–149). Oxford University Press, Oxford.

RALL, W., and RINZEL, J. (1971a). "Dendritic Spines and Synaptic Potency Explored Theoretically," *Proceedings of the International Congress Physiological Society (XXV International Congress)* **9,** 466.

RALL, W., and RINZEL, J. (1971b). "Dendritic Spine Function and Synaptic Attenuation Calculations," *Society for Neuroscience Abstracts* **1,** 64.

SHEPHERD, G. M. (1979). *The Synaptic Organization of the Brain.* Oxford University Press, Oxford.

STRATFORD, K., MASON, A., LARKMAN, A., MAJOR, G., and JACK, J. J. B. (1988). "The Modeling of Pyramidal Neurones in the Visual Cortex," in R. Durbin, C. Miall, and G. Mitchison (eds.), *The Computing Neuron* (pp. 296–321). Addison-Wesley, Workingham, U.K.

Chapter 2   Canonical Neurons and Their Computational Organization

GORDON M. SHEPHERD

Section of Neurobiology
Yale University School of Medicine
New Haven, Connecticut

A hallmark of the nervous system is that its constituent neurons are cells with complex structures and complex physiological properties. One of the main objectives of neuroscience is to understand the relations between these cellular structures and properties and the operational tasks of the nervous system. The premise of this review is that the main task of the nervous system is to process information, and that the way this is done at the systems level is heavily dependent on the rules for information processing by individual neurons functioning as complex operational units.

Some clues to these rules have emerged from investigations of specific types of neurons in different neural regions. The aim of this chapter is to review progress on this problem, using neurons in the olfactory pathway as models for analysis. The results of this work not only shed some light on the processing of information in this relatively simple sensory system, but also have a broader significance in pointing toward a better understanding of the principles of organization of complex neurons.

It will be useful to begin by considering the theoretical and computational foundations that underlie our current approach to the analysis of complex neurons. From a consideration of different neuronal types, we will develop the concept of the canonical neuron. To illustrate this concept, we will consider the functional organization of several types of canonical neurons in the olfactory pathway. This will lead to a discussion of the canonical pyramidal neuron of the cerebral cortex, and its significance for the construction of realistic cortical networks.

I. Historical Background for the Complex Neuron

The classical studies by Cajal and his contemporaries revealed the basic morphology of most of the main types of neurons in different regions that we recognize today. It is notable that in this centennial year of the announcement of the *Neuron Doctrine* (von Waldeyer-Hartz, 1891), several new works make available the classical papers in English translations so that there is a better opportunity to appreciate the magnitude of the achievement of these workers, using mainly the Golgi stain. These include translations of Cajal's papers on the cerebral cortex (DeFelipe and Jones, 1989); a translation by Swanson and Swanson of Cajal's first book summarizing his studies, entitled *New Ideas on the Structure of the Nervous System of Man and Vertebrates* (Cajal, 1991); a critical evaluation of the articles on which the Neuron Doctrine is based (Shepherd, 1991); and a translation in preparation by Swanson and Swanson of Cajal's monumental *Histologie du Système Nerveux de l'Homme et des Vertébrés* (Cajal, 1911).

These classical studies showed clearly the extraordinary elaborateness of the dendritic trees and axonal arborizations of most neurons. They also showed that the neuronal population in a given region can be characterized in terms of a few distinct neuronal types, based on the dendritic and axonal branching patterns (*cf.* Figs. 1A, B). To make sense of the functions of neurons in the face of this diversity of structure, the classical workers in the 1890s generally agreed that one had to look past the particularities of the different neuronal shapes and sizes to infer some common functions shared by all neurons. From this arose the idea of the *dynamic polarization* of the neuron, that is, that all neurons receive impulse inputs from other neurons by means of axon terminals on their dendrites and cell bodies, and send outputs to other neurons by means of impulses in their axons and axon terminals (*cf.* Cajal, 1911).

This idea, that the neuron has a holistic input–output function, coupled with the idea that this function is similar in all neurons, provided a very useful shorthand method for making sense of the overall flow of information through neurons. However, it de-emphasized the significance of dendritic branching, except as pathways for the flow of activity toward the cell body and/or the origin of the axon. As earlier pointed out by Rall (1962) (Fig. 1), it was only a small step to assume that the dendritic branches were largely immaterial to the overall

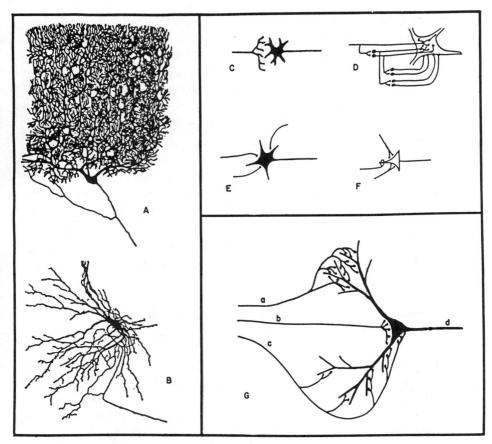

FIGURE 1.   Examples of different types of dendritic branching and different simplified neuronal representations: (A) Cerebellar Purkinje cell (B) Motoneuron. (C, E) Reduced representations of neurons and input axons. (D) Feedforward pathways onto a neuron. (F) Formal node of McCulloch and Pitts. (G) Bipolar neuron with different inputs (a–c) and output axon (d). See text for explanation. From Rall (1964).

characterization of the input–output functions of the neuron, and from there to reduce the representation of the neuron to a single node, at which different inputs are summed and from which output arises to be sent to other nodes (Figs. 1C, E). This tendency is clearly seen in the representation of reverberating cortical circuits by Lorente de No (1938) (Fig. 1D), and reached its ultimate in the paper of McCulloch and Pitts (1943) (Fig. 1F), in which neurons were represented as simple summing nodes, arranged to form circuits that could perform varieties of logic operations. This paper by McCulloch and Pitts marked the beginning of the analysis of information processing at the neuronal level; at the same time, by ignoring the dendrites, it divorced from that analysis the very features most critical to those functions.

The first intracellular recordings from neurons (Eccles, 1953; Eyzaguirre and Kuffler, 1955) showed dendrites receiving synaptic or sensory inputs, implying that the resulting synaptic or receptor potentials must spread through the dendrites to the site of impulse initiation, in line with the hypothesis of Cajal and others. Nonetheless, the representation of the cell body as a simple summing node encouraged the idea that synaptic inputs near the cell body would be most effective in exerting immediate control of impulse initiation in the axon hillock and initial axon segment. Despite the early recognition by Bullock (1959), Bodian (1962), and a few others of the *complexity within unity* of the neuron, this *soma-centric* hypothesis of neuronal function has been a powerful idea in neuroscience, perhaps the most important hindrance to appreciating the importance of dendrites as the basis for complex information processing by the neuron.

II.   Development of the Computational Representation of the Complex Neuron

Modern neurobiology, at the membrane and cellular level, rests on three main theoretical pillars. The first is the quantitative model of Hodgkin and Huxley (1952) for the nerve impulse. Second is the quantitative characterization of synaptic transmission at the neuromuscular junction by Katz and his colleagues (*cf.* del Castillo and Katz, 1954; Katz, 1969). Application of these models, based on simple and accessible peripheral structures, to the central nervous system was hindered by the dendritic complexity of the central neuron. A third pillar of neurobiological theory thus was necessary.

An understanding of the contributions of dendrites to the functions of neurons required the development of methods for characterizing the generation and distribution of electric current in complex branching systems. The methods introduced by Wilfrid Rall for this purpose have provided much of our present theoretical framework for the study of the complex neuron. Since this contribution is generally unrecognized by new workers coming into the field, it will be necessary to summarize it briefly.

The first step was to adapt cable models, originally applied to simple unmyelinated and unbranched axons (*cf.* Hodgkin and Rushton, 1946), to branching systems, by developing mathematical methods for representing current flow through these systems. In the simplest case, a branching dendritic tree could be represented as an *equivalent cylinder,* in which there is impedance matching between a parent stem and the daughter branches. This has become known as the *Rall model*. This is only the simplest representation of a dendritic tree; the methods, in fact, are completely general, and allow for *equivalent cables* of any arbitrary dendritic diameter and taper for particular cases (see Rall, 1959, 1977). These methods went far beyond simple passive cable analysis, laying the basis for understanding the complex properties of impedance matching that dominate the generation and flow of all types of signals through branching dendritic trees. The mathematical basis for electrotonic current flow in passive and active neurons is now well-developed (*cf* Jack *et al.*, 1975; Rall, 1977; Tuckwell, 1988).

The next crucial step in the analysis of dendritic function was the development of computational, rather than analytical, models of dendrites. This was necessary because the analytical mathematical methods were laborious and could only be applied to the simplest cases of electric current spread through passive membranes. Rall (1964), therefore, introduced compartmental modeling for computational representations of neuronal form and function. With these methods, any given branch could be modeled with any arbitrary combinations of passive membrane properties, synaptic conductances, or voltage-gated conductances (Fig. 2), and the compartments could be connected in any arbitrary branching pattern at any desired level of detail. The first applications of these methods were to motoneurons (Rall, 1964, 1967), which provided the first quantitative analysis of the generation, spread, and integration of synaptic potentials in branching dendritic trees, and to olfactory neurons (Rall and Shepherd, 1965, 1968; Rall

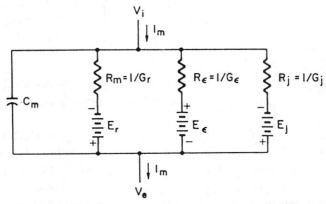

FIGURE 2. Representation of a locus on a neuron by the compartmental method. Abbreviations: V_i, intracellular potential; V_e, extracellular potential; C_m, membrane capacitance; I_m, membrane current; E_r, resting potential, E_e, excitatory equilibrium potential; E_j, inhibitory equilibrium potential, R_m and G_r, membrane resistance and conductance, respectively; R_e and G_e, excitatory resistance and conductance, respectively; R_j and G_j, inhibitory resistance and conductance, respectively. From Rall (1964).

TABLE I.   Compartmental Neuronal Modeling Programs.

general	1964, Rall	SAAM
Motoneuron	1967, Rall	SAAM
Olfactory mitral and granule cells	1968, Rall and Shepherd	SAAM
Motoneuron	1973, Dodge and Cooley	
Axons	1975, Ramon, Joyner and Moore	
Motoneuron	1976, Traub	
Purkinje cell	1976, Pellionicz and Llinás	
Renshaw cell	1977, Traub	
general	1978, Perkel and Mulloney	MANUEL
general	1978, Shepherd and Brayton	ASTAP
Olfactory mitral and granule cells	1979, Shepherd and Brayton	ASTAP
Pyramidal neuron	1979, Traub and Llinás	
Invertebrate neurons	1979, Graubard and Calvin	
Retinal ganglion cell	1982, Koch, Poggio, and Torre	
general	1984, Guthrie	NEUROS
general	1984, Hines	CABLE/NEURON
Dendritic spine	1984, Wilson	
Dendritic spine	1985, Perkel and Perkel	MANUEL
general	1985, Bunow *et al.*	SPICE
Dendritic spine	1985, Miller *et al.*	SPICE
Dendritic spines	1985, Shepherd *et al.*	ASTAP
Dendritic spines	1987, Segev and Rall	SPICE
Motoneuron	1988, Fleshman *et al.*	SPICE
general	1987, 1990, Carnevale *et al.*	SABER
Dendritic spines	1989, Shepherd *et al.*	SABER
general	1990, Wilson and Bower	GENESIS

et al., 1966), which extended the analysis to both synaptic potentials and action potentials in dendrites, and began the analysis of synaptic interactions between neurons.

For these early studies, Rall utilized a modeling program (SAAM) that had been developed for kinetic analysis of radioactively labeled tracer distribution in body compartments by Mones Berman at the National Institutes of Health, adapting it for the specific representation of the electrical properties of neurons and their branching dendrites. Table I traces the development of neural modeling programs since then. They cover a range of kinds of approaches, from the general to the specific. The general approach is represented by ASTAP, SPICE, and SABER, which are general-purpose circuit analysis programs used in electrical engineering and the computer and electronics industry for electronic circuit design. Similar programs, but designed specifically for neuron modeling, are PEDRO,

CABLE, NEURON, and GENESIS. Programs written for modeling specific neurons, neuronal circuits, or neuronal functions are represented by the work of Traub, Pellionicz, and many others.

It is important to emphasize that there is no one modeling approach or program that is better than all the rest. Each has its advantages and disadvantages (see Koch and Segev, 1989). The general circuit analysis programs have the advantages that they are written in high-level languages so that they are easy to use and one does not have to be a programmer to construct and explore any arbitrary model; in addition, most of these programs are well-documented and well-supported. The main disadvantage is reduced efficiency, since only a part of the program is used to model neuronal properties. At the other extreme, programs written expressly for specific neuronal models have the advantage of greater efficiency, permitting construction of larger systems that can run within a given time frame; however, they require considerable programming expertise, and considerable time and expense for development and support; in addition, the more computationally efficient they are, the less likely they are to permit easy explorations of model parameters. The general-purpose neural modeling programs combine ease of model construction with relative efficiency. All the types provide accessible means for incorporating neural modeling methods into local laboratory settings. These programs presently run on workstations such as the DEC 5000 and the SUN 4.

In our own research, we use several types of programs for neural modeling, including ASTAP (IBM), SPICE, SABER (ANALOGY), GENESIS, and NEURON. This has several advantages over being limited to one approach: First, it enables us to baseline any application of a particular program to a new model; second, since documentation for the public domain programs is often limited and support sometimes unavailable, it ensures continuity in pursuing a problem by the well-documented commercial programs (i.e., SABER); third, it provides the ability to carry out exploratory phases of the work on one program optimal for a particular problem (i.e., NEURON), then do the display phase on the program with the best graphics (i.e., GENESIS); some programs (i.e., SABER) are particularly effective for modeling diffusional processes (e.g., second messengers) combined with electrical properties. Finally, since we collaborate with a number of other laboratories, we find that different collaborators prefer different types of software; for example, computer scientists and electrical engineers prefer the larger commercial general purpose programs (ASTAP, SABER), whereas neuroscientists prefer the more specialized programs (GENESIS, NEURON). We stress these aspects of software because of the recent tendency of some neuroscientists to focus on using only one specialized program for all modelling; we feel this is less effective than a broader approach, for the reasons discussed above. A broader approach also encourages the effort to relate neural modelling to some of the simulation techniques and strategies used routinely in electrical engineering and computer science.

III.   Strategies for Neuronal Modeling

In summary, the current generation of modeling programs, combined with the availability of powerful and affordable machines to run them in a laboratory setting, means that the experimental and theoretical analysis of complex neurons, so slow and laborious in its early period during the 1960s, can now proceed closely in parallel. Increasingly, experimenters can routinely subject electrophysiological recordings to immediate theoretical analysis in model neurons, and the model neurons can suggest hypotheses for immediate experimental testing, as envisaged by the early pioneers (see Rall, 1970).

There are several important warnings for those pursuing these studies. The available programs vary significantly in their optimization for different types of models. For example, speed of computation may be gained by numerical algorithms that preclude certain circuit configurations such as feedback loops, or only permit representation of the cable properties of dendrites but not of axons. Thus, a slower but more general program may be preferable, at least for initial characterization of a given neuron, to one that is faster but less flexible. Another point is that the programs at present differ markedly in their degree of development, documentation, support, and ability to run on different platforms. There is a pressing need for better support for the new programs, particularly those that have been developed in laboratory environments through heroic efforts by an individual scientist.

Finally, the new computational power means that neurons may now be represented by an unlimited number of compartments. However, this temptation should be resisted, because as a model is increased in size it quickly runs the risk of becoming underconstrained. The point has been addressed specifically by Rall (1990):

> . . . the apparent flexibility of the large multicompartment model must be reduced by the imposition of severe constraints before there is any hope of matching its parameters to an experiment. When it is so constrained, it becomes equivalent, in many respects, to one of the simple (theoretical) models. . . . It is very important to remember that when there are too many degrees of freedom, it may be easy to find parameter sets that fit the data, but such solutions are not unique and may be far from optimal. To avoid such pitfalls, it is better to reduce the degrees of freedom severely at first, and to make use of anatomical information as well as both steady-

state and transient electrophysiological data; then the problem becomes similar to that (covered by the simpler theoretical models).

Thus, the critical significance of the mathematical and computational methods is that they provide a theoretical foundation for reducing neuronal complexity to the minimum needed to capture the essential properties relevant to a particular functional operation or set of operations. The rules of reduction are available (e.g., Fleshman *et al.*, 1988), and need to be applied so that models can be used to carry out critical tests of competing hypotheses based on experimental data.

IV.   The Concept of the Canonical Neuron

As noted previously, the classical studies with the Golgi stain showed that, within a given region of the nervous system, dendritic branching patterns are usually not continuously variable over the entire population, but rather there are a few distinct types. Most of these neuronal types were summarized in Cajal's (1911) great work on the histology of the nervous system, and have withstood the test of time.

Modern studies have greatly extended the analysis of these types, by a variety of methods. These include: analysis of fine structure and synaptic connections under the electron microscope; examination of neurons injected with intracellular dyes, ions and enzymes; staining with monoclonal antibodies; analysis of uptake of different neurotransmitters and other bioactive molecules; localization of receptor binding of neurotransmitter agonists and antagonists; and recording of different physiological properties. The results of these studies have shown that the classical types of neurons may have multiple subtypes. The main subtypes are related to details of axonal or dendritic branching pattern, specificity of synaptic connections, expression of specific antigens, uptake and release of different neurotransmitters, and generation of different physiological properties.

A critical question for both experimental neuroscience and neural networks is how to deal with this diversity. Are the numbers of types, subtypes, and sub-subtypes virtually limitless? If so, deduction of general principles underlying neuronal structure and function would seem almost impossible, and neural network modelers may feel continued justification in ignoring most of the results being generated by modern neuroscience.

A more useful strategy is to designate each of the main types as a *canonical neuron*. The relevant definition of *canonical* in Webster's dictionary is "reduced to the simplest or clearest schema possible." Thus, we may define a canonical neuron as *the simplest type of a particular pattern or motif of neuronal structure and function.* In practice, it is one of the irreducible few types into which all neurons of a given region are divided.

Canonical neurons by these criteria may seem equivalent to the traditional neuronal types identified by the Golgi stain; in fact, attempts have been made to reduce all branching patterns to a few basic types (*cf.* Ramón-Moliner, 1962). However, there are several additional considerations that warrant the new term.

First, there is a convenient analogy with the concept of molecular families that has arisen in the study of membrane proteins. It is now recognized that diverse membrane proteins are grouped in families on the basis of homologies in primary amino acid sequences and related aspects of secondary and tertiary structure. As an example, a seven-membrane-spanning segment motif is shared by receptor proteins that are activated by diverse neurotransmitter molecules but have in common that they all activate second messengers through GTP binding proteins (e.g., Ross, 1990). In a similar way, a canonical neuron is a consensus pattern or motif embracing both structural and functional properties; this motif is diversified into a family of subtypes for different but related functional operations. Thus, a canonical neuron is an idealized representation of the properties that distinguish one neuronal type from another.

A second reason for the new term is related to the construction of computational neuronal models. The essence of this task is to identify the properties that are critical for the particular input–output operations under study, and represent the neuron by the minimum number of compartments containing the minimum set of properties sufficient to *capture* those operations. In this process, one in fact converts a descriptive list of the structural and functional properties of a neuron, as gained from experimental analysis, into a biophysical model in a computational form. Although somewhat idealized, the model is mathematically precise and representationally accurate, up to the limits of the operations that it has been designed to simulate. We thus have an enlarged definition: a canonical neuron is *the simplest type of a particular pattern or motif of neuronal structure and function that can represent a given neuron in computational form as a basis for simulating a functional operation or set of operations essential for that type.* In this context, the canonical neuron is a necessary concept for creating a unified theoretical, biophysical, and computational foundation for the simulation of neuronal properties. This foundation is a critical base for bridging the gap between the domain of experimental data and the domain of artificial neural nets.

A final need for the new term is discovered when one moves from the level of simulation of individual neurons to the level of neuronal circuits. The concept of the *basic circuit* has been used to characterize *the minimum set of synaptic connections and neuronal properties that can carry out the basic input–output operations of a given region of the nervous system* (Shepherd, 1979). This concept has been given a computational form by the term *canonical circuit* (Douglas and Martin, 1990). We will discuss canonical circuits shortly. For the present, it is sufficient to note that canonical neurons are obvious building blocks for canonical circuits.

V.   Hierarchical Organization of Canonical Neurons in the Olfactory System

To illustrate the concept of a canonical neuron, we will consider several examples in a model system, the olfactory pathway (see Fig. 3).

A.   The Olfactory Receptor Neuron

The first neuron in the olfactory pathway is the receptor neuron within the olfactory epithelium. In a macrosomatic mammal such as the rabbit, there are

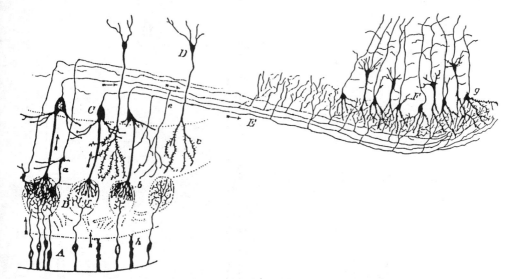

FIGURE 3.   Neuronal organization of the mammalian olfactory pathway. The main outlets of the organization, as portrayed in the classical work of Cajal, have been confirmed by modern work: (A) olfactory receptor neurons in the olfactory epithelium; (B) olfactory glomeruli in the olfactory bulb; (C) mitral cell in the olfactory bulb; (D) granule cell in the olfactory bulb; (E) lateral olfactory tract; (F) pyramidal neurons of the olfactory cortex; (a) primary dendrite of a tufted cell; (b) glomerular tuft of a mitral cell; (c) peripheral spiny dendritic branches of a granule cell; (e) mitral axon collateral; (g) small pyramidal neuron; (h) supporting cell. From Cajal (1991).

as many as 50 million of them. This exceeds the number of any other neuronal type except for cerebellar granule cells (10–100 billion) and retinal rods (100 million). Olfactory receptor neurons vary in the depth of their cell body within the epithelium, and consequently in the length of their dendrites. Electrophysiological experiments have shown that these cells have very high input resistances, and are very tightly coupled electrotonically. They thus are extremely sensitive to sensory input. There is evidence that a single membrane channel opening can give rise to an impulse (*cf.* Lynch and Barry, 1985; Hedlund *et al.,* 1987; Firestein and Werblin, 1987).

Compartmental models have been constructed to model these properties. These have varied from detailed models including representation of the dendrite and individual cilia (20–30 compartments) to reduced models in which the cilia are lumped together (13 and 7 compartments) (see Fig. 4). The tight electrotonic coupling suggests that for many purposes the reduced models will be sufficient for simulating the main input–output working range of these cells in transducing their odor ligand inputs in their cilia into impulse output in these axons (Pongracz

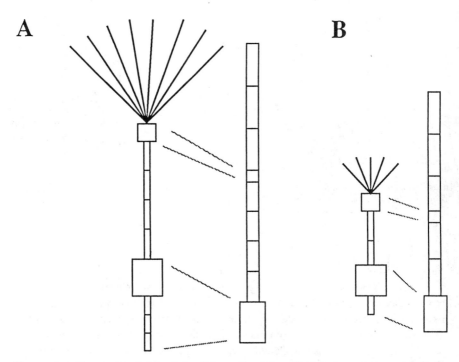

FIGURE 4.   Compartmental models of the salamander olfactory receptor neuron. Left, full model representing cilia, knob, dendrite, cell body, and axon. Right, reduced model of same. See text. From Pongracz *et al.* (1991a).

et al., 1991a). This basic canonical representation will then be adapted to explore more detailed effects of dendrites and cilia on such properties as detection threshold, distribution of different molecular receptors on different cilia, signal-to-noise relations, and linear and nonlinear response summation. Our strategy in this work is to use the olfactory neuron as a relatively simple model for working out some of the rules for reducing complex neurons to their simplest representation that still capture their essential operating model. This will then serve as a basis for replicating these models to build networks of neurons.

B. The Mitral/Tufted Cell

The output neurons of the olfactory bulb consist of mitral cells and smaller versions called tufted cells. Although these two main types are distinct in many ways from each other, and themselves consist of subtypes (*cf*. Scott and Harrison, 1987; Mori, 1987), they all share certain common features: a distal dendritic tuft of branches that receives the peripheral input within regions called glomeruli; one or more primary radial dendrites that serve as the link between the distal tuft and the cell body; and several secondary lateral dendrites that branch and terminate in the external plexiform layer, where they interact with granule cell dendrites.

This extreme differentiation of the dendrites expresses an important concept: *the hierarchical organization of the complex neuron*. This concept is illustrated diagrammatically in Fig. 5. Branch compartments of this nature have been called *topological functional units* (Shepherd, 1972); that is, they are large structural subdivisions within the overall topology of the dendritic tree of a neuron. Other examples are the cilia of sensory neurons, the apical and basal dendrites of pyramidal neurons, and the two oppositely directed branches of bipolar neurons (*cf*. Fig. 1G). This differentiation of branch structure can be significant in several ways: It can provide for the separation of different inputs to different compartments of the dendritic tree; it can enable different branch compartments to connect to different output sites in the case of presynaptic dendrites; it can isolate one set of inputs or outputs from another along the same dendrite. In the case of the mitral/tufted cells, the compartment is so extreme that the cell can be regarded from a functional point of view as effectively subdivided into the equivalent of three virtual cells: the glomerular tuft for receiving the olfactory nerve input; the primary dendrite for conveying the glomerular response to cell body and axon; and the secondary dendrite for gating and modulating the output in the axon. Each of these findings is anatomically, physiologically, and computationally distinct from the others.

Topological branch units in turn contain several layers of organization. As

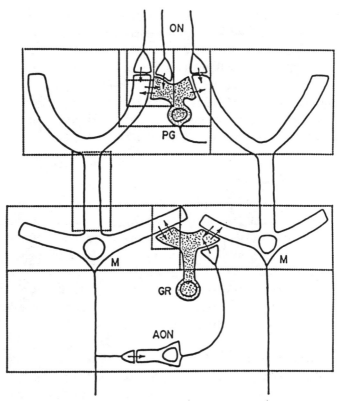

FIGURE 5.   Hierarchical organization of the mitral cell and related synaptic circuits in the olfactory bulb. Dotted lines frame different levels of organization, from single synapse, synaptic microcircuits, dendritic topological units, and interregional circuits. See text. From Shepherd (1977).

indicated in Fig. 5, at the next lower level are individual dendritic branches; below this are the local patterns of synaptic connections associated with a given branch; at the lowest level is the individual synapse. The synapse can be regarded as the basic unit underlying the construction of the higher levels of organization. For this reason, from the point of view of information processing, the most fundamental unit of the nervous system is the synapse, rather than the neuron (von Neumann, 1958; Shepherd, 1979). The traditional Neuron Doctrine, with its emphasis on the whole neuron as the basic anatomical and functional unit of the nervous system, does not do justice to this view; indeed, it led inexorably to the reduced view of the whole neuron represented by the McCulloch–Pitts model. To move away from this view, it has been argued that we need to develop a *Synaptic Doctrine*, in which the synapse is regarded as the basic structural and functional unit for information processing in the nervous system (Shepherd, 1979, 1990).

C.   The Olfactory Granule Cell

The granule cell is one of the two main types of interneurons in the olfactory bulb, and is therefore intimately involved in regulating mitral/tufted cells. However, the synaptic organization, and with it the computational organization, of the granule cell stands in marked contrast to the mitral/tufted cells. The key feature of the granule cell recognized in classical times is that it lacks an axon (*cf.* Fig. 3). The main contributions of modern work (Fig. 6) have been that the

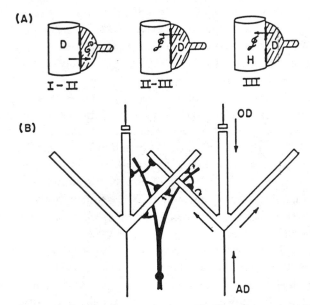

FIGURE 6.   Local synaptic input–output operations of olfactory neurons. (a) Diagrammatic representation of reciprocal synapses between mitral dendrite and granule dendritic spine. Time periods I–II: Sequence begins with mitral dendrite depolarization (D) activating the excitatory synapse, causing excitatory (E) depolarization of the granule dendritic spine. Time periods II–III: During transition from time periods I to II, the excited granule cell spine activates the inhibitory synapse, eliciting feedback inhibition (J) of the mitral dendrite. Time period III: During this period, there is continued long-lasting hyperpolarizing (H) inhibition of the mitral dendrite. (b) Diagram showing circuit relations between mitral (open profiles) and granule (filled profiles) cells. Large arrows indicate direction of orthodromic (OD) and antidromic (AD) activity. Small arrows indicate how reciprocal synapses provide for both reciprocal (feedback) and lateral inhibition. From Rall and Shepherd (1968).

major synaptic input to the granule cell comes through dendrodendritic synapses
from mitral cell secondary dendrites onto granule cell spines; all of the granule
cell output is by way of reciprocal dendrodendritic synapses from the spines
onto mitral cell dendrites; these output synapses are inhibitory; input–output
operations through the dendrodendritic synapses can run in the absence of impulse
activity; individual granule cell spines can function as local input–output units
operating semi-independently of the rest of the neuron. These properties were
established a number of years ago (Rall et al., 1966; Rall and Shepherd, 1968),
and have been confirmed by numerous studies subsequently. (See especially
Nicoll, 1969; Price and Powell, 1970; Jackowski et al., 1978; Mori et al., 1981;
Jahr and Nicoll, 1982.) It is also known that the granule-to-mitral synapses are
GABAergic (Ribak et al., 1978). There is increasing evidence that the mitral-
to-granule synapses are glutamatergic (McLennan, 1971; Nicoll, 1971; Trombley
and Westbrook, 1990). There is also evidence that these synapses have NMDA
receptors (Cotman and Iverson, 1987) and may be involved in olfactory learning
(cf. Brennan et al., 1990).

From a computational point of view, the dendritic spine is of special interest.
Its properties enable it to be "the smallest compartment in the nervous system
capable of carrying out a complete synaptic input–output function" (Shepherd
and Greer, 1989). This point is well-illustrated by the granule cell spine, which
has both a synaptic input and synaptic output, but it also applies to other types
of spines whose output is by means of current spread into a dendritic branch.
These considerations support the notion discussed previously regarding the importance of the synapse as the basic building block for information-processing
circuits in the nervous system.

The connection from the spine head to the dendritic branch depends on the
effective resistance through the spine neck, which provides a sensitive means
for controlling the linkage between the two, as pointed out by Rall (1974).
This has been explored in granule cells by detailed reconstructions, as illustrated in Figs. 7 and 8. The simulations carried out on these compartmental
models have shown that the spread of potential from an activated spine is critically dependent on the local pattern of spine organization and spine clustering, as well as on the value for the specific membrane resistance (R_m). Different values of R_m create different sizes of populations of neighboring activated
spines, equivalent to the *dendritic subunits* characterized in a similar way by
Koch et al. (1983) in starburst amacrine cells. These dendritic subunits are at
an intermediate level of intraneuronal organization, and show that some functional levels are not rigidly determined by fixed synaptic sites but can shift in
a dynamic fashion.

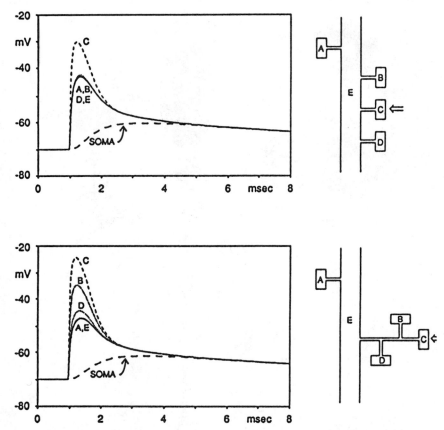

FIGURE 7.   Computational models of the electrotonic spread of synaptic potentials between spines of olfactory granule cells. Above, right: Synaptic excitation is delivered to
spine C; left, graph shows EPSP in spine C, and relatively similar decremental synaptic
potentials in other spines (A, B, D), which are only slightly smaller than the response
in the dendrite (E). This illustrates the asymmetrical impedance relation between a spine
and its parent branch; there is significant decrement of a potential spreading from spine
to dendrite, because of the relatively large conductance load, but insignificant decrement
spreading from dendrite to spine because of the high impedance boundary condition
represented by the spine and the relatively short electrotonic length of the spine stem.
Below, another configuration: synaptic excitation delivered to spine C, which is part of
a complex spine (B–D). Note separation of amplitudes of electrotonically spreading EPSPs
in A–D. Dashed line shows slowed smooth transit in end dendritic compartment, indicating
the kind of transient seen in the soma. Synaptic output from the granule cell occurs at
the spines in the lateral plexiform layer, as represented in these models. From Woolf *et
al.* (1991).

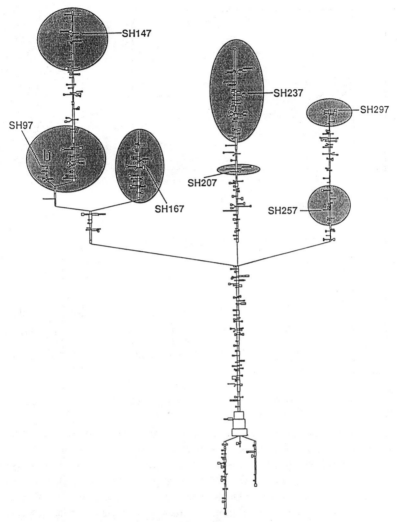

FIGURE 8.   Computational model simulating the electrotonic properties of an olfactory granule cell in the rat. The model is a quantitative simulation of all the dendrites and dendritic spines. The model was explored functionally by injecting an excitatory synaptic current in the spine heads (SH) as indicated, and determining the extent of spread of the EPSP. The shaded regions indicate the extent of spines depolarized above 10 mV. These regions define dendritic subunits relating a given input to the extent of output from neighboring spine populations; these subunits can vary in size depending on passive as well as active properties of the dendrites. From Woolf *et al.* (1991).

VI. The Cortical Pyramidal Neuron

We now consider the cortical pyramidal neuron, a type that is obviously of central importance for understanding the basis of cortical function.

The mitral/tufted cells send the olfactory bulb output to the olfactory cortex. There the axon terminals make excitatory synapses onto the spines of distal branches of apical dendrites of pyramidal neurons.

The distinguishing anatomical features of olfactory pyramidal neurons were identified in the classical studies of Cajal (Fig. 3), and have been confirmed and greatly extended by modern studies (reviewed in Haberly, 1990). These may be summarized as follows (Fig. 9):

1. The dendritic tree is divided into apical and basal compartments.
2. The apical dendrite is oriented radially toward the cortical surface, across non-repeating cortical layers.
3. Both apical and basal dendrites bear spines, which are sites of excitatory synaptic input.
4. The axon gives off collaterals, which are excitatory either to other pyramidal neurons or to inhibitory interneurons within the cortex.

Together, these features may be considered to defined anatomically the olfactory pyramidal neuron (*cf.* Shepherd, 1974; Shepherd and Koch, 1990; Haberly, 1990).

The *olfactory cortex* is considered to be the simplest example of forebrain cortex in the mammal. It is similar to the hippocampus in being three-layered, with pyramidal cell bodies located in a single intermediate layer. By contrast, the *neocortex* has five or six layers. The additional layers are due to several factors: Pyramidal cells have different sizes and their cell bodies have different locations; nonpyramidal cell populations are interposed, spiny stellate cells particularly in layer IV, nonspiny cells in other layers; afferent fibers have different levels of termination; intrinsic fiber systems are grouped at different levels. These features have been thoroughly documented for many cortical areas (see Peters and Jones, 1984; Rakic *et al.*, 1988). Although a sharp distinction traditionally has been made between "primitive" three-layered and more highly evolved six-layered cortex, it has been argued that the three-layer cortex may represent in simplest form the basic circuit for the neocortex (Shepherd, 1974). The pyramidal cell is the main cortical output neuron, and is therefore the key element in cortical circuits. It is this element that is represented in simple neural network simulations of cortical function (*cf.* Rumelhart and McClelland, 1986). It is therefore critical to consider the extent to which the pyramidal neuron in the different kinds of cortices can be regarded as representing variations on a canonical form.

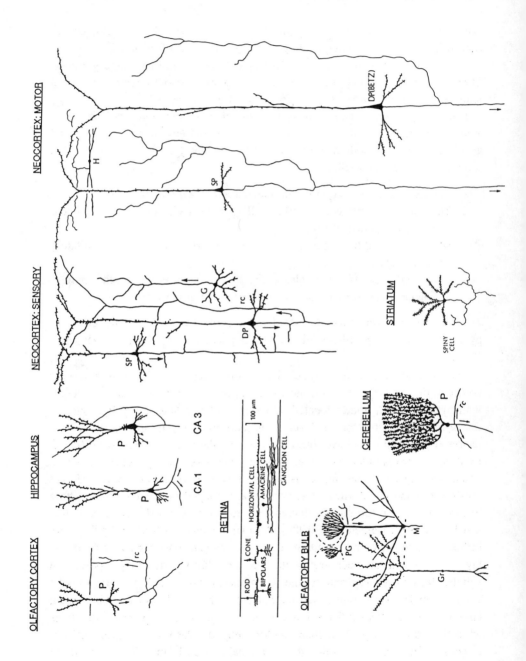

OLFACTORY CORTEX

HIPPOCAMPUS

NEOCORTEX: SENSORY

NEOCORTEX: MOTOR

P

rc

P

P

CA 1

CA 3

SP

DP

rc

G

H

SP

DP (BETZ)

STRIATUM

SPINY
CELL

RETINA

ROD   CONE   HORIZONTAL CELL

BIPOLARS   AMACRINE CELL

GANGLION CELL

100 μm

CEREBELLUM

P

rc

OLFACTORY BULB

PG

M

Gr

Although there has been varied evidence that supports this notion, closer consideration requires attention to the absolute size of the neurons in the different types of cortices. Surprisingly, relatively little attention has been given to this question. In Fig. 9, slightly simplified drawings of the main varieties of pyramidal neurons in the three main types of cortices—olfactory cortex, hippocampus, and neocortex—are shown. Two main conclusions may be drawn from this comparison. First, the pyramidal neurons in all areas share in the main features described previously for the olfactory pyramidal neuron: apical and basal dendrites; radial orientation of the apical dendrite; excitatory inputs on spines; excitatory axon collaterals providing for feedback; and lateral excitation and inhibition. Second, this similarity of form is maintained despite the large range in cell size. The smallest pyramidal cells are found in the olfactory cortex and in the most superficial part of neocortical layer II, where the apical dendrites extend 200–400 μm. By contrast, the largest pyramidal neurons are found in layers V and VI of the motor cortex, where the apical dendrite extends up to 4 mm in the human; even in layer III, the lengths may be up to 2 mm. Thus, despite a range of x5–10 in apical dendritic length, there is nonetheless a retention of the overall pattern. This suggests rather strongly that the pyramidal neuron type represents a canonical form that scales well over different cell sizes.

These conclusions based on anatomy are supported by results of electrophysiological studies of pyramidal cell properties. Sufficient analysis has been carried out in the different types of cortices to provide basic parameters of membrane properties, synaptic responses, and intrinsic membrane properties. An exhaustive account of the increasing number of studies of these properties is beyond the scope of this review. Our interest is first in the consensus properties across different types of pyramidal neurons. This question has been addressed by several workers, and a recent view is provided by Table II, from Connors and Gutnick (1990).

It can be seen that, in terms of basic physiological properties, most neocortical neurons fall into one of three main groups. Pyramidal neurons are represented in two of these groups: those that show regular spiking (RS) in response to injected current, and those that are intrinsically bursting (IB). RS cell bodies are found in all layers except layer I, whereas IB cell bodies are found in layer IV or V. Interestingly, spiny stellate cells in layer IV also fall into these two groups.

FIGURE 9.   Comparisons between pyramidal neurons in different types of cerebral cortex. Examples of several other types of neurons are shown for comparison. Note the constancy of form of the pyramidal neuron despite large differences in cell size, type of input fibers, and type of output targets, and in different cortical areas. The individual diagrams are drawn to the same scale. From Shepherd (1979).

From these data, it appears that spiny stellate cells resemble pyramidal neurons with only basal dendrites emanating from their cell bodies. This view suggests that the basal dendritic tree has similar functional properties in both anatomical types, and that the apical dendrite confers an additional set of functional properties in the pyramidal neuron type. A crucial property resulting from this arrangement is that a single neuron can receive and integrate a number of lamina-specific inputs (see White, 1989). From this emerge a number of consequences for the computational organization of the neuron, as discussed next. The third class of physiological properties is characterized by the generation of a fast-spiking (FS) response to injected current, and has been correlated with nonspiny cortical interneurons (Table II). These results have been reported from a number of laboratories. (See, for example, Mountcastle *et al.*, 1969; Stafstrom *et al.*, 1984; McCormick *et al.*, 1985.) Similar classes of regular firing and bursting pyramidal neurons have been described in the CA1 and CA3 areas, respectively, of the hippocampus (*cf.* Schwartzkroin and Mueller, 1987), and in the turtle dorsal cortex (Kriegstein and Connors, 1986).

The operational significance of these different impulse firing patterns for cor-

TABLE II.   General Classification Scheme for Neocortical Neurons (From Connors and Gutnick, 1990).

Characteristics	Physiological class		
	Regular-spiking (RS)	Intrinsically bursting (IB)	Fast-spiking (FS)
Single-spike:			
rate of rise	+ +	+ +	+ +
rate of fall	+	+	+ +
Single-spike afterpotential	Complex, AHPS, and ADP	Complex, AHPs, and ADP	Simple, AHP alone
Frequency adaptation	+ +	Variable	−
Spike bursts during injected current	−	+ +	−
Laminar location of soma	II to VI	IV or V	II to VI
Presumed morphology	Pyramidal or spiny	Pyramidal or spiny	Aspiny or sparsely spiny non-pyramidal
Presumed synaptic function	Excitatory (glutamate or aspartate)	Excitatory (glutamate or aspartate)	Inhibitory (GABA)

Symbols: +, the presence and relative strength of a characteristic; −, the absence of a characteristic. Abbreviations: AHP, afterhyperpolarizatoin; ADP, afterdepolarization. Table adapted from Ref. 15, and compiled from data in the following Refs.: 7–12, 16, 18, Gutnick, M.J., and White, E. L., unpublished observations, and Agmon, A., Ph.D. Thesis, Stanford University, CA, 1988.

tical functions is only beginning to be analyzed. At present, it can be surmised (*cf*. Connors and Gutnick, 1990) that rapidly adapting responses will tend to extract information about phasic rather than tonic inputs; that bursting cells will amplify weak inputs; and that bursting cells must be important in generating oscillatory cortical activity. It is also clear that actions of specific transmitter systems can shift a cell from one firing mode to another, which is believed to underlie different behavioral states (*cf*. McCormick and Pape, 1990). Thus, one of the cardinal physiological features of the canonical pyramidal neuron may be its wide range of functional properties. The set of common properties is large, giving pyramidal neurons a wide range of possibilities for synaptic integration of diverse inputs and for generating different impulse firing patterns. Another way to put it is that pyramidal neurons are not restricted to the highly stereotyped activity patterns characteristic of cerebellar Purkinje cells, nor the monotonic firing patterns of motoneurons; the population contains both types. This supports the idea of the pyramidal neuron as a canonical type, which is adapted for different specific processing operations dependent on the type of cortex, the area of cortex, and the layer within an area.

Recently, it has been possible to test these conclusions, based on studies of laboratory animals, in human cortical tissue obtained during neurosurgical procedures. These studies have shown that human cortical pyramidal neurons stained with intracellular dyes resemble their counterparts in subhuman spines. These neurons also share in most of the electrophysiological properties listed in Table II, with the exception of burst firing properties. However, cells that fire action potential doublets have been recorded, particularly in deeper layers (Foehring and Wyler, 1989; *cf*. also Strowbridge *et al.*, 1987).

Computational models: The idea of the canonical pyramidal neuron has been supported by studies utilizing computational models. The first compartmental model of a pyramidal neuron incorporating realistic representations of both synaptic inputs and intrinsic membrane conductances was that of Traub and Llinás (1979) for the hippocampal pyramidal cell. Similar multicompartment models have been developed for olfactory pyramidal cells (Wilson and Bower, 1990). A recent version of the model for the hippocampal pyramidal cell is shown in Fig. 10. This model (Pongracz *et al.*, 1991b) contains simulated apical and basal dendrites; an axon; excitatory synaptic conductance with an NMDA component; $GABA_A$ and $GABA_B$ inhibitory conductances; active Na, K, and Ca conductances; as well as a simulated extracellular compartment for K movements affecting cell excitability. With appropriate adjustments of parameters, this canonical model can be adapted for all specific types of pyramidal neurons (*cf*. also Llinás, 1988).

The canonical models of Traub and Wilson and Bower (see also Granger *et*

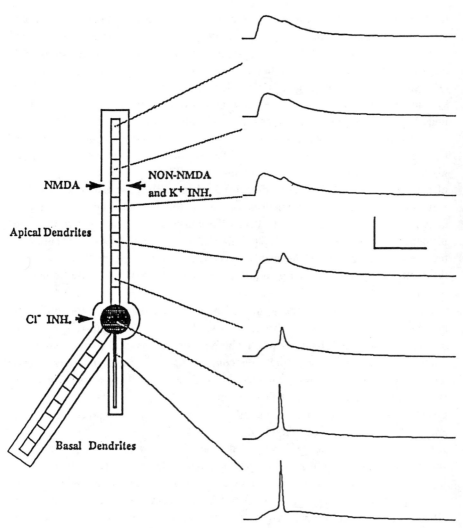

FIGURE 10.   Compartmental model of a pyramidal neuron in area CA1 of the hippocampus. On the left are indicated the types of physiological properties included in this model. Active conductances: Hodgkin–Huxley impulse with G_{Na} and G_{DR} (filled compartments at soma and initial axon segment); $G_{Ca}2+$ and $G_{K(Ca)}$. Synaptic conductances: glutamatergic excitatory conductances ($G_{non-NMDA}$ and G_{NMDA}) and GABAergic inhibitory conductances (G_{Cl} and G_K). Ionic concentrations: internal calcium and external potassium. On the right, excitatory dendritic input gives rise to an EPSP, which leads to generation of an impulse. This model showed potentiation of the EPSP by the NMDA synaptic conductance during slow repetition activation, as has been found in experimental studies. From Pongracz et al. (1991b).

al., 1989) have been the basis for the development of cortical networks containing up to 10,000 pyramidal neurons interconnected by means of associational fibers and including representation of inhibitory neurons as well. Recognition of the canonical pyramidal neuron is thus the necessary first step in the development of canonical cortical circuits (*cf.* Douglas and Martin, 1990).

The canonical models described thus far lack a very important level of organization, that of the dendritic spine. In an effort to explore properties at this level, we have constructed compartmental models aimed at simulating interactions between spines on distal dendrites. For this purpose, we have used the paradigm of logic operations to test the hypothesis that distal dendrites carry out highly specific operations involved in information processing in pyramidal neurons. The results, such as those illustrated in Fig. 11, show that with active properties in the spines or in the dentate, the system readily generates the basic logic operations of AND, NOT, and NOT-AND (Shepherd and Brayton, 1987; Shepherd *et al.*, 1989). Furthermore, active spines provide a means for boosting the responses spreading from distal dendrites to the same, to control impulse output (Shepherd *et al.*, 1985). Even more complex operations are possible through interactions between active spine clusters (Rall and Segev, 1987).

Distal dendritic branches with spines that receive excitatory synaptic inputs are a basic component of olfactory and hippocampal pyramidal neurons, and we

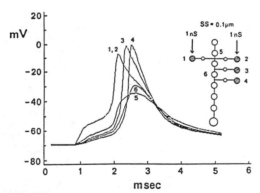

FIGURE 11.   Interactions between distal dendritic spines with active membrane. Inset shows model of a distal dendrite of a cortical pyramidal neuron with 4 spines, each with a spine stem (SS) diameter of 0.1 μ. An excitatory synaptic conductance of 1 nS is activated simultaneously in spines 1 and 2 and maintained for 1 msec. An impulse is generated in both spines 1 and 2, which spreads to generate impulses in spines 3 and 4. The transients generated in the different compartments are indicated in the graph. The spine system under these conditions functions as an AND gate. See text. Shepherd and Brayton (1987).

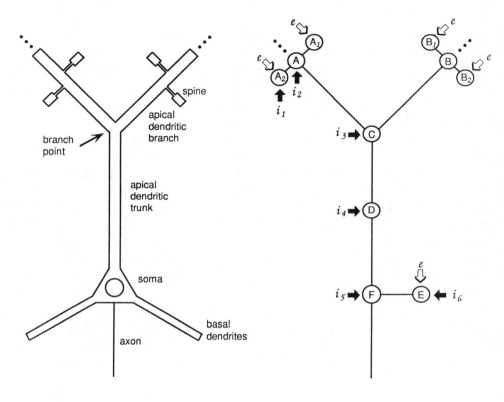

A. Schematic representation of
reduced pyramidal neuron

B. Pyramidal neuron as a branching
system of nodes for logic operations

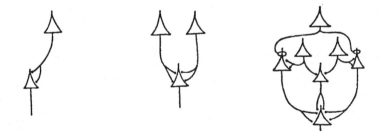

C. Original McCulloch-Pitts representation of neurons
as oversimplified nodes for logic operations

postulate that they are a component of neocortical pyramidal neurons as well. They thus may be considered as a canonical subunit of the canonical pyramidal neuron. They are another example of information processing based on the synapse as the functional unit rather than the neuron.

How do we represent these complexities of spine and dendritic function in a canonical model of the pyramidal neuron that, on the one hand, will contain the minimum of essential properties revealed by the experimental data while, on the other hand, being sufficiently reduced so that it can be incorporated into network simulations? A formal representation of the canonical pyramidal neuron that can serve this purpose is shown in Fig. 12. This builds on the models of Traub and Llinás (1979) and Traub *et al.* (1990), which have represented apical and basal dendrites, adding to those models the properties of dendritic spines and spine interactions. Key features are the presence of multilevel synaptic inhibition, and multilevel intrinsic membrane properties. These features give the dendritic tree the ability to generate nonlinear interactions and logic operations at every level along the distal-proximal extent of the dendritic tree. To pursue the logic paradigm, it has been pointed out that a dendritic tree can compute a complicated logic function, and that this can be proved to include any threshold function (Shepherd and Brayton, 1987). Such a threshold logic function has the form of

$$f(x) = 1 \text{ if } \sum w(i)x(i) \geq \theta, \text{ otherwise O},$$

where $x(i)$ are the inputs; $w(i)$ are the nonnegative weights, and θ is the threshold value. An overall function of this type can thus represent the sum of the separate logic functions carried out by a system. In the past, this formalism has

---

FIGURE 12.   The *canonical cortical pyramidal neuron:* (a) The minimum architecture necessary to represent the essential functional information-processing features of the cortical pyramidal neuron. These anatomical features include apical and basal dendrites; apical trunk; distal apical branches; dendritic spines. Physiological features include excitatory synaptic inputs to dendritic spines; inhibitory synaptic inputs to branches, trunk, and soma; Hodgkin–Huxley impulse in the soma and initial axon segment; various intrinsic membrane conductances (G_{Na}, $G_{K(CA)}$, G_{Ca}) distributed in the dendrites. (b) Representation of the canonical form in terms of functional compartments for carrying out a hierarchy of logic operations: A, B, spines and branches for synaptic logic operations; C, branch point for summing and gating of branch units; D, trunk sites for gating of transfer from sites of distal logic operations to soma; E, operations similar to A, B, in basal dendrites; F, global summation at the soma. Arrows i_1–i_6 indicate possible sites of synaptic inhibition, which can provide for AND-NOT operations at all hierarchical levels. (c) Examples of the representation of neurons as simple nodes for logic operations, in the original paper of McCulloch and Pitts (1943). From Shepherd *et al.* (1989).

been applied to the operation of networks, as in the original McCulloch–Pitts (1943) model, and more recent simulations (Hopfield, 1982). We suggest that cortical pyramidal neurons compute complicated logic functions due to nonlinear interactions within their dendritic trees, and that incorporation of these properties into network nodes will confer more realistic and much enhanced computational capacity onto those networks, as originally predicted by Rall (1970).

These considerations are of interest from the point of view of integrated circuits; in particular, the design of macro-cell chips. According to C. Sechen and T. Stanion (personal communication): "A common design method for a digital logic cell is for the inputs to feed the output in a converging tree structure. Given a particular tree structure, it is straightforward to enumerate the possible functions which that structure can implement, using a set of inhibitors which act at different levels of the tree structure." We are pursuing the development of circuits of this nature that can implement in silicon the computational design features of the canonical neuron illustrated in Fig. 12.

A final direction for the future is to inquire into the specific relevance of these models for cognitive functions of humans. The availability of human cortical tissue presents an unprecedented opportunity to pursue a direct analysis of the neuronal basis of human cortical function. We have postulated that there may be properties or sets of properties unique to human cortical cells. At present, we are analyzing preliminary evidence that human cortical neurons show a tendency toward more sustained impulse firing in response to injected current depolarization or excitatory synaptic inputs (Williamson *et al.*, 1991). We suggest that if human cortical neurons have such properties, it is likely that they are relevant to the generation of properties at the network level that are crucial for human cognitive function. This gives added emphasis to the need for developing models of cortical neurons that represent more adequately their computational organization along the lines discussed in this chapter.

Acknowledgments

This work was supported by research grants from the National Institute of Neurological Disease and Stroke, the National Institute of Deafness and other Communicative Disorders, and the Office of Naval Research. I am grateful to Anne Williamson, Ben Strowbridge, John Kauer, Charles Greer, Tom Woolf, Carl Sechen, and Ted Stanion for valuable discussions.

References

BODIAN, D. (1962). "The Generalized Vertebrate Neuron," *Science* **137**, 323–326.

BRENNAN, P., KABA, H., and KEVERNE, E. B. (1990). "Olfactory Recognition: A Simple Memory System," *Science* **250**, 1223–1226.

BULLOCK, T. H. (1959). "Neuron Doctrine and Electrophysiology," *Science* **129**, 997–1002.

CAJAL, S. RAMÓN y (1911). *Histologie du Système Nerveux de l'Homme et des Vertébrés Maloine, Paris. Translated by L. Azoulay.*

CAJAL, S. RAMÓN y (1991). *New Ideas on the Structure of the Nervous System of Man and Vertebrates.* Translated by N. Swanson and L. M. Swanson from *Les Nouvelles Idées sur la Structure du Système Nerveux ches l'Homme et chez les Vertébrés* (1894). MIT Press, Cambridge, Massachusetts.

CONNORS, B. W. and GUTNICK, M. J. (1990). "Intrinsic Firing Patterns of Diverse Neocortical Neurons," *Trends Neurosci.* **13**, 99–104.

COTMAN, C. W., and IVERSON, L. L. (1987). "Excitatory Amino Acids in the Brain—Focus on NMDA Receptors," *Trends Neurosci.* **7**, 263–265.

DEFELIPE, J., and JONES, E. G. (1988). *Cajal on the Cerebral Cortex.* Oxford University Press, New York.

DE CASTILLO, J., and KATZ, B. (1954). "Quantal Components of the End-Plate Potential," *J. Physiol. (London)* **124**, 500–573.

DOUGLAS, R. J., and MARTIN, K. A. C. (1990). "Neocortex," in Shepherd, G. M. (ed.), *The Synaptic Organization of the Brain* (pp. 389–438). Oxford University Press, New York.

ECCLES, J. C. (1953). *The Physiology of Nerve Cells.* John Hopkins, Baltimore.

EYZAGUIRRE, C., and KUFFLER, S. W. (1955). "Processes of Excitation in the Dendrites and in the Soma of Single Isolated Sensory Nerve Cells of the Lobster and Crayfish," *J. Gen. Physiol.* **39**, 87–119.

FIRESTEIN, S., and WERBLIN, F. S. (1987). "Gated Currents in Isolated Olfactory Receptor Neurons of the Larval Tiger Salamander," *Proc. Natl. Acad. Sci.* **84**, 6292–6296.

FIRESTEIN, S., and WERBLIN, F. (1989). "Ionic Mechanisms Underlying the Olfactory Response," *Science* **244**, 79–82.

FLESHMAN, J. W., SEGEV, I., and BURKE, R. E. (1988). "Electrotonic Architecture of Type-Identified Alpha-Motoneurons in the Cat Spinal Cord," *J. Neurophysiol.* **60**, 60–85.

FOEHRING, R. C., and WYLER, A. R. (1990). "Two Patterns of Firing in Human Neocortical Neurons," *Neurosci Letters.* **110** , 279–285.

GRANGER, R., AMBROS-INGERSON, J., and LYNCH, G. (1989). "Derivation of Encoding Characteristics of Layer II Cerebral Cortex," *J. Cog. Neurosci.* **1**, 61–87.

HABERLY, L. B. (1990). "Olfactory Cortex," in G. M. Shepherd (ed.), *The Synaptic Organization of the Brain* (pp. 317–345). Oxford University Press, New York.

HEDLUND, B., and SHEPHERD, G. M. (1983). "Biochemical Studies on Muscarinic Receptors in the Salamander Olfactory Epithelium," *FEBS Letters* **162**, 428–431.

HEDLUND, B., MASUKAWA, L. M., and SHEPHERD, G. M. (1987). "Excitable Properties of Olfactory Receptor Neurons," *J. Neurosci.* **7,** 2338–2344.

HODGKIN, A. L., and HUXLEY, A. F. (1952). "A Quantitative Description of Membrane Current and Its Application to Conduction and Excitation in Nerve," *J. Physiol. (London)* **117,** 500–544.

HODGKIN, A. L., and RUSHTON, W. A. H. (1946). "The Electrical Constants of a Crustacean Nerve Fibre," *Proc. Roy. Soc. London B* **133,** 444.

HOPFIELD, J. J. (1982). "Neural Networks and Physical Systems with Emergent Collective Computational Properties," *Proc. Natl. Acad. Sci. USA* **79,** 2554–2558.

JACK, J. J. B., NOBLE, D., and TSIEN, R. W. (1975). *Electric Current Flow in Excitable Cells.* Oxford University Press, Oxford.

JACKOWSKI, A., PARNEVALAS, J. G., and LIEBERMAN, A. R. (1978). "The Reciprocal Synapse in the External Plexiform Layer of the Mammalian Olfactory Bulb," *Brain Res.* **195,** 17–28.

JAHR, C. E., and NICOLL, R. A. (1982). "An Intracellular Analysis of Dendrodendritic Inhibition in the Turtle in Vitro Olfactory Bulb," *J. Physiol. (Lond.)* **326,** 213–234.

KATZ, B. (1969). *The Release of Neural Transmitter Substances.* Charles C. Thomas, Springfield, Illinois.

KOCH, C., and SEGEV, I. (1989). *Methods in Neuronal Modeling: From Synapses to Networks.* MIT Press, Cambridge, Massachusetts.

KOCH, C., POGGIO, T., and TORRE, V. (1983). "Nonlinear Interactions in a Dendritic Tree: Localization: Timing and Role in Information Processing," *Proc. Natl. Acad. Sci. USA,* **80,** 2799–2802.

KRIEGSTEIN, A. R., and CONNORS, B. W. (1986). "Cellular Physiology of the Turtle Visual Cortex: Synaptic Properties and Intrinsic Circuitry," *J. Neurosci.* **6,** 178–191.

LLINÁS, R. (1988). "The Intrinsic Electrophysiological Properties of Mammalian Neurons: Insights into Central Nervous System Function," *Science* **242,** 1654–1664.

LORENTE DE NO, R. (1938). "The Cerebral Cortex: Architecture, Intracortical Connections and Motor Projections," in J. F. Fulton (ed.), *Physiology of the Nervous System* (pp. 291–339). Oxford University Press, London.

McCORMICK, D. A., CONNORS, W. B., LIGHTHALL, J. W., and PRINCE, D. A. (1985). "Comparative Electrophysiology of Pyramidal and Sparsely Spine Stellate Neurons of the Neocortex," *J. Neurosci.* **54,** 782–806.

McCORMICK, D. A., and PAPE, H.-C. (1990). "Properties of a Hyperpolarization-Activated Cation Current and Its Role in Rhythmic Oscillation in Thalamic Relay Neurons," *J. Physiol.* **431,** 291–318.

McCULLOCH, W., and PITTS, W. (1943). "A Logical Calculus of Ideas Immanent in Nervous Activity," *Bull Math Biophys.* **5,** 115–133.

McLENNAN, H. (1971). "The Pharmacology of Inhibition of Mitral Cells in the Olfactory Bulb," *Brain Res.* **29,** 177–187.

MORI, K. (1987). "Membrane and Synaptic Properties of Identified Neurons in the Olfactory Bulb," *Progress in Neurobiol.* **29,** 275–320.

MORI, K., NOWYCKY, M. C., and SHEPHERD, G. M. (1981). "Electrophysiological Analysis of Mitral Cells in Isolated Turtle Olfactory Bulb," *J. Physiol.* (*Lond.*) **314,** 281–294.

MOUNTCASTLE, V. B., TALBOT, W. H., SAKATA, H., and HYVÄRINEN, J. (1968). "Cortical Neuronal Mechanisms in Fluter-Vibration Studied in Unanesthetized Monkeys, Neuronal Periodicity and Frequency Discrimination," *J. Neurophysiol.* **32,** 452–484.

NICOLL, R. A. (1969). "Inhibitory Mechanisms in the Rabbit Olfactory Bulb: Dendrodendritic Mechanisms," *Brain Res.* **14,** 157–172.

NICOLL, R. A. (1971). "Pharmacological Evidence for GABA as the Transmitter in Granule Cell Inhibition in the Olfactory Bulb," *Brain Res.* **35,** 137–149.

PETERS, A., and JONES, E. G. (eds.), (1984). *Cerebral Cortex, Vol. 1: Cellular Components of the Cerebral Cortex.* Plenum, New York.

PONGRACZ, F., FIRESTEIN, S., and SEPHERD, G. M. (1991a). "Electrotonic Structure of Olfactory Sensory Neurons Analyzed by Intracellular and Whole Cell Patch Techniques," *J. Neurophysiol.* **65,** 747–758.

PONGRACZ, F., POOLOS, N. P., KOCSIS, J. D., and SEPHERD, G. M. (1991b). "A Model of NMDA Receptor-Mediated Activity in Hippocampal Pyramidal Cells" (submitted).

PRICE, J. L., and POWELL, T. P. S. (1970). "The Morphology of Granule Cells of the Olfactory Bulb," *J. Cell Sci.* **7,** 91–123.

RAKIC, P., GOLDMAN-RAKIC, P. S., and GALLAGER, D. (1988). "Quantitative Autoradiography of Major Neurotransmitter Receptors in the Monkey Striate and Extrastriate Cortex," *J. Neurosci.* **8**(10), 3670–3690.

RALL, W. (1959) "Branching Dendritic Trees and Motoneuron Membrane Resistivity," *Exp. Neurol.* **1,** 491–527.

RALL, W. (1962) "Electrophysiology of a Dendritic Neuron Model," *Biophys. J.* **2,** 145–167.

RALL, W. (1964) "Theoretical Significance of Dendritic Trees and Motoneuron Input–Output Relations," in R. F. Reiss (ed.), *Neural Theory and Modelling* (pp. 73–97). Stanford University Press, Stanford, California.

RALL, W. (1967) "Distinguishing Theoretical Synaptic Potentials Computed for Different Soma-Dendritic Distributions of Synaptic Input," *J. Neurophysiol.* **30,** 1138–1168.

RALL, W. (1970) "Cable Properties of Dendrites and Effects of Synaptic Location," in P. Anderson and J. K. S. Jansen (eds.), *Excitatory Synaptic Mechanisms* (pp. 175–187). Universitetsforlag, Oslo, Norway.

RALL, W. (1974) "Dendritic Spines, Synaptic Potency in Neuronal Plasticity," in C. D. Woody, K. A. Brown, T. J. Crow, and J. D. Knispel (eds.), *Cellular Mechanisms Subserving Changes in Neuronal Activity* (pp. 13–21). Brain Information Service, Los Angeles.

RALL, W. (1977) "Core Conductor Theory and Cable Properties of Neurons," in E. R. Kandel (ed.), *The Nervous System, Vol. I: Cellular Biology of Neurons, Part 1* (pp. 39–97). American Physiological Society, Bethesda, Maryland.

RALL, W. (1990) "Perspectives on Neuronal Modeling," in M. D. Binder and L. M. Mendell (eds.), *The Segmental Motor System* (pp. 129–149). Oxford University Press, New York.

RALL, W., and RINZEL, J. (1971). "Dendritic Spine Function and Synaptic Attenuation Calculations," *Soc. Neurosci. Abstracts.* **1,** 64.

RALL, W. and RINZEL, J. (1973) "Branch Input Resistance and Steady Attenuation for Input to One Branch of a Dendritic Neuron Model," *Biophys. J.* **13,** 648–688.

RALL, W. and SEGEV, I. (1987) "Functional Possibilities for Synapses on Dendrites and on Dendritic Spines," in G. M. Edelman, W. F. Gall, and W. M. Cowan (eds.), *Synaptic Function (Neurosci. Res. Found.)* (pp. 605–636). New York, John Wiley.

RALL, W. and SHEPHERD, G. M. (1965) "Theoretical Reconstruction of Olfactory Bulb Potential," *XXIII Int. Physiol. Congr.*

RALL, W. and SHEPHERD, G. M. (1968) "Theoretical Reconstruction of Field Potentials and Dendro-Dendritic Synaptic Interactions in Olfactory Bulb," *J. Neurophysiol.* **31,** 884–915.

RALL, W. SHEPHERD, G. M., REESE, T. S., and BRIGHTMAN, M. W. (1966) "Dendrodendritic Synaptic Pathway for Inhibition in the Olfactory Bulb," *Exp. Neurol.* **14,** 44–56.

RAMÓN-MOLINER, E. (1962). "An Attempt at Classifying Nerve Cells on the Basis of Their Dendritic Patterns," *J. Comp. Neurol.* **119,** 211–227.

RIBAK, C. E., VAUGHN, J. E., SAITO, K., BARBER, R., and ROBERTS, E. (1977). "Glutamate Decarboxylase Localization in Neurons in the Olfactory Bulb," *Brain Res.* **126,** 1–18.

ROSS, E. (1990). "Signal Sorting and Amplification through G Protein-Coupled Receptors," *Neuron* **3,** 141–152.

RUMELHART, D. E., and MCCLELLAND, J. L. (1986). *Parallel Distributed Processing: Explorations in the Microstructure of Cognition, Vol. 1: Foundations.* MIT Press, Cambridge, Massachusetts.

SCHWARTZKROIN, P. A., and MUELLER, A. L. (1987). "Electrophysiology of Hippocampal Neurons," in E. G. Jones and A. Peters (eds.), *Cerebral Cortex: Further Aspects of Cortical Function, including Hippocampus, Vol. 6.* Plenum Press, New York.

SCOTT, J. W., and HARRISON, T. A. (1987). "The Olfactory Bulb: Anatomy and Physiology," in T. E. Finger and W. L. Silver (eds.), *Neurobiology of Taste and Smell* (pp. 157–178). John Wiley, New York.

SECHEN, C., and STANION, T. (1991). (personal communication).

SHEPHERD, G. M. (1972). "The Neuron Doctrine: A Revision of Functional Concepts," *Yale J. Biol. Med.* **45,** 584–589.

SHEPHERD, G. M. (1974). *The Synaptic Organization of the Brain.* Oxford University Press, New York.

SHEPHERD, G. M. (1977). "The Olfactory Bulb: A Simple System in the Mammalian Brain," in E. R. Kandel (ed.), *Handbook of Physiology, Sect. 1: The Nervous System, Part 1: Cellular Biology of Neurons* (pp. 945–968). American Physiological Society, Bethesda, Maryland.

SHEPHERD, G. M. (1979). *The Synaptic Organization of the Brain,* 2nd ed. Oxford University Press, New York.

SHEPHERD, G. M. (1990). "The Significance of Real Neuron Architectures for Neural Network Simulations," in E. Schwartz (ed.), *Computational Neuroscience.* MIT Press, Cambridge, Massachusetts.

SHEPHERD, G. M. (1991). *Foundations of the Neuron Doctrine*. Oxford University Press, New York.

SHEPHERD, G. M. and BRAYTON, R. K. (1987). "Logic Operations are Properties of Computer-Simulated Interactions between Excitable Dendritic Spines, *Neuroscience* **21,** 151–166.

SHEPHERD, G. M. and KOCH, C. (1990). "Introduction to Synaptic Circuits," in G. M. Shepherd (ed.), *The Synaptic Organization of the Brain* (pp. 3–31). Oxford University Press, New York.

SHEPHERD, G. M., BRAYTON, R. K., MILLER, J. F., SEGEV, I., RINZEL, J., and RALL, W. (1985). "Signal Enhancement in Distal Cortical Dendrites by Means of Interactions between Active Dendritic Spines," *Proc. Natl. Acad. Sci. USA* **82,** 2192–2195.

SHEPHERD, G. M., CARNEVALE, N. T., and WOOLF, T. B. (1989). "Comparisons between Computational Operations Generated by Active Responses in Dendritic Branches and Spines," *J. Cognit. Neurosci.* **1,** 273–286.

SHEPHERD, G. M. and GREER, C. A. (1988). "The Dendritic Spine: Adaptations of Structure and Function for Different Types of Synaptic Integration," in R. Lasek, and M. Black (eds.), *Intrinsic Determinants of Neuronal Form and Function* (pp. 245–314). Alan R. Liss, New York.

STAFSTROM, C. E., SCHWINDT, P. C., and CRILL, W. E. (1984). "Cable Properties of Layer V Neurons from Cat Sensorimotor Cortex *in vitro*," *J. Neurophysiol.* **52,** 278–289.

STROWBRIDGE, B. W. SHEPHERD, G. M., SPENCER, D. D., and MASUKAWA, L. M. (1987). "Intracellular Responses of Human Cortical Biopsies Maintained *in vitro*," *Soc. Neurosci. Absts.* **13,** 941.

TRAUB, R. D. and LLINÁS, R. (1979). "Hippocampal Pyramidal Cells: Significance of Dendritic Ionic Conductances for Neuronal Function and Epileptogenesis," *J. Neurophysiol.* **59,** 1352–1376.

TRAUB, R. D., MILES, R., and WONG, R. K. S. (1989). "Models of the Origin of Rhythmic Population Oscillations in the Hippocampal Slice," *Science* **243,** 1319–1325.

TROMBLEY, P. Q., and WESTBROOK, G. L. (1990). "Excitatory Synaptic Transmission in Cultures of Rat Olfactory Bulb," *J. Neurophysiol.* **64,** 598–606.

TUCKWELL, H. C. (1988). *Introduction to Theoretical Neurobiology, Vol. 1: Linear Cable Theory and Dendritic Structure; Vol. 2: Nonlinear and Stochastic Theories*. Cambridge University Press, Cambridge.

VON NEUMANN, J. C. (1958). *The Computer and the Brain*. Yale University Press, New Haven, Connecticut.

VON WALDEYER-HARTZ, H. W. G. (1958). "Über Einige Neuere Forschungen im Gebiete der Anatomie des Centralnervensystems," *Deutsch. Med. Wschr.* **17,** 1213–1218, 1244–1246, 1267–1269, 1287–1289, 1331–1332, 1352–1356.

WHITE, E. L. (1989). *Cortical Circuits, Synaptic Organization of the Cerebral Cortex: Structure, Function, and Theory*. Birkhüser, Boston.

WILLIAMSON, A., SHEPHERD, G. M., and SPENCER, D. D. (1991). "Electrophysiological Comparison of Human and Rodent Dentate Granule Cells Maintained *in vitro*," *Soc. Neurosci. Abst.,* **17,** 1253.

WILLIAMSON, A., McCORMICK, D. A., SHEPHERD, G. M., and SPENSER, D. S.

(1990). "Intracellular Recording from Human Dentate Granule Cells: Evidence for Hyperexcitability," *Epilepsia* **31,** 625.

WILSON, M., and BOWER, J. (1989). "The Simulation of Large-Scale Neural Networks," in C. Koch and I. Segev (eds.), *Methods in Neuronal Modelling* (pp. 291–334). MIT Press, Cambridge, Massachusetts.

WOOLF, T. B., SHEPHERD, G. M., and GREER, C. A. (1991). "Serial Reconstitutions of Granule Cell Spines in the Mammalian Olfactory Bulb," *Synapse* **7,** 181–192.

Chapter 3

Computational Models of Hippocampal Neurons

BRENDA J. CLAIBORNE[1], ANTHONY M. ZADOR,[2]
ZACHARY F. MAINEN,[2] AND THOMAS H. BROWN[2]

[1]Division of Life Sciences
University of Texas at San Antonio
San Antonio, Texas
[2]Department of Psychology
Yale University
New Haven, Connecticut

There is now substantial evidence that the hippocampal formation plays a pivotal role in certain aspects of learning and memory in humans and other animals (Press *et al.*, 1989; Squire, 1987), but almost nothing is known at the cellular or circuit level about how the hippocampus computes its suspected mnemonic functions (Brown and Zador, 1990). As part of an overall effort designed to address this issue, we have begun to create realistic models of the principal classes of neurons and their synapses. We decided to explore the computational capabilities of single neurons before considering more complex properties that may emerge at the circuit level, although the latter is the ultimate target of this effort.

In this chapter, we first describe and illustrate the methods and their rationale, focusing on pyramidal neurons of the CA1 region of the hippocampus. In particular, we show how morphometric and neurophysiological properties can be combined and incorporated into compartmental models that enable efficient simulations. We also describe the graphical display system, which we have found to be essential for interpreting the results of the simulations. For those unfamiliar with the history of this field, we briefly review some key technical advances—

in the areas of neuromorphometry, neurophysiology, and neurocomputing—that have made this kind of research possible.

Having laid the methodological foundation, we then illustrate the application of the simulator to the problem of understanding intracellular signalling of synaptic potentials. A model of a CA1 hippocampal neuron is used to illustrate the decrement of synaptic potentials as they spread throughout the dendritic arbor. Additional applications are presented in the following chapter, where we examine these voltage gradients in more detail and show that they have important implications both for experimental neurobiologists and theoreticians. Of special interest to us is the consequence of these gradients on a kind of synaptic plasticity, called long-term potentiation (LTP), that occurs in the hippocampus and many other brain structures (Brown *et al.*, 1988b, 1991).

I. Neuromorphometry

Since the introduction of the Golgi staining technique in the late 1800s, it has been known that neurons in the mammalian cortex exhibit complex dendritic branching patterns (Cajal, 1911; Golgi, 1886). Obtaining quantitative three-dimensional data about these branching patterns from Golgi material, however, has presented several major problems. It is not always possible to visualize the three-dimensional structure of the dendrites from a single section of Golgi-impregnated material—the dendritic trees of cortical neurons often extend through several hundred microns, while Golgi-impregnated sections are rarely thicker than 150 μm and are often as thin as 75 μm. To capture an entire Golgi-stained neuron requires that the dendritic branches be traced through several or more serial sections. In the past, this was most often accomplished by tracing the neurons with the aid of a camera lucida attached to a compound microscope.

This tracing technique, although useful for qualitative studies, was not adequate for quantitative studies. Even when dendrites were traced in their entirety, it was not possible to incorporate accurately the third dimension, or depth, into the data. Quantitative analyses, therefore, were based primarily on two-dimensional measurements taken from camera-lucida drawings. Only with the introduction of computer-assisted reconstruction systems in the 1970s was it possible to obtain accurate three-dimensional measurements of complete dendritic trees (Capowski and Sedivec, 1981; Lindsey, 1977; Macagno *et al.*, 1979; Wann *et al.*, 1973). Computer–microscope systems allowed an operator to digitize a Golgi-stained neuron from serial sections and include depth measurements in the digitization. A number of dendritic parameters, including branch length and number, could then be obtained from the three-dimensional data files.

Reconstructing Golgi-impregnated neurons from serial sections, even with the aid of a computer, can be error-prone and time-consuming. (See Desmond and Levy, 1982, for a technical discussion.) For example, impregnated dendrites often break at the surface of a section, increasing the difficulty of tracing and aligning branches from one section to the next. Further, because many neurons are impregnated in Golgi sections, it is often difficult to trace the dendrites of a specific neuron through the maze of stained processes.

The use of intracellular labeling techniques in the mammalian cortex was the first step in alleviating these problems (Kelly and Van Essen, 1974). By injecting a single neuron with dye, the experimenter was able to control the number of stained neurons in any one cortical region, thereby eliminating the problem of dendritic overlap. For detailed studies of dendritic structure, horseradish peroxidase (HRP) soon became the preferred label because, unlike the more common fluorescent dyes, it resulted in a dense product that was not subject to photobleaching (Bishop and King, 1982; Jankowska et al., 1976, Kitai et al., 1976). However, the use of HRP for labeling cortical neurons did not immediately eliminate the need to reconstruct labeled neurons from serial sections. When neurons were filled either in vivo or in the in vitro brain slice preparation, the tissue had to be sectioned and then reacted with diaminobenzidine to produce a dense product from the injected HRP. Again, reconstructive techniques were needed for the analysis of complete dendritic trees.

This final problem was solved by filling neurons in brain slices in vitro (Dingledine, 1984) and visualizing the labeled neurons in whole mounts of the thick slices (Claiborne et al., 1986, 1990). Slices, 350 to 400 μm thick, of an appropriate brain region were maintained in vitro and individual neurons filled with HRP. With some modifications of the histological procedures (Claiborne et al., 1986), it was possible to leave the slices intact during the processing (i.e., not resection them). The slices were cleared with glycerol and complete dendritic trees were then visible in whole mounts of the thick slices. These labeled trees could be digitized in three dimensions directly from the slice, thus eliminating the need to reconstruct dendritic arbors from serial sections.

The combination of the morphological techniques—intracellular labeling in the in vitro slice preparation—and computer-assisted analysis of dendritic trees was first used to analyze the structure of granule cells in the dentate gyrus of adult (Claiborne et al., 1986, 1990) and developing rats (Rihn and Claiborne, 1990), and to describe the CA1 and CA3 pyramidal neurons in the hippocampus of young adult animals (Amaral et al., 1990; Ishizuka, 1988). These studies showed that distal dendrites, those most often neglected in Golgi studies, account for a significant proportion of the total dendritic length of hippocampal neurons. For example, approximately 40% of the total length of the granule cells is located

in the distal third of the molecular layer (Claiborne *et al.*, 1990), and about 20%
of the total length of a CA1 pyramidal cell is found in stratum lacunosum-
moleculare, that portion of the dendritic layer most distant from the cell body
(Amaral *et al.*, 1990). The basal dendrites located in stratum oriens account for
another 35% of the total length of a CA1 pyramid. Results also showed previous
estimates of the total dendritic length of CA1 pyramidal cells based on the Golgi
staining technique had greatly underestimated their size. Reports of the total
dendritic length of Golgi-impregnated CA1 neurons have ranged from 4,359 μm
(Englisch *et al.*, 1974) to 5,613 μm (Minkwitz, 1976), whereas the total length
of HRP-filled CA1 neurons was found to average about 13,000 μm (Amaral *et
al.*, 1990).

To quantify the structure of CA1 neurons for the simulation studies described
next, we are currently using a computer–microscope system designed by John
P. Miller of the University of California, Berkeley (with software written by
R. H. W. Nevin of the University of California, Berkeley, and IBM). This
system provides three-dimensional data that has sufficient resolution and accuracy
to be used for the neuronal modeling described shortly. It has two significant
advantages over earlier systems (Miller *et al.*, 1990; Nevin, 1989). First, linear
encoders, mounted on the microscope stage, are used to monitor stage position.
They provide measurements with an accuracy of 0.2 μm in the X and Y dimensions. Second, the filled neuron is viewed on a video monitor, rather than through
the microscope. The neuron is digitized directly from the monitor by the operator,
thereby reducing operator error and fatigue, and allowing dendritic diameters to
be recorded with greater precision.

Using this computer–microscope system, individual CA1 neurons are digitized
into a series of 3,000 to 4,000 data points. Each point consists of an $X, Y,$ and
Z coordinate (referenced to an arbitrary zero) and a diameter measurement. Other
dendritic parameters, such as branch points and termination points, are also
encoded. Dendritic spines are not included in these morphometric measurements,
but data from other studies (Harris and Stevens, 1989) can be incorporated into
the single-neuron model (Brown *et al.*, 1991). The resulting three-dimensional
data files are incorporated directly into the compartmental models, as described
next.

II.   Electrotonic Structure

Accurate morphometric data alone are not sufficient to construct a computational
model. To create such a model, the structural data must be combined with
information about the electrical properties of the neuron. In the present study,

we have only considered the *passive* (linear) electrical properties of the cells, although the simulator can easily accommodate nonlinear membrane properties as well as consequences of second messengers. In analyzing a system this complex, it is useful to gain some insights into the linear case before introducing the varieties of nonlinearities that are known to be present in these cells.

Thus, we will consider only what is termed the *electrotonic structure* of the neurons. Electrotonic structure refers to the combined effects of morphometry and linear electrical properties on the spread of voltage and charge throughout the neuron. To create an electrotonic model, three electrical properties of the cell must be specified: R_m, the specific membrane resistivity; C_m, the specific membrane capacitance; and R_i, the specific resistivity of the cytoplasm. If these are assumed to be homogeneous throughout the neuron, and if one neglects the problem of a somatic shunt (*cf.* Stratford *et al.*, 1989), then these three parameters, in conjunction with the morphology, uniquely determine the neuron's electrotonic structure.

Usually, C_m is assumed to be a biological constant with a value of about 1 μF/cm^2 (Brown *et al.*, 1988a; Brown *et al.*, 1981a; Haydon *et al.*, 1980; Jack *et al.*, 1975; Johnston and Lam, 1981; Takashima, 1976). The widely accepted value for R_i is 75–100 Ω-cm, but recently there have been suggestions that the value could be 200 Ω-cm or even larger (Shelton, 1985; Stratford *et al.*, 1989). The values of R_m were taken from measurements of the membrane time constant τ_m, using $R_m = \tau_m/C_m$ (Barrionuevo and Brown, 1983; Brown *et al.*, 1981b; Johnston, 1981; Spruston and Johnston, 1990). As a check on internal consistency among the combination of electrical constants and morphometry, we compared the input resistance R_N of the model, measured at the soma, with experimentally determined values.

Two technical advances over the last decade have improved the quality of the electrophysiological data required for measuring the steady-state and transient current and voltage responses of hippocampal neurons. The first was the development and application to brain slices of the single-microelectrode clamp (SEC) in conjunction with relatively low-resistance (20–50 MΩ) microelectrodes (Barrionuevo *et al.*, 1986; Brown and Johnston, 1983; Johnston and Brown, 1981, 1983). These so-called *sharp electrodes* have tip diameters of 0.2 μm or less. The SEC is a time-share system that allows a single microelectrode to alternate (at 3–12 kHz) between measuring membrane potential and passing current. When used in the current-clamp mode, it allows more accurate and more convincing determinations of transient and steady-state responses to current injections than the previous approach, which involved the use of a bridge circuit to cancel the electrode resistance.

When used in the voltage-clamp mode, the SEC permitted, for the first time, measurements of synaptic and other conductances (Barrionuevo *et al.*, 1986; Brown and Johnston, 1983; Johnston and Brown, 1981, 1983). In previous voltage-clamp systems, it was necessary to use two electrodes (Cole, 1968)—one for the measuring membrane potential and the other for passing current. A feedback circuit then minimized the difference between the desired and the actual voltage. Attempts to apply this two-electrode method directly to cortical neurons met with little success for several reasons. First, it was difficult to obtain dual intrasomatic impalements on small cortical cells. Second, even when this was possible, the likelihood of cell injury was great. Finally, there was the problem of cross talk between the two microelectrodes, which usually had to be glued together to assure that the tips were adjacent in the soma. The SEC eliminated these difficulties by using a single microelectrode to serve the function of two electrodes. The feedback circuit is similar to that used in a two-electrode system, except that it is discontinuous because of the switching back and forth between measuring current and passing current.

A second important technical advance was the development and application to ordinary brain slices of the whole-cell, patch-clamp method (Blanton *et al.*, 1989; Malinow and Tsien, 1990; Sakmann and Neher, 1983). A large (1–3 μm) electrode is advanced to the surface of the membrane, and then gentle suction is applied to form a gigaohm (GΩ) seal between the electrode and the membrane. The membrane within the lumen of the patch electrode is then ruptured with further suction, resulting in a very low access resistance. There are several advantages, including a much lower noise level and the ability to pass more current. In addition, by forming a GΩ seal, one eliminates the problem of leakage resistance associated with sharp electrodes. Because of the large diameter of the patch pipette, it is also much easier to inject the cell with dyes or other agents. We find that the noise level can be reduced about an order of magnitude lower than even the best recordings obtained using sharp electrodes with the SEC (Xiang *et al.*, 1990).

There is, however, one major disadvantage of this method. Because of the large diameter of the patch pipette, the contents of the electrode mix readily with the cell. As mentioned previously, this is an advantage for injecting substances, but the dialysis can produce unwanted effects on the cellular physiology. One possible solution to this problem of dialysis is to use the nystatin (or perforated) patch technique (Bashir *et al.*, 1991; Horn and Marty, 1988; Spruston and Johnston, 1990, 1992). With this method, the cell membrane is not ruptured, and the low access resistance is produced by incorporating nystatin, which is present in the electrode tip, into the part of the membrane that is within the lumen of the pipette. The disadvantage of the perforated patch method is that the access resistance is higher than with the conventional whole-cell method.

We have used data from experiments that utilize both sharp and patch-clamp electrodes to place limits on the values of R_m and R_N. The input resistance R_N obtained using the whole-cell patch can be almost an order of magnitude higher than that obtained using sharp electrodes. Using sharp electrodes, the value of R_N obtained with healthy CA1 pyramidal cells is typically 30–60 MΩ (Nobre and Brown, unpublished). Using the whole cell method, R_N can be greater than 200 MΩ (Malinow and Tsien, 1990). Obviously, it is important to know which of these is closer to the true value—the value that would have been obtained if the ideally non-obtrusive experiment could have been performed. Both methods begin with what can only be considered a major insult to the integrity of the neuron—penetration or dialysis. However, the direction of the error introduced by these insults is different. Penetration may produce a somatic shunt—a low-resistance pathway to ground caused by rupture of the membrane around the electrode. This would result in an underestimate of the true value of R_N. Using the whole-cell patch method, no such shunt would be expected. However, dialysis of the cell may increase the effective membrane resistivity by closing a variety of membrane channels, resulting in an overestimate of R_N. The perforated-patch method tends to give intermediate values. It may furnish the best single estimate because it may minimize the problems associated with the other two methods (Spruston and Johnston, 1990, 1992).

III.   Computer Simulations

A.   Varieties of Models

The electrotonic models that have been applied to cortical neurons over the past three decades span a wide range of complexity. At the one extreme, a model can be as simple as a single compartment. At the other extreme, it can seek to capture the full morphological and physiological complexity of the neuron. All of these models can be seen as reformulations and special cases of the system of coupled partial differential equations with the appropriate boundary conditions (Eqs. (1) and (2) in this chapter; Hines, 1984; Jack et al., 1975; Rall, 1977). These equations become nonlinear if voltage-dependent conductances are considered. The form of the model is limited only by the availability of computer power and data. More complex models have more parameters that must be determined experimentally.

In one class of model, which was popular in the 1970s and early 1980s, the dendritic tree was reduced to a single *equivalent cylinder* (Brown et al., 1981a; Jack et al., 1975; Johnston, 1981; Rall, 1977). Equivalent-cylinder models can be viewed as a first-order correction of the simplest single-compartment models. They assume that a special relationship—the so-called *3/2 power law*—holds at each branch point between parent and daughter processes:

$$d_{\text{parent}}^{3/2} = \sum_{\text{daughters}} d_{\text{daughter}}^{3/2},$$

where d is the diameter of the process. The equivalent-cylinder model also assumes that all of the dendritic processes terminate at the same electrotonic distance from the soma. If these and certain other conditions are met, then it is easy to show (Rall, 1977) that, for some purposes, the original branched structure can be replaced by a mathematically equivalent single cylinder. The appeal of models belonging to this class is two-fold. First, they permit the closed-form analytical solution for voltage as a function of current under a wide range of conditions (Rall, 1977), thereby reducing the requisite numerical calculations to the evaluation of a few hyperbolic trigonometric functions easily performed on a pocket calculator. Second, these models reduce the number of parameters required to a very manageable size. This reduced parameter set can be computed from electrophysiological data by either the method of the peeling of exponentials (Rall, 1977) or what has been called the *direct fitting* method (Stratford *et al.*, 1989).

The simplicity and elegance of the equivalent-cylinder model comes at a steep price. One limitation is that the preceding assumptions may not be appropriate. In the case of hippocampal pyramidal neurons, the 3/2 power law may not be a reasonable approximation. In addition, the different dendritic processes certainly do not terminate at the same electrotonic distance from the soma. The basilar dendrites are electrically much shorter than the apical dendrites, and the apical dendrites themselves terminate at different electrotonic distances (Claiborne *et al.*, 1991). A much more serious problem is that there are a number of important scientific issues that are entirely outside the scope of the equivalent cylinder approach. Often, for example, the nature of the research question requires a full representation of the dendritic arbor. The reason is that for some problems—including most of those that we are pursuing—it is necessary to understand the spread of synaptic potentials from one part of the arbor to every other part. A further limitation concerns nonlinear aspects of the physiology— the equivalent-cylinder approach does not lend itself to the inclusion of active membrane properties and second messengers.

Recognition of the need to incorporate this complexity has led to two complementary approaches. In one approach, analytical insights have been obtained by recasting traditional cable models in terms of two-port electrical networks (Carnevale and Johnston, 1982; Koch *et al.*, 1982). This formalism is sufficiently general that interesting statements can be formulated about neurons with arbitrary dendritic morphologies. For example, within this framework it is easy to show that, in the case of a passive neuron model, the attenuation of voltage from a dendritic site to the soma is always exactly equal to the attenuation of current from soma to synapse (Carnevale and Johnston, 1982). Many similar statements, applicable to all passive models, can be demonstrated. The second approach, which we have adopted, involves using numerical methods to simulate specific cases of interest on a computer. Selected examples are illustrated next and in the following chapter.

B.   Algorithms

Any simulation of electrical activity in a neuron must begin with the system of partial differential equations governing this activity (Hines, 1984; Jack et al., 1975; Rall, 1977). Each branch of the neuron is assumed governed by the one-dimensional cable equation:

$$\frac{1}{2\pi a}\frac{\partial}{\partial x}\left(\frac{\pi a^2}{R_i}\frac{\partial V}{\partial x}\right) = C_m\frac{\partial V}{\partial t} + a\frac{V}{R_m} + I_{SYN} + I_{HH}, \tag{1}$$

where $a(x)$ is the branch radius; $V(x,t)$ is the membrane potential; C_m, the specific membrane capacitance; R_m, the specific membrane resistivity; R_i, the specific cytoplasmic resistivity; $I_{SYN}(x,t)$, the membrane current due to synaptic conductances; and $I_{HH}(x,t)$, the membrane current due to any voltage-dependent channels.

For simplicity, here we consider only the passive case. At the junction of separate branches, conservation of charge requires that the axial current I_{axial} into the node, given by the term on the left of Eq. (1), equal the membrane current out of the node, given by the term on the right:

$$\frac{1}{A}\sum_{branches}\left(\frac{\pi a^2}{R_i}\frac{\partial V}{\partial x}\right) = C_m\left(\frac{\partial V}{\partial t}\right)\bigg|_{node} + a\frac{V}{R_m} + I_{SYN} + I_{HH}, \tag{2}$$

where A is the area of the node and the summation is over the branches to a node. In the special case of the *sealed end* boundary conditions, Eq. (2) can be written as $(\partial V/\partial x)|_{node} = 0$.

Typically, it is not these equations themselves that are solved. The reason is that they are formulated in terms of continuous temporal and spatial derivatives of voltage, whereas digital computers can deal only with discrete values. The computer simulations thus use a finite difference approximation (Press et al., 1988) in which the value of voltage at points on a discrete spatio-temporal lattice is used. Each partial differential equation is replaced with a system of difference equations. It has become traditional in the field of neural modeling to refer to a finite difference approximation as a compartmental model because each point in the spatial lattice can be thought of as a compartment.

A great deal of research in computer science and numerical analysis has been devoted to developing efficient methods for the solutions of partial differential equations (Press *et al.*, 1988). There are many possible finite difference approximations to a given partial differential equation, and the solutions of these different discretizations have different properties. A key requirement of any discretization scheme is stability, meaning that the solutions must not diverge to infinite values if the corresponding continuous equations do not. In what follows, we will discuss a wrong and a right way to simulate Eqs. (1) and (2) for a branched neuron. First, we show how the naive approach to discretization gives rise to an unstable system of equations unless certain stringent criteria are satisfied. Satisfying them produces a method in which the solution time scales with the third power of the number of compartments, resulting in unmanageably long simulation times for even moderately large structures. We then describe an alternative method that scales linearly with the number of compartments. For a more complete exposition, the interested reader is referred to Hines (1984), who first applied this second method to neuron equations, and Mascagni (1989), who has analyzed these methods in detail.

Beginning with a discrete approximation to the first derivative with respect to time in Eq. (1):

$$\frac{\partial V_x}{\partial t} \rightarrow \frac{V_x^{t+1} - V_x^t}{\Delta t},$$

where the superscript and subscript indicate, respectively, the points in time and space at which V is evaluated. Similarly, we can write the second spatial derivative as:

$$\frac{\partial^2 V}{\partial x^2} = \frac{\partial}{\partial x}\left(\frac{\partial V}{\partial x}\right) \rightarrow \frac{\left(\frac{\partial V}{\partial x}\bigg|_{x+1} - \frac{\partial V}{\partial x}\bigg|_x\right)}{\Delta x} \rightarrow \frac{V_{x+1} - 2V_x + V_{x-1}}{\Delta x^2}.$$

Combining these two discrete equations gives rise to a coupled system of algebraic (difference) equations; but at which time—t or $t + 1$—should one evaluate the spatial derivative? This seemingly trivial choice determines the properties of the solution.

If we evaluate the spatial derivative at t, then the resulting equation,

$$\frac{a}{2R_i} \frac{V_{x+1}^t - 2V_x^t + V_{x-1}^t}{\Delta x^2} = C_{\mathrm{m}} \frac{V_x^{t+1} - V_x^t}{\Delta t} + a \frac{V_x^t}{R_{\mathrm{m}}} + I',$$

is the forward Euler approximation. This equation can be solved explicitly for V_x^{t+1} in terms of V^t, as follows:

$$V_x^{t+1} = (V_{x+1}^t + V_{x-1}^t) \left(\frac{a\Delta t}{2\Delta x^2 C_{\mathrm{m}} R_i} \right)$$

$$+ V_x^t \left(1 - \frac{a\Delta t}{R_{\mathrm{m}} C_{\mathrm{m}}} - \frac{2a\Delta t}{\Delta x^2 C_{\mathrm{m}} R_i} \right) - \frac{\Delta t I'}{C_{\mathrm{m}}}, \quad (3)$$

so it is an explicit method.

This naive method is easy to implement, but it is unstable—V diverges to infinity unless $\Delta t \leqslant (R_i C_{\mathrm{m}} \Delta x^2)/a$ (Mascagni, 1989). That means that increasing spatial accuracy through an n-fold reduction in Δx requires an n^2 reduction in Δt, or an n^3 increase in computer time. For example, increasing the number of compartments from 10 to 100 might require a reduction of the time step from 50 microseconds to 50 nanoseconds. For large simulations, this monumental inefficiency makes the method unusable regardless of increases in computer speed.

An alternative is the backward Euler approximation, in which the spatial derivative is evaluated at $t + 1$:

$$\frac{a}{2R_i} \frac{V_{x+1}^{t+1} - 2V_x^{t+1} + V_{x-1}^{t+1}}{\Delta x^2} = C_{\mathrm{m}} \frac{V_x^{t+1} - V_x^t}{\Delta t} + a \frac{V_x^{t+1}}{R_{\mathrm{m}}} + I^{t+1}. \quad (4)$$

This implicit method is unconditionally stable for any combination of Δx and Δt. In the semi-implicit Crank–Nicolson scheme, the average of Eqs. (3) and (4) is used. Both methods share the disadvantage that V_x^{t+1} depends on $V_{x\pm1}^{t+1}$, which is also unknown. Solution requires the inversion at each time step of the matrix resulting from the coupling of each spatial compartment to the previous and next spatial compartments. The structure of the matrix is tridiagonal, since only the main diagonal and two off-diagonal elements are nonzero, as can be seen from the fact that V is evaluated only at those three locations in the equation. Inversion of an arbitrary matrix requires about order n^3 operations, where again n is the number of compartments, so on first analysis there is little apparent advantage to this method. However, the special structure of the matrix permits a tremendous increase in efficiency, because inversion of a tridiagonal matrix requires only order n operations. By taking advantage of the sparse structure of the matrix, the backward Euler method can be implemented so that it scales linearly with the number of compartments.

The matrix, corresponding to Eq. (4), is strictly tridiagonal only in the case of an unbranched neuron. In the case of a branched neuron, Eq. (2) creates far off-diagonal nonzero elements at each branch point. This complication prevented the widespread application of the implicit method, since it appeared that a full matrix inversion was required at each time step. While it is intuitively clear that the complexity of the matrix is not substantially increased by the inclusion of a few far off-diagonal elements—the matrix remains essentially sparse—it was not until Hines (1984) described a systematic method for numbering in compartments that this sparseness was exploited. Using this scheme, the advantages of the implicit method can be retained even when simulating branched dendritic structures.

C.   Simulator

Algorithmic efficiency is not the only requirement for a successful simulator. The ease of use is also critical. The amount of time required to specify a simulation and analyze the results can be longer than the time required to execute it. The challenge is to design a *user-friendly* simulator that is sufficiently general to accommodate a very wide range of hypothetical conditions, and one that is also convenient enough to permit interactive exploration of a large parameter space. We have adopted a two-fold strategy that combines a general-purpose neuron description language with a graphical front-end for analysis.

The neuron description language is based on the program NEURON, written by Michael Hines (1989). NEURON is a language, itself coded in C, that is tailored to the specification of a wide range of compartmental simulations. It incorporates the efficient algorithms discussed previously—as well as others for simulating active membranes—in a way that remains transparent to the user. We have extended NEURON so that morphological data from hippocampal neurons can be conveniently incorporated and combined with electrophysiological data.

To analyze the results of simulations, and to develop an intuitive understanding of the results, we have implemented a graphical display that permits the data to be viewed in a number of different ways. For example, the voltage in each compartment due to some stimulus can be color-coded, with *hot* colors like red and yellow indicating depolarization and *cold* colors like blue indicating hyperpolarization. Examples are illustrated in the next chapter. Snapshots of the neuron's state can be viewed individually, or the simulation can be played back in *animation mode*. Using these methods, the user can quickly grasp in a qualitative manner the tremendous amount of information—often on the order of 10 Mbytes—that constitutes a simulation. Alternatively, a more quantitative appreciation of simulation results can be obtained by plotting current, voltage, or some other variable as a function of time at just a few locations. An example of this approach, illustrated in Fig. 1, is discussed in the following section. The front-end of our simulator is sufficiently general that almost any display mode can be rapidly programmed. The practical limit is the imagination of the user.

IV. Methods and Results

In the next chapter, we discuss how dendrites might contribute to the computational capabilities of hippocampal neurons by allowing the synapses to self-organize across the dendritic arbor. Critical to this self-organization is the existence of steep, transient voltage gradients within the dendritic tree. In what follows, we will compare the voltage gradients due to distal and medial inputs to illustrate how we are combining morphological and electrophysiological data into the simulator.

We began by choosing a mature CA1 pyramidal neuron from our library of digitized neurons. In this case, the neuron had only a single main apical branch, although other neurons have primary apicals that bifurcate. This neuron was represented by about 3,000 compartments. The electrical constants were as follows: $R_m = 50$ $k\Omega$-cm^2, $C_m = 1\mu F$/cm^2, and $R_i = 100$ Ω-cm. Spines were incorporated into the membrane at a density of 2/μm (Harris and Stevens, 1989). We then selected two sites for synaptic inputs, which were located on two-compartment spines (Brown *et al.*, 1991; Zador *et al.*, 1990). Several sites were then selected to monitor the voltage response. The entire time required to specify these simulations was under one hour, and parametric variations could be specified in seconds.

The run-time for the simulations was short. Using our implementation of NEURON on a Sun 4/330 Unix workstation (16 MIPS/2.6 MFLOPS), the typical simulation speed is about 10,000 compartment steps/second; that is, the solution for a 3,000-compartment neuron can be advanced about 10 discrete steps/second. Using a time step of $\Delta t = 0.5$ msec, the 200 time steps were executed in about one minute of real time. Because the implicit scheme is unconditionally stable for all time steps and relatively accurate for even large ones, a very large time step (>10 msec) could have been used if the fine details of the evolution of the solution had not been of interest. This is useful, for example, when calculating the steady-state input resistance R_N.

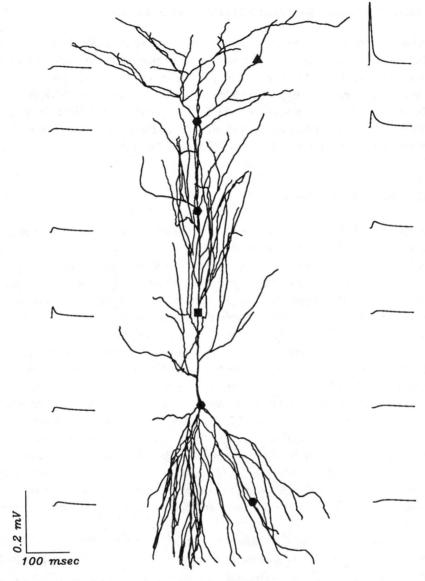

FIGURE 1. A distal synapse (*triangle*) and a proximal synapse (*square*) synapse were discharged in separate simulations and the resulting voltage transients at different locations in the dendritic tree measured. Activation consisted of a single presynaptic impulse, represented by a standard alpha function (Jack *et al.*, 1975; Brown and Johnston, 1983):

$$g_{SYN}(t) = g_{PEAK} \kappa t e^{-\alpha t},$$

with $E_{rev} = 0$ mV, $\alpha^{-1} = 1.5$ msec, $g_{peak} = 50$ pS, and the normalization factor $\kappa = \alpha e$. Voltage traces show the resultant activity (from distal synapse on the right, proximal on left) at different locations in the tree (hexagons).

74

Figure 1 shows the results from two separate simulations. The simulations consisted of activation of a single distal synapse (triangle) or a medial synapse (square). The synaptic waveform was represented by an α-function (Brown and Johnston, 1983; Jack *et al.*, 1975) with a peak conductance change of 50 pS, a time-to peak of 1.5 ms, and an equilibrium potential of 0 mV (Zador *et al.*, 1990). The traces show the voltage at each point marked with a symbol. The traces on the right column are responses to activation of the distal synapse, while those on the left column are responses to activation of the more proximal synapse. This method of representing the data permits a direct quantitative comparison between the two simulations. These same simulations can also be viewed in the *color animation mode* to obtain a more qualitative feeling for the data.

At least two results are immediately obvious from Fig. 1. First, steep voltage gradients result from transient synaptic inputs. In the case of the distal synapse, the peak voltage response even halfway down the dendritic tree is much smaller than at the site of the activity. In fact, the response is more than an order of magnitude smaller at the soma than at the site of the synapse. As discussed in the next chapter, these voltage gradients might play an important role in single-neuron computations. Second, the site of the input contributes to both the magnitude and shape of the response. Thus, the peak response at the distal site is much larger than at the medial site. While this is no surprise in principle—it is well-known that the higher input impedance at a distal dendritic branch causes a larger response to the same conductance change—the magnitude of the difference could not have been calculated without a simulation. These features are robust with respect to the range of uncertainty of the values of the electrical constants (Claiborne *et al.*, 1991).

V.   Summary and Conclusions

The last decade has witnessed great advances in neuromorphometry, neurophysiology and biophysics, and methods for creating and comprehending simulations of realistically represented neurons. In the current decade, the fruits of this progress will be a rapid growth in our understanding of the nature of the computations and information processing that occurs in single neurons and large circuits of interacting neurons. The field of computational neuroscience (Sejnowski *et al.*, 1988) can be expected to furnish the theoretical glue that is needed to integrate and comprehend the explosion of experimental knowledge that is now being generated. Some preliminary results that are pertinent to our particular fields of interest—learning, memory, and development—are presented in the following chapter. Once we create a more detailed library of the types of hippocampal neurons and their synapses, it will be possible to explore adaptive computations at the circuit level. These circuit models should furnish useful insights into both the development and mnemonic functions of the hippocampus.

The utility of this computational approach is obviously not limited to our particular interests. Other areas of psychology and neurobiology that are likely to benefit include studies of motor control, sensation and perception, and even emotion and cognition. Ultimately, we expect computational neuroscience also to yield important insights that will impact the fields of engineering and computer science—particularly in the areas of robotics, pattern recognition, adaptive process control, industrial inspection, learning theory, and parallel computation.

Acknowledgment

This work was supported by grants from the National Institute on Aging, the Office of Naval Research and the Defense Advanced Research Projects Agency.

References

AMARAL, D. G., ISHIZUKA, N., and CLAIBORNE, B. J. (1990). "Neurons, Numbers and the Hippocampal Network," in O. P. Ottersen, J. Storm-Mathiesen, and J. Zimmer (eds.), *Progress in Brain Research, Vol. 83,* (pp. 1–11). Elsevier Science Publishers B. V., Amsterdam.

BARRIONUEVO, G., and BROWN, T. H. (1983). "Associative Long-Term Potentiation in Hippocampal Slices," *Proceedings National Academy of Science USA* **80,** 7347–7351.

BARRIONUEVO, G., KELSO, S., JOHNSTON, D., and BROWN, T. H. (1986). "Conductance Mechanism Responsible for Long-Term Potentiation in Monosynaptic and Isolated Excitatory Inputs to Hippocampus," *Journal of Neurophysiology* **55,** 540–550.

BASHIR, Z. I., ALFORD, S., DAVIES, S. N., RANDALL, A. D., and COLLINGRIDGE, G. L. (1991). "Long-Term Potentiation of NMDA Receptor-Mediated Synaptic Transmission in the Hippocampus," *Nature (London)* **349,** 156–158.

BISHOP, G. H., and KING, J. S. (1982). "Intracellular Horseradish Peroxidase Injections for Tracing Neural Connections," in M.-M. Mesulam (ed.), *Tracing Neural Connections with Horseradish Peroxidase* (pp. 185–247). John Wiley, Chichester, New York.

BLANTON, M. G., LoTURCO, J. J., and KREIGSTEIN, A. R. (1989). "Whole Cell Recording from Neurons in Slices of Reptilian and Mammalian Cerebral Cortex," *Journal of Neuroscience Methods* **30,** 203–210.

BROWN, T. H., CHANG, V. C., GANONG, A. H., KEENAN, C. L., and KELSO, S. R. (1988a). "Biophysical Properties of Dendrites and Spines that May Control the Induction and Expression of Long-Term Synaptic Potentiation," in S. A. Deadwyler and P. W. Landfield (eds.), *Long-term Potentiation: From Biophysics to Behavior* (pp. 197–260). Liss, New York.

BROWN, T. H., CHAPMAN, P. F., KAIRISS, E. W., and KEENAN, C. L. (1988b). "Long-Term Synaptic Potentiation," *Science* **242,** 724–728.

BROWN, T. H., FRICKE, R. A., and PERKEL, D. H. (1981b). "Passive Electrical Constants in Three Classes of Hippocampal Neurons," *Journal of Neurophysiology* **46**, 812–827.

BROWN, T. H., and JOHNSTON, D. (1983). "Voltage-Clamp Analysis of Mossy Fiber Synaptic Input to Hippocampal Neurons," *Journal of Neurophysiology* **50**, 487–507.

BROWN, T. H., PERKEL, D. H., NORRIS, J. C., and PEACOCK, J. H. (1981a). "Electrotonic Structure and Specific Membrane Properties of Mouse Dorsal Root Ganglion Neurons," *Journal of Neurophysiology* **45**, 1–15.

BROWN, T. H., and ZADOR, A. M. (1990). "The Hippocampus," in G. Shepherd (ed.), *The Synaptic Organization of the Brain* (pp. 346–388). Oxford University Press, New York.

BROWN, T. H., ZADOR, A. M., MAINEN, Z. F., and CLAIBORNE, B. J. (1991). "Hebbian Modifications in Hippocampal Neurons," in J. Davis and M. Baudry (eds.), *Long-Term Potentiation: A Debate of Current Issues* (pp. 357–389). MIT Press, Cambridge, Massachusetts.

CAJAL, S. RAMÓN Y (1911). *Histologie du Système Nerveux de l'Homme et des Vertébrés*, (*Vol. 2*). Maloine, Paris.

CAPOWSKI, J. J., and SEDIVEC, M. J. (1981). "Accurate Computer Reconstruction and Graphics Display of Complex Neurons Utilizing State-of-the-Art Interactive Techniques," *Computers and Biomedical Research* **14**, 518–532.

CARNEVALE, N. T., and JOHNSTON, D. (1982). "Electrophsiological Characterization of Remote Chemical Synapses," *Journal of Neurophysiology* **47**, 606–621.

CLAIBORNE, B. J., AMARAL, D. G., and COWAN, W. M. (1986). "A Light and Electron Microscopic Analysis of the Mossy Fibers of the Rat Dentate Gyrus," *Journal of Comparative Neurology* **246**, 435–458.

CLAIBORNE, B. J., AMARAL, D. G., and COWAN, W. M. (1990). "Quantitative Three-Dimensional Analysis of Granule Cell Dendrites in the Rat Dentate Gyrus," *Journal of Comparative Neurology* **302**, 206–219.

CLAIBORNE, B. J., ZADOR, A. M, MAINEN, Z. F., and BROWN, T. H., (1991). "Electrotonic Structure of Hippocampal Neurons," in preparation.

COLE, K. S. (1968). *Membranes, Ions and Impulses*. University of California Press, Berkeley California.

DESMOND, N. L., and LEVY, W. B. (1982). "A Quantitative Anatomical Study of the Granule Cell Dendritic Fields of the Rat Dentate Gyrus Using a Novel Probabilistic Method," *Journal of Comparative Neurology* **212**, 131–145.

DINGLEDINE, R. (1984). *Brain Slices*. Plenum Press, New York.

ENGLISCH, H.-J., KUNZ, G., and WENZEL, J. (1974). "Zur Spines-Verteilung an Pyramidenneuronen der CA-1-Region des Hippocampus der Ratte," *Zeitschrift für mikroskopisch-anatomische Forschung* **88**, 85–102.

GOLGI, C. (1886). *Sulla Fina Anatomia degli Organi Centrali del Sistems Nervoso*. U. Hoepli, Milano.

HARRIS, K. M., and STEVENS, J. K. (1989). "Dendritic Spines of Rat Cerebellar Purkinje Cells: Serial Electron Microscopy with Reference to Their Biophysical Characteristics," *Journal of Neuroscience* **9**, 2982–2997.

HAYDON, D. A., REQUENA, J., and URBAN, B. W. (1980). "Some Effects of the Aliphatic

Hydrocarbons on the Electrical Capacity and Ionic Currents of the Squid Giant Axon Membrane," *Journal of Physiology* **309**, 229–245.

HINES, M. (1984). "Efficient Computation of Branched Nerve Equations," *International Journal of Bio-Medical Computation* **15**, 69–76.

HINES, M. (1989). "A Program for Simulation of Nerve Equations with Branching Geometries," *International Journal of Bio-Medical Computation* **24**, 55–68.

HORN, R., and MARTY, A. (1988). "Muscarinic Activation of Ionic Currents Measured by A New Whole-Cell Recording Method," *Journal of General Physiology* **92**, 145–159.

ISHIZUKA, N. (1988). "On the Length of the Dendrites of the CA3 Pyramidal Neurons in the Rat Hippocampus," *Acta Anatomica Nipponica* **63**, 376.

JACK, J., NOBLE, A., and TSIEN, R. W. (1975). *Electrical Current Flow in Excitable Membranes*. Oxford Press, London.

JANKOWSKA, E., RASTAD, J., and WESTMAN, J. (1976). "Intracellular Application of Horseradish Peroxidase and Its Light and Electron Microscopic Appearance in Spinocervical Tract Cells," *Brain Research* **105**, 557–562.

JOHNSTON, D. (1981). "Passive Cable Properties of Hippocampal CA3 Pyramidal Neurons," *Cellular and Molecular Neurobiology* **1**, 41–55.

JOHNSTON, D. and BROWN, T. H. (1981). "Giant Synaptic Potential Hypothesis for Epiletiform Activity," *Science* **211**, 294–297.

JOHNSTON, D., and BROWN, T. H. (1983). "Interpretation of Voltage-Clamp Measurements in Hippocampal Neurons," *Journal of Neurophysiology* **50**, 464–484.

JOHNSTON, D., and LAM, D. (1981). "Regenerative and Passive Membrane Properties of Isolated Horizontal Cells from a Teleost Retina," *Nature* **292**, 451–454.

KELLY, J. P., and VAN ESSEN, D. C. (1974). "Cell Structure and Function in Visual Cortex of the Cat," *Journal of Physiology (London)* **238**, 515–547.

KITAI, S. T., KOCSIS, J. D., PRESTON, R. J., and SUGIMORI, M. (1976). "Monosynaptic Inputs to Caudate Neurons Identified by Intracellular Injection of Horseradish Peroxidase," *Brain Research* **109**, 601–606.

KOCH, C., POGGIO, T., and TORRE, V. (1982). "Retinal Ganglion Cells: A Functional Interpretation of Dendritic Morphology," *Proceedings Royal Society, London B* **298**, 227–264.

LINDSEY, R. D. (1977). *Computer Analysis of Neuronal Structures*. Plenum Press, New York.

MACAGNO, E. R., LEVINTHAL, C., and SOBEL, I. (1979). "Three-Dimensional Computer Reconstruction of Neurons aned Neuron Assemblies," *Annual Review of Biophysics and Bioengineering* **8**, 323.

MALINOW, R., and TSIEN, R. W. (1990). "Presynaptic Enhancement Shown by Whole-Cell Recordings of Long-Term Potentiation in Hippocampal Slices," *Nature* **346**, 177–180.

MASCAGNI, M. V. (1989). "Numerical Methods for Neuronal Modeling," in C. Koch and I. Segev (eds.), *Methods in Neuronal Modeling: From Synapses to Networks* (pp. 439–484). MIT Press, Cambridge, Massachusetts.

MILLER, J. P., TROMP, J. W., and NEVIN, R. H. W. (1990). "Strategies for the Analysis and Modeling of Current Transients in Neurons with Complex Morphology," in *Neural Computation, 1990 Short Course 3 Syllabus, Society for Neuroscience* (pp. 10–22).

MINKWITZ, H.-G. (1976). "Zur Entwicklung der Neuronenstruktur des Hippocampus während der prä- und postnatalen Ontogenese der Albinoratte, III. Mitteilung: Morphometrische Erfassung der ontogenetischen Veränderungen in Dendritenstruktur und Spinebesatz an Pyramidenneuronen (CA1) des Hippocampus," *Journal für Hirnforschung* **17**, 255–272.

NEVIN, R. H. W. (1989). *Morphological Analysis of Neurons in the Cricket Cercal System.* Unpublished doctoral dissertation, University of California, Berkeley, California.

PRESS, G. A., AMARAL, D. G., and SQUIRE, L. R. (1989). "Hippocampal Abnormalities in Amnesic Patients Revealed by High-Resolution Magnetic Resonance Imaging," *Nature* **341**, 54–57.

PRESS, W. H., FLANNERY, B. P., TEUKOLSKY, S. A., and VETTERLING, W. T. (1988). *Numerical Recipes: The Art of Scientific Computing.* Cambridge University Press, Cambridge, U.K.

RALL, W. (1977). "Core Conductor Theory and Cable Properties of Neurons," in E. R. Kandel (ed.), *Handbook of Physiology: The Nervous System, Vol. 1, Sect. 1* (pp. 39–97). American Physiological Society, Bethesda, Maryland.

RIHN, L. L., and CLAIBORNE, B. J. (1990). "Dendritic Growth and Regression in Rat Dentate Granule Cells during Late Postnatal Development," *Development Brain Research* **54**, 115–124.

SAKMANN, B., and NEHER, E. (eds.) (1983). *Single Channel Recordings.* Plenum Press, New York.

SEJNOWSKI, T. J., KOCH, C., and CHURCHLAND, P. S. (1988). "Computational Neuroscience," *Science* **241**, 1299–1306.

SHELTON, D. P. (1985). "Membrane Resistivity Estimated for the Purkinje Neuron by Means of a Passive Computer Model," *Neuroscience* **14**, 11–131.

SPRUSTON, N., and JOHNSTON, D. (1990). "Whole-Cell Patch Clamp Analysis of the Passive Membrane Properties of Hippocampal Neurons," *Society for Neuroscience Abstract* **16**, 1297.

SPRUSTON, N. and JOHNSTON, D. (1992). "Perforated Patch-Clamp Analysis of the Passive Membrane Properties of Three Classes of Hippocampal Neurons," *Journal of Neurophysiology,* in press.

SQUIRE, L. R. (1987). *Memory and the Brain.* Oxford University Press, New York.

STRATFORD, D., MASON, A., LARKMAN, A., MAJOR, G., and JACK, J. (1989). "The Modeling of Pyramidal Neurons in the Visual Cortex," in R. Durbin, C. Miall, and G. Mithchison (eds.), *The Computing Neuron* (pp. 296–321). Addison-Wesley, London.

TAKASHIMA, S. (1976). "Membrane Capacity of Squid Giant Axon during Hyper- and Depolarizations," *Journal of Membrane Biology* **27**, 21–39.

WANN, D. F. WOOLSEY, T. A., DIERKER, M. L., and COWAN, W. M. (1973). "An

On-Line Digital Computer System for the Semiautomative Analysis of Golgi-Impregnated Neurons," *IEEE Transactions on Biomedical Engineering* **20,** 233–247.

WILSON, W., and GOLDNER, M. A. (1975). "Voltage Clamping with a Single Microelectrode," *Journal of Neurobiology* **6,** 411–422.

XIANG, Z., KAIRISS, E. W., KEENAN, C. L., LANDAW, E., and BROWN, T. H. (1990). "Quantal Model of Synaptic Current Fluctuations in Hippocampal Neurons," *Society for Neuroscience Abstract* **16,** 492.

ZADOR, A., KOCH, C., and BROWN, T. H. (1990). "Biophysical Model of a Hebbian Synapse," *Proceedings National Academy of Science USA* **87,** 6718–6722.

Chapter 4   Hebbian Computations in Hippocampal Dendrites and Spines

THOMAS H. BROWN,[1] ANTHONY M. ZADOR,[1]
ZACHARY F. MAINEN,[1] AND BRENDA J. CLAIBORNE[2]

[1]*Department of Psychology*
Yale University
New Haven, Connecticut
[2]*Division of Life Sciences*
University of Texas at San Antonio
San Antonio, Texas

I. Introduction

Neurobiologists are often critical of connectionist studies that attempt to understand *brain-style* computations using *artificial neural networks*. These networks consist of interconnected sets of summation nodes, also called *processing elements* (PEs), which usually have very simple input–output relationships. The criticism is that these artificial neural networks fail to take into account the complexity that we know to characterize neurons and neuronal circuits. The rebuttal, of course, is that neurobiologists have failed to produce a computational theory that demonstrates or at least furnishes some guidance regarding which of the biological details are computationally important and which can be safely ignored or abstracted. The complexity issue thus cuts both ways, providing some interesting and potentially productive challenges.

    In this chapter, we begin to address this issue at the single-cell level, focusing on obvious differences between neurons and the PEs that are commonly used in studies of artificial neural networks. We consider three abstract categories of differences that could be computationally significant—space, time, and probability. The main emphasis is on spatial differences. PEs are points, meaning that they have no internal structure analogous to the dendritic tree structure of a neuron. To appreciate the consequences of this omission, we need to know what dendrites do. We must understand the computational significance of the fact that synapses are located on a tree structure that has certain passive and active electrical properties.

This understanding requires a theory of dendritic computation that takes into account the adaptive and dynamic characteristics of neuronal information processing. To be applicable to hippocampal and many other cortical neurons, such a theory should acknowledge two facts. First, some of the synapses display use-dependent modifications that are controlled in part by the voltage at the point of the synaptic input. This is true for some synapses that undergo a *Hebbian* form of long-term synaptic potentiation (LTP) (Kelso *et al.*, 1986; Zador *et al.*, 1990; Brown *et al.*, 1990). Second, at least for transient events, such as individual synaptic responses, the dendrites may not be isopotential (Claiborne *et al.*, 1991a,b). In pyramidal neurons, the voltages can be quite different in different parts of the dendritic arbor. Thus, synaptic modifications can be determined in part by the local dendritic potential, and the dendritic potential can be affected by the outcome of the modifications. Depending on the properties of the dendrites, this reciprocal relationship can easily be imagined to produce categories of adaptive computations that have not yet been captured in simple PEs. What follows is a preliminary elaboration of this possibility.

II.   Nodes and Neurons

A.   *Nodes*

Consider the architecture of a typical neural network node or PE (Fig. 1). A PE is usually a simple summation junction whose output is a weighted sum of its inputs passed through a transfer function:

$$a_i^{t+1} = \theta \left(\sum_{j=1}^{N} w_{ij}^t a_j^t \right), \tag{1}$$

where the output a_i of the ith element at time $t + 1$ is equal to the sum of N inputs a_j at time t weighted by associated *connection weights* w_{ij}. This sum, which is denoted I and often called an *input function,* can be negative if some of the weights are negative. The value of I is passed through a nonlinear transfer function $\theta(I)$, commonly represented by a sigmoid:

$$\theta(I) = \frac{1}{1 + e^{-I}}, \tag{2}$$

which is bounded between 0 and 1. The popularity of Eq. (1) and its variants stems in large part from their mathematical tractability.

B.   Neurons

The summation node (Fig. 1 and Eqs. (1) and (2)) does not capture the *stochastic, spatial,* and *temporal* complexity inherent in neuronal information processing. Let us review these three categories of differences between neurons and PEs that might prove to be computationally important.

1.   Random Fluctuations.   Stochastic fluctuations may affect the output of a neuron in several ways. One important source of these fluctuations occurs at the level of individual synapses as a result of the quantal nature (Katz, 1969) of neurotransmitter release. (See Fig. 2.)

Neurotransmitter is discharged in the form of integral numbers of multimolecular packets, where each packet is called a *quantum*. Quantization is thought to reflect the fact that neurotransmitter is packaged within vesicles before its exocytotic release. The postsynaptic response to the release of a quantum of transmitter can be measured in terms of the peak conductance change, the peak current, the net charge transfer, or the peak voltage response.

The magnitude of the postsynaptic response to the discharge of a single quantum is termed the *quantal size*. Even at a single synapse, the quanta are not all

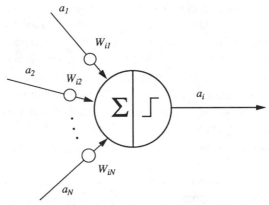

FIGURE 1.   Typical processing element. The summation node a_i receives inputs from other processing elements $a_1 \ldots a_N$ weighted by connections $w_{i1} \ldots w_{iN}$. The output of this a_i is the sum of the inputs (Eq. (1)) passed through a sigmoid (Eq. (2)).

the same size. There is a certain quantal variance σ^2 around the *mean quantal size* μ. Sometimes the quantal size distribution is approximately symmetrical, in which case it can often be reasonably well approximated by a Gaussian probability density function, which has two parameters, μ and σ^2. When the distribution is positively skewed, a two- or three-parameter gamma density function usually furnishes a better fit.

The number x of quanta released with each presynaptic nerve impulse, which is called the *quantal content,* is also a random variable. The average number of quanta released (the *mean quantal content*) is usually denoted m. In those cases in which it has been most accurately determined (chiefly the peripheral nervous system), m has been found to range—among different types of synapses—from a few tenths to a few hundred. Relatively little comparative data are available regarding the value of m in synapses in the mammalian brain, but preliminary evidence suggests that the value of m in the CNS may commonly be less than 10 and possibly less than 1 (Bekkers and Stevens, 1990; Malinow and Tsien, 1990).

The fluctuations in x can sometimes be described in terms of a Poisson probability function, which has the single parameter m. In other cases, a more accurate fit can be obtained with a binomial probability function, which has parameters n and p whose product is m. The Poisson probability function is the limiting case of a binomial probability function in which p becomes small and n becomes large. A binomial model can be derived by assuming that there are n presynaptic release sites, each associated with probability p of discharging a quantum in response to the arrival of a nerve impulse. If there is reason to suspect that n and/or p vary over time or that p is not the same at all n sites, then it may be necessary to consider a compound binomial release function, which is more complex (Brown *et al.,* 1976). A stochastic model of quantal transmission therefore must account for two sources of variance—one associated with the number of quanta released and the other with differences among individual quanta.

One might combine these two sources of variance into a single random variable κ. In the simplest case, the probability that κ falls within the range between a and b is given by a 3–5 parameter quantal probability function that consists of the integral between a and b of the sum of the product of a Poisson (or binomial) probability function for release and a Gaussian (or gamma) density function for quantal size (Fig. 2).

If the quanta are released nearly synchronously, the stochastic fluctuations in the observed synaptic responses can be approximated as the product of this random variable κ and an expression that describes the time course of the individual quantal conductance change. In some synapses, the conductance waveform can be adequately represented by a so-called *alpha function* (Jack *et al.,*

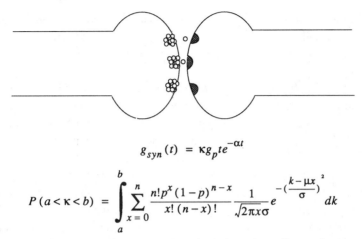

$$g_{syn}(t) = \kappa g_p t e^{-\alpha t}$$

$$P(a < \kappa < b) = \int_a^b \sum_{x=0}^{n} \frac{n! p^x (1-p)^{n-x}}{x!(n-x)!} \frac{1}{\sqrt{2\pi x \sigma}} e^{-\left(\frac{k-\mu x}{\sigma}\right)^2} dk$$

FIGURE 2. Probabilistic basis of quantal transmission. According to the quantum hypothesis of synaptic transmission, neurotransmitter is discharged in the form of integral numbers of multimolecular packets called quanta. The magnitude of the postsynaptic response is a probability distribution given by $P(\kappa)$. (See text.) The time course of the synaptic response has a time course given by the alpha function shown.

1975; Brown and Johnston, 1983), in which case the observed conductance fluctuations are approximated by:

$$g_{syn}(t) = \kappa g_p t e^{-\alpha t}, \tag{3}$$

where g_p is a constant and α^{-1} is the time-to-peak of the conductance waveform. (See Fig. 2.)

It is clear from the preceding discussion that the quantity $g(t)$ is more complex than the connection strength w_{ij}. The connection strength is dimensionless; it does not have a temporal component (on a single trial); and its value does not fluctuate randomly across trials. It is natural to wonder how one should think about the relationship between the scalar quantity w_{ij} and the synaptic conductance waveform $g(t)$. One possibility is to regard w_{ij} as an abstract variable that is proportional to the mean peak value of $g(t)$. Other possibilities include the mean net charge transfer or the mean voltage change over some interval. The appropriateness of these abstractions obviously depends on whether the time course of the synaptic response and the inherent randomness in the process are computationally significant.

In addition to fluctuations associated with quantal transmission, there are numerous other sources of variance that can also affect neuronal function. For example, the active and passive membrane properties of the neuron can vary over time as a consequence of circulating neuromodulators or the recent activity in the neuron. For this reason, the amount of synaptic current required to trigger an action potential can also fluctuate. In a PE, the corresponding effect might be a fluctuating *bias* term that is applied to the input function.

2. Electrotonic Structure. A summation node is a point. There is no spatial component to the processing of information by a PE. By contrast, neurons have a three-dimensional structure—owing to the complex dendritic tree—that may be computationally significant. This spatial aspect of neuronal structure affects not only the integration and transmission of information, but possibly also the synaptic modifications that underlie learning. The reason for this last suggestion, as we shall see, is that certain kinds of synaptic modifications that may be important in learning or activity-dependent aspects of development depend critically on the voltage across the dendritic membrane at the site of synaptic contact.

Anatomical structure is one of the determinants of electrotonic structure. In a passive neuron, electrotonic structure determines the manner in which the electrical signal spreads from one part of the dendritic tree to any other part of the tree and from the dendrites to the cell body (Claiborne *et al.*, 1991b). The passive spread of charge is one vital factor that determines the voltage at each part of the anatomical structure. In addition, the voltage is also a function of active membrane responses, as we will discuss in the next section. Before considering the active membrane, however, it is important to understand the passive (linear) case. In any complex system, it is often helpful to comprehend the linear case before introducing complex nonlinearities.

The manner in which the electrotonic structure is specified and modeled is elaborated in detail in the companion chapter (Claiborne *et al.*, 1991a). As indicated there, under special circumstances it is possible to use simple analytical expressions that describe limited aspects of the steady-state and transient spread of voltage (Jack *et al.*, 1975; Rall, 1977; Johnston and Brown, 1983, 1984). However the application domain for these analytical approaches is too limited for the kinds of questions with which we are concerned here. In practice then, we usually employ numerical methods (Claiborne *et al.*, 1991a) to represent the electrotonic structure in terms of a set of electrically coupled compartments (Perkel *et al.*, 1981; Hines, 1984; Mascagni, 1989). The amount of biological detail that can be represented in these compartmental models and the range of questions that can be addressed is currently only limited by the speed and memory of the computer that is used for the simulations (Claiborne *et al.*, 1991a).

In brief, the electrotonic structure of a neuron results from the combination of the anatomical structure plus three electrical parameters—the specific membrane resistance (R_m, units of Ω-cm^2), the specific membrane capacitance (C_m, units of μF/cm^2), and the specific resistivity of the cytoplasm (R_i, units of Ω-cm). It is perhaps easiest to understand the relationship between anatomical structure and electrotonic structure operationally—in terms of the method by which the first is transformed into the second. The initial step in producing this transform is usually to convert the morphology into a set of idealized shapes that can be used to calculate areas and volumes. Thus, the soma might be approximated by a sphere or an ovoid or a cone. The dendritic tree is usually approximated by a series of short, interconnected cylinders. The size of each of these geometric structures is made to be so small that each can be assumed to approximate an isopotential region of the neuron.

The second step is to convert these idealized structures into their equivalent electrical circuits. In the simplest case, we represent the membranes of each isopotential compartment as a capacitance that is in parallel with a resistance in series with a battery. The battery represents the electrochemical equilibrium potential associated with the *resting membrane conductance* (the reciprocal of the membrane resistance). The values of the membrane resistance and capacitance are obtained from R_m and C_m, respectively, and the area of the membrane. In this way, the plasma membrane associated with each anatomical region is converted into a series of electrically equivalent RC compartments. These RC compartments are connected to each other via coupling resistances that are calculated from R_i plus the geometry of each set of connected structures.

Figure 3 illustrates, in a highly simplified form, the nature of the relationship between the RC network and the anatomy. The number of compartments needed to represent the electrotonic structure of a neuron depends on the neuron itself as well as the intended use of the electrotonic model. When spines are not included, the digitization of CA1 pyramidal cells yields about 3,000 cylinders. When including spines at a realistic density, there may be as many as 64,000 compartments. Usually, it is possible to collapse all of this data into just a few hundred compartments. For example, in most of the simulations presented in this chapter, we found that the neurons could be modeled by 300 to 400 compartments without introducing any serious inaccuracy.

The RC properties of the dendritic arbor can have two major effects on the passive spread of transient potentials. The time constant of the cell membrane, given by

$$\tau_m = R_m C_m, \tag{4}$$

is about 15 to 70 msec in hippocampal pyramidal neurons. This long-time constant has several significant effects on the passive spread of signals. (See Fig. 9 later in chapter.) First, because the neuronal membrane acts as a low-pass filter, the voltage waveform begins to rise and decay more slowly as it spreads passively throughout the neuron. This has the effect of limiting the bandwidth

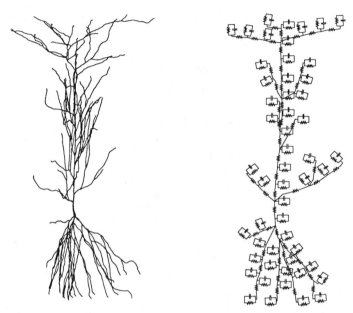

FIGURE 3.   Compartmental representation of structure. The three-dimensional positions and diameters of 2,714 locations of a CA1 pyramidal neuron (left) were digitized using the methods described in Claiborne *et al.* (1991a). The morphology was then reduced to an equivalent circuit representation for use with the neural simulator NEURON (Hines, 1989). As indicated in a highly schematic form (right), the equivalent circuit of a passive neuron consists primarily of coupled *RC* compartments. In typical simulations, we used several hundred compartments. Not shown are the time- and/or voltage-varying elements.

of the cell. Thus, it limits the amount of information that can be transmitted from point to point. In addition, it allows the cell to integrate information over significant intervals. Finally, with increasing distance from the initiation site, there is a progressive attenuation of the amplitude of the response.

3.   Nonlinear Membrane Dynamics.   A convenient starting point for under-standing neuronal dynamics is the idealized passive neuron shown in Fig. 3. This neuron is a linear system: The response—the change in membrane volt-ages—at any point is directly proportional to the stimulus—an injected current. If the injected current doubles, then the change in voltage across a point in the membrane also doubles. This is a direct consequence of assuming an ohmic membrane and a stimulus consisting of current injection. In almost any case more complex than this, however, nonlinearities become a factor.

a. Conductance Increase versus Current Injection.   It is important to note that even the passive response of a neuron to a synaptic input is not adequately described in terms of a time-invariant linear system. The reason is that a synaptic input is better modeled as a variable conductance than as a variable current source. The current through the synapse is given by Ohm's Law:

$$I_{syn}(t) = (V(t) - E_{syn})\, g_{syn}(t), \tag{5}$$

where $V(t)$ is the membrane potential, E_{syn} is the synaptic driving force, and $g_{syn}(t)$ is the time-varying synaptic conductance. In the steady-state, the potential across the equivalent circuit element shown in Fig. 4, excluding the voltage-dependent components, is given by

$$V = \frac{-I_{inj} + g_L\, E_L + g_{syn}\, E_{syn}}{g_L + g_{syn}} \tag{6}$$

where the three terms in the numerator are related to the injected, leak, and synaptic currents, respectively, and the two terms in the denominator to the leak and synaptic conductances, respectively. The difference between a synaptic stimulus and current injection is clear from Eq. (6). While a synaptic stimulus causes a change in both the numerator and the denominator, injected current changes only the numerator.

This has two implications. First, the synaptic stimulus causes a transient change in the electrotonic structure of the neuron—the synaptic stimulus is equivalent to a transient change in the value of g_L at the site of the stimulus. The importance of this change may be relatively minor for a fast synaptic event. Second, if the conductance is large, V approaches E_{syn}, and the response I_{syn} saturates (see Eq. (5)). Under conditions of a large synaptic input leading to a large conductance change, this saturation can have important implications, including nonlinear summation of synaptic stimuli, which is observed experimentally. In the other limit, a small conductance change that does not cause a large change in V is little different from an injected current.

b. Voltage-Dependent Membrane Conductances. The intrinsic voltage dependence of the membrane itself causes further deviations from the approximation of a neuron as a linear system. In many neurons, this voltage dependence is the source of powerful nonlinearities that completely dominate dynamics in certain voltage ranges. These nonlinearities can lead to sharp threshold phenomena such as the action potential. Even far from threshold, many neurons exhibit non-ohmic behavior for voltage excursions greater than 5–10 mV. In many neurophysiological experiments, attempts are made to block these nonlinearities pharmacologically.

The mechanism underlying the initiation and propagation of the action potential was elucidated by Hodgkin and Huxley (1952). They postulated that, near threshold, a rapid positive feedback process depolarizes the membrane, which in turn

Outside

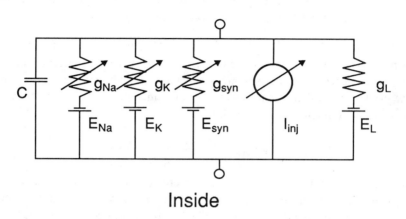

Inside

$$i_m(V,t) = (V - E_{Na})\,g_{Na} + (V - E_K)\,g_K + (V - E_{syn})\,g_{syn} + (V - E_L)\,g_L + I_{inj} + C\frac{dV}{dt}$$

FIGURE 4. Nonlinear membrane circuit element. The equivalent circuit for an idealized patch of nonlinear membrane with Hodgkin–Huxley channels and a current source is shown. Six components are in parallel. Each of the four ionic currents is modeled as a conductance in series with a battery. The battery corresponds to the driving force produced by the ionic gradient. The capacitor C and the leak conductance $g_L = 1/R_m$ are time-invariant and constitute the purely passive portion of the circuit. The synaptic conductance g_{syn} is a function only of time, while the two Hodgkin–Huxley conductances (g_{Na} and g_K) are functions of time and voltage. Other voltage-dependent conductances, such as calcium channels, are not shown.

FIGURE 5. Hodgkin–Huxley formalism. The Hodgkin–Huxley equations (A) provide a quantitative account of the initiation and propagation of the action potential in the squid axon. Their voltage-dependent time-constants τ_{HH} and their steady-state values HH_∞ are illustrated in (B). The continuous Hodgkin–Huxley equations can be interpreted (C) in terms of voltage-dependent transitions between open and closed states. These two states can be observed experimentally, as shown in an idealized record showing holding current as a function of time (D).

A

$$g_{Na}(V,t) = \bar{g}_{Na}m^3h \qquad\qquad g_K(V,t) = \bar{g}_K n^4$$

$$\frac{dm}{dt} = (1-m)\,\alpha_m(V) - m\beta_m(V) \qquad \frac{dn}{dt} = (1-n)\,\alpha_n(V) - n\beta_n(V) \qquad \frac{dh}{dt} = (1-h)\,\alpha_h(V) - h\beta_h(V)$$

$$= \frac{m_\infty(V) - m}{\tau_m(V)} \qquad\qquad = \frac{n_\infty(V) - n}{\tau_n(V)} \qquad\qquad = \frac{h_\infty(V) - h}{\tau_h(V)}$$

B

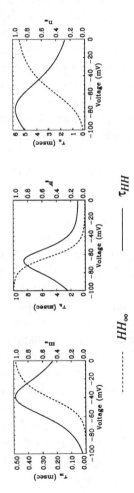

- - - - - HH_∞ —— τ_{HH}

C

$$\overset{\alpha_n(V)}{\underset{\beta_n(V)}{\rightleftharpoons}}$$

D

I

t

activates a slower negative feedback process that repolarizes the membrane. The positive feedback process consists of activation of an excitatory sodium conductance, while the negative feedback process consists of both inactivation of the sodium conductance and activation of an inhibitory potassium conductance. In a series of experiments for which they received the Nobel Prize, Hodgkin and Huxley were able to provide a quantitative account of the action potential in the squid axon in terms of the voltage-dependent activation and inactivation rates of these two membrane conductances (Fig. 5).

The physical substrate of these variable membrane conductances consists of ion channels (Fig. 5)—transmembrane proteins that behave as selectively permeable ion pores (Hille, 1984). These ion channels undergo conformational transitions between open, conducting states and closed, nonconducting states (Fig. 5c). Single-channel recording (Neher and Sakmann, 1983) permits direct study of the opening and closing of individual channels. The kinetics of many channels have been modeled as a Markoff random process in which the voltage-dependent rate constants correspond to the transition probabilities between open and closed states. The Hodgkin–Huxley equations can be derived as the continuum limit of many discrete channels.

The role of voltage-dependent channels depends on their distribution and their kinetics. Threshold phenomena, such as spiking, require both fast channel kinetics and a relatively high channel density. In the many neurons that show somatic action potentials, both conditions are satisfied. Many neurons also have a variety of voltage-dependent channels distributed across the dendrites, often at densities lower than at the soma. In hippocampal pyramidal neurons, for example, there is experimental evidence for at least a dozen voltage-dependent conductances, and at least some of these are distributed in a heterogeneous fashion in the dendrites (reviewed in Brown and Zador, 1990). Recent advances in electrophysiological and optical methods may provide a better quantitative basis for modeling of voltage-dependent channels in dendrites.

c. *Logic Operations in Dendrites.* The computational role of voltage-dependent dendritic channels is not well understood. One early proposal (Perkel and Perkel, 1986; Shepherd et al., 1985) is that high-density patches of sodium and potassium channels—with Hodgkin–Huxley (1952) kinetics—located at the base of dendritic spines might perform a logical AND or OR on inputs to adjacent spines. In this scheme, an XOR could be produced by appropriately timed extrinsic inhibitory inputs. The appeal of this approach lies in its clear demonstration of the computational power of suitably designed tree structure. All logical operations can be generated by combinations of AND, OR, and XOR. If neurons contained such membrane, then the dendritic tree might be regarded, in some

Outside

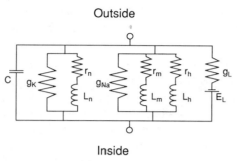

Inside

FIGURE 6.   Linearized membrane. The nonlinear components in the equivalent circuit of a patch of a nonlinear membrane (Fig. 4) can be linearized about a particular voltage to obtain the equivalent RLC circuit shown. Each gating particle (m, h, and n) of the Hodgkin–Huxley equations is replaced with a resistor and inductor in parallel. The values of these components depend on the parameters of the Hodgkin–Huxley equations and the voltage about which they are linearized. (See DeFelice, 1981, and Koch, 1985, for details). Synaptic components and batteries are not shown.

sense, as a universal computer. The drawback of this approach, however, is that it disregards both the analog and temporal characteristics of dendritic signals.

As noted previously, if channel density is below a critical value, threshold behavior—required for digital logical operations—will not be observed. Even above this critical value, the response to subthreshold inputs must be accounted for in analog terms. For small inputs, the voltage-dependent components of the circuit in Fig. 4 can be linearized at about the resting membrane voltage (Chandler *et al.*, 1962; Sabah and Leibovic, 1969; Koch, 1985) to produce an RLC circuit (Fig. 6). Koch (1985) has shown that under certain conditions, this circuit can perform temporal differentiation. Simulations have shown that a patch of potassium channels in the dendrites of a hippocampal pyramidal neuron can perform a computation that might be characterized as an analog temporal XOR (Zador *et al.*, 1992). More generally then, heterogeneous distributions of different voltage-dependent channels can implement a wide range of operations that are not well characterized in terms of digital operations.

III.   Voltage Gradients in Dendrites and Spines

Most of what follows will be concerned only with the spatial aspect of neuronal function, focusing on the effects of the electrotonic structure on synaptic plasticity. We hasten to point out that, in the case of hippocampal neurons, few if any of the membrane conductances are strictly voltage-independent. However, within limited voltage ranges, and for small perturbations, the membrane may behave in a nearly ohmic (linear) fashion. The linear case that we will be describing can be viewed as a useful simplification, a platform upon which to build intuitions that will be helpful in considering the more general circumstance that includes nonlinear membrane responses. Even if we had detailed information about the density and distribution of all the active conductances that are thought to reside in hippocampal membranes (McCormick, 1990)—which is not the case—it would still be valuable to explore and understand the linear case before incorporating nonlinear membrane dynamics.

The concern here is with voltage gradients along the plasma membrane. This is motivated in large part by the fact, mentioned earlier, that certain synaptic modifications that are thought to be important in various forms of associative learning and self-organization are dependent on the voltage across the membrane at the site of the synaptic input. The synaptic modification process thus is presumably affected by the extent that the cell is not isopotential. The effects of nonisopotentiality become even more interesting when we introduce various kinds of nonlinear membrane responses. Even in the purely passive case, however, we shall see that voltage gradients across the space of the neuron may result in patterns of synaptic changes that could not arise in a simple processing element of the sort described earlier.

A.   *Voltage Attenuation along the Dendritic Spine*

There are likely to be voltage gradients even on a very small spatial scale. To illustrate this, consider the *dendritic spine,* which is the specialized postsynaptic structure associated with most excitatory synaptic inputs to a hippocampal neuron. The dendritic spine has a diameter in the submicron range and a length in the micron range. Owing to its small diameter, it affords a substantial resistance to synaptic current flow. Although precise estimates are not available, the spine axial resistance has been calculated—based on the value of R_i and the geometry— to be of the order of 100–200 MΩ. From Ohm's law, we know that a synaptic current of 5 pA will cause a 1mV voltage drop along a spine that has a 200 MΩ axial resistance.

Simulated effects of spines on electrical signaling are discussed at length elsewhere (Brown *et al.,* 1988a, 1989). The results of one such simulation are illustrated in Fig. 7. This figure illustrates a general point, which is sometimes not appreciated, that is relevant to the subsequent discussion. In an asymmetrical neural structure, such as the spine and its associated dendrite, the voltage attenuation from point A on the structure to some other point B is generally not the same as the voltage attenuation from B to A. Note that the voltage attenuation

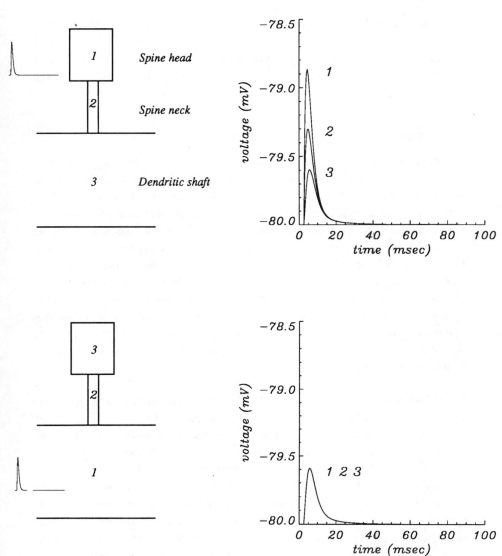

FIGURE 7. Voltage gradients in a dendritic spine. A synaptic stimulus (Eq. (3); g_p = 50 pS) was applied at a spine head (top) or the dendritic shaft (bottom). The voltage at the spine head, the spine shaft, and the dendritic shaft are plotted as a function of time. There is attenuation of the peak transient for a stimulus applied at the spine head (top), but almost none for the stimulus applied at the shaft (bottom). Note that the peak transient is larger for the stimulus applied at the spine head.

from the head of the spine to the dendritic shaft (Fig. 7, top) is greater than from the shaft to the spine head (Fig. 7, bottom). A theoretical discussion of asymmetric voltage transfer is provided in Carnevale and Johnston (1982). This general principle will be important when we consider the spread of voltage throughout the dendritic tree.

B.   Voltage Attenuation along the Dendritic Shaft

Several years ago, Brown and colleagues (Brown *et al.*, 1981) suggested that hippocampal neurons are *electrotonically compact*. What was meant is that the steady-state spread of voltage from the soma to the tips of the dendrites suffers much less attenuation than had been previously assumed for some other neurons, such as spinal motor neurons. From the preceding discussion of dendritic spines, we should not be surprised to learn that the steady-state attenuation of voltage from the soma to the tips of the dendrites is not the same as the steady-state attenuation in the opposite direction, which is much larger. Usually, we are less interested in steady-state attenuation than in the attenuation of transient signals, such as synaptic potentials, voltage-dependent membrane responses, or sudden voltage commands presented by the experimenter. The attenuation of transient signals is larger than for steady-state signals. The amount of attenuation increases as the frequency increases.

1.   What is the True Value of R_m?   One of the factors that determines the amount of voltage attenuation along the dendritic shaft is the value of R_m. Recent studies that have made use of *whole cell* recordings have suggested that the value of R_m may be much larger than suggested from previous estimates that were based on intracellular recordings using *sharp electrodes*. The explanation for this disparity has been that sharp electrodes inevitably cause a leakage between the intracellular microelectrode and the surrounding plasma membrane. The argument is that this artificial leakage produces an artificially low apparent (measured) cell resistance and time constant. By contrast, the whole-cell recording electrode eliminates this leakage conductance by forming a giga-ohm seal with the plasma membrane.

This explanation is probably partly correct, but other factors may also be important. For example, the whole-cell recording mode has a greater tendency to perfuse the cell with the contents of the recording electrode. This internal perfusion of the cell can itself reduce some of the normal membrane conductances. One cannot assume therefore that the results of whole-cell recordings necessarily give the correct resistance.

Although the actual value of R_m remains uncertain, several investigators have suggested that it may be so high that hippocampal neurons are effectively iso-potential. The implication is that there are no voltage gradients along the dendritic shaft. If true, this would be reason for celebration, for at least three reasons. First, it means that neurons can be represented by a single isopotential com-partment. The algorithm that we use to simulate neuronal responses scales as $O(N)$, where N is the number of compartments (Claiborne *et al.*, 1991a). If a neuron can be adequately represented by a single compartment, this could entail an enormous economy in computation time and memory—by two or three orders of magnitude. This becomes particularly significant when one considers modeling complex circuits of neurons.

Second, it implies that voltages and currents recorded in the soma do not suffer any attenuation or distortion. Thus, the signals that are observed in intra-somatic recordings can be assumed to be identical to what actually occurs at the site of the conductance change in the dendrites. If the cell is, in fact, a single isopotential compartment, then a voltage command applied at the soma will be produced instantly and uniformly throughout the entire plasma membrane of the neuron. Isopotentiality implies a perfect space clamp, which is the ideal condition for voltage-clamp studies.

Third, it provides an important constraint on the possible computational role(s) of dendrites. If the cells are always isopotential (if the computations performed by the neuron can be adequately captured by a single isopotential compartment), then we can dismiss the idea that some types of logic operations occur in the dendrites as a consequence of the spatial inhomogeneity of various active mem-brane conductances. Isopotentiality thus limits the range of possible answers to the question: What do dendrites do? In regard to this important question, any major limits that can be placed on the range of possible answers can be seen as progress.

2.   Are Hippocampal Neurons Isopotential?   Are there, in fact, steep voltage gradients along the plasma membrane of hippocampal neurons? For reasons already mentioned, our interest in this question concerns the transient case. To approach this question, we modeled hippocampal pyramidal neurons assuming various values for R_m. These results will be presented in more detail elsewhere (Claiborne *et al.*, 1991b). Here, we will illustrate briefly two kinds of simulations. In both, we let $R_m = 15,000$ Ω-cm^2 or $150,000$ Ω-cm^2. These values include the range that might be inferred from studies using either sharp electrodes (Brown *et al.*, 1981) or whole-cell recordings (Spurston and Johnston, 1992).

a. Somatic Voltage Step.   The first type of simulation was a voltage-clamp experiment (Claiborne *et al.*, 1991b). The results are specifically relevant to experimental neurophysiologists and also illustrate some theoretical principles. The voltage in the soma was instantaneously jumped from its resting value of -80 mV to a new steady value of 0 mV. The resulting voltage changes were observed in the dendrites over time. If the entire cell could be represented as a single isopotential compartment, then the dendritic potential should be identical to the somatic potential at all times. At the risk of redundancy, we repeat that what is of interest, from the point of view of a voltage-clamp experiment, is almost always the transient case, because most of the events that one wants to study—synaptic conductances and active membrane conductances—have rapid kinetics. The conductances often peak within a few milliseconds or less.

Based on previous simulations, we were not surprised to find that the low-R_m cell was not isopotential. After a full 10 msec following the jump in the somatic potential, there were still steep voltage gradients in the dendrites, as illustrated in Fig. 8 (lower left). Interestingly, the gradients in the high-R_m cell were also relatively steep (Fig. 8, lower right). The reason of course, is that much of early current is capacitative. Naturally, in the steady-state, there is less voltage attenuation in the high-R_m cell. The steady-state is approached after about three time constants—almost half a second in this case. The fact that the steady-state voltage control is much better in the high-R_m cell is of little comfort to neurophysiologists interested in measuring the peak current or conductance produced by a distal synapse. The excitatory synaptic currents typically peak within 2–3 msec (Brown and Johnston, 1983). This means that there would be no hope of accurately controlling the transient subsynaptic membrane potential.

Because there seems to be some confusion on this last point, it might be worthwhile to elaborate. The confusion arises from the following argument. Suppose one wants to measure the conductance increase at a distal synapse. If enough time is allowed—following the application of a voltage step in the soma—the voltage at the subsynaptic membrane should eventually approach the desired value. If R_m is large enough, this inference is reasonable. To appreciate the problem with this argument, we need to consider the actual dynamics of voltage clamping. Suppose that in the steady-state the voltage applied at the soma is, in fact, closely approximated in the subsynaptic membrane at some distal site on the dendritic tree. Unless the subsynaptic potential is at the electrochemical equilibrium potential, as soon as the synapse is activated, the resulting synaptic current will produce a transient local voltage change at the subsynaptic membrane. The goal is to prevent this transient voltage change. If this goal cannot be achieved, it is impossible to obtain an accurate measurement of the synaptic current and conductance waveform.

There are two problems. First, even if the somatic voltage clamp could immediately sense the first instant of the local voltage perturbation at the synaptic

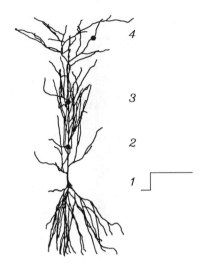

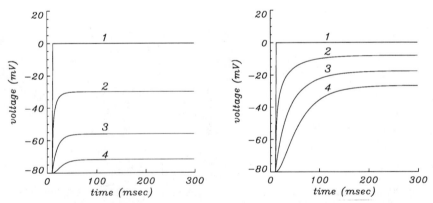

FIGURE 8.   Voltage gradients following a somatic voltage step. The membrane potential at the soma was stepped instantaneously to 0 mV. The membrane potential was plotted as a function of time at the four points indicated. The results were compared for a low (R_m = 15 kΩ-cm^2; left) and a high (R_m = 150 kΩ-cm^2; right) value of membrane resistivity. The neuron with the higher value of membrane resistivity charges more completely but more slowly.

site, it cannot produce the required instantaneous correction at the synaptic site. As we have already noted, it takes a long time—relative to the speed of the synaptic conductance—for the full amount of a command delivered at the soma to reach a distal site on the dendritic tree. Second, the voltage clamp does not, in fact, immediately sense the magnitude of the voltage perturbation at the

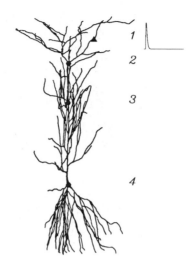

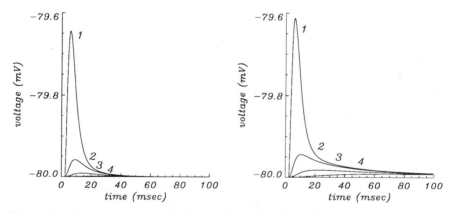

FIGURE 9.   Voltage gradients following a distal synaptic input. A synaptic stimulus was applied on the head of a spine (location 1). The membrane potential was plotted as a function of time at three other locations (locations 2–4) on the path from the synapse to the soma. The results were compared for a low (R_m = 15 kΩ-cm^2, left) and a high (R_m = 150 kΩ-cm^2; right) value of membrane resistivity. In both cases, the peak voltage shows significant attenuation even relatively near the site of the input. The early transient components of the response are almost indistinguishable, but the decay time for the high-R_m cell is substantially longer.

subsynaptic membrane. Just as it takes time for voltage to spread from the soma to the synapse, it also takes time for the peak synaptic potential to reach the soma (Fig. 9). As we shall see in a moment, the problem is actually more serious than this because there is considerable attenuation and distortion of the synaptic potential as it passively spreads to the soma (Fig. 9). Thus, the voltage clamp does not, in fact, sense the actual voltage waveform at the synaptic site and will not then produce the appropriate correction.

Although this is not the place to expound on the theory and practice of voltage-clamping hippocampal neurons, it is important to note that—depending on the purpose, logic, and outcome of the experiment and on the required measurement precision—these considerations need not pose insurmountable obstacles. For two examples, we refer the reader to the experiments of Brown and Johnston (1983) and Kelso et al. (1986).

b. Spread of Synaptic Potentials.   The simulated voltage clamp experiment illustrated the fact that hippocampal neurons are not isopotential. This fact has practical implications for the experimental neurophysiologist. For the purpose of the present chapter, however, we are actually more interested in voltage gradients produced by synaptic inputs. These attenuations are directly relevant to some of the kinds of synaptic modifications that might naturally occur on the dendrites of hippocampal neurons. We will elaborate later about the biophysical mechanisms underlying the plasticity at this kind of synapse. For now, though, it is only important to appreciate that the synaptic strengthening is a joint function of presynaptic activity and the voltage at the site of the synaptic input.

In the second type of simulation, we activated synapses at various parts of the dendritic tree and examined gradients in the resulting potential over space and time (Fig. 9). In these same experiments, we also compared the currents injected at the synaptic site with those that would have been recorded with a voltage clamp applied to the soma (data not shown). As in the previous simulations, we examined two values of R_m—15,000 Ω-cm^2 and 150,000 Ω-cm^2. Figure 9 shows the potentials resulting from a synaptic input to a distal branch of the dendritic tree. Note that the synaptic potential suffers considerable attenuation as it spreads toward the soma. The peak voltages were remarkably similar in the low- and high-R_m cells.

c. Summary.   What is common to both kinds of simulations is that hippocampal neurons are not isopotential in the transient case even when R_m is assumed to be large. There can be steep voltage gradients along the dendrites. Later, we will suggest that these gradients could be computationally significant in terms of synaptic plasticity. A more detailed presentation is currently being prepared (Claiborne *et al.*, 1991b).

IV.   Spatial Representation of Electrotonic Structure

The voltage gradients that we have been discussing arise because of the electrotonic structure of the neurons. Electrotonic structure is not a popular subject in neurobiology for several reasons. First, the concepts are often buried in mathematical treatments that are not easily accessible to most neurobiologists. Second the mathematical formalisms sometimes require neurobiologically unrealistic assumptions. Third, the formalisms often do not lend themselves to some of the kinds of questions that are of most interest to neurobiologists. We have been developing graphical and simulation methods that circumvent some of these problems.

A.   Electrotonic Distance

The *electrotonic distance X* between two points is defined as:

$$X = \frac{x}{\lambda}, \tag{7}$$

where x is the distance that separates the points and the *space constant* λ is defined as:

$$\lambda = \left(\frac{R_m}{R_i}\frac{d}{4}\right)^{1/2}, \tag{8}$$

where R_m is the membrane resistivity; R_i, the axial resistance; and d, the process diameter. The motivation for these definitions lay in the particularly simple form they provide for the spread of potential in an infinite cylinder. Specifically, if the voltage at some point is V_0, then the voltage at any other point is simply:

$$V = V_0 e^{-X}. \tag{9}$$

In this idealized structure then, the voltage decays to $1/e$ of its initial value with each space constant. The electronic distance X between two points thus is the number of space constants separating the two points. Discussion of electrotonic distance thus far has been restricted to the steady-state (or dc) case. The basic concept expressed in Eq. (8) can be readily generalized to include the frequency-dependent (or ac) case. This requires that we consider the frequency-dependent space constant λ_{ac}. Responses to sinusoidal stimuli are discussed elsewhere (Jack et al., 1975; Rinzel and Rall, 1974; Koch et al., 1982; Carnevale and Johnston, 1982; Brown et al., 1988a).

Electrotonic distance is a useful and intuitive concept when considering infinite or semi-infinite cables. For electrotonically short processes, such as dendrites, more complicated expressions relate λ to voltage decay, and some of the intuitive appeal of Eq. (9) is lost. Further complications must be introduced to account for voltage spread in branched structures. As the complexity of the structure increases, there quickly comes a point where there are no convenient expressions for the electrotonic distance between any two points.

B. Electrotonic Structure

An understanding of electrotonic distance as defined in semi-infinite and terminated cables is a useful first step toward appreciating the electrotonic structure of morphologically complex neurons. It is important to appreciate, however, that hippocampal pyramidal neurons are *electrotonically compact* (Brown et al., 1981), meaning that the electrotonic distance from the soma to the tips of the dendrites does not approach infinity. The average electrotonic distance from the soma to the tips of the dendrites is only about 0.5 to 1.5, depending on assumptions and the particular neuron (Claiborne et al., 1991b).

A full account of the electrotonic structure of hippocampal neurons requires recognition of two important principles. First, the electrotonic structure is viewpoint-specific. It depends on the observation point and the direction of voltage or current transfer of interest. Second, the electrotonic structure depends on the temporal nature of the stimulus. As stated earlier, neuronal membrane acts as a low-pass filter, so that fast transient events are more filtered than slow or steady-state events. A familiar consequence of this filtering is the long, slow voltage response observed at the soma in response to a distal, fast synaptic event (Fig. 9).

These two principles suggest that a physiologically relevant measure of electrotonic distance must be a function of at least three parameters; the stimulus site, the observation site, and the temporal nature of the stimulus. Thus, one might, for example, tabulate the voltage response at the soma due to a synaptic current at every point in the dendritic tree. We have developed a graphical method of appreciating this tremendous amount of data. This method is based on a measure, L_{ij}, that can be thought of as the appropriate generalization of the concept of electrotonic distance to arbitrary branched structures. For the case of an infinite cable, L_{ij} reduces to X.

The method has its roots in the well-established theory of two-port electrical circuits, which has only relatively recently been applied to electrotonic structure (Butz and Cowan, 1974; Carnevale and Johnston, 1982; Koch *et al.*, 1982). The formulation is in the frequency domain—in keeping with its electrical engineering roots. Let $V_i(f)$ be the response at some point i to an injected sinusoidal current $I_i(f)$ of frequency f, and $V_j(f)$ the response at any other point j to that same injected current. The frequency of the response at all points is equal to the frequency of the stimulus, f.

The frequency-dependent voltage attenuation $A_{ij}(f)$ between any two points i and j is defined as (Rinzel and Rall, 1974; Koch *et al.*, 1982; Carnevale and Johnston, 1982):

$$A_{ij}(f) \equiv \frac{V_i(f)}{V_j(f)}. \tag{10}$$

The steady-state response is the special case of $f = 0$. The response is maximal at i, so:

$$A_{ij}(f) \geq 1, \tag{11}$$

with equality if and only if $i = j$. Finally, we introduce L_{ij}, defined as:

$$L_{ij}(f) \equiv \log\,(A_{ij}(f)). \tag{12}$$

By Eq. (11), L_{ij} is a nonnegative number. By combining Eqs. (9), (10), and (12), it can be seen that L_{ij} exactly equals the electrotonic distance X in the case of an infinite cable.

C.   Morphoelectrotonic Transform

The attenuation and its logarithm have many interesting properties, two of which are relevant to the present discussion. First, the attenuation is not in general symmetric:

$$L_{ij}(f) \neq L_{ji}(f). \tag{13}$$

In fact, it can be shown that the log-attenuation of *voltage* from i to j is exactly equal to the log-attenuation of *current* from j to i. A rule of thumb applicable to most neurons is that the voltage attenuation from a dendritic synapse to the soma is larger than the attenuation from the soma to the synapse.

A second, convenient property of the log-attenuation is that if j is any point between i and k, then the log-attenuation from i to k is just the sum of the log-attenuation from i to j and the log-attenuation from j to k:

$$L_{ik}(f) = L_{ij}(f) + L_{jk}(f). \tag{14}$$

In this respect, the log-attenuation behaves as a distance measure. As with a true distance, the log-attenuation between two points is equal to the sum of the log-attenuations between each of them and a collinear intermediate point. Further, by Eq. (10), $A_{ii} = 1$ and $L_{ii} = 0$, which is consistent with the requirement that the distance from a point to itself be 0. Because of the asymmetry noted in Eq. (13), L_{ij} is not, however, a distance measure in the true mathematical sense, since the log-attenuation from i to j is not equal to the log-attenuation from j to i. Nevertheless, with the appropriate interpretation, Eq. (14) provides the basis for a convenient graphical transformation of the data.

Each segment of the 2,714-compartment neuron pictured in color insert Fig. 1A (left) can be specified by a triplet of parameters: its diameter, the angle of its attachment to its parent process, and its length. In our *morphoelectrotonic transform*, the first two of these parameters are preserved, but some measure of that electrotonic length is substituted for the anatomical length. In color insert Fig. 1A (middle) the length of each compartment has been replaced by the value of the steady-state log-attenuation of voltage from the soma. Each segment of the anatomical neuron has been color-coded according to its physical distance from the soma—the color changes for each 100 μm of distance. This same color-coding—in terms of physical distance—has been applied to the transformed neuron so that the correspondences between the two representations can be readily appreciated. In the transformed representation, each unit of distance represents an n-fold decrement of voltage.

At least three features of the transformed neuron are noteworthy. First, the apical structure is dominated by the primary apical stalk. The length of the side branches is greatly reduced relative to the main stalk, indicating that most of the attenuation occurs across the main stalk, and very little across the smaller process. This is consistent with the well-known principle that voltage transfer from a large process to a small process is effective (Brown *et al.*, 1988a). Second, relatively more of the attenuation along the primary apical branch occurs prox-imally than distally, as is indicated by the very small region of the transformed neuron with blue. Finally, the basal portion of the tree is relatively small, indicating very little attenuation within the basal dendrites.

The log-attenuation of voltage in response to a 250 Hz sinusoidal current injected at the soma (color insert Fig. 1A, right) provides a very different per-spective on electrotonic structure. Most striking is its increased size relative to the steady-state representation, reflecting the much greater attenuation of the higher frequency components of any signal. There is also an interesting restruc-turing of the overall form, with a relatively greater fraction of the attenuation occurring across some of the secondary branches and the basal tree.

We have explored many other aspects of electrotonic structure through morphoelectrotonic transforms. Different pictures of the neuron emerge as a function of the point-of-view determined by the site of the stimulus, the frequency of the input, and the variable represented (current, voltage, or power). Temporal aspects can be explored by substituting phase-delay or time-to-peak for physical distance. We expect that these transforms will provide a useful tool for understanding electrotonic structure.

V.   Voltage-Dependent Synaptic Modification

The significance of voltage gradients in hippocampal neurons derives in part from recent discoveries regarding a fascinating form of synaptic plasticity in these neurons. Long-term potentiation (LTP) is a rapid and persistent synaptic enhancement that can be induced in many of the excitatory synapses of the hippocampus. (For reviews, see Brown *et al.*, 1988a,b, 1989, 1990.) There are several varieties of LTP, including one that resembles the synaptic modification proposed by the famous neuropsychologist, Donald Hebb (1949).

A.   Hebbian Form of Long-Term Potentiation

LTP has been studied most extensively in the Schaffer collateral/commissural input to pyramidal neurons of hippocampal region CA1. There are three key features of LTP in these synapses (Kelso *et al.*, 1986):

1. Conjunctive pre- and postsynaptic activity is necessary to induce the enhancement.
2. To be effective in producing the enhancement, the pre- and postsynaptic activity must occur at about the same time.
3. The enhancement is specific to just those synaptic inputs that were co-activated with the postsynaptic cell.

The Hebbian nature of LTP induction in these synapses has been clearly demonstrated (Fig. 10). In particular, tetanic stimulation of an afferent input during strong postsynaptic depolarization induced LTP in the activated synapses but not in unstimulated synapses (Kelso *et al.*, 1986; Malinow and Miller, 1986; Wigstrom *et al.*, 1986). In contrast neither postsynaptic depolarization alone nor synaptic stimulation performed while voltage-clamping the postsynaptic soma to -80 mV resulted in LTP induction (Kelso *et al.*, 1986). The same study also demonstrated that the elicitation of sodium spikes in the postsynaptic cell is not necessary for LTP induction. The conclusion was that some consequence of strong depolarization other than the elicitation of sodium spikes is necessary to

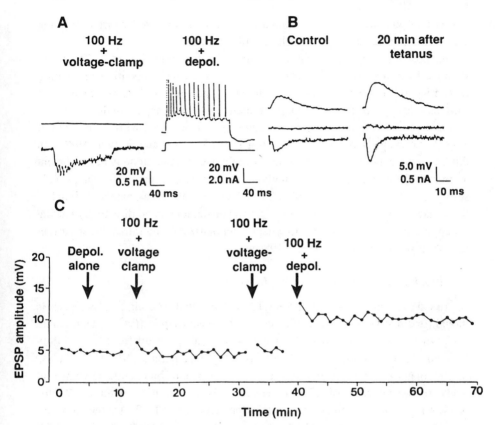

FIGURE 10.   Demonstration of Hebbian modification. The experiment demonstrates that the induction of LTP in the Schaffer/commissural synapses is controlled by a Hebbian mechanism. Current- and voltage-clamp recordings were done in a pyramidal cell of region CA1. (A) left; voltage-clamp record of inward synaptic currents (lower trace) and a membrane potential (upper trace) during a high-frequency train of synaptic stimulations. Right: Current clamp recording of postsynaptic action potentials (upper trace) produced by an outward current step (lower trace), which was sometimes paired with the synaptic stimulation train. (B) Current-clamp (top trace) and voltage-clamp (bottom trace) records before and 20 minutes after pairing synaptic stimulation with the outward current step. The middle trace is the membrane potential during the voltage clamp. (C) The excitatory postsynaptic potential (EPSP) amplitudes are plotted as a function of the time. Arrows denote the occurrence of three manipulations: an outward current step alone (depol. alone) or synaptic stimulation trains delivered while applying either voltage clamp (100 Hz + voltage clamp) or an outward current step (100 Hz + depol.). Each point is the average of five consecutive EPSP amplitudes. From Kelso *et al.*, 1986.

enable LTP induction at synapses that are eligible to change by virtue of being active at about the same time as the depolarization.

The postsynaptic depolarization required for the induction of LTP can be supplied by several types of experimental manipulations. Repetitive stimulation of a strong afferent pathway (one containing a large number of fibers) can provide sufficient postsynaptic depolarization to induce LTP at synapses in that pathway. Alternatively, stimulation of a strong pathway can be paired with stimulation of a weak pathway (Barrioneuvo and Brown, 1983; Kelso and Brown, 1986). In this case, the depolarization resulting from the weak input alone is not sufficient to induce LTP, but coactive stimulation of the strong pathway can provide sufficient depolarization to enable LTP induction in the weak pathway. Finally, the voltage dependence of LTP induction can be demonstrated directly by pairing synaptic activation and direct postsynaptic depolarization by intracellular current induction (Fig. 10) (Kelso et al., 1986).

B.   Biophysical Mechanism of LTP Induction

A critical step in the induction of LTP in the Schaffer collateral synapses is thought to involve a transient increase in the postsynaptic $[Ca^{2+}]$ (Dunwiddie and Lynch, 1979). This increase, which is thought to result from Ca^{2+} influx through channels associated with the N-methyl-D-aspartate (NMDA) subclass of glutamate receptor, apparently triggers a sequence of biochemical events that leads to the induction of LTP (reviewed in Brown et al., 1988a,b, 1991b). NMDA receptors play an essential role in the induction of LTP. Another subclass of glutamate receptor, the AMPA receptor, participates in the expression of LTP and in normal synaptic transmission.

The NMDA receptor–ionophore complex can respond to the conjunction of presynaptic activity and postsynaptic depolarization that is characteristic of a Hebbian synapse. The channel allows Ca^{2+} influx when there is glutamate binding to the receptor and simultaneous depolarization of the membrane containing the channel (Nowak et al., 1984; Mayer and Westbrook, 1987; Jahr and Stevens, 1987). The requirement for postsynaptic depolarization arises from a block of the NMDA receptor-gated channel by extracellular Mg^{2+} at negative membrane potentials. This block is gradually relieved with membrane depolarization. Thus, the NMDA receptor–ionophore complex can act as a *coincidence detector* for pre- and postsynaptic activity.

C.   Biophysical Model of a Hebbian Synapse

Most of the excitatory synaptic inputs to hippocampal neurons are located on dendritic spines (color insert Fig. 2), and there is reason to suspect that the

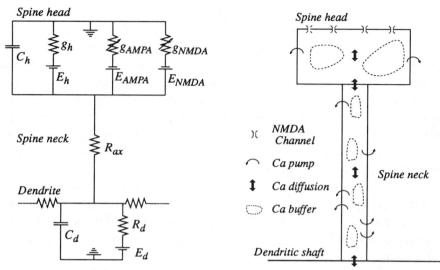

FIGURE 11.   Biophysical model of spine. The model has an electrical component (left) and a chemical component (right). The electrical component incorporates both an NMDA and an AMPA conductance (Eq. (3)). The chemical component of the model simulated Ca^{2+} influx, transport, and buffering. Ca^{2+} entered the spine through channels located only on the distal spine head and then diffused along the length of the spine to the dendritic shaft. Diffusion was limited by binding to saturable membrane pumps and cytosolic buffers. From Zador *et al.*, 1990.

microphysiology of the spine plays a significant role in the induction of LTP. To explore this possibility, Zador *et al.* (1990) have developed a model of Ca^{2+} influx, transport, and buffering following activation of NMDA receptor-gated channels located on the spine head (Fig. 11). The amount of synaptic enhancement was assumed to be a monotonic function of the free Ca^{2+} or Ca^{2+} bound to substrates such as calmodulin. The model accounts for the three key aspects of the phenomenonology of LTP induction that were mentioned previously: the conjunctive requirement, temporal specificity, and input specificity. A simplified version of the model was used in the simulations of hippocampal neurons presented next.

VI.   Self-Organization and Pattern Association

Up to this point, we have discussed two potentially interrelated facts about hippocampal pyramidal neurons from the CA1 region. First, there can be steep voltage gradients along the dendritic arbor. Second, many or most of the excitatory synapses onto these neurons display a Hebbian form of LTP that depends on the subsynaptic membrane potential. Our intuition was that the voltage gradients could affect the synaptic modification process. The finding that different parts of these neurons can be at different potentials suggests that differential modifications of synapses can take place across the space of the dendritic tree.

To evaluate this possibility, we simulated hippocampal neurons containing several hundred synapses that underwent modification according to a Hebbian learning rule. The simultaneous or nearly simultaneous activation of a subset of these synapses constituted an input vector or *feature*. A set of input vectors made up the *environment* of the neuron. Initial simulations involved the sequential presentation of a small number of temporally correlated, non-overlapping input vectors chosen according to a simple probability distribution.

The Hebbian modification algorithm (Brown *et al.*, 1990) related changes in synaptic strength to presynaptic activity $(a_j(t))$, postsynaptic activity $(a_i(t))$, and the current strength of the synapse $(w_{ij}(t))$. In general, the modifications could be expressed in the form:

$$F[a_i(t),a_j(t)] = F[\alpha(a_j(t) - A_j)(a_i(t) - A_i)], \qquad (15)$$

where A_i, A_j, and α are constants and $a_i(t)$ and $a_j(t)$ are defined as before. If we let $F[\bullet]$ be a linear functional, then Eq. (15) becomes:

$$F[a_i(t),a_j(t)] = \alpha F[a_i(t)a_j(t)] - \beta F[a_i(t)] - \gamma F[a_j(t)] - \delta, \qquad (16)$$

where the constants β, γ, and δ depend upon A_i, A_j, and α. In initial simulations (Brown *et al.*, 1991a,b), we related each of the terms in this equation to various known biological processes, as explained elsewhere (Brown *et al.*, 1990). The term $F[a_i(t)a_j(t)]$ can be compared to an interactive process for synaptic enhancement. In the simulations, its actual form was an abstraction of our biophysical model (Fig. 11) of LTP induction in the CA1 region of the hippocampus. The other three terms represented forms of synaptic depression. Specifically, $F[a_i(t)]$ represented heterosynaptic depression, which was proportional to depolarization from the resting membrane potential; $F[a_j(t)]$ represented homosynaptic depression, which was made to be proportional to the AMPA conductance; and the constant δ represented a passive decay process. A squashing function was used to maintain the final synaptic strength within specified bounds.

A.   Feature Selection

At the beginning of the simulations, the total postsynaptic conductance produced by each of the several features was the same. The synapses associated with the different features differed only with respect to their location on the dendritic tree. The location of each synapse was randomly and independently assigned. In spite of the fact that the synaptic conductances all began the same, over trials the neurons typically became more responsive to some features and less responsive to others. Some features would *win* in the sense that the average postsynaptic conductance (or the postsynaptic potential (PSP) amplitude in the soma) markedly increased beyond its initial value. Other features would *lose* in the sense that the average conductance (or somatic PSP) dramatically decreased from its initial value. In this way, the neuron became selectively tuned to a subset of its original set of inputs. The specificity of this tuning to particular features was dependent on the parameter values of the neuronal model, the learning rule, and the statistical structure of the environment.

As indicated, winning and losing can be measured by just looking at the mean synaptic conductance or the PSP amplitude in the soma. However, this measure does not furnish much insight into the spatiotemporal process that gives rise to the end result. To observe more readily the process of self-organization, we developed a color graphic display that made it easy to grasp in a semi-quantitative manner any changes in the conductance associated with each synapse of each feature. (See color insert Fig. 1B.)

B.   Cluster Formation

This graphical system made it obvious that the conductances of the synapses within a feature did not usually change at the same rate. Often they did not even change in the same direction; that is, some of the synapses within a winning feature might become weaker and some of the synapses within a losing feature might become stronger. It was common to observe spatial clusters of synapses with similar conductances. This visual impression, which could be quantified in any of several ways, was reflected in an initial increase in the variance of the conductances of the synapses within a pattern. Recall that at the beginning of each simulation the variance was zero. In some simulations, we observed a single large cluster of strengthened or weakened synapses and a smaller cluster of the opposite kind. (See color insert Fig. 1B.) In other cases, there were multiple clusters of smaller size. Across trials, it was often possible to observe some feature clusters form and then dissappear.

What causes these clusters to form, and why do some features win while others lose? Since the synapses initially differ only with respect to dendritic location, we assume that, for a fixed environment, it is the initial electrotonic location of the inputs that determines their fate. There are possibly two principles at work. First, for a given number of coactive synapses, the likelihood of strength-

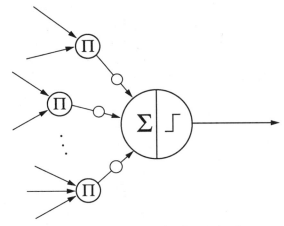

FIGURE 12.   Sigma-pi unit. The computations of a sigma-pi unit are more complex than those of the simplest processing element described in Eqs. (1) and (2). In a sigma-pi unit, subsets of of the inputs are multiplied. The output of the unit is the weighted sum of these products, passed through a threshold.

ening should increase if the synapses within the feature are electrotonically near each other. This follows from the fact that the synaptic enhancement is a joint function of presynaptic activity and the local dendritic depolarization and the fact that the local depolarization will increase if the coactive synapses are electrotonically closer together. Second, for any given synapse, the likelihood of depression should increase if the synapse is electrotonically closer to those of a strengthened cluster from another feature. This follows from the fact that the modification algorithm contains a term for heterosynaptic depression, which causes weakening of synapses that are not active during a strong local depolarization.

The full significance of this self-organization may only become apparent when dendritic voltage-dependent channels are included in our model. Evidence is accumulating for the heterogeneous distribution of voltage-gated Ca^{2+} and possibly other channels in the dendrites of hippocampal neurons (Jones *et al.*, 1989; Regehr *et al.*, 1989). The local nonlinearities introduced by these channels could allow regions of dendrites to make decisions about subsets of the inputs previously organized into a feature cluster. The existence of nonlinear membrane responses could allow the neurons to extract higher-order information from the environment. The closest analogy that one might find in the connectionist literature might be the so-called *sigma-pi* or product units, an example of which is illustrated in Fig. 12. We are currently exploring the computational consequence of such nonlinearities in dendritic membranes.

VII. Summary and Conclusions

Hippocampal pyramidal neurons differ from the PEs that are commonly used in connectionist studies along at least three dimensions—space, time, and probability. For some purposes, it may be inappropriate to view these PEs as being *neuron-like*. Our model of the dendritic spine is computationally considerably more complex than a typical PE, and there are more than 10^5 spines on each pyramidal neuron. (See color insert Fig. 2.) In this chapter, we have focused on the implications of spatial differences between neurons and PEs. Preliminary simulations suggest that the electrotonic structure of neurons adds a computational complexity that is not present in PEs. Electrotonic structure may be an important determinant of the manner in which synaptic strengths self-organize in response to a structured environment. This intuition is based on two facts. First, hippocampal pyramidal neurons are not isopotential. Synaptic inputs result in transient voltage gradients along the dendritic arbor. Second, some of the synapses onto these cells undergo a Hebbian modification that depends on the local voltage—that is, the voltage in the postsynaptic dendritic spine. We are currently exploring the consequences of adding voltage-dependent conductances to the dendritic membrane. The addition of such nonlinearity could influence the process of self-organization and alter the effects of this self-organization on neuronal output.

Acknowledgments

This research was supported by grants from the Office of Naval Research and the Defense Advanced Research Projects Agency.

References

Barrionuevo, G., and Brown, T. H. (1983). "Associative Long-Term Potential in Hippocampal Slices," *Proceedings National Academy of Science USA* **80**, 7347–7351.

Bekkers, J. M., and Stevens, C. F. (1990). "Presynaptic Mechanism for Long-Term Potentiation in the Hippocampus," *Nature (London)* **346**, 724–729.

Brown, T. H., Chang, V. C., Ganong, A. H., Kennan, C. L., and Kelso, S. R. (1988a). "Biophysical Properties of Dendrites and Spines that May Control the Induction and Expression of Long-Term Synaptic Potentiation," in S. A. Deadwyler and P. W. Landfield (eds.), *Long-Term Potentiation: From Biophysics to Behavior* (pp. 197–260). Liss, New York.

Brown, T. H., Chapman, P. F., Kairiss, E. W., and Keenan, C. L. (1988b). "Long-Term Synaptic Potentiation," *Science* **242**, 724–728.

BROWN, T. H., FRICKE, R. A., and PERKEL, D. H. (1981). "Passive Electrical Constants in Three Classes of Hippocampal Neurons," *Journal of Neurophysiology* **46**, 812–827.

BROWN, T. H., GANONG, A. H., KAIRISS, E. W., KEENAN, C. L., and KELSO, S. R. (1989). "Long-Term Potentiation in Two Synaptic Systems of the Hippocampal Brain Slice," in J. H. Byrne and W. O. Berry (eds.), *Neural Models of Plasticity* (pp. 266–306). Academic Press, San Diego, California.

BROWN, T. H., and JOHNSTON, D. (1983). "Voltage-Clamp Analysis of Mossy Fiber Synaptic Input to Hippocampal Neurons," *Journal of Neurophysiology* **50**, 487–507.

BROWN, T. H., KAIRISS, E. W., and KEENAN, C. L. (1990). "Hebbian Synapses: Biophysical Mechanisms and Algorithms," *Annual Review of Neuroscience* **13**, 475–512.

BROWN, T. H., MAINEN, Z. F., ZADOR, A. M., and CLAIBORNE, B. J. (1991a). "Self-organization of Hebbian Synapses in Hippocampal Neurons," *Neural Information Processing Systems* **3**, 39–45.

BROWN, T. H., PERKEL, D. H., and FELDMAN, M. W. (1976). "Evoked Transmitter Release: Statistical Effects of Nonuniformity and Nonstationarity," *Proceedings of the National Academy of Science USA* **73**, 2913–2917.

BROWN, T. H., and ZADOR, A. M. (1990). "The Hippocampus," in G. Shepherd (ed.), *The Synaptic Organization of the Brain, Vol. 3* (pp. 346–388). Oxford University Press, New York.

BROWN, T. H., ZADOR, A. M., MAINEN, Z. F., and CLAIBORNE, B. J. (1991b). "Hebbian Modifications in Hippocampal Neurons," in J. Davis and M. Baudry (eds.), *Long Term Potentiation: A Debate of Current Issues* (pp. 357–389). MIT Press, Cambridge, Massachusetts.

BUTZ, E. G., and COWAN, J. D. (1974). "Transient Potentials in Dendritic Systems of Arbitrary Geometry," *Biophysical Journal* **14**, 661–689.

CARNEVALE, N. T., and JOHNSTON, D. (1982). "Electrophysiological Characterization of Remote Chemical Synapses," *Journal of Neurophysiology* **47**, 606–621.

CHANDLER, W. K., FITZHUGH, R., and COLE, K. S. (1962). "Theoretical Stability Properties of a Space Clamped Axon," *Biophysical Journal* **2**, 105–127.

CLAIBORNE, B. J., ZADOR, A. M., MAINEN, Z. F., and BROWN, T. H. (1991a). "Computational Models of Hippocampal Neurons," in T. McKenna, J. Davis, and S. F. Zornetzer (eds.), *Single Neuron Computation*.

CLAIBORNE, B. J., ZADOR, A. M., MAINEN, Z. F., and BROWN, T. H. (1991b). "Electrotonic Structure of Hippocampal Neurons," in preparation.

DEFELICE, L. J. (1981). *Introduction to Membrane Noise*. Plenum Press, New York.

DUNWIDDIE, T., and LYNCH, G. (1979). "The Relationship between Extracellular Calcium Concentrations and the Induction of Long-Term Potentiation," *Brain Research* **169**, 103–110.

HEBB, D. O. (1949). *The Organization of Behavior*. Wiley and Sons Ltd., New York.

HILLE, B. (1984). *Ionic Channels of Excitable Membranes*. Sinauer, Sunderland, Massachusetts.

HINES, M. (1984). "Efficient Computation of Branched Nerve Equations," *International Journal of Bio-Medical Computation* **15**, 69–76.

HINES, M. (1989). "A Program for Simulation of Nerve Equations with Branching Geometries," *International Journal of Bio-Medical Computation* **24**, 55–68.

HODGKIN, A. L., and HUXLEY, A. F. (1952). "A Quantitative Description of Membrane Current and Its Application to Conduction and Excitation in the Nerve," *Journal of Physiology* **117**, 500–544.

JACK, J., NOBLE, A., and TSIEN, R. W. (1975). *Electrical Current Flow in Excitable Membranes.* Oxford, London.

JAHR, C. E., and STEVENS, C. F. (1987). "Glutamate Activates Multiple Single Channel Conductances in Hippocampal Neurons," *Nature (London)* **325**, 522–525.

JOHNSTON, D., and BROWN, T. H. (1983). "Interpretation of Voltage-Clamp Measurements in Hippocampal Neurons," *Journal of Neurophysiology* **50**, 464–484.

JOHNSTON, D., and BROWN, T. H. (1984). "Biophysics and Microphysiology of Synaptic Transmission in Hippocampus," in R. Dingledine (ed.), *Brain Slices* (pp. 51–86). Plenum Press, New York.

JONES, O. T., KUNZE, D. L., and ANGELIDES, K. J. (1989). "Localization and Mobility of ω-conotoxin-sensitive Ca^{2+} Channels in Hippocampal CA1 Neurons," *Science* **244**, 1189–1193.

KATZ, B. (1969). *The Release of Neural Transmitter Substances.* Charles C. Thomas, Springfield, Massachusetts.

KELSO, S. R., and BROWN, T. H. (1986). "Differential Conditioning of Associative Synaptic Enhancement in Hippocampal Brain Slices," *Science* **232**, 85–87.

KELSO, S. R., GANONG, A. H., and BROWN, T. H. (1986). "Hebbian Synapses in Hippocampus," *Proceedings of the National Academy of Science USA* **83**, 5326–5330.

KOCH, C. (1985). "Cable Theory in Neurons with Active, Linearized Membranes," *Biological Cybernetics* **50**, 15–33.

KOCH, C., POGGIO, T., and TORRE, V. (1982). "Retinal Ganglion Cells: A Functional Interpretation of Dendritic Morphology," *Philosophical Transactions of the Royal Society of London (Biology)* **298**, 227–264.

MALINOW, R., and MILLER, J. P. (1986). "Postsynaptic Hyperpolarization during Conditioning Reversibly Blocks Induction of Long-Term Potentiation," *Nature (London)* **320**, 529–530.

MALINOW, R., and TSIEN, R. W. (1990). "Presynaptic Enhancement Shown by Whole-Cell Recordings of Long-Term Potentiation in Hippocampal Slices," *Nature (London)* **346**, 177–180.

MASCAGNI, M. V. (1989). "Numerical methods for neuronal modeling," in C. Koch and I. Segev (eds.), *Methods in Neuronal Modeling: From Synapses to Networks* (pp. 439–484). MIT Press, Cambridge, Massachusetts.

MAYER, M. L., and WESTBROOK, G. L. (1987). "The Physiology of Excitatory Amino Acids in the Vertebrate Central Nervous System," *Progress of Neurobiology* **28**, 197–276.

McCORMICK, D. (1990). "Membrane properties in neurotransmitter actions," in G. Shepherd (ed.), *The Synaptic Organization of the Brain* (pp. 32–66). Oxford University Press, New York.

NEHER, E., and SAKMANN, B. (1983). *Single Channel Recording.* Plenum Press, New York.

NOWAK, L., BREGESTOVSKI, P., ASCHER, P., HERBET, A., and PROCHIANTZ, A. (1984). "Magnesium Gates Glutamate-Activated Channels in Mouse Central Neurons," *Nature* **307**, 462–465.

PERKEL, D. H., and PERKEL, D. J. (1985). "Dendritic Spines: Role of Active Membrane in Modulating Synaptic Efficacy," *Brain Research* **325**, 331–335.

PERKEL, D. H., MULLONEY, B., and BUDELLI, R. W. (1981). "Quantitative Methods for Predicting Neuronal Behavior," *Neuroscience* **6**, 823–838.

RALL, W. (1977). "Core Conductor Theory and Cable Properties of Neurons," in E. R. Kandel (ed.), *Handbook of Physiology: The Nervous System, Vol. 1 Sect. 1* (pp. 39–97). American Physiology Society, Bethesda, Maryland.

REGEHR, W. G., CONNOR, J. A., and TANK, D. W. (1989). "Optical Imaging of Calcium Accumulation in Hippocampal Pyramidal Cells during Synaptic Activation," *Nature (London)* **341**, 533–536.

RINZEL, J., and RALL, W. (1974). "Transient Response in a Dendritic Neuron Model for Current Injected at One Branch," *Biophysical Journal* **14**, 759–790.

SABAH, N. H., and LEIBOVIC, K. N. (1969). "Subthreshold Oscillatory Responses of the Hudgkin–Huxley Cable Model for the Squid Giant Axon," *Biophysical Journal* **9**, 1206–1222.

WIGSTROM, H., GUSTAFSSON, B., HUANG, Y.-Y, and ABRAHAM, W. C. (1986). "Hippocampal Long-Term Potentiation Is Induced by Pairing Single Afferent Volleys with intracellularly Injected Depolarizing Pulses," *Acta Physiology Scandinavia* **126**, 317–319.

ZADOR, A., KOCH, C., and BROWN, T. H. (1990). "Biophysical Model of a Hebbian Synapse," *Proceedings of the National Academy of Science USA* **87**, 6718–6722.

ZADOR, A., CLAIBORNE, B. J., and BROWN, T. H. (1992). "Nonlinear Processing in Single Hippocampal Neurons with Dendritic Hot and Cold Spots," *Neural Information Processing Systems* **4**. In press.

Chapter 5   Synaptic Integration by Electro-Diffusion in Dendritic Spines

TERRENCE J. SEJNOWSKI

Computational Neurobiology Laboratory
The Salk Institute
San Diego, California

NING QIAN

Department of Brain and Cognitive Sciences
Massachusetts Institute of Technology
Cambridge, Massachusetts

I. Introduction

Many vertebrate and invertebrate neurons receive synaptic inputs on spines (Coss and Perkel, 1985). In hippocampal pyramidal cells, almost all synapses found on spines are excitatory (Harris and Stevens, 1989), but in the cat primary visual cortex, 7% of synapses on spines are inhibitory (Beaulieu and Colonnier, 1985), which comprise almost one-third of the total number of inhibitory synapses on a pyramidal cell. It has been suggested that shunting inhibition on a spine could perform a selective AND-NOT-like operation (Koch and Poggio, 1983b). The nonlinear effects would be localized to the spine, making it an effective computational module. In this chapter, we compare the predictions made by the cable model with those made by the electro-diffusion model for this problem.

The conduction of action potentials in axons can be accurately modeled by the Hodgkin–Huxley equation, which is based on the cable equation (Hodgkin and Huxley, 1952). The integration of postsynaptic signals in dendrites has also been studied with analytic solutions to passive cables (Rall, 1977), and recently, several investigators have used the cable model to examine the possibility of more complex signal processing in dendrites with complex morphologies, multiple synaptic inputs, and passive or excitable membranes (Shepherd *et al.*, 1985;

Koch and Poggio, 1983; Koch *et al.*, 1983; Rall and Segev, 1985; Perkel and Perkel, 1985; Wathey *et al.*, 1989). A central assumption of the cable model is that ionic concentrations do not change appreciably, so that the driving forces can be approximated by fixed batteries. However, if the intracellular volume is relatively small, as in dendritic spines, then ionic concentrations can change rapidly following a transient change in ionic conductances. Moreover, a sudden change in concentration at one location can lead to gradients of ionic concentration within a thin process, such as the neck of a spine, which violates another fundamental assumption of the cable model. Under these circumstances, it is necessary to consider the fundamental laws governing the movements of ions, as given by the Nernst–Planck equations for electro-diffusion (Jack *et al.*, 1975).

In this chapter, we summarize the results of applying an electro-diffusion model based on the Nernst–Planck equation to dendritic spines (Qian and Sejnowski, 1989, 1990). The electro-diffusion model gives more accurate predictions than the cable model, especially for small structures. The electro-diffusion model provides a unified framework for the computation of both the membrane potentials and the intracellular ionic concentrations during synaptic activation. We have also developed a modified cable model that is a better approximation to the electro-diffusion model than the standard cable model and is less demanding computationally.

II. Cable Model Predictions

Consider first the case where the excitatory and inhibitory synapses are very close to each other. According to the cable model, the excitatory and the inhibitory synaptic currents are, respectively, given by:

$$I_e(t) = G_e(t)(V(t) - E_e) \tag{1}$$

and

$$I_i(t) = G_i(t)[V(t) - E_i] \sim G_i(t)[V(t) - V_{\text{rest}}], \tag{2}$$

where E_e and E_i are the reversal potentials of the excitatory and the inhibitory synapses, G_e and G_i are the transient synaptic conductances, V is membrane potential at the synapse, and V_{rest} is the resting membrane potential (Rall, 1977).

We have assumed in Eq. (1) that the reversal potential for shunting inhibition is very close to the resting membrane potential. In contrast, the excitatory inputs usually cause conductance increases to ions like Na^+ or Ca^{++} that have a reversal potential, E_e, well above resting membrane potential. The inhibition will be effective if $|I_i|$ is comparable to $|I_e|$. This requires, first, that G_i be larger than

G_e. Second, V should be well above the resting membrane potential so that the driving force for the inhibition $(V - V_{rest})$ is comparable to the driving force for the excitation $(V - E_e)$. This in turn requires that G_e be large and/or that the synapses are on small structures, such as spines or thin dendrites, where input resistances are large and small synaptic conductance change can cause a large depolarization. (Large inhibitory driving forces can also be achieved when the cell is firing an action potential.) In summary, for shunting inhibition to be effective when excitation and inhibition are located close to each other, the cable model requires that the synapses should be on small structures and $G_i > G_e \gg G_{rest}$, where G_{rest} is resting conductance of the membrane at the synapse.

In their analysis and simulations of shunting inhibition, Koch et al., (1983) mainly considered large synaptic conductances on spines and distal (thin) dendrites that satisfy the inequalities discussed in the preceding. Their G_e was as large as 10 nS and G_i was 100 nS, but more recent physiological data suggest that G_e should be about 1 nS (Higashima et al., 1986; Brown et al., 1988). They also found that, for large excitatory conductances, inhibition on the direct path to the cell body was also effective, and that the most effective location for the inhibition moves toward the soma as the excitatory conductance increases (Koch et al., 1982). Two opposing factors explain the phenomenon: When the inhibition is on the direct path from excitation to the soma, I_i is smaller because, at the site of inhibition, the membrane is less depolarized; but I_e is also smaller because, at the site of inhibition, the membrane is less depolarized; but I_e is also smaller because, at the site of excitation, the membrane is more depolarized. They also found that when the inhibition was more distal than the excitatory synapse, the inhibition was no longer effective. In this case, the resistance from the excitatory synapse to the cell body is much less than the resistance to the inhibitory synapse at the distal tip, so less current is shunted. Finally, Koch et al., (1983) mentioned that increasing the value of the cytoplasmic resistivity and the membrane resistance increased the effectiveness of inhibition. This occurred because the membrane depolarization was larger, which made the driving force for the inhibitory current larger and the driving force of the excitatory current smaller.

III. Limitations of the Cable Model

In the cable model, the membrane potential, $V(z, t)$, at distance z and time t along a cable obeys the equation (Jack et al., 1975):

$$\frac{d}{4R_i} \frac{\partial^2 V}{\partial z^2} = C_m \frac{\partial V}{\partial t} + I_m, \tag{3}$$

where d is the diameter of the cable. R_i (Ωcm) is the total intracellular cytoplasmic resistivity, C_m (μF/cm^2) is the membrane capacitance per unit area, and I_m (mA/cm^2) represents the total non-capacitative membrane current density, which is the summation of all non-capacitative membrane current densities for each ionic species, $I_{m,k}$. If we assume that the movement of ionic species k across the membrane can be described by a membrane resistance of unit area $R_{m,k}$ (Ωcm^2) in series with a battery whose electromotive force E_k is equal to the ionic equilibrium potential, then

$$I_{m,k} = \frac{V - E_k}{R_{m,k}} \tag{4}$$

and

$$I_m = \sum_k I_{m,k} = \frac{V - V_{rest}}{R_m}. \tag{5}$$

where the resting membrane potential V_{rest} and the total membrane resistance R_m are given by:

$$V_{rest} = R_m \sum_k (E_k/R_{m,k}), \tag{6}$$

$$\frac{1}{R_m} = \sum_k (1/R_{m,k}). \tag{7}$$

Through these definitions, the electrical circuit can be reduced to a simpler equivalent circuit that has a single battery in series with a leak resistance. The standard equation for the cable model is obtained by substituting Eq. (5) into Eq. (3), assuming that V is measured from the resting potential V_{rest} (Rall, 1977):

$$\lambda^2 \frac{\partial^2 V}{\partial z^2} = \tau_m \frac{\partial V}{\partial t} + V, \tag{8}$$

where the space and time constants are defined as

$$\lambda = (d\, R_m/4R_i)^{1/2}, \tag{9}$$

$$\tau_m = R_m\, C_m. \tag{10}$$

The electromotive forces of the membrane batteries (equilibrium potentials) in the cable model are usually obtained from the Nernst equation and are considered constants. This is a good approximation in the squid giant axon or other large neurons, but may introduce errors if the concentrations of some ions change significantly. This applies to Ca^{++} in many situations and to synaptic events in small structures such as dendritic spines (Rall, 1978; Koch and Poggio, 1983a; Zador *et al.*, 1990).

A second limitation of the cable model is in the treatment of longitudinal spread of current within neurons. In the cable model, the gradient of the electrical potential in the cytoplasm is the driving force for the ionic current, but there is no provision for the driving forces due to concentration gradients. This is usually a good assumption, but it may not be valid for small structures like dendritic spines where the spatial concentration gradients can be very large.

Finally, different ions may have different concentration-dependent cytoplasmic resistivities, but the cable model only incorporates the total cytoplasmic resistivity. This may not be a valid approximation when the concentration of ions are changing differentially. In summary, one expects that the cable model may not be appropriate when spatial and/or temporal ionic concentration changes are large and, especially when ionic concentration changes need to be determined.

IV.   Electro-Diffusion Model Predictions

The cable model fails for small structure and large conductance changes, precisely the conditions required for effective shunting inhibition by the cable model (Qian and Sejnowski, 1988, 1989). the electro-diffusion model predicts that the shunting inhibition cannot be effective on small structures for the following reasons. Consider first the case when the conductance changes are large. If the inhibitory current is carried by Cl^- ions, then during a large conductance change the Cl^- concentration in a small structure such as a spine or a thin dendrite will very rapidly increase. The Nernst potential for Cl^- becomes more positive and the inhibition is ineffective. Changes in the Cl^- Nernst potential have been reported (Griffith *et al.*, 1986; Huguenard and Alger, 1986). If the conductance changes are small, then the concentration changes for Cl^- are small and the electro-diffusion model will reduce to the cable model. Thus, shunting inhibition will not be effective because the membrane depolarization is small and the driving force for the inhibitory current is much smaller than that for the excitatory current, as discussed in section II. As a consequence, the electro-diffusion model predicts that the shunting inhibition can never be very effective in small structures.

A similar analysis can be applied for hyperpolarizing inhibition carried by K^+. When both the excitatory and the inhibitory synaptic conductances are large on a small structure, K^+ hyperpolarizing inhibition is just as ineffective as the Cl^- shunting inhibition because of large ionic concentration changes. However, the situation for small synaptic conductances is different. The reversal potential for K^+ is sufficiently below the resting potential that the driving force for the inhibition can be large even at the resting potential. In addition, the intracellular K^+ concentration is much higher than Cl^- and, therefore, the percentage change is usually smaller. These statements will be made quantitatively precise in the model and the numerical simulations presented next.

A. Electro-Diffusion Model

The movement of ions in neurons is governed by the Nernst-Planck equation (Jack *et al.*, 1975):

$$\bar{J}_k = -D_k(\bar{\nabla}n_k + (n_k/\alpha_k)\bar{\nabla}V), \tag{11}$$

where V is the potential, $\bar{J}_k$ is the flux of ionic species k (number of particles per unit area), D_k is the diffusion constant, n_k is the concentration, and the constant α_k is defined as

$$\alpha_k = \frac{RT}{Fz_k}, \tag{12}$$

where z_k is the valence of ionic species k, R is the gas constant, F is the Faraday constant, and T is the absolute temperature. The ionic concentrations and ionic currents must additionally satisfy the continuity equation:

$$\bar{\nabla} \cdot \bar{J}_k + \frac{\partial n_k}{\partial t} = 0. \tag{13}$$

Consider a cylinder of diameter d and assume that the longitudinal current and ionic concentrations are uniform across the transverse cross section of the cylinder. Assume also that transverse currents occur only at the surface of the cylinder and are independent of angle around the axis of the cylinder. These assumptions reduce the problem of electro-diffusion to a one-dimensional problem along the axis of the cylinder. The equations can be written in cylindrical coordinates and reduced to a single equation for the concentration as a function of the distance along the z axis of a cylinder:

$$\frac{\partial n_k}{\partial t} = D_k \frac{\partial^2 n_k}{\partial z^2} + \frac{D_k}{\alpha_k} \frac{\partial}{\partial z}\left(n_k \frac{\partial V}{\partial z}\right) - \frac{4}{d}J_{m,k}, \tag{14}$$

where $J_{m,k}$ is the membrane flux of ionic species k, positive for outgoing flux.

Equation (14) must be supplemented by an additional constraint between the membrane potential and the ionic concentrations. We adopt the same capacitative model of the membrane used in the cable model; that is, we assume that the potential change in a short segment of a process is equal to the change of the total charge in the segment divided by its membrane capacitance:

$$V(z, t) = V_{rest} + (Fd/4C_m) \sum_k [n_k(z, t) - n_{k,rest}]z_k, \qquad (15)$$

where V_{rest} is the initial potential and $n_{k,rest}$ is the initial ionic concentration of species k.

Neuronal processes often branch and change their diameters. If branches are allowed, then these equations must be solved on a tree rather than a line. Continuous diameter changes can be approximated by segments having piecewise constant diameters. At points where the diameter jumps and/or branches occur, the solutions can be matched using the continuity of flux at that point. The continuity constraint at a branch point can be derived from Eq. (11). The continuity constraint at a branch point where three processes join is given by:

$$d_1^2 \left(\frac{\partial n_k}{\partial z} + \frac{n_k}{\alpha_k} \frac{\partial V}{\partial z} \right) \Bigg|_1$$
$$= d_2^2 \left(\frac{\partial n_k}{\partial z} + \frac{n_k}{\alpha_k} \frac{\partial V}{\partial z} \right) \Bigg|_2 + d_3^2 \left(\frac{\partial n_k}{\partial z} + \frac{n_k}{\alpha_k} \frac{\partial V}{\partial z} \right) \Bigg|_3, \qquad (16)$$

where d_i is the diameter of the ith branch.

The square of the diameter enters into this equation because the flux through the areas of each branch must be matched. There is an analytic solution of the cable model for branching dendrites having passive membranes if Rall's *3/2 power law* (Rall, 1977) is satisfied. This law for an equivalent cylinder does not hold for our electro-diffusion model except in the limit when the concentration gradients go to zero and Eq. (4) is used to compute the membrane currents. A compartment approximation for the solution of Eq. (14) is inaccurate for large ionic fluxes if the continuity constraint in Eq. (16) is not used to match solutions on the two sides of a diameter jump or at a branch point.

B.   Simulations of Postsynaptic Potentials in a Dendritic Spine

For simplicity, only three types of ions, K^+, Na^+ and Cl^-, will be considered in this chapter. Excitation was modeled by a combination of transient conductance changes to Na^+ and K^+, with the K^+ conductance equal to one-tenth of Na^+ conductance (Hille, 1984). The synaptic reversal potential under this combination is about 50 mV. We also made simulations with the reversal potential of the excitatory synapse equal to 0 mV and similar conclusions were obtained. Silent shunting inhibition and hyperpolarizing inhibition were modeled by transient Cl^-

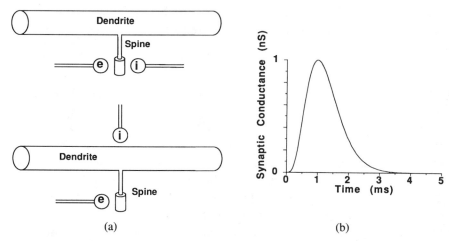

(a)   (b)

FIGURE 1.   (a) Geometry of the dendrite and spine for the simulations showing excitation and inhibition on the spine head (top) and inhibition on the dendritic shaft at the base of the spine (bottom). The spine was located in the center of a dendrite with total length 300 μm and diameter 1 μm; the spine neck was 1 μm long and 0.1 μm in diameter; the spine head was 0.69 μm long and 0.3 μm in diameter. In the simulations of the electro-diffusion model, sample points in the dendrite were 10 μm apart and the integration time step was 10^{-7} sec; in the spine head and neck, the spacing was 0.173 μm and 0.167 μm, respectively, and the time steps were 2×10^{-9} sec. The model had a total of 41 sample points: 31 in the dendrite, six in the spine neck, and four in the spine head. In the conventional cable model, only 33 lumped compartments were used (one for head, one for neck, and 31 for dendrite) due to the large spatial constant. The time step for spine head and neck was 10^{-7} sec and that for the dendrite was 10^{-6} sec. (b) Excitatory and inhibitory synaptic conductance changes were modeled by $G(t) = G_M (et/t_{peak})^4 e^{-4t/t_{peak}}$, where t_{peak} was the time to reach the peak conductance, G_M. A graph of this expression is shown with $t_{peak} = 1$ ms, and $G_M = 1$ nS. Parameters used in our simulations were: $t_{peak} = 1$ ms; membrane capacitance $C_m = 1$ μF/cm^2; diffusion coefficients $D_K = 1.96 \times 10^{-5}$ cm^2/sec, $D_{Na,M} = 1.33 \times 10^{-5}$ cm^2/sec, and $D_{Cl} = 2.03 \times 10^{-5}$ cm^2/sec; resting membrane conductances of unit area $g_{K,rest} = 1.95 \times 10^{-4}$ Scm^2, $g_{Na,rest} = 1.63 \times 10^{-5}$ Scm^2, and $g_{Cl,rest} = 3.89 \times 10^{-5}$ Scm^2; initial intracellular concentrations $n_{K,0} = 140$ mM, $n_{Na,0} = 12$ mM, and $n_{Cl,0} = 5.5$ mM; extracellular concentrations $n_{K,out} = 4$ mM, $n_{Na,out} = 145$ mM, and $n_{Cl,out} = 120$ mM. With this set of parameters, the resting membrane potential was -78 mV. The Nernst potentials for K$^+$, Na$^+$, and Cl$^-$ were -90 mV, 63 mV, and -78 mV, respectively. Total membrane resistivity R_m was 4000 Ωcm^2. Total cytoplasmic resistivity R_i at rest was calculated (Qian and Sejnowski, 1989) to be 87 Ωcm. Total surface area of the spine head was 0.65 μm^2. Some of these parameters were varied, as explained in the relevant figures or tables. The sources for these parameters are given in Qian and Sejnowski (1989).

and K^+ conductance changes, respectively. Ionic driving forces similar to Eqs. (1) and (2) were used for the electro-diffusion model rather than the constant-field approximation used in Qian and Sejnowski (1989). The Nernst Potentials were updated at each time step according to the instantaneous ionic concentrations. We varied the magnitudes and durations of the conductance changes and the spine neck dimensions. We also compared the effectiveness of inhibition on the spine with inhibitory input on the dendrite at the base of the spine. The standard parameters used and details of the simulations are summarized in the caption of Fig. 1; any variation will be explicitly mentioned.

A measure of the effectiveness of shunting inhibition is the ratio of the maximum depolarization at a reference point in the neuron caused by an excitatory input alone to the depolarization when both the excitatory and the inhibitory inputs are present. This ratio, called the *F factor* (Koch and Poggio, 1983a), is equal to 1 if the inhibition has no effect on the excitation. One obvious requirement for effective inhibition is that its time course should overlap substantially with the excitatory synaptic conductance change. We modeled a spine located in the middle of a 300 μm-long dendrites and the response at the spine head was used to calculate F factors. Our simulation results based on both the cable model and the electro-diffusion model are shown in Table I. The cable model indeed showed strong veto effects, especially when the conductances were large, as predicted. However, our electro-diffusion model showed no significant veto effect over a wide range of conductances. Note also that when the $G_{Cl,M}/G_{Na,M}$ ratio was increased, there were cases where the F factor decreased slightly. This occurred because the Cl^- Nernst potential shifted so much that it depolarized the membrane away from its resting level. A Cl^- conductance change alone, however, did not cause any depolarization because there was no driving force and, therefore, no concentration change.

TABLE I.   *F* Factors at the Spine Head when Both Excitatory and Cl^- Mediated Inhibitory Synaptic Inputs are Located on the Same Spine Predicted by Both the Electro-Diffusion Model and the Cable Model.

$G_{Cl,M}/G_{Na,M}$	$G_{Na,M} = 0.1$ nS		$G_{Na,M} = 1.0$ nS		$G_{Na,M} = 10$ nS	
	diffusion	cable	diffusion	cable	diffusion	cable
1	1.02	1.02	1.10	1.17	1.23	1.65
10	1.10	1.20	1.20	2.74	1.26	7.56
10^2	1.16	3.04	1.19	18.63	1.25	66.20
10^3	1.14	20.35	1.19	163.86	1.25	602.19

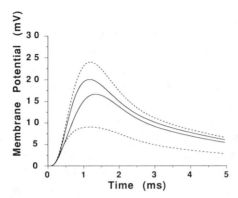

FIGURE 2.   Postsynaptic responses relative to the resting level at the spine head calculated with the electro-diffusion model (solid lines) and the cable model (dotted lines). Two traces are shown for each model: The top trace is the response with excitatory synaptic input alone and the lower trace is the response to both excitatory and shunting inhibitory inputs. Excitatory synaptic input: $G_{Na,M} = 1$ nS; inhibitory synaptic input: $G_{Cl,M} = 10$ nS.

The difference between the cable model and the electro-diffusion model decreased as the synaptic conductances decreased. For $G_{Na,M} = 0.1$ nS, the two models were essentially identical and both predicted that the inhibition was ineffective. However, for longer durations of the synaptic input, the two models may not agree even for synaptic conductance changes as small as 0.1 nS. (See Section VI.) The details of the postsynaptic responses on the spine head are shown in Fig. 2.

Quantal analysis on excitatory postsynaptic potentials in area CA3 of the rat hippocampus (Higashima *et al.*, 1986; Brown *et al.*, 1988) gave a quantal conductance of about 1 nS at mossy fiber synapses. Therefore, the synaptic conductance change due to a single presynaptic action potential should be about a few nS. Similar measurements of unitary inhibitory conductance performed on CA3 pyramidal cells of guinea-pig hippocampus obtained a value of 5–9 nS (Miles and Wong, 1984). However, the conductance of synapses on pyramidal cells in cerebral cortex may be much smaller. In the following simulations, we fixed $G_{Na,M}$ at 1 nS and varied $G_{Cl,M}$ unless otherwise indicated.

The morphologies of spines vary greatly. The critical parameters for our simulations were the diameter and length of spine neck, which were varied from 0.1 μm to 0.25 μm and 0.4 μm to 1.0 μm, respectively, with the neck membrane area kept constant. We also considered the case where there was no spine neck and the spine head was connected directly to dendrite. The cable model gave large F factors when the neck was long and narrow and/or $G_{Cl,M}$ was large, but the electro-diffusion model produced no F factor larger than 2 over the entire range. The effectiveness of inhibition was not very sensitive to the dimensions of the spine neck because of two competing effects that cancel: As the spine neck length was decreased and the diameter increased, the concentration changes in the spine were reduced, making the inhibition more effective. However, the input resistance of the spine head was also decreased, resulting in a smaller depolarization and a reduced driving force for the inhibition.

1.   On-Path Inhibition.   Inhibitory synapses on pyramidal neurons are commonly located on the dendritic shaft at the base of the spine (Beaulieu and colonnier, 1985; Martin, 1984). The simulations in Fig. 3 show that dendritic on-path inhibition is much more effective than inhibition on the spine head. The ionic concentration changes were much smaller for the dendritic inhibition because the dendrite had a diameter of 1 μm, and hence the cable equation was a good approximation. Also, the driving force for the inhibition was strong because the spine was electrically coupled to the dendrite well enough that the excitation of the spine caused a large depolarization at the dendritic shaft.

2.   Duration of Inhibition.   Table II shows the F factors at the spine for a range of durations of the synaptic conductance (determined by t_{peak} in Fig. 1). A longer duration produced a larger membrane depolarization that increased the driving force for the inhibition and made the inhibition more effective. This

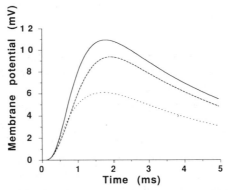

FIGURE 3.   Responses relative to the resting level at the dendritic shaft at the base of a spine under three conditions: no inhibition and the excitatory input ($G_{Na,M} = 1$ nS) on the spine head alone (solid), the excitatory input ($G_{Na,M} = 1$ nS) and the inhibitory input ($G_{Cl,M} = 10$ nS) both on the same spine head (dashed), and the excitatory input ($G_{Na,M} = 1$ nS) on the spine head and the inhibitory input ($G_{Cl,M} = 10$ nS) on the dendritic shaft at the base of the spine (dotted).

TABLE II. F Factors at the Spine for a Range of
Durations of the Synaptic Conductance. The time
to peak (t_{peak}, explained in caption to Fig. 1a) was
varied for a synapse with $G_{Na,M} = 1$ nS and $G_{Cl,M}$
$= 11$ nS.

t_{peak}(ms)	diffusion	cable
0.5	1.32	2.46
1.0	1.20	2.73
2.0	1.17	3.01
3.0	1.16	3.30
4.0	1.17	3.53

explains the increase of F factors with t_{peak} predicted by the cable model. However, the longer the conductance change, the larger the concentration change,
which makes inhibition less effective according to the electro-diffusion model.
The latter factor predominated for inhibition on the spine head, as shown in
Table II.

Repetitive stimulation of inhibitory interneurons in hippocampal slices for tens
of seconds can cause disinhibition of inhibitory potentials in pyramidal cells
(Thompson and Gahwiler, 1989). The most likely explanation is the intracellular
accumulation of Cl$^-$. Our simulations suggest that a similar disinhibition can
occur on spines within milliseconds. The difference between the time scales can
be attributed to the difference between the intracellular volumes of cell bodies
and dendrites compared with spine heads.

3.   Interactions between Synapses on Dendrites.

We next studied interac-
tions between excitatory and inhibitory synapses at adjacent sites on dendrites
ranging in diameter from 0.1 to 2.0 μm. The predicted F factors for the cable
model, given in Table III, were very large when the dendritic diameter was small
and the G_{Cl} was large. For the electro-diffusion model, two competing factors
determined the effectiveness of inhibition: For dendrites with large diameters,
the concentration effects were small, so the Nernst potential did not change very
much and the inhibition was effective. However, the polarization of the membrane from the resting level was smaller in larger dendrites, which made inhibition
less effective. The F factors in Table IIIa were not monotonically increasing
with increasing dendritic diameter because of these two factors and their interaction. For $G_i = G_e$, the first factor dominated and the inhibition was compar-
atively more effective on small dendrites. When $G_i \geq 100\, G_e$, the second factor
dominated and the inhibition was more effective on large dendrites. In any case,

TABLE III. *F* Factors for Both Excitatory and Inhibitory Synapses on Dendrites at the Same Site, with Different Dendritic Diameters. $G_{Na,M} = 1$ nS. Top: Electro-Diffusion Model; Bottom: Cable Model.

	dendritic diameter (μm)				
$G_{Cl,M}/G_{Na,M}$	0.1	0.25	0.5	1.0	2.0
0.1	1.07	1.04	1.02	1.01	1.00
1	1.47	1.45	1.18	1.07	1.03
10	1.87	1.66	2.26	1.66	1.31
10^2	1.91	2.67	3.43	3.10	3.25

	dendritic diameter (μm)				
$G_{Cl,M}/G_{Na,M}$	0.1	0.25	0.5	1.0	2.0
0.1	1.08	1.06	1.04	1.02	1.01
1	1.72	1.39	1.19	1.08	1.04
10	8.31	5.16	3.03	1.88	1.38
10^2	73.07	43.15	22.46	11.01	5.65

when the dendritic diameter was 0.1 μm, the *F* factors were always less than 2, similar to the previous results for inhibition on spines.

4. K$^+$-Mediated Inhibition.

The equilibrium potential for K$^+$ is generally below the resting membrane potential (12 mV below in our model), so that an increase in K$^+$ conductance leads to a hyperpolarization. In a previous study, we found that inhibition on spines mediated by K$^+$ was not effective for excitatory conductances greater than 10 nS (Qian and Sejnowski, 1989). In this section, we consider excitatory conductances that are lower and more realistic for pyramidal neurons. We find that for smaller excitatory conductances, hyperpolarizing inhibition can be quite effective. Synaptic responses to $G_{Na,M} = 0.1$ nS are shown in Fig. 4, which also shows that an inhibition of $G_{K,M} = 1$ nS is very effective in reducing the response. In comparison, the inhibition due to a similar or much larger conductance change for Cl$^-$ was not effective. For large excitatory conductance changes, the K$^+$ inhibition became as ineffective as Cl$^-$ because of the large K$^+$ concentration changes that rapidly shift the K$^+$ Nernst potential, as shown in Table IV.

Inhibition mediated by K$^+$ in cortical neurons has a time course that can last for a significant fraction of a second when it is activated by GABA$_B$ receptors

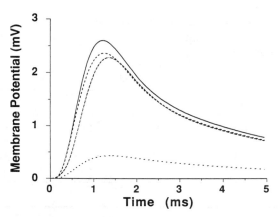

FIGURE 4.   Response relative to the resting level at the spine head to an excitatory input of $G_{Na,M} = 0.1$ nS and one of the following four different inhibitory synaptic inputs: no inhibition (solid), K^+ inhibitory synaptic input with $G_{K,M} = 1$ nS (dotted), Cl^- inhibitory input with $G_{Cl,M} = 1$ nS (dashed), and $G_{Cl,M} = 100$ nS (long-dashed).

through G proteins. We therefore studied the steady-state behavior following a step change in conductances of $G_{Na} = 0.1$ nS and $G_K = 1$ nS and found that the response at the spine head was about 6.9 mV with excitation alone and 1.3 mV with both excitation and inhibition. In steady state, the K^+ efflux from the spine head was balanced by the K^+ diffusion from the dendritic shaft to the head. Thus, the inhibition mediated by K^+ conductances remained effective for slow inhibitory synaptic potentials when the excitatory conductances were small. Since an excitatory synaptic conductance typically lasts for only a few ms, an excitatory input arriving a few ms earlier than the inhibitory input will not be affected by the inhibition. Once an inhibitory input is active, there is a long time window during which arriving excitatory inputs are inhibited.

TABLE IV. F Factors at the Spine Head when Both Excitatory and K^+ Mediated Inhibitory Synaptic Inputs are Located on the Same Spine Predicted by both the Electro-Diffusion and the Cable Models.

$G_{K,M}/G_{Na,M}$	$G_{Na,M} = 0.1$ nS		$G_{Na,M} = 1.0$ nS		$G_{Na,M} = 10$ nS	
	diffusion	cable	diffusion	cable	diffusion	cable
0.1	1.01	1.07	1.02	1.07	1.07	1.07
1	1.11	1.18	1.24	1.33	1.58	1.79
10	6.05	8.03	7.35	17.65	1.70	47.53
10^2	*	*	*	*	1.66	*

*F factors undefined because the responses were hyperpolarizing, indicating very effective inhibition.

V.   The Cable Model for Electro-Diffusion

The electro-diffusion model is highly computation-intensive and cannot be used routinely for large-scale simulations of complex dendritic trees (Wathey *et al.* 1989). It would be desirable to have a model that was as efficient as the cable model. We will prove here that a simple extension of the cable model can, in fact, provide an accurate approximation to the predictions of the electro-diffusion model.

The following modifications to the discrete approximation of the standard cable model should be made at each time step:

(1) Calculate the intracellular concentration of each ionic species explicitly in each compartment from the membrane currents and the ionic currents flowing between compartments.

(2) Compute the new membrane equilibrium potentials for each compartment using the intracellular and extracellular ionic concentrations according to the Nernst equation and update the membrane batteries:

$$E_k = \frac{RT}{Fz_k} \ln \frac{n_k(\text{out})}{n_k(\text{in})}, \tag{17}$$

where $n_k(\text{out})$ is the ionic concentration of species k outside the membrane and $n_k(\text{in})$ is the ionic concentration inside the membrane. This makes the membrane current expressions identical to those used in the electro-diffusion model.

(3) Replace the single longitudinal resistance between compartments with parallel resistances $R_{i,k}$ (Qian and Sejnowski, 1989).

$$\frac{1}{R_{i,k}} = (F^2/RT)D_k n_k z_k^2, \tag{18}$$

in series with batteries (step 4) for each ionic species.

(4) Determine the longitudinal batteries by the Nernst potential for ionic concentrations of the two compartments they connect and update in the same way as the membrane batteries. The potential of the battery for species k between compartments j and $j + 1$ is:

$$E_{i,k} = \frac{RT}{Fz_k} \ln \frac{n_k(j)}{n_k(j + 1)}, \tag{19}$$

where a positive value for $E_{i,k}$ means that the positive terminal of the battery is pointing to the $j + 1$ compartment.

Steps 3 and 4 account for the effects of the longitudinal component of electro-diffusion in Eq. (11). We show here that this approximation is exact in limit that the compartments approach zero length.

Consider two compartments at x and $x + dx$. The longitudinal battery for ionic species k between these two compartments, according to step 4, is:

$$dE_{i,k}(x) = \frac{RT}{Fz_k} \ln\frac{n_k(x)}{n_k(x + dx)} = -\frac{RT}{Fz_k} \ln\left[1 + \frac{1}{n_k}\frac{dn_k}{dx}dx\right] = -\frac{RT}{Fz_k}\frac{dn_k}{n_k dx}dx.$$

For a nerve fiber of cross-sectional area A, the ionic current of species k between these two compartments is:

$$I_k(x) = \frac{dE_{i,k}(x) + V(x) - V(x + dx)}{R_{i,k}dx/A}$$

$$= -AFz_kD_k\frac{dn_k}{dx} - \frac{AFz_kD_kn_k}{\alpha_k}\frac{dV}{dx}. \quad (20)$$

Equation (18) and the definition of α_k in Eq. (12) were used in this derivation. The flux $J_k(x)$ is defined as the number of ions moving across a unit area in unit time, of ionic species k between the two compartments. Thus, from Eq. (20),

$$J_k(x) = \frac{I_k(x)}{AFz_k} = -D_k\left[\frac{dn_k}{dx} + \frac{n_k}{\alpha_k}\frac{dV}{dx}\right], \quad (21)$$

Modified Cable Model

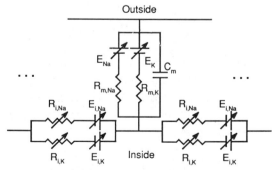

FIGURE 5.   Electrical circuit for a single compartment of the modified cable model. The batteries in the membrane and between compartments are calculated from the Nernst potentials and change value during a response as concentrations change. The resistivities in the cytoplasm are also updated at each time step.

which is identical to the longitudinal component of the Nemst–Planck electro-diffusion equation for ionic species k. The continuity equation of the electro-diffusion model (Eq. (13)) is also satisfied in cable model because of Kirchoff's law. Therefore, results predicted by the modified cable model should be equivalent to those of the one-dimensional electro-diffusion model.

A schematic view of the modified cable model is represented in Fig. 5. This model was applied to the dendritic spine model in Qian and Sejnowski (1989). The results were in excellent agreement with the electro-diffusion model except when the conductances and ionic concentrations changed very rapidly. The differences were mainly due to the fact that we used the constant-field approximation for the membrane currents in that version of the electro-diffusion model, but used ohm's law for the batteries in the modified cable model.

VI.   Discussion

In most circumstances, the cable model of electrical conduction in neurons gives accurate predictions for membrane potentials during transient electrical events. A tiny amount of charge is enough to cause a substantial change in the membrane potential because the membrane capacitance is small, and as a consequence, the ionic concentrations usually are nearly constant (Qian and Sejnowski, 1989). However, the *individual* concentrations of certain ions could change significantly as long as changes in the total charge are nearly balanced out. This is more likely to happen in very small structures such as dendritic spines and for an ion such as Ca^{++}, which is normally maintained at a very low concentration inside a cell. Thus, the predictions from the cable model should not be used without a careful analysis.

We have developed a one-dimensional electro-diffusion model of electrical conduction, which reduces to the cable model when ionic concentrations are approximately constant. This model was used to study changes in ion concentrations and membrane potentials in dendritic spines in response to synaptic inputs. We found that ionic concentrations changed considerably in many circumstances. Thus, significant errors can be made in estimating membrane potentials and concentration changes using the cable model if the effects of diffusion and the changes in the driving forces for membrane current are not taken into account.

The major conclusion of this study is that Cl^- shunting inhibition on spines cannot be very effective regardless of how large the synaptic conductance changes are. Shunting inhibition is significantly more effective when it is on the dendritic shaft on-path to the cell body. This may partly explain the anatomical findings that most synapses on spines are putatively excitatory and that the majority of the putative inhibitory synapses are found on dendritic shafts. Shunting inhibitory synapses on spines may have other functions. Although they may not contribute significantly to the electrical responses of the cell, they can certainly cause large local ionic concentration changes that may be important in regulating certain cellular functions.

The inhibitory synapses on spines may contribute to the electrical responsiveness of a cell if they are mediated through K^+ currents. Our simulations predict that K^+ hyperpolarizing inhibition on a spine head can be very effective when the excitatory synaptic conductance changes are less than 10 nS. The major inhibitory neurotransmitter in the visual cortex is GABA; GABA$_A$ receptors are coupled to Cl$^-$ channels and GABA$_B$ receptors are linked to K^+ channels. Therefore, we specifically predict that the inhibition on spines is mediated by the GABA$_B$ receptors. This prediction is consistent with the finding that GABA$_B$ input to hippocampal pyramidal cells is preferentially dendritic (Janigro and Schwartzkroin, 1988), where the majority of inputs are onto spines. Another way to have effective inhibition on a spine is through conductance decreases of either Na^+ or Ca^{++}, although this type of inhibitory mechanism has not been found in cortical neurons.

All these results can be understood as a consequence of changing ionic concentrations and shifting equilibrium potentials. The postsynaptic responses are mainly determined by the ionic species with the largest transient membrane permeability. Regardless of the equilibrium potential for this species (excitatory or inhibitory), the opening of channels with sufficient duration will shift the equilibrium potential of the synapse toward zero because of the large concentration changes. Therefore, if the equilibrium potential of the ionic species is greater than zero (Na^+ and Ca^{++}), the cable model will overestimate the response. For ions whose equilibrium potential is less than zero (K^+ and Cl$^-$), the cable model makes predictions that can be qualitatively incorrect. For example, under certain conditions an inhibitory synaptic input that causes a large K^+ conductance increase may, in fact, produce a postsynaptic depolarization.

Our simulations have shown that discrepancies between the cable model and our electro-diffusion model increase with the magnitude and duration of the synaptic conductance changes. Since the cable model is valid only when the concentration changes are small, we can derive a condition under which the cable model is self-consistent. The intracellular concentration change of the kth ionic species caused by membrane current I_k within time duration Δt is $\Delta n_k = I_k \Delta t / v z F$, where v is the effective intracellular volume, z is the valence of the ion involved, and F is the Faraday's constant. (Of course, n_k will eventually stop changing with time when the membrane current is balanced by the intracellular diffusion.) The criteria for the self-consistency of the cable model is simply $|\Delta n_k|/n_{k,0} \ll 1$, where $n_{k,0}$ is the initial intracellular ionic concentration, or $\Delta t \ll |vzFn_{k,0}/I_k|$.

When the synaptic conductance G_k is small, say, 0.1 nS, $I_k \sim G_kE_k$, where E_k is the reversal potential relative to the resting potential. The preceding condition gives $\Delta t \ll 10$ ms for Na^+ in a spine, assuming that the effective volume v is equal to twice the volume of the spine. Therefore, the cable model may not be valid for spines if the duration of the conductance change is longer than 10 ms, even for synaptic conductance changes as small as 0.1 nS. the inclusion of ionic pumps would not alter the preceding conclusions for a typical Na–K pump current density of $1 \sim \mu A/cm^2$ (Weer and Rakowski, 1984), in which case the total pump current of the spine head is about 10^{-14} A, three orders of magnitude smaller than the synaptic current. Even when the pump molecules are close-packed in the membrane, the maximum possible pump current density is 100 $\mu A/cm^2$, and the total pump current of the spine head is still about 10 times smaller than the synaptic current for a 0.1 nS conductance. The effect of the Na–K pump would be significant if we assume that the spine apparatus is also densely packed with pump molecules and its surface area is about 10 times that of the spine.

Ionic concentration changes are usually not explicitly considered in the cable model. Although ionic currents in the cable model can be integrated to yield concentration changes, this usually gives erroneous results (Qian and Sejnowski, 1989), even when the membrane potentials are predicted fairly well. Often, the cable model is solved first to find the membrane potentials and then diffusion processes are introduced to determine the ionic concentration changes (Gamble and Koch, 1987; Yamada *et al.*, 1989). Our model, however, considers the membrane potential and the ionic concentration changes at the same time and thus provides a more natural and accurate way for solving the problem.

In our simulations, we have assumed that extracellular ionic concentrations were constant to simplify our calculations. This may not be a valid approximation in restricted extracellular spaces for the same reasons that the cable model broke down in restricted intracellular spaces. If the extracellular space around a spine head that is effectively available for exchange in 0.5 ms is about the same as the volume of spine, then a change of concentration in the spine head would cause an equal change with opposite sign outside the spine head. For $t_p = 0.25$ ms, the extracellular $[Na^+]$ would change from 140 mM to about 110 mM, and $[K^+]$ from 4 mM to about 34 mM in about 0.5 ms. (See Yamada *et al.*, 1989 for a similar estimate.) Although the extracellular K^+ concentration would increase by a larger factor, the maximum value of an excitatory synaptic response is mainly determined by the Nernst potential of Na^+ because the Na^+ permeability is much larger during an excitatory synaptic input. Also, glial cells are very effective in maintaining K^+ homeostasis on a longer time scale so that the actual change during maintained activity is probably less.

Thus, the main effect of a limited extracellular space is on the Nernst potential of Na^+. Based on the preceding estimates, this would reduce the peak response of the postsynaptic potential by about 6% at the spine head. For large t_p and multiple synaptic inputs, the Na^+ concentration change is greater but is achieved over a longer period of time. The corresponding effective extracellular space around the spine head would then be larger because more time allows ions to diffuse further. Thus, the modification would not be much greater. For an excitatory input driven by a Na^+ current, the effects of restricted extracellular space always reduce the amplitude of response and thus will tend to make the differences with the cable model even greater.

For synapses on large dendrites rather than on spines, we do not expect any significant difference between the electro-diffusion model and the cable model because the ionic concentration changes are negligible. Synaptic inputs on thick dendrites should not suffer the saturation caused by shifts of the Nernst potential, the absence of temporal summation, and lack of an inhibitory veto effect that we have demonstrated for synapses on spines. On the other hand, large compartments are also harder to depolarize, which is needed to increase the driving force of the inhibitory currents. This second factor may not be as important if a compartment also receives a large number of convergent excitatory synapses. Note that the more depolarized a cell, the more effective the inhibitory synapses become and the less effective the excitatory synapses. Indeed, if the depolarization is large enough to trigger an action potential, the driving forces for the inhibitory currents on soma and dendrites reach their maximum and the driving forces for the excitatory currents reach their minimum, especially if the effects of the action potentials propagate up the dendritic tree. Thus, inhibitory synapses on the cell body and proximal dendrites could control the effects of action potentials propagating up dendritic trees and the temporal firing patterns of the neuron (Lytton and Sejnowski, 1991).

This raises the interesting possibility that otherwise identical synapses could have different functions depending on their location. It has been reported that during learning and development, the spine neck shortens and merges into the dendrite (Rausch and Scheich, 1982; Coss and Globus, 1978; Brandon and Coss, 1982; Coss et al., 1980). The concentration changes in spines are caused not just by the small volume of the spine head but also by the long, narrow spine neck, which helps to maintain the large concentration changes that occur in the spine head. Thus, synapses on spines with long necks could switch to a different functional state if the neck were to shorten sufficiently for the spine to merge with the dendritic shaft.

Acknowledgments

We are grateful to Drs. Richard Cone, Francis Crick, and Christof Koch for helpful discussions. This research was supported by the Mathers Foundation, the Drown Foundation, and the Office of Naval Research.

References

BEAULIEU, C., and COLONNIER, M. (1985). "A Laminar Analysis of the Number of Round-Asymmetrical and Flat-Symmetrical Synapses on Spines, Dendritic Trunks, and Cell Bodies in Area 17 of the Cat," *J. Comp. Neuro.* **231,** 180–189.

BRANDON, J. G., and COSS, R. G. (1982). "Rapid Dendritic Spine Stem Shortening during One-Trial Learning: the Honeybee's First Orientation Flight," *Brain Research* **252,** 51–61.

BROWN, T., CHANG, V., GANONG, A., KEENAN, C., and KELSO, S. (1988). In P. Landfield and S. Deadwyler (eds.) *Long-Term Potentiation: From Biophysics to Behavior* (pp. 201–264). Alan R. Liss, New York.

COSS, R. G., BRANDON, J. G., and GLOBUS, A. (1980). "Changes of Morphology of Dendritic Spines on Honeybee Calycal Interneurons Associated with Cumulative Nursing and Foraging Experiences," *Brain Research* **192,** 49–59.

COSS, R. G., and GLOBUS, A. (1978). "Spine Stems on Tectal Interneurons in Jewelfish Are Shortened by Social Stimulation," *Science* **200,** 787–789.

COSS, R. G., and PERKEL, D. H. (1985). "The Function of Dendritic Spines," *Behavioral and Neural Biology* **44,** 151–185.

GAMBLE, E., and KOCH, C. (1987). "The Dynamics of Free Calcium in Dendritic Spines in Response to Repetitive Synaptic Input," *Science* **236,** 1311–1315.

GRIFFITH, W., BROWN, T., and JOHNSTON, D. (1986). "Voltage-Clamp Analysis of Synaptic Inhibition during Long-Term Potentiation in Hippocampus," *J. Neurophysiol.* **55,** 767–775.

HARRIS, K. M., and STEVENS, J. K. (1989). "Dendritic Spines of CA1 Pyramidal Cells in the Rat Hippocampus: Serial Electron Microscopy with Reference to Their Biophysical Characteristics," *J. Neurosci.* **9,** 2982–2997.

HIGASHIMA, M., SAWADA, S., and YAMAMOTO, C. (1986). "A Revised Method for Generation of Unitary Postsynaptic Potentials for Quantal Analysis in the Hippocampus," *Neuroscience Letter* **68,** 221–226.

HILLE, B. (1984). *Ionic Channels of Excitable Membranes*. Sinauer Associates, Sunderland, Massachusetts.

HODGKIN, A. L., and HUXLEY, A. F. (1952). "Currents Carried by Sodium and Potassium Ions through the Membrane of the Giant Axon of *Loligo*," *J. Physiol.* **116,** 449–472.

HUGUENARD, J., and ALGER, B. (1986). "Whole-Cell Voltage-Clamp Study of the Fading of the GABA-Activated Currents in Acutely Dissociated Hippocampal Neurons," *J. Neurophysiol.* **56,** 1–18.

JACK, J. J. B., NOBLE, D., and TSIEN, R. W. (1975). *Electrical Current Flow in Excitable Cells.* Oxford University Press, Oxford.

JANIGRO, D., and SCHWARTZKROIN, P. (1988). "Effects of GABA on CA3 Pyramidal Cell Dendrites in Rabbit Hippocampal Slices," *Brain Research* **453**, 265–274.

KOCH, C., POGGIO, T., and TORRE, V. (1982). "Retinal Ganglion Cells: A Functional Interpretation of Dendritic Morphology," *Phil. Trans. R. Soc. Lond.* **298**, 227–264.

KOCH, C., and POGGIO, T. (1983a). "A Theoretical Analysis of Electrical Properties of Spines," *Proc. Roy. Soc. Lond.* **B218**, 455 –477.

KOCH, C., and POGGIO, T. (1983b). "Electrical Properties of Dendritic Spines," *Trends in Neuroscience* **3**, 80–83.

KOCH, C., POGGIO, T., and TORRE, V. (1983). "Nonlinear Interaction in a Dendritic Tree: Location, Timing, and Role in Information Processing," *Proc. Natl. Acad. Sci. USA* **80**, 2799–2802.

LYTTON, W. W., and SEJNOWSKI, T. J. (1991). "Simulations of Cortical Pyramidal Neurons Synchronized by Inhibitory Interneurons." *J. Neurophys.* **66**, 1059–1079.

MARTIN, L. F. (1984). "Morphology of the Neocortical Pyramidal Neurons," A. Peters and E. G. Jones (eds.), *Cerebral Cortex, Vol. 1* (pp. 123–200). Plenum, New York.

MILES, R., and WONG, R. (1984). "Unitary Inhibitory Synaptic Potentials in the Guinea-Pig Hippocampus *in vitro*," *J. Physiol.* **356**, 97–113.

PERKEL, D. H., and PERKEL, D. J. (1985). "Dendritic Spines: Role of Active Membrane Modulating Synaptic Efficacy," *Brain Research* **325**, 331–335.

QIAN, N. and SEJNOWSKI, T. J. (1988). "Electro-Diffusion Model of Electrical Conduction in Neuronal Processes," in C. W. Woody, D. L. Alkon, and J. L. McGaugh (eds.), *Cellular Mechanisms of Conditioning and Behavioral Plasticity.* Plenum, New York.

QIAN, N. and SEJNOWSKI, T. J. (1989). "An Electro-Diffusion Model for Computing Membrane Potentials and Ionic Concentrations in Branching Dendrites, Spines and Axons," *Biol. Cybernetics* **62**, 1–15.

QIAN, N. and SEJNOWSKI, T. J. (1990). "When is an Inhibitory Synapse Effective?" *Proc. Natl. Acad. Sci USA* **87**, 8145–8149.

RALL, W. (1977). "Core Conductor Theory and Cable Properties of Neurons," in E. R. Kandel (ed.), *Handbook of Physiology: The Nervous System:* (pp. 39–97). American Physiological Society, Bethesda, Maryland.

RALL, W. (1978). "Dendritic Spines and Synaptic Potency." in R. Porter (ed.), *Studies in Neurophysiology.* Cambridge University Press, Cambridge.

RALL, W., and SEGEV, I. (1987). "Functional Possibilities for Synapses on Dendrites and Dendritic Spines," in G. M. Edeleman, W. F. Gall, and W. M. Cowan (eds.), *New Insights into Synaptic Function.* John Wiley, New York.

RAUSCH, G., and SCHEICH, H. (1982). "Dendrite Spine Loss and Enlargement during Maturation of the Speech Control System in Mynah Bird *Gracula religiosa*," *Neurosci. Lett.* **29**, 129–133.

SHEPHERD, G. M., BRAYTON, R. K., MILLER, J. P., SEGEV, I., and RALL, W.

(1985). "Signal Enhancement in Distal Cortical Dendrites by Means of Interactions between Active Dendritic Spines," *Proc. Natl. Acad. Sci. USA* **82,** 2192–2195.

THOMPSON, S. M., and GAHWILER, B. H. (1989). "Activity-Dependent Disinhibition, I. Repetitive Stimulation Reduces IPSP Driving Force and Conductance in the Hippocampus *in vitro*," *J. Neurophysiol.* **61, 501–511.**

WATHEY, J., LYTTON, W., JESTER, J., and SEJNOWSKI, T. (1989). "Simulations of Synaptic Potentials Using Realistic Models of Hippocampal Pyramidal Neurons," *Soc. Neuroscience Abstracts* **15.**

WEER, P. D., and RAKOWSKI, R. F. (1984). "Current Generated by Backward-Running Electrogenic Na$^+$ Pump in Squid Giant Axons," *Nature* **309,** 450–452.

YAMADA, W. M., KOCH, C., and ADAMS, P. R. (1989). "Multiple Channels and Calcium Dynamics," in C. Koch and I. Segev (eds.), *Methods in Neuronal Modeling*. MIT Press, Cambridge, Massachusetts.

ZADOR, A., KOCH, C., and BROWN, T. H. (1990). "Biophysical Model of a Hebbian Synapse," *Proceedings of the National Academy of Sciences USA* **87,** 6718–22.

Chapter 6   Dendritic Morphology, Inward Rectification, and the Functional Properties of Neostriatal Neurons

CHARLES J. WILSON

Department of Anatomy and Neurobiology
University of Tennessee, Memphis
Memphis, Tennessee

I. Introduction

Recent advances in cellular neurophysiology have revealed a wealth of very nonlinear membrane properties that greatly enrich the landscape of possible neuronal computation. At the same time, the complexity introduced by these nonlinearities has made the task of analyzing their functional properties enormously challenging. With relatively simple linear systems, it is a reasonable strategy to attempt a complete description of the dynamic behavior of the system that can be used to predict the output for any input signal. Given the complexity and nonlinearity displayed by neurons, it may be expedient to use a less ambitious strategy, restricting the analysis to a limited set of inputs that are already known to be important for the operation of the system. To the experimental neurobiologist, the most natural approach is to ask what new computational capability is introduced when a particular ion channel is inserted into the membrane of the neuron, or a new synapse is attached to its surface. The problem is less straightforward than it sounds, however, because the functional impact of new inputs or new membrane properties is strongly dependent upon the nature of the input signals and interactions with other membrane nonlinearities. In the absence of a more general solution, it may be best to examine specific neurons with their own characteristic sets of nonlinearities, arrangements of inputs, and patterns of afferent activity. The computational facilities endowed upon those neurons by particular nonlinearities can then be examined in their natural setting. This increases the likelihood that the results will fall in a relevant part of the complicated parameter space that represents the potential interactions of large numbers of inputs with all possible manner of nonlinear synaptic and membrane properties.

The spiny projection neuron of the neostriatum is a good candidate for this kind of analysis. The neostriatum and related structures represent a large proportion of the forebrain, second only to the cerebral cortex itself. Understanding the role of the neostriatum in the context of the overall organization of the brain requires a knowledge of the transformation applied by the neostriatum upon its input. This is especially critical because the roles of the neostriatum in experience and behavior are somewhat obscure, or at least not as well-understood as those of the cortical regions that give rise to its input fibers. By studying the modifications of cortical signals that occur as they traverse the neostriatum, we can expect to gain information of fundamental importance on the functional properties of this prominent region of the brain. In addition, the neostriatal projection neuron shares many of its morphological and physiological characteristics with cells in a variety of other regions of the forebrain, and it is likely that the principles that govern its function will be similar to those other neurons.

Like the cerebral cortex, the neostriatum contains a very large number of neurons whose axons exit the nucleus, and that differ according to the target of their axons. Although there are several different types of cells known to be present, the majority of the neurons are of one morphological type characterized by its long axon and spine-laden dendrites and called the *spiny projection neuron* (e.g., DiFiglia *et al.*, 1976, Chang *et al.*, 1981). Among these cells, there are a number of subtypes, but they do not differ greatly in either their overall morphological features or their physiological properties (Kawaguchi, *et al.*, 1990a, b). The electrical properties of these neurons, their anatomical connections, and their cellular morphology have all been studied and described in some detail. Importantly, these same cells receive the majority of synaptic inputs from afferent fibers. Thus, the circuit consisting of the spiny projection neurons and their afferent fibers comprises the simplest and the numerically predominant through-pathway for information entering the neostriatum. The computation that can be performed by a single neuron in this circuit is the fundamental functional operation of the neostriatum.

II.   Firing Pattern of Neostriatal Spiny Projection Neurons

The firing pattern of the spiny projection neurons is characterized by episodes of firing separated by long periods of silence. In animals deeply anesthetized with barbiturates, practically no spontaneous firing can be observed at all. In awake animals, episodes of spontaneous activity are associated with the initiation and execution of planned or learned movements (e.g., Evarts *et al.*, 1984). Intracellular recording experiments in unanesthetized animals (Hull *et al.*, 1970; Wilson and Groves, 1981) and in lightly anesthetized animals (Sedgwick and Williams, 1967; Calabresi *et al.*, 1990) have shown that firing during these episodes arises from periods of sustained (0.1–3 sec) membrane depolarization. Separating these periods of depolarization are longer episodes of membrane hyperpolarization and greatly decreased synaptic noise. An example showing this firing pattern, and the underlying membrane potential changes, is shown in Fig. 1. The transitions between the depolarized and hyperpolarized states underlie the pattern of action potentials generated by the striatal neuron, and the patterns of firing of neurons in their target structures, the *globus pallidus,* and *substantia nigra.*

III.   Distribution of Synaptic Inputs on the Spiny Projection Neuron

Afferent inputs to the spiny projection neurons terminate mostly on the dendrites, with more proximal inputs arising from interneurons within the nucleus. The

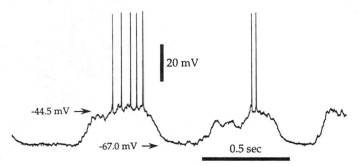

FIGURE 1.   Spontaneous membrane potential fluctuations of a neostriatal spiny neuron, and the spontaneous firing pattern caused by those fluctuations in an animal anesthetized with urethane. Noisy depolarizing episodes arise from a hyperpolarized baseline with relatively little spontaneous noise. Action potentials arise from membrane potential fluctuations that are superimposed upon the depolarizing episodes. Transitions between the depolarized and hyperpolarized states are abrupt and occur at irregular intervals.

structural features of the neostriatal spiny neuron are well-known from studies using intracellular staining, as well as the classical Golgi technique. An example of the dendritic and local axonal arborizations of a spiny projection cell is shown in Fig. 2, as reconstructed from serial sections through an intracellularly stained cell. Somatic diameter ranges from 10 to 15 μm, and the dendritic tree fills a volume ranging from 250–500 μm in diameter. Twenty to 30 dendrites arise from approximately half that number of short spine-free dendritic trunks that branch within the first 20–30 μm to give rise to less branched, gradually tapering dendrites that are densely covered with dendritic spines. The local axonal arborization usually stays in the general vicinity of the dendritic field. Electron microscopic studies have shown that the synapses made by the axons are primarily on the somata, dendrites, and dendritic spines of other spiny projection neurons (Wilson and Groves, 1980).

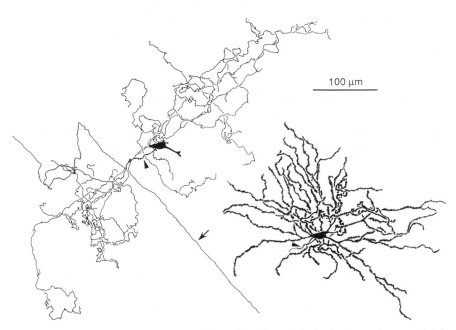

100 μm

FIGURE 2.   Reconstruction of the dendritic and local axonal arborizations of a neostriatal spiny neuron stained by intracellular injection of biocytin. The initial segment of the axon is indicated by an arrowhead, and the main axonal branch is indicated by the arrow. After emitting the local collaterals, the axons of these cells descend through the internal capsule bundles to innervate the globus pallidus, entopeduncular nucleus and substantia nigra. From Kawaguchi *et al.* (1990b).

An adequate representation of the morphological features of the neuron requires information that is not available in the light microscope. The dendrites are of small caliber (on the order of 1 μm) at their origin, and become very fine at their tips. Accurate estimates of dendritic surface area cannot be obtained in the light microscope, both because the measurement of diameter is not sufficiently accurate and because the dendrites are not cylindrical in cross section. The dendritic spines have cross-sectional diameters on the order of 0.1 μm, making estimation of their surface area even more problematical. Even the density of dendritic spines cannot be accurately judged from the light microscope, as they lie so closely together that they often fall into each others' circles of confusion.

More accurate estimates of the distribution of membrane on the neostriatal neuron have been made using electron microscopy of thin sections, and a more specialized technique, high-voltage electron microscopy (HVEM) of thick (3–5 μm) sections (Wilson *et al.*, 1983b). An example showing the improved resolution gained by the HVEM technique is shown in Fig. 3. Intracellular staining is responsible for nearly all the contrast in the image, as it is with light microscopic examination, and so the image can be interpreted in exactly the same way as it would be in the light microscope. The resolution is approximately 50 times that of the light microscope, however, allowing for very accurate measurement of somatic and dendritic diameters, spine densities, and the dimensions of dendritic spines. Because the stained neurons used for HVEM analysis are also visible in the light microscope, systematic sampling of different parts of the dendritic tree, required for quantitative analysis, is a simple matter. Analysis of spiny neurons using this technique has allowed accurate measurement of the linear dimensions of the cells, but because dendrites are not made up of symmetrical components such as cylinders and spheres (as is obvious from examination of the cell in Fig. 3), surface area cannot be estimated from these measurements in a straightforward manner. Systematic sampling and serial electron microscopic reconstruction of samples from the dendrites have provided corrections that can be applied to the HVEM measurements to yield more accurate surface area estimates (Wilson *et al.*, 1983b). Although very reliable, this approach requires three different kinds of data using three different microscopic techniques, and is expensive in time and effort. A more direct route to surface area measurements has been introduced recently, using axial tomographic reconstruction of dendrites from series of HVEM images taken over a range of specimen tilt angles. This technique has not yet been systematically applied, but in a preliminary test it yielded estimates of spine and dendrite surface area that closely matched those obtained previously from serial section data (Mastronarde *et al.*, 1989).

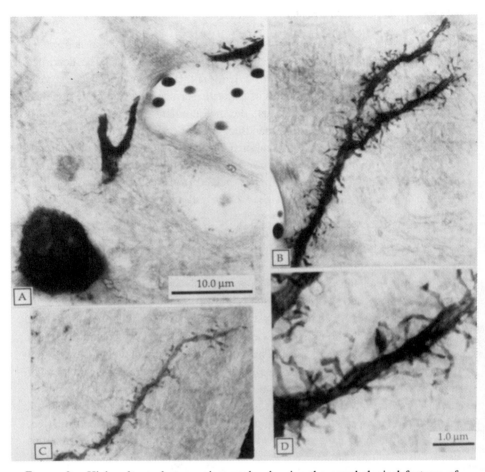

FIGURE 3. High-voltage electron micrographs showing the morphological features of neostriatal spiny projection neurons at high resolution. The neuron was stained by intracellular injection of horseradish peroxidase, and prepared for examination in the HVEM as described previously (Wilson, 1987). Section thickness is 5 μm. (a) The soma, a portion of the aspiny initial part of a dendritic tree with a branch point, and a part of the spiny part of the same dendritic tree are shown. The soma was only partly contained in the section, but was sectioned through its center, so its apparent diameter is representative of spiny neurons. The diameters of the aspiny part of the dendrites, and their noncylindrical shapes, are evident. (b) A segment of dendrite taken from the most densely spiny region of the dendritic tree. The high density of spines and the variety of their shapes is evident, as is the irregular dendritic contour. (c) A dendritic tip from a spiny neuron, shown at the same magnification as in a and b. Note the reduced spine density and rapid tapering of dendritic diameter. (d) High-magnification view of a portion of the densely spiny dendrite shown in b, to illustrate the high resolution available in the HVEM for measurement of dendritic spine shapes. The texture of the background tissue is due to treatment of the section with osmium tetroxide.

146

IV.   A Model of the Spiny Neuron

The distribution of surface area on the neostriatal spiny neuron as estimated using the combination of light, HVEM, and conventional electron microscopy is shown in Fig. 4. Rather than the distribution of surface area for a single neuron, this is a composite based on microscopic analysis of many neurons. The surface area is combined for all the dendrites, and is expressed in square micrometers per micrometer of dendritic length as a function of distance from the soma. The total dendritic surface area decreases with distance from the soma because of the tapering of the dendrites and the sparse branching of their arborizations. The surface area due to dendritic spines increases rapidly between 20 and 50 μm from the soma, and this reflects the changes in spine density that occur over this region in the dendritic field. Spine density tapers off gradually toward the tips of the dendrites, and this is reflected in the surface area as well. None of the variation of spine surface area is due to any systematic change in spine shape on the neurons. Despite great variation in the shapes and sizes of

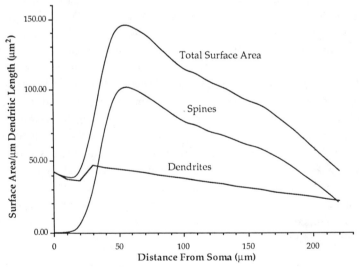

FIGURE 4.   Surface area profile of the neuronal model used in the simulations. This profile was based on the average dendritic spine surface areas measured from serial section electron microscopic analysis, spine density, and dendritic diameter data obtained from HVEM analysis of intracellularly stained neurons, and dendritic branching data obtained using the light microscope. The surface areas do not represent any one neuron, but a composite of the spiny projection neurons analyzed in a number of experiments. The dendritic branching patterns have also been simplified so that all the dendritic trees arising from the soma are identical.

dendritic spines as seen in electron micrographs of striatal neuropil, these features were not found to be correlated with position on the dendritic tree (Wilson *et al.*, 1983b).

In the model used for the simulations, the surface area represented in Fig. 4 was distributed onto six identical dendritic trees, each of which branched twice. The first branch was placed 10 μm from the soma, in the spine-free portion of the dendrites, and the second 30 μm from the soma, in the first part of the spiny portion of the dendritic field. The model of the cell thus had 24 dendritic tips. Spine density distribution was based on a HVEM correction of counts of the mean spine density from five neurons whose spines were counted in an earlier study (Wilson *et al.*, 1983b). It was maximal at a distance of 50 μm from the soma, where there were three spines per micron of dendrite, and tapered in an approximately linear manner to a value of 0.6 spines per micron at the dendritic tip. The primary dendrites were 10μm-long tapered cylinders with diameters of 2.25 μm and no spines; the secondary dendrites were 20 μm long, with a diameter of 1.0 μm initially and tapering to 0.96 μm. The tertiary dendrites were 190 μm in length, and their diameters were tapered from 0.63 μm at their origin to 0.29 μm at the tips using data from HVEM analysis. Dendritic spines not activated synaptically had a mean surface area of 1.46 μm², corresponding to the mean value determined from serial section electron microscopic analysis (Wilson *et al.*, 1983b). Synaptically quiet spines in the simulations were represented by increasing the membrane capacitance and conductance of the dendritic segment attached to the spine by an amount that represented the increased surface area associated with the spine. Detailed simulations of charge redistribution in the limited environment of an active synapse in a segment of dendrite showed that this approximation was accurate for the dimensions used for the spiny neuron and the time course of its synaptic conductances used in these simulations. Synaptically activated spines were represented using one isopotential compartment to represent the head, and three compartments to represent the spine neck. Dimensions of these spines were varied over the range found in striatal spiny neurons.

V.   Input Resistance and Electrotonic Length of the Passive Model

Although much is already known about the electrical characteristics of distributed neuronal membrane, the large contribution of dendritic spines to the dendritic surface area in spiny neostriatal neurons deserves some analysis as a basis for comparison with the same neuron after addition of the anomalous rectification conductance. Figure 5 shows the input resistance and electrotonic dendritic length of this model as a function of membrane resistivity. To illustrate the effect of dendritic spines on these parameters, a model of the neuron without dendritic spines is compared with the results for the spiny dendrites. These results show that the additional surface area represented by dendritic spines greatly reduces the input resistance of the neuron at any value of membrane resistivity. Likewise, the additional dendritic surface area is added without any additional increase in the cross-sectional profile of the dendrite available for longitudinal current flow. Thus, the spines act to reduce membrane resistance of the dendrite without reducing the axial resistance. This is similar in effect to a decrease in the membrane resistance that occurs with no alteration of membrane time constant (as described earlier). The effective electrotonic length of the dendrite is increased by the presence of dendritic spines, as would happen if membrane resistivity were reduced. This effect is very pronounced when the membrane resistivity is low, but continues to be felt even with very high membrane resistivity.

The range of input resistances and electrotonic lengths normally attributed to spiny neurons in neurophysiological experiments (Kita et al., 1984; Bargas et al., 1988) are shown using dashed lines (Fig. 5). One remarkable feature of these empirical data is their wide range. This variability from cell to cell can be seen to correspond with an even larger range of membrane resistivity, from about 3,500 Ω-cm^2 to over 10,000 Ω-cm^2. If the spines were not present, this same variability would correspond to a narrower range of membrane resistivity, between 1,500 and 6,500 Ω-cm^2, and the resistivity calculated from input resistances would make a much poorer match with those predicted from the measurement of electrotonic length. If the range of input resistances and electrotonic lengths observed in the neurophysiological measurements were due to differences in membrane resistivity among neurons, it might be expected from the curves in Fig. 5 that membrane time constant would vary even more, reflecting the very wide range of membrane resistivity. If the membrane capacitance were 1 μF/cm^2, the time constant might be expected to vary from about 3.5 ms to greater than 10 ms.

VI.   Effect of Fast Anomalous Rectification on Input Resistance and Time Constant

A very fast anomalous rectification in neostriatal neurons has been described in most studies of the membrane properties of neostriatal neurons (Kita et al., 1984; Calabresi, et al., 1987; Kawaguchi et al. 1990a). If this conductance could alter membrane resistivity over a wide range, it could be responsible for the range of observed values shown in Fig. 5. In a recent study of the membrane properties

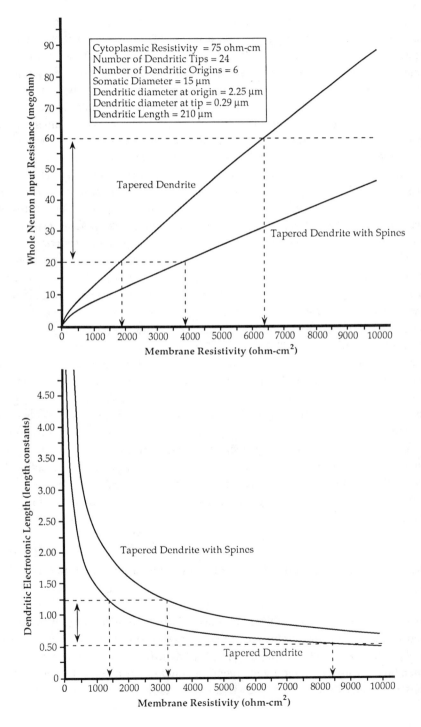

of neostriatal neurons in slices, Kawaguchi *et al.*, (1990a) measured the variation in input resistance and time constant over the normal range of membrane potentials. An example from the results of that study is shown in Fig. 6. The very rapid time course of activation of the inward rectification is evident in the traces at the top of Fig. 6. The non-exponential nature of the charging curves is evident in the responses to the large current steps, and from the asymmetry in the responses to onset and offset of the curves. The rapid response of membrane resistivity to changes in membrane potential can explain most of the features of these charging curves. Because the change in membrane resistivity follows changes in membrane potential very rapidly (i.e., much faster than the rate of decay of charge through the membrane resistance), the time constant may be considered to be varying continuously during the response to current. Thus, deviation of the responses from an exponential curve are greatest in the first part of the transients when the rate of change of membrane potential is highest. The later parts of both the onset and offset transients are associated with smaller rates of change of membrane potential, and these segments of the response are very nearly exponential in form. If an exponential curve is fit to the late portions of the responses to onset and offset of the response to current pulses in spiny neurons, as is the usual way of measuring the time constant of the cell, very different values are obtained depending upon whether the onset or the offset of the response is used. Fitting exponentials to the offset curves gives a single value for the time constant regardless of the amplitude of the current pulse. This time constant reflects the membrane resistivity at the resting membrane potential. If the time constant is measured from the late portion of the transient at the onset of a current pulse, the result depends very much on the amplitude of the current pulse. This is because the time constant obtained this way is measured at a membrane potential near the steady-state value that will be achieved by the pulse, and so reflects the membrane resistivity of the cell at this membrane potential.

FIGURE 5.   Effect of dendritic spines on the steady-state parameters of the neuron model. In the upper graph, input resistance is shown as a function of membrane resistivity for the neuron model with and without dendritic spines. The range of input resistances usually reported for spiny projection neurons, and the corresponding membrane resistivity values, are indicated by dotted lines. In the lower graph, dendritic electrotonic length is shown as a function of membrane resistivity for both neuron models. Again, the range of reported values for spiny projection neurons is shown with dotted lines. The presence of dendritic spines greatly lowers the input resistance and raises the dendritic electrotonic length expected at any value of membrane resistivity. Likewise, the membrane resistivity predicted from estimates of these parameters is increased several-fold due to the presence of spines on the neurons.

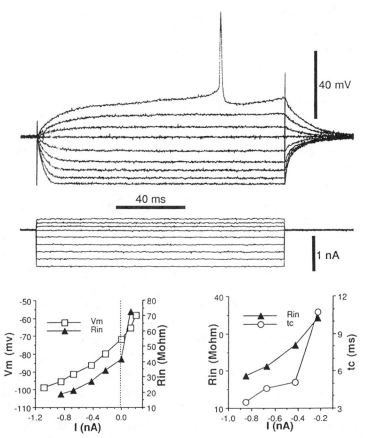

FIGURE 6.   Responses of a striatal spiny projection neuron to intracellularly applied current pulses, and the variation of input resistance and time constant as a function of the amplitude of the injected current. In the upper panel, the current pulses and transients are shown. In the graph at the lower left, the steady-state membrane potential is plotted against the amplitude of the current for all transients except the one evoking the action potential. The anomalous rectification is indicated by the curvature of this steady-state current voltage relationship. This change in the steady-state input resistance is indicated in the same graph. At the lower right, the variation in input resistance is superimposed on the change in time constant of the charging curve for each current step. The time constant was calculated from the late portion of the charging curve, in which the membrane potential was near its final value. The response to the most hyperpolarizing current step (in which a time-dependent anomalous rectification is evident) and the depolarizing responses were not used in this analysis to avoid confusion of the fast inward rectification with other voltage-dependent currents. Data from Kawaguchi *et al.* (1990a).

Using the membrane potential responses to the onset of current pulses, it thus is possible to estimate the time constant of the neuron as a function of membrane potential. The input resistance likewise can be estimated (although somewhat less rigorously) by taking the slope of the steady-state voltage-current curve at each voltage point. These measures of input resistance and time constant are shown for a spiny neuron in the lower part of Fig. 6. The time constant and input resistance of this one neuron could be made to vary over the entire range reported for the population of neostriatal neurons, simply by adjusting the membrane potential.

Because the anomalous rectification acts so quickly, it is permissible to continue to use the conventional model of the neuron, but with the membrane resistivity made a function of instantaneous membrane potential. The range of time constants and input resistances obtained from analyses of individual spiny neurons like the one shown in Fig. 6 can be compared with the input resistance curve in Fig. 5. This comparison suggests that the membrane resistivity of spiny neurons varies between about 3,000 Ω-cm^2 and 20,000 Ω-cm^2, over the rather modest range of membrane potentials from about -100 mV to -60 mV. These values were used to simulate the anomalous rectification of neostriatal spiny neurons using the model of the fast inward rectification described by Hagiwara and Takahashi (1974) for the starfish egg. The potassium equilibrium potential was fixed at -90 mV. This is important for the anomalous rectifier because the conductance, as well as the current, has been shown to be dependent upon the deviation of the membrane potential from E_k. The maximum conductance and the activation curve was varied somewhat in these simulations, and the curve best matching the input resistance and time constant data for neostriatal spiny neurons is shown in Fig. 7, along with the current density associated with that conductance. Because the fast anomalous rectification approaches zero conductance as the cell is depolarized, a constant leak conductance (20,000 Ω-cm^2) was placed in parallel with it, limiting the whole neuron input resistance (at the soma) to 60 MΩ (not shown). A constant leak conductance of this type is probably not what determines the input resistance of neostriatal neurons when depolarized, but rather other voltage-sensitive potassium currents. (See following.) To prevent confusion with the effects of the anomalous rectifier, such currents were excluded from these simulations, however.

The effect of the anomalous rectification on the responses of the model cell to current steps is shown in Fig. 8 (top). The characteristic asymmetry of the responses of neostriatal spiny neurons is evident in the simulated responses, as is the ramp-like depolarization that is seen with moderate depolarizing pulses.

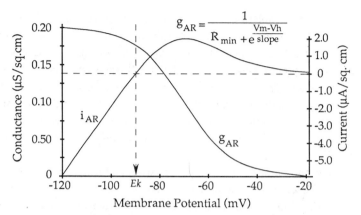

FIGURE 7.   Activation curve for the inwardly rectifying conductance used in the model of the spiny neuron, and the equation used to generate it. R_{min} determines the maximum conductance, while the shape of the activation curve is determined by the half-maximal voltage (Vh), the potassium equilibrium potential (Ek), and the slope factor (slope). The current density associated with this conductance also is shown.

This almost linear ramp depolarization is caused by the positive feedback inherent in the anomalous rectification. Small depolarizations increase the input resistance of the neuron, thereby increasing the depolarization caused by the constant applied current, which again raises the input resistance. This is active over the entire range of the activation curve for the anomalous rectification, and ultimately

FIGURE 8.   Responses of the neuron model to current steps. The graph at the top shows the transient response to somatic current steps of various amplitudes. (Amplitudes of current steps are shown at the right of each transient.) The graph at the middle left shows the steady-state membrane potential recorded both at the soma and at the tip of a dendrite. The effect of the inward rectifier on the somatic resistance matches that of the real neuron in Fig. 6. The decay of voltage from the soma to the dendritic tip is also seen to be dependent upon the size of the current step. This is due to the change in electrotonic length of the dendrite as a function of membrane potential. This change is shown in the bottom panel. Dendritic length, calculated by peeling exponentials from the transient, is shown as a function of the size of the current step, along with the electrotonic length obtained from the steady-state voltage difference between soma and dendritic tip. Both show a strong dependence on injected current, which is due to the variation of membrane resistivity with changes in membrane potential. At depolarized membrane potentials, the dendrites become electrotonically compact. Similarly, the membrane time constant (obtained from the last part of the transient response) is greatly lengthened when the membrane is depolarized.

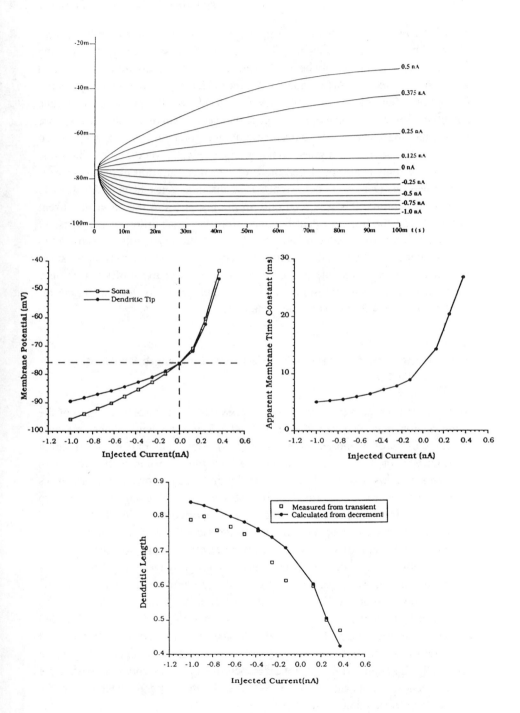

causes the membrane potential to saturate at a level determined (in these simulations) by the leak conductance. In a more realistic simulation, this limiting conductance would probably be due to voltage-sensitive potassium currents, and would be a complicated function of the recent history of membrane potential fluctuations (Surmeier et al., 1988, 1989, 1991). Over the range of membrane potentials used here, which did not engage the mechanism that limits the response to depolarizing currents, the current–membrane potential relation was a good match for that seen in spiny neostriatal neurons (Fig. 8, middle left). The same range of currents also reproduced the time constant variation observed in neostriatal neurons (Fig. 8, middle right). Like the time constant variation seen in neostriatal neurons, this was seen only in the responses to the onset of current injection, and the time constant obtained from the last portion of the transient was appropriate for the membrane resistivity at the asymptotic (steady-state) level of depolarization of the neuron for each pulse. Time constants obtained from examination of the late portion of the transient to current offset were very nearly the same for all transients, and were appropriate for the membrane resistivity of the neuron at its resting potential.

VII.   If the Time Constant Is Not Constant, the Length Constant Is Not Either

This variation in membrane resistivity implies that the electrotonic length of the spiny neuron may also vary between about 1.3 and 0.5 length constants over the same range of membrane potential (Fig. 5). This could explain the apparent contradiction in the literature concerning both the time constants and length constants of spiny neostriatal neurons. In the first study of membrane properties of neostriatal neurons, performed by Sugimori et al. (1978) using cat neostriatum in vivo, the time constant of the neurons was reported to be about 15 ms, and transients recorded in response to current steps could be fit with a single exponential, suggesting that the dendrites were electrotonically too short to measure. In subsequent studies in slices of neostriatum, a range of dendritic lengths and time constants was observed, but generally the time constants were shorter, around 5 ms, and at least a second exponential decay could be detected in the responses to current steps (Kita et al., 1984; Bargas et al., 1988). This discrepancy has been suggested to be due to species differences, or to the local potassium concentration, but a more simple explanation is suggested by these results. In the slightly more depolarized neurons of the in vivo cat neostriatum (with synaptic inputs intact), time constants are longer and dendritic lengths shorter than those obtained in slices.

In the simulations shown in Fig. 8, electrotonic length was examined as a function of the injected current, using two approaches. Because of the non-cylindrical distribution of membrane and the nonlinearity of the membrane, the equivalent cylinder model cannot be guaranteed to yield an accurate estimate of dendritic electrotonic length from a decomposition of the transient into exponential components (Rall, 1977; Fleshman, *et al.*, 1983; Durand, 1984; Kawato, 1985). Nonetheless, this analysis was performed on the transients, to estimate the error introduced by these deviations from the ideal case. For comparison, the effective dendritic length was also measured using the difference between the average response at the dendritic tips and that at the soma, measured at the end of the 100 ms current pulses (Fig. 8, middle left). Because the response to the larger depolarizing currents had not achieved steady state at the end of the current pulse, this estimate was somewhat approximate. This failure to achieve steady state with depolarizing pulses is an essential feature of the model with anomalous rectification, and could not be avoided. The model will not achieve steady state until the input resistance is limited by some additional mechanism (in this case, the leak conductance). The same behavior is observed in intracellular recordings from neostriatal neurons (Kawaguchi *et al.*, 1990a), but probably for somewhat more complicated reasons.

The variation in dendritic electrotonic length measured by peeling exponential components from the transient (calculated using Kawato's 1985 variant of Rall's method) was in reasonably close agreement with the values obtained using the steady-state dendritic decrement (Fig. 8, bottom), and both of them reproduced the entire range of dendritic lengths reported for neostriatal neurons in intracellular recording experiments. Thus, the dendritic length of a neostriatal neuron may not be constant, but may vary from less than half of one length constant to more than one length constant, depending upon membrane potential.

VIII.   Synaptic Integration in the Spiny Neuron

The variation of length and time constants of the neurons could also have profound effects upon synaptic integration in these neurons. Many excitatory synaptic inputs are located in the distal dendrites. When the dendrite is relatively polarized, due to an overall low level of excitatory input, the anomalous rectifier acts like a distributed shunting inhibition, reducing the input resistance, increasing the effective dendritic length, and reducing the time constant. These decrease the ability of excitatory synaptic inputs to sum effectively. If, however, enough inputs are active to depolarize the dendritic membrane somewhat, they can remove a portion of the conductance of the inward rectifier, and thereby increase their own effectiveness and that of any synaptic excitation arriving during the time course of the depolarization. The spatial extent of this cooperative effect among synapses would correspond to the spatial spread of the depolarization. One interesting feature of this mechanism is that it has the same time course and spatial domain as the classical synaptic nonlinearity, but it acts in the opposite direction. Synapses with similar actions are known to sum in a less than arithmetic way because of the classical synaptic nonlinearity, in which the voltage change generated by each synapse decreases the driving force for synaptic current at the others. Like the cooperative effect that might be expected to result from anomalous rectification, the synaptic nonlinearity is very fast in onset. It therefore can be thought of as strictly dependent upon membrane potential.

To determine the effect of the synaptic nonlinearity and inward rectification when acting simultaneously, a simulation of synaptic activation of the model neostriatal spiny neuron was performed, comparing activation of large populations of synapses distributed in the dendritic tree in a neuron with and without the inward rectification. Because the placement of the synapses on dendritic spines and the distribution of the synapses on the dendritic tree proved to be important influences on the results, the effects of these morphological parameters will be addressed first.

IX.   Dendritic Spines and Synaptic Strength

Many theoretical studies have emphasized the potential role of dendritic spines in the regulation of synaptic strength (e.g., Diamond et al., 1970; Rall, 1974; Koch and Poggio, 1984; Perkel and Perkel, 1985; Wilson, 1984; Shepherd et al., 1985). In many of these, the dendritic spine is postulated to act as a series impedance to current flow from the synapse to the dendrite, and to thereby produce a very large local synaptic potential in the spine. This large local potential has different implications in the various models, but it always relies upon some nonlinearity of the postsynaptic neuron. If the nonlinearity is a fast voltage-gated inward current, the local potentials in spines are enough to trigger spikes in the spines that may propagate to other spines (e.g., Perkel and Perkel, 1985; Shepherd et al., 1985, Miller et al., 1985; Rall and Segev, 1988). If the nonlinearity is nothing more than the classical synaptic nonlinearity, it should act to decrease the effectiveness of the synapse by limiting synaptic current flow (e.g., Rall, 1974; Koch and Poggio, 1984; Kawato et al., 1984; Wilson, 1984). There are a variety of other possibilities, based on other kinds of membrane nonlinearities. The possible synaptic attenuation effect of the dendritic spine that arises from the classical synaptic nonlinearity has been emphasized in studies of neostriatal neurons, because it helps to explain why the cells are not continuously depolarized and firing due to the spontaneous activity in the thousands of afferent synapses that are formed onto each cell (Wilson *et al.*, 1983b). Computer simulations of dendritic spines in the size range observed in the neostriatum have shown that these may act to limit synaptic current and so attenuate synaptic inputs (Wilson, 1984). In such computer models, however, there is only a single synaptically activated spine on the neuron, and the effect of the spine on that synapse is studied. When this is done, the synaptic effect of an axospinous synapse can be seen to be attenuated, and to make a relatively small EPSP (<10 mV) in the dendrite, despite the presence of a very large (>40 mV) EPSP amplitude in the spine head. When the influence of this small synaptic input is seen at the soma, it is further diminished and appears appropriately small; for example, less than 0.1 mV.

In the model of the spiny neuron used here, the spine density in the dendrites is often 2–3, and occasionally reaches 4–5 per micrometer. An examination of the effect of anomalous rectification on synaptic transmission in this neuron requires some additional data on the linear model in the case of large numbers of active synapses. For example, if only 10% of synapses are active within the time period that allows their nonlinear interaction, there may still be as many as 5 active synapses active per 0.1 length constant of dendrite (about 25 μm at −90 mV membrane potential). The resulting dendritic EPSP may be as large as the EPSP in the spine head, and so might engage nonlinear properties in the dendrites as well as those in the spines, if these are present. As an extreme case, consider the case when all synapses are simultaneously active with identical synapses, on an infinitely long, densely spiny neuron. Because the currents injected into the dendrite by each spine will be equal, each spine would behave as if it were placed on a short segment of dendrite terminated at each end with an open circuit. The length of the segment of dendrite will be equal to the distance between dendritic spines. If this is very short (less than 1 μm in the example used here), the dendritic input resistance will be as great as that of the spine, and any special privilege of the spine in triggering voltage-sensitive nonlinearities may be lost (unless this privilege is gained by something more than simply the morphology of the spine).

Simulation of simultaneous activation of synapses clustered in one spot on the spiny neuron model showed that this effect occurs even with relatively low densities of activated spines. This is illustrated in Fig. 9. In this simulation, the dendritic spines were made to be rather long and thin (3.0 μm long and 0.1 μm neck diameter) and the synaptic conductance change rather short ($\alpha = 100$), and high amplitude (peak conductance 1.0 nS). This combination favors the current attenuation effect of the dendritic spine simulation, and produced

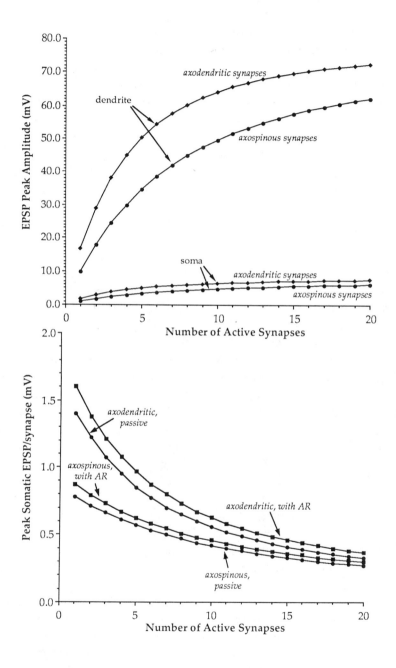

an approximately 50% attenuation of the synaptic potential as seen at the soma when compared with an axodendritic synapse at the same location (Fig. 9, bottom, traces marked *passive*). The active spines were placed 120–140 μm from the soma. This 50% attenuation was obtained when only one synapse was tested in isolation. Additional synapses were then synchronously activated within $\pm$ 10 μm ($\sim$0.05 length constants) of that active synapse, and the effect of those other inputs on the effectiveness of the original one was examined. The effect of clustered synchronous activation on the peak amplitude of the composite synaptic potential, as it appeared locally in the dendrite, and as it appeared in the soma, is shown in Fig. 9 (top). The current attenuation effect of the dendritic spine depends upon the existence of a large difference between the input resistances at the spine head and the dendrite. Thus, the current injected by the synapse cannot move the dendrite as near the reversal potential of the synapse as can the current injected at the higher resistance site at the spine head. Fig. 9 (top) shows, however, that clustered synaptic activity can depolarize the dendrite very effectively, and thereby reduce the proportional difference between axodendritic and axospinous synaptic inputs. This is clearly seen in Fig. 9 (bottom), which shows the EPSP contribution per synapse as a function of the number of synapses active. The effectiveness of axodendritic and axospinous synapses approach each other quickly as the number of active synapses approaches 1 per μm of dendrite (only about 25% of the spine density at 80 μm from the soma of the spiny neuron).

FIGURE 9. Integration of spatially clustered synaptic input on the model neuron. The peak amplitude of the EPSP evoked by synchronous activation of varying numbers of synapses along a single 25μm segment of dendrite on one dendrite of a model neuron is shown in the upper graph. The two upper curves show the amplitude of the local EPSP in the dendrite at the level of the synapses. Anomalous rectification has little effect on these curves, and so only the ones from the model with the inward rectifier are shown. The two lower curves show the resulting somatic EPSPs. Axodendritic synapses lead to larger local and somatic EPSPs, but the proportional difference between axodendritic and axospinous synapses decreases as the number of active synapses increases. This reduction in the difference between axospinous and axodendritic synapses as the synaptic density increases is more clearly shown in the lower graph, in which the somatic EPSP contribution of a single synapse is shown as a function of the total number of active synapses in the cluster. As the dendritic EPSP grows due to summation of EPSPs, it becomes large enough to engage the synaptic nonlinearity that attenuates synaptic current in the axospinous synapse. Addition of the inwardly rectifying nonlinearity has little effect on these differences because the local EPSPs in all cases are large enough to drive the local membrane into its high-resistance region.

It is not yet known exactly what proportion of afferent synapses are active on one neostriatal spiny neuron at any one moment. This is not because we do not know the firing patterns of the afferent neurons in the cortex, thalamus, and elsewhere. It is because we do not know the divergence of these pathways in the neostriatum. Thus, we do not know how many different afferent neurons contribute synaptic input to each neostriatal cell (although there are indirect arguments for the number being quite large). Similarly, we do not know the degree of correlation in the firing patterns of those cells whose influences do converge in the neostriatum. However, these simulations do not support the view that there is a very high level of synaptic input that is attenuated simply by the electrotonic properties of the dendritic spines. If the level of activity in converging afferents is more than a few percent, the effect of these synapses must be weakened by another mechanism.

X.   Effect of Fast Anomalous Rectification on Synaptic Integration

The result of adding the fast anomalous rectification to the simulation of spatially clustered synaptic input is evident in Figure 9 (bottom). In this simulation, the passive neuron had a membrane resistivity near the low end of the range for spiny neurons (5,000 Ω-cm^2). The anomalous rectification simulation allowed a voltage-dependent variation of membrane resistivity between 3,500 and 10,000 Ω-cm^2, matching that of the simulations shown in Fig. 8. Because local EPSPs (at the site of the synapse) were always large enough to engage the membrane resistance increasing effect of the anomalous rectification, the membrane in the dendrite had an effectively higher resistivity in the anomalous rectification model. The result of this was an increased length constant in the synaptically activated dendrite, and more effective spread of synaptic current to the soma. Thus, the synaptic effectiveness for the anomalous rectification model in Figure 9 (bottom) is generally higher than that for the passive model. This is not a very important effect of anomalous rectification. Of course, the comparison would have been more realistic if the passive neuron had an input resistance at the high end of the range, as the fast anomalous rectification in neostriatal neurons is probably due to an increase in potassium conductance with hyperpolarization rather than a decrease in membrane conductance caused by depolarization (e.g., Hagiwara and Takahashi, 1974). In any case, this simply shifts the curves vertically, and the addition of anomalous rectification has little or no practical effect on the function of the dendritic spine or the removal of the current attenuation effect of dendritic spines with clustered synaptic excitation.

When synaptic inputs are distributed on the dendrites, addition of anomalous rectification to the neuron model might be expected to have a larger effect. In this case, synaptic inputs produce more modest depolarizations throughout the dendritic tree, increasing the membrane resistivity in a general fashion. This might have the effect of reducing the electrotonic extent of the dendritic tree, and so the synapses could act in a cooperative way over a range of membrane potentials that correspond to that for the voltage sensitivity curve of the anomalous rectification. This possibility was tested by simulating synaptic excitation distributed uniformly in the dendritic field, and adjusting the density of this excitation. To correct for the effects of synaptic location when only a few synapses were active, the results for a number of different randomly selected locations were averaged. The results of these simulations are shown in Fig. 10. The synaptic effectiveness obtained with a single synapse is less in this simulation than in Fig. 9 because a smaller conductance change was used (peak = 0.5 nS), and because the effective location on the dendrite is slightly more distal. (The datum in Fig. 10 for one synapse is an average of many simulations at different dendritic locations.) In the absence of the fast anomalous rectification, distributed synaptic inputs interfered destructively, but more weakly than in the case of clustered synaptic input. As in the case of clustered synapses, increasing the number of synapses active decreased the average effectiveness of any one, because the depolarization of the neuron moved all of the postsynaptic sites nearer to the reversal potential for the synapses. The degree to which this occurred was closely related to the membrane potential achieved during the synaptic activation, as can be appreciated by comparing the curves in Figs. 9 and 10; but this depolarization, and the decrease in synaptic strength, occurs over a much larger scale. While synaptic strength was decreased to less than half, with 10 active synapses when clustered, more than 250 distributed synapses were required to get the same decrement in synaptic strength. With such a large number of active synapses, the somatic depolarization is several times greater than that achieved by clustered synaptic inputs.

These simulations of the passive neuron suggest that despite the extended nature of the dendritic field and the low input resistance of the spiny neuron, only a small fraction of its inputs (perhaps less than 10%) may be required to depolarize the neuron to the tonic level usually observed *in vivo* (about 15–20 mV above the resting potential for the membrane). They also show that the most effective distribution of these synapses is the most uniform possible one, with synapses placed as far apart as possible.

Addition of anomalous rectification to the neuron in this simulation had a profound effect on synaptic effectiveness. When few synaptic inputs were active, the membrane remained relatively polarized despite the synaptic activation, and it remained in its low resistance state. Somewhat larger groups of distributed

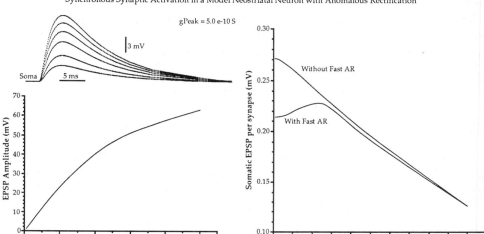

Synchronous Synaptic Activation in a Model Neostriatal Neuron with Anomalous Rectification

FIGURE 10.   Integration of spatially distributed synaptic input on the model neuron. Synapses were activated at very low density uniformly over the surface of the spiny neuron model. At low values of synaptic density, points represent an average of responses obtained with synapses placed at various regions of the dendritic field. The traces at the upper left show the somatic EPSP obtained with varying numbers of active synapses. The graph at the left shows the peak somatic EPSP as a function of the number of simultaneously active synapses. Note that the somatic EPSP grows more linearly with the number of synapses than it does with clustered synaptic inputs. This is due to the deemphasis of the synaptic nonlinearity when synapses are distributed on the neuron. The somatic EPSP contribution of one synapse is shown as a function of the total number of active synapses, in the graph at the right. Synaptic effectiveness in the linear neuron model decays more slowly with numbers of synapses than in the clustered synapse case. When the inward rectifier is added to the neuron model, an increase in synaptic effectiveness is seen with low synaptic numbers, and a maximal synaptic effectiveness is observed when approximately 150 synapses are active at the same time. This number of synapses depolarizes the soma by about 30 mV, and so corresponds roughly to the size of the depolarizing episodes, as seen in Fig. 1.

synaptic inputs could lift the membrane slightly into the higher resistance state, decreasing electrotonic decay and so boosting effective synaptic strength as seen from the soma. Because the half-activation voltage for the anomalous rectification is much nearer the baseline membrane potential than is reversal potential for the synaptic potential, the cooperative effect produced by such moderate depolarizations can overcome the opposite influence exerted by the synaptic nonlinearity. Thus, in the curve showing synaptic strength in the case of a model neuron with fast anomalous rectification in Fig. 10, increasing the number of active synapses produces an initial increase in synaptic strength. When the number of synapses becomes great enough to produce larger EPSPs and the effect of the anomalous rectification becomes maximal, the trend reverses, and the synaptic nonlinearity again undermines synaptic strength. This did not occur in the case of clustered synaptic inputs because of the large size of the local synaptic input (which turned the inward rectifying conductance off and raised the input resistance locally to its maximum) and the small region of the dendritic tree that was affected. The regions of the neuron that were not involved in the synaptic events remained mostly in the low-input-resistance voltage region, and the EPSP was degraded by electrotonic decay.

This effect of anomalous rectification on the strength of distributed synapses resembles a bandpass filter, in which synapses are most effective if they arrive in concert with a limited number of others. The peak of this curve—that is, the preferred number of simultaneous synapses—is determined by the initial membrane potential, the strength of the synapses, and especially by the slope of the activation curve and the half-activation voltage of the anomalous rectification. Increasing the half-activation voltage, either directly or by changing the potassium equilibrium potential, will raise the number of synapses preferred by the neuron, and increase the voltage range over which synapses act cooperatively. If it becomes too positive, however, the strength of the cooperative effect is diluted by the powerful effect of the synaptic nonlinearity. Increasing the slope of the activation curve strengthens the cooperative effect, but at the expense of narrowing the range of synaptic densities over which the cooperative effect can be observed.

The current attenuating effect of dendritic spines is not required for this cooperative interaction between synapses, which can be seen within simulations of axodendritic synapses as well. The reduced dendritic EPSP amplitudes obtained with current attenuating dendritic spines, however, increases the magnitude of the cooperative effect, and the preferred number of active synapses, as does decreasing the magnitude or duration of the synaptic conductance change.

XI. Implications for Neostriatal Function

One of the most obvious features of the neostriatal spiny neuron is its relative silence. Even in awake animals, these cells often fire only in rare episodes. Because the cells contain and release GABA, and because they have dense local axonal arborizations, much of the early work on neostriatal neurons concentrated on the idea that the silence of these cells is due to active inhibition generated within the neostriatum (e.g., Hull *et al.*, 1973; Bernardi *et al.*, 1975, 1976).

More recent studies have failed to reveal this inhibition, and have suggested the opposite view, i.e., the neurons do not fire because they have insufficient excitatory synaptic input, despite the large number of excitatory synaptic contacts formed onto them by cortical and thalamic afferent fibers (Wilson and Groves, 1981; Wilson et al., 1983a; Wilson, 1986; Calabresi et al., 1990). If corticostriatal or thalamostriatal neurons fired very rarely, there would be no paradox in this view, but this is not the case. Studies of corticostriatal neurons, at least, show them to be spontaneously active under conditions in which neostriatal neurons are mostly silent (e.g., Landry et al., 1984; Cowan and Wilson, 1990). In addition, neostriatal spiny neurons have been shown to be chronically depolarized by 10–20 mV by tonic but subthreshold afferent synaptic activity (Wilson et al., 1983b). These experiments suggested that neostriatal neurons do not lack synaptic input, but rather that individual synapses are so weak that large numbers of correlated synaptic inputs are required to depolarize the neuron sufficiently to fire action potentials (Wilson and Groves, 1981; Wilson et al., 1983b; Calabresi et al., 1990).

The reasons for the ineffectiveness of afferent synapses on the neostriatal spiny neurons remain unavailable for direct study. On the basis of the findings presented here, the current attenuation effect of dendritic spines, as previously suggested (Wilson et al., 1983a), seems unlikely to be a large factor in determining the strength of the synapses in the neostriatum. Unless synaptic conductances on neostriatal neurons have unusually short durations or spine necks have unexpectedly high cytoplasmic resistances, this effect is not likely to account for much more than a 50% attenuation of EPSP amplitudes at the soma. A potentially larger effect may be due to the effect of the additional membrane of the dendritic spines on the effective electrotonic length of the dendrites, coupled with the strong anomalous rectification, which both lowers the overall input resistance of the neuron, and exaggerates the effects of the dendritic spines on the length constant. The reduction of input resistance in the dendrite would also promote the operation of dendritic spines as current attenuators. The current attenuation effect of dendritic spines requires conductance changes large enough to produce saturating EPSPs in the spine heads, and relies upon the low-input impedance of the dendrite to insure that the local dendritic EPSP will be small. The linear model of the spiny neuron, combined with current estimates of the electrical characteristics of the spine membrane and cytoplasm, predicts that the EPSP generated from such an axospinous synapse on a neostriatal neuron would actually be quite large; that is, the input resistance of the dendrite is so high that saturating EPSPs in the spine head produce dendritic and somatic EPSPs that are larger than would be consistent with neurophysiological data. If the conductance change is made small enough to get the appropriate-sized EPSP in the soma, it fails to produce a saturating EPSP in the spine, and the current attenuation effect is lost. This is similar to the situation described for hippocampal neurons, and that has led to the conclusion that dendritic spines probably do not substantially attenuate synaptic current in those cells (Turner, 1984; Harris and Stevens, 1989). In neostriatal neurons, the extremely low membrane resistivities generated by the fast inward rectifier can result in low enough dendritic input impedances to reconcile the simulations and the physiological data. Thus, the relative ineffectiveness of individual afferent synapses in neostriatal neurons when the membrane potential is near its resting level can be accounted for by a combination of the fast inward rectifier, the short length constant of the spiny dendrite, and perhaps also the current attenuating effect of dendritic spines.

Dendritic spines have also been proposed to act to linearize synaptic interactions. (See references in Wilson, 1984.) Because the spine neck acts as a series resistance between the conductance change on a dendritic spine head and the rest of the dendritic tree, a large conductance decrease in a spine head will not as effectively act to shunt currents generated elsewhere on the dendrites. In addition, the small dendritic EPSPs that are generated by axospinous synapses may reduce the effect of the classical synaptic nonlinearity. The possible effect of the spine neck resistance on synaptic shunting does not play a role when the interactions are between excitatory afferents with the same synaptic reversal potentials, as considered here. The effect on local dendritic EPSPs, and thus the driving force for synaptic currents, was large in the simulations of synchronous afferent synaptic activity, but only when the density of synaptic inputs was low. When axospinous synaptic inputs were closely spaced on the dendrite, either because they were clustered or because the overall level of synaptic input was large, the dendritic EPSPs became large enough to engage the synaptic nonlinearity. The difference between axospinous and axodendritic synapses was greatly reduced when the local density of activated synapses exceeded only a few percent of the density of afferent synapses present on the spiny neurons, and was practically absent with activation of 50% of the synapses. One result of this is that the effectiveness of synapses on the spiny neuron is greatly decreased if the synapses are spatially clustered on the dendrites. A single axon that makes several synaptic contacts on a neostriatal neuron will have a much greater potential for summing with other active inputs if it distributes its synapses as far apart as possible. Afferent fibers in the neostriatum probably do make spatially distributed synaptic contacts, because they branch infrequently and make synapses *en passant* as they cross over the dendritic trees of the spiny neuron.

The dependence of the membrane resistance of the spiny neuron on membrane potential introduces the possibility for cooperative synaptic interactions. The conditions that favor this cooperative effect are:

1. Widely distributed synaptic input.
2. Adjustment of the activation curve of the inwardly rectifying conductance so that it is almost fully on at the potassium equilibrium potential and disengages over a range of potentials corresponding to the local (dendritic) amplitudes of the EPSPs that should interact cooperatively, but is far from the reversal potential of the excitatory inputs.
3. Placement of the inwardly rectifying conductance on the dendrites.

This latter condition is necessary because the effect of the inward rectification on dendritic length constant is critical to the synaptic cooperativity.

This cooperative effect among synapses may contribute to the existence of the episodic depolarizations that characterize the *in vivo* pattern of activity in neostriatal projection neurons. In the hyperpolarized state of the membrane that separates depolarizing episodes, the input resistance is low and the effect of an individual synaptic input is negligible. If there are correlated synaptic inputs distributed over the dendritic field, these can depolarize the neuron to some degree, raising the input resistance and decreasing the electrotonic distance between the interacting synapses. This will make the neuron more sensitive to these and other excitatory inputs. This positive feedback effect of the inward rectification will tend to give an abrupt onset and offset to episodes of synaptic depolarization, as has been reported for neostriatal neurons (Wilson *et al.*, 1982). The number of active synapses required to produce a depolarizing episode would, according to this view, depend upon the activation curve for the inward rectification, the background level of unpatterned synaptic input present but not correlated with the inputs whose correlated activity will determine the timing of the episode, and the uniformity of the distribution of synaptic inputs whose activity is correlated. Thus, the conductance responsible for the fast anomalous rectification in neostriatal neurons is well-situated to adjust the sensitivity of the neostriatal neuron to different patterns of synaptic input and would make an ideal target for modulation by other neurotransmitters and peptides.

Acknowledgments

This work was supported by NIH grant NS20743 and ONR grant N00014-89-J-3179. High-voltage electron microscopy was supported by NIH grant RR00592 to the Laboratory for High Voltage Electron Microscopy at Boulder, Colorado. Thanks to Dr. D. J. Surmeier for thoughtful comments and helpful suggestions.

References

BARGAS, J, GALARRAGA, E., and ACEVES, J. (1988). "Electrotonic Properties of Neostriatal Neurons Are Modulated by Extracellular Potassium," *Brain Res.* **72,** 390–398.

BERNARDI, G., MARCIANI, M. G., MOROCUTTI, C., and GIACOMINI, P. (1975). "The Action of GABA on Rat Caudate Neurones Recorded Intracellularly," *Brain Res.* **92,** 511–515.

BERNARDI, G., MARCIANI, M. G., MOROCUTTI, C., and GIACOMINI, P. (1976). "The Action of Picrotoxin and Bicuculline on Rat Caudate Neurons Inhibited by GABA," *Brain Res.* **102,** 397–384.

BROWN, T. H, CHANG, V. C., GANONG, A. H., KEENAN, C. L., and KELSO, S. R. (1988). "Biophysical Properties of Dendrites and Spines that May Control the Induction and Expression of Long-Term Potentiation," *Neurol. Neurobiol.* **35,** 201–264.

CALABRESI, P., MERCURI, N., STANZIONE, P., STEFANI, A., and BERNARDI, G. (1987). "Intracellular Studies on the Dopamine-Induced Firing Inhibition of Neostriatal Neurons *in vitro:* Evidence for D1 Receptor Involvement," *Neuroscience* **20,** 757–771.

CALABRESI, P., MERCURI, N. B., STEFANI, A., and BERNARDI, G. (1990). "Synaptic and Intrinsic Control of Membrane Excitability of Neostriatal Neurons, I. An *in vivo* Analysis," *J. Neurophysiol.* **63,** 651–662.

CHANG, H. T., WILSON, C. J., and KITAI, S. T. (1981). "Single Neostriatal Efferent Axons in the Globus Pallidus: A Light and Electron Microscopic Study," *Science* **213,** 915–918.

COWAN, R. L., and WILSON, C. J. (1990). "Contralateral Neostriatal and Ipsilateral Pyramidal Tract Stimulation Produce Synaptic Responses in Pyramidal Tract and Crossed Corticostriatal Neurons," *Soc. Neurosci. Abst.* **16,** 417.

DIAMOND, J., GRAY, E. G., and YASARGIL, G. M. (1970). "The Function of the Dendritic Spine: An Hypothesis," in P. Andersen and J. K. S. Jansen (eds.), *Excitatory Synaptic Mechanisms* (pp. 213–222). Universitet Forlaget, Oslo, Norway.

DIFIGLIA, M., PASIK, P., and PASIK, T. (1976). "A Golgi Study of Neuronal Types in the Neostriatum of Monkeys," *Brain Res.* **114,** 245–256.

DURAND, D. (1984). "The Somatic Shunt Cable Model for Neurons," *Biophys. J.* **46,** 645–653.

EVARTS, E. V., KIMURA, M., WURTZ, R. H., and HIKOSAKA, O. (1984). "Behavioral Correlates of Activity in Basal Ganglia Neurons," *Trends Neurosci.* **7,** 447–453.

FLESHMAN, J. W., SEGEV, I., CULLHEIM, S., and BURKE, R. E. (1983). "Matching Electrophysiological Measurements in Cat Motoneurons," *Neurosci. Abst.* **9,** 341.

HAGIWARA, S., and TAKAHASHI, K. (1974). "The Anomalous Rectification and Cation Selectivity of a Starfish Egg Cell," *J. Membrane Biol.* **18,** 61–80.

HARRIS, K. M., and STEVENS, J. K. (1989). "Dendritic Spines of CA1 Pyramidal Cells in the Rat Hippocampus: Serial Electron Microscopy with Reference to Their Biophysical Characteristics," *J. Neurosci.* **9,** 2982–2997.

HULL, C. D., BERNARDI, G., and BUCHWALD, N. A. (1970). "Intracellular Responses of Caudate Neurons to Brain Stem Stimulation," *Brain Res.* **22,** 163–179.

HULL, C. D., BERNARDI, G., PRICE, D. D., and BUCHWALD, N. A. (1973). "Intracellular Responses of Caudate Neurons to Temporally and Spatially Combined Stimuli," *Exp. Neurol.* **38,** 324–336.

KAWAGUCHI, Y., WILSON, C. J., and EMSON, P. C. (1990a). "Intracellular Recording of Identified Neostriatal Patch and Matrix Spiny Cells in a Slice Preparation Preserving Cortical Inputs," *J. Neurophysiol.* **62**, 1052–1068.

KAWAGUCHI, Y., WILSON, C. J., and EMSON, P. C. (1990b). "Projection Subtypes of Rat Neostriatal Matrix Cells Revealed by Intracellular Injection of Biocytin," *J. Neurosci.* **10**, 3421–3438.

KAWATO, M. (1985). "Cable Properties of a Neuron Model with Non-Uniform Membrane Resistivity," *J. Theor. Biol.* **111**, 149–169.

KAWATO, M., HAMAGUCHI, T., MURAKAMI, F., and TSUKAHARA, N. (1984). "Quantitative Analysis of Electrical Properties of Dendritic Spines," *Biol. Cybern.* **50**, 447–454.

KITA, T., KITA, H., and KITAI, S. T. (1984). "Passive Electrical Membrane Properties of Rat Neostriatal Neurons in an *in vitro* Slice Preparation," *Brain Res.* **300**, 129–139.

KOCH, C., and POGGIO, T. (1984). "A Theoretical Analysis of Electrical Properties of Spines," *Proc. R. Soc. Lond. B* **218**, 455–477.

LANDRY, P., WILSON, C. J., and KITAI, S. T. (1984). "Morphological and Electrophysiological Characteristics of Pyramidal Tract Neurons in the Rat," *Exp. Brain Res.* **57**, 177–190.

MASTRONARDE, D. N., WILSON, C. J., and McEWEN, B. (1989). "Three-Dimensional Structure of Intracellularly Stained Neurons and Their Processes Revealed by HVEM and Axial Tomography," *Soc. Neurosci. Abst.* **15**, 256.

MILLER, J. P., RALL, W., and RINZEL, J. (1985). "Synaptic Amplification by Active Membrane in Dendritic Spines," *Brain Res.* **325**, 325–330.

PERKEL, D. H., and PERKEL, D. J. (1985). "Dendritic Spines: Role of Active Membrane in Modulating Synaptic Efficacy," *Brain Res.* **325**, 331–335.

RALL, W. (1977). "Core Conductor Theory and Cable Properties of Neurons," in J. M. Brookhart, V. B. Mountcastle, and E. R. Kandel (eds.), *Handbook of Physiology, Vol. 1., Sect. 1* (pp. 39–97). American Physiological Society, Bethesda, Maryland.

RALL, W. (1974). "Dendritic Spines, Synaptic Potency and Neuronal Plasticity," in C. D. Woody, K. A. Brown, T. J. Cros, and J. D. Knispel (eds.), *Cellular Mechanisms Subserving Changes in Neuronal Activity* (pp. 13–21). Brain Information Service, UCLA, Los Angeles.

RALL, W., and SEGEV, I. (1988). "Synaptic Integration and Excitable Dendritic Spine Clusters: Structure/Function," *Neurol. Neurobiol.* **37**, 263–282.

SEDGWICK, E. M., and WILLIAMS, T. D. (1967). "The Response of Single Units in the Caudate Nucleus to Peripheral Stimulation," *J. Physiol. (London)* **189**, 281–298.

SHEPHERD, G. M., BRAYTON, R. K., MILLER, J. P., SEGEV, I., RINZEL, J., and RALL, W. (1985). "Signal Enhancement in Distal Cortical Dendrites by Means of Interactions Between Active Dendritic Spines," *Proc. Natl. Acad. Sci. USA* **87**, 2192–2195.

SUGIMORI, M., PRESTON, R. I., and KITAI, S. T. (1978). "Response Properties and Electrical Constants of Caudate Neurons in the Cat," *J. Neurophysiol.* **41**, 1662–1675.

SURMEIER, D. J., BARGAS, J., and KITAI, S. T. (1988). "Voltage-Clamp Analysis of a Transient Potassium Current in Rat Neostriatal Neurons," *Brain Res.* **473**, 187–192.

SURMEIER, D. J., BARGAS, J., and KITAI, S. T. (1989). "Two Types of A-Current

Differing in Voltage Dependence Are Expressed by Neurons of the Rat Neostriatum," *Neurosci. Lett.* **103,** 331–337.

SURMEIER, D. J., STEFANI, A., FOEHRING, R. C., and KITAI, S. T. (1991). "Developmental Regulation of a Slowly Inactivating Potassium Conductance in Rat Neostriatal Neurons," *Neurosci. Lett.* **122,** 41–46.

TURNER, D. A. (1984). "Conductance Transients onto Dendritic Spines in a Segmental Cable Model of Hippocampal Neurons," *Biophys. J.* **46,** 85–96.

WILSON, C. J. (1984). "Passive Cable Properties of Dendritic Spines and Spiny Neurons," *J. Neurosci.* **4,** 281–297.

WILSON, C. J. (1986). "Postsynaptic Potentials Evoked in Spiny Neostriatal Projection Neurons by Stimulation of Ipsilateral or Contralateral Neocortex," *Brain Res.* **367,** 201–213.

WILSON, C. J. (1987). "Three-Dimensional Analysis of Neuronal Geometry Using HVEM," *Journal of Electron Microscopic Technique* **6,** 175–183.

WILSON, C. J., and GROVES, P. M. (1980). "Fine Structure and Synaptic Connections of the Common Spiny Neuron of the Rat Neostriatum: A Study Employing Intracellular Injection of Horseradish Peroxidase," *J. Comp. Neurol.* **194,** 599–615.

WILSON, C. J., and GROVES, P. M. (1981). "Spontaneous Firing Patterns of Identified Spiny Neurons in the Rat Neostriatum," *Brain Res.* **220,** 67–80.

WILSON, C. J., CHANG, H. T., and KITAI, S. T. (1982). "Origins of Postsynaptic Potentials Evoked in Identified Rat Neostriatal Neurons by Stimulation in Substantia Nigra." *Exp. Brain Res.* **45,** 157–167.

WILSON, C. J., CHANG, H. T., and KITAI, S. T. (1983a). "Disfacilitation and Long Lasting Inhibition of Neostriatal Neurons in the Rat," *Exp. Brain Res.* **51,** 217–226.

WILSON, C. J., GROVES, P. M., KITAI, S. T., and LINDER, J. C. (1983b). "Three-Dimensional Structure of Dendritic Spines in the Rat Neostriatum," *J. Neurosci.* **3,** 383–398.

Chapter 7 Analog and Digital Processing in Single Nerve Cells: Dendritic Integration and Axonal Propagation

IDAN SEGEV, MOSHE RAPP,
YAIR MANOR, and YOSEF YAROM
Department of Neurobiology
Institute of Life Science
Hebrew University
Jerusalem, Israel

I. Introduction

The basic information-processing unit of the nervous system, the *nerve cell,* is a complicated element, both morphologically (large dendritic and axonal trees) and physiologically (regional specialization of electrical properties along the cell surface). Whether this complexity is of functional significance or is just the result of various biological constraints is still an unresolved question. One extreme approach, inspired by successful attempts to model large physical systems, is that in such systems the single element performs a simple operation and its exact details thus have no functional consequences. According to this approach, the information processing function of the nervous system depends mainly on the *interaction between the neurons* rather than on the properties of the single nerve cell. The other extreme approach, cherished by the biologist, is that one should explore the detailed morphology and physiology of the single cell in depth, since the overall performance of the system does depend critically on these details. This line of though has been strengthened by the success of biologically realistic models, such as those of Hodgkin and Huxley for the action-potential initiation and of Katz and Miledi for synaptic transmission.

This chapter reflects a midway approach, arguing that one should start by studying the computational capability of the individual nerve cells that comprise the network, and that such a study requires the construction of detailed models that incorporate knowledge of neuronal morphology and physiology as well as synaptic architecture. Only the investigation of such detailed models will eventually lead to the construction of more simplified model units that still capture the essential characteristics of the actual cells. These units then can be used to build realistic network models of the systems under study. Presently we are at the stage where novel experimental and theoretical techniques enable one to tailor such detailed models for both the input region (the *dendritic tree*) and the output region (the *axonal tree*). This is demonstrated in the present chapter for a morphologically and physiologically characterized dendritic tree of a *cerebellar Purkinje* cell, as well as for a reconstructed axon from the *cat somatosensory cortex*. Examples of the insights that were gained from exploring these models are given in Section III. Several integrative properties that are likely to be important for the formulation of the more simplified (but yet biologically realistic) units are elaborated on in Section IV.

II.   Methods

The different stages in constructing realistic models of morphologically and physiologically characterized neurons is given elsewhere (Segev *et al.*, 1989; Stratford *et al.*, 1989). As will be elaborated on shortly, the goal is to end up with an electrical model that represents the detailed morphology of the cell and, at the same time, faithfully mimics all the electrophysiological measurements obtained from that same cell. Exploring such a model enables one to ask specific questions regarding the input–output properties of the neuron. One hopes that eventually, in the process of this exploration, the principles that govern the integrative function of the cell will be understood. In the following sections, only the main features of the different stages in constructing such as model are elucidated.

A.   Morphology

Recent development of new techniques for intracellular staining and graphic reconstruction have enabled researchers to accurately quantize the detailed morphology of nerve cells. Morphometric measurements are usually made at the

LM level after staining single cells with either HRP or Biocytin. The morphological measurements of the stained cell are obtained using computer-driven tracing systems; they typically include information on the X, Y, Z coordinates of the sampled points, the diameter of the process at these points, and the distance between the points. An example of such a reconstruction is shown in Fig. 2 for the dendritic tree of a Purkinje cell from the guinea pig cerebellum. Although this level of resolution is sufficient for many purposes, one should remember that the dimensions of fine dendritic and axonal processes (such as dendritic spines or axonal varicosities) can not be faithfully measured, nor can the type (symmetrical or nonsymmetrical) and the location of the different synaptic inputs be identified. When this information is required, measurements at the EM level should be performed (Segev et al., 1989).

B. Physiology

Complete characterization of the electrical properties of the cell requires knowledge of the distribution of the different channel types along the cell surface. This is beyond the capability of present techniques, although a significant step towards this end is provided by recent developments of antibodies against specific voltage-dependent ion channels, together with voltage- and ion-dependent dyes (Tank et al., 1988; Westenbroek et al., 1990; also Chapter 9 in this volume). Usually, however, physiological data from single neurons are collected from intracellular stimulating and recording at the cell body. The development of the slice technique made possible the study of the properties of neurons from the mammalian CNS. Basic properties for the *passive* model, such as the input resistance of the cell (R_N), the system time constant (τ_0), the electrotonic length of the dendrites (L), and the dendritic-to-soma conductance ratio (ρ) can be estimated from such measurements. (See Jack et al., 1975; Rall, 1977; Segev et al., 1989; also Section III of this chapter.) When possible, selective synaptic inputs are stimulated, and the somatic response can be used to refine the estimates obtained from direct somatic current stimulation (Rall, 1967; Segev et al., 1990). Limited information regarding the kinetics and distribution of various *excitable* channels along the dendritic surface can also be derived from somatic recordings (e.g., Llinás and Yarom, 1981).

Although nerve cells are inherently nonlinear, their passive models (and the corresponding biophysical parameters) serve as the basis for more extended models which incorporate nonlinear properties (both synaptic and excitable) as data become available. Indeed, many of the insights that were obtained from passive models are applicable also to the excitable case. (See Segev and Rall, 1988.) Constructing passive models for the cell under study is, therefore, the primary goal for the modeler.

C.   Matching Dendritic Morphology and Physiology through Passive Cable and Compartmental Models

A *faithful model* of a morphologically and physiologically characterized dendritic tree is hereby defined as a model whose responses to physiologically applied inputs (e.g., a short current pulse injected into the soma) fit the corresponding experimental responses. As noted previously, one typically starts by estimating the values of the biophysical parameters for the *passive* model, namely, the values for R_m and C_m (the specific membrane resistance and capacitance, respectively), as well as the value for R_i (the specific axial resistance). The general procedure is to first construct a model that describes the dendritic morphology, using some initial values for R_m, C_m, and R_i. Then some (or all) of these parameters are allowed to vary and the cable properties of the model (e.g., R_N or τ_0), or its whole transient voltage response, are compared to the corresponding experimental results. Eventually, estimates for the values of R_m, C_m, and R_i that produce a close match between the model behavior and the experimental data are obtained.

Two complementary theoretical approaches were developed, both by W. Rall, for the purpose of constructing realistic models of dendritic neurons. The first employs analytical *cable* models for describing the flow of electrical currents in arbitrarily complex *passive* trees (Rall, 1959, 1977, 1989). In this approach, the dendritic tree is decomposed into a connected set of passive cylinders (cables), each with a given morphology (length and diameter) and specific electrical properties (R_m, C_m, and R_i). Rall has shown that, for a passive membrane the input–output relation between any pair of points in the tree can be analytically computed for both current input and an input in the form of a transient (synaptic) conductance change (Barret and Crill, 1974; Butz and Cowan, 1974; Holmes, 1986; Horwitz, 1981; Koch and Poggio, 1985; Poggio and Torre, 1977; Rall, 1959; Rall and Rinzel, 1973; Rinzel and Rall, 1974; Turner, 1984). Cable models for dendritic neurons have proven very useful for estimating the values of biophysical parameters and for analyzing the response of the dendritic tree to linear inputs. These become computationally cumbersome, however, when many synaptic inputs are involved. When the membrane consists of voltage-gated channels, this approach is no longer valid and *compartmental* models should be employed (Rall, 1964).

In the compartmental approach, the neuron is decomposed into a set of (linear or nonlinear) R–C compartments, each of which is a lumped representation of a sufficiently small patch of membrane to be regarded as isopotential. Adjacent compartments are connected by an axial resistance, representing the cytoplasm resistivity (and thus, the voltage difference) between these compartments. Mathematically, the compartmental approach is a finite difference discretization of the continuous cable equation; it involves the solution of a set of a coupled ordinary differential equations. Typically, numerical methods are employed to solve the corresponding set of equations (e.g., Cooley and Dodge, 1966; Segev *et al.*, 1985; Shelton, 1985; Traub, 1977), although for some cases analytical methods are available (Perkel *et al.*, 1981; Holmes *et al.*, 1991). Compartmental models are very flexible and can incorporate any type of membrane nonlinearity and an unlimited number of compartments. With the recent advance of powerful computers, realistic models of morphologically and physiologically complicated neurons were constructed. (See review on compartmental modeling by Segev *et al.*, 1989.) It seems, therefore, that this approach is here to stay.

D. Incorporating Dendritic Spines into the Neuron Model

The dendritic trees of many types of neurons are studded with a large number of short appendages—the *dendritic spines* (approximately 10,000 in cortical pyramidal cells and 100,000 in cerebellar Purkinje cells). Typically, each spine is composed of a thin neck that emerges from the dendrite and ends with a bulbous head. In most cases, a single synaptic contact, usually of the asymmetrical type, is attached to the spine head membrane.

Since a large percentage of the dendritic surface is in the spines, they must be incorporated into any faithful model of the cell. However, because of their large number, it is computationally very costly to represent each dendritic spine as a short cylinder (in cable models) or as one (or several) compartments (in the compartmental models). Some global method for including the dendritic spines in the neuron model should be devised.

Three such methods are schematically shown in Fig. 1. All are based on the assumption that, when current flows *from the dendrite into the spine,* the membrane area of the spine can be incorporated into the membrane of the parent dendrite. Indeed, for plausible physiological values of R_m, R_i, and C_m, this approximation is valid, since (for this direction of current flow) the spine base and the spine head membranes are essentially isopotential (negligible voltage attenuation from the spine base into the spine head). (See Jack *et al.*, 1975; Rall, 1974; Segev and Rall, 1988.) This approximation does not hold when the current flows *from the spine head membrane into the dendrite,* as is the case when the spine receives a synaptic input. In this case, a significant voltage attenuation does exist between the spine head and the spine base. (They are not isopotential.) The implication is that spines receiving synaptic inputs should be represented in full (Fig. 1d).

The first method, suggested by Stratford *et al.* (1989), transforms the original spiny parent dendrite into a single "equivalent" cylinder whose length and

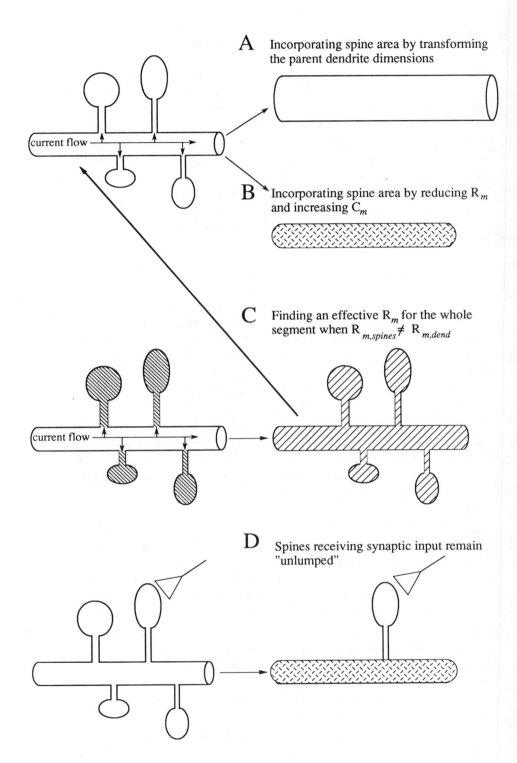

A Incorporating spine area by transforming the parent dendrite dimensions

current flow

B Incorporating spine area by reducing R_m and increasing C_m

C Finding an effective R_m for the whole segment when $R_{m,spines} \neq R_{m,dend}$

current flow

D Spines receiving synaptic input remain "unlumped"

diameter are both larger than those of the parent dendrite. If l and d are the length and diameter of the spiny dendrite, respectively, then the length l' and the diameter d' of the equivalent cylinder are:

$$l' = l \cdot F^{2/3} \quad \text{and} \quad d' = d \cdot F^{1/3},$$

and

$$F = \frac{\text{area}_{\text{dend}} + \text{area}_{\text{spines}}}{\text{area}_{\text{dend}}}, \tag{1}$$

where $\text{area}_{\text{dend}}$ is the area of the parent dendrite without the spines and $\text{area}_{\text{spines}}$ is the membrane area of the spines at that dendritic segment. This transformation preserves the area, input resistance, axial resistance, and effective electrotonic length of the spiny segment (Fig. 1a). It holds, however, only if the specific properties of the membrane of the parent dendrite and of the spines are the same.

A second method (Fig. 1b), suggested by Holmes (1989), is to preserve the original dimensions of the parent dendrite, but to alter its specific membrane properties to effectively include the spine area in the model. (See also Shelton, 1985.) If R_{m} and C_{m} are the values for the spiny segment, the transformed values are:

$$R'_{\text{m}} = \frac{R_{\text{m}}}{F} \quad \text{and} \quad C'_{\text{m}} = C_{\text{m}} \cdot F,$$

where F is as defined in Eq. (1). Note that the time constant of the original spiny dendrite is preserved under this transformation. Also, as in the first method, this transformation holds only when the parent dendrite and the spines have the same

---

FIGURE 1.   Dendritic spines that do not receive direct inputs can be incorporated into the membrane of the parent dendrite. In A, the spiny segment shown at the left is transformed into an "equivalent" cable whose length and diameter are larger than those of the original dendritic segment. In B, the spine's membrane is effectively incorporated into the parent branch membrane by reducing R_{m} and, accordingly, increasing C_{m}, keeping the dimensions of the parent branch unchanged. In both a and b, the specific membrane properties of the dendrite and the spines are assumed to be identical. In c, the case where $R_{\text{m,spines}} \neq R_{\text{m,dend}}$ is treated. The first step is to find an effective, uniform R_{m} value for the whole segment, taking into account the relative areas of the spines and the parent dendrite. Then either one of the methods shown in a and b could be employed. Frame d shows that spines can be incorporated into the dendrite only when the current flows from the dendrite *into* the spines. Spines receiving direct synaptic input are modeled in full.

specific membrane properties. It is easy to show that the two methods shown in Figs. 1a and 1b are mathematically equivalent.

In practice, however, the membrane of the dendritic spines and that of the parent dendrite may be different. (Typically, spine membranes bear synaptically activated channels.) The third method, shown schematically in Fig. 1c, extends the two previous methods to include the possibility that $R_{m,dend}$ (the specific membrane resistivity of the parent dendrite) is different than $R_{m,spines}$ (the specific membrane resistivity of the spines). In this case, the time constant of the spiny segment (dendrite and spines) lies between the time constant of the dendrite's membrane and that of the spines' membrane (assuming an identical C_m value for the two membranes). The first step, therefore, is to find an effective R_m (denoted here by R_m^*) for the whole segment so that ($R_m^* \cdot C_m$) is approximately the time constant of the original segment. Our calculations suggest that, for an electrically short dendritic segment, R_m^* is given by:

$$R_m^* = \frac{1}{\left(\dfrac{1}{R_{m,dend}}\right)\left(\dfrac{area_{dend}}{area_{total}}\right) + \left(\dfrac{1}{R_{m,spines}}\right)\left(\dfrac{area_{spines}}{area_{total}}\right)}, \qquad (2)$$

where $area_{total} = area_{dend} + area_{spines}$. (See also Fleshman et al., 1988.) Having a single R_m^* value as calculated from Eq. (2) for the whole segment, one can utilize either one of the first two methods shown earlier for incorporating the dendritic spine membrane area into the dendritic membrane.

Alternative methods to those shown in the preceding were employed by Turner (1984). Recently, a novel analytical approach to incorporate dendritic spines (both passive and excitable) into passive dendritic trees was developed by Baer and Rinzel (1991).

E. Compartmental Models of Axonal Trees

Because of their complicated morphology and excitable (nonlinear) membrane properties, axons pose a difficult task for neuronal modelers. Membrane excitability is usually described by partial differential equations (PDEs), which can not be solved analytically; numerical (compartmental) methods should, therefore, be employed. Cooley and Dodge (1966) were the first to show how to apply compartmental techniques for modeling the propagating of an action potential along an axonal tree, utilizing the Hodgkin et al. (1952) PDEs. Their approach was further employed to study the theoretical consequence of a local geometrical change (a diameter change or a branch point) or a change in electrical properties of the membrane and cytoplasm for the processing of action potentials along the axon. (See review by Khodorov and Timin, 1975; also Moore et al., 1983;

Parnas and Segev, 1979; Stockbridge, 1989.) Only recently, with the advance of powerful computers and fast integration methods for solving PDEs for axonal branching (Hines, 1984; Manor *et al.,* 1991a; Mascagni, 1989), has it become possible to model realistically complicated axonal trees. An example of this model is given in the present chapter. Details on AXONTREE, the computer program that we developed specifically for this purpose, is given elsewhere (Manor *et al.,* 1991a).

It is important to note that, although the branching patterns of many axonal trees have been studied in detail during the last several years (e.g., Gilbert & Wiesel, 1979; Humphrey *et al.,* 1985; Sereno and Ulinski, 1987), in most cases the *diameters* of the different axonal branches were not measured, nor was information regarding myelination available in these studies. Critical electrical information is still missing for those interested in modeling signal processing along axonal trees, including, specifically, the properties (kinetics, density) of the excitable channels along the tree. Until these morphological and electrical data become available, ranges of possible values have to be theoretically explored.

III.   Results

A.   Modeling Purkinje Cell Dendrites

Figure 2 shows a Purkinje cell (PC) from the guinea pig cerebellum that was reconstructed following intracellular staining with HRP. The reconstruction of this cell was performed utilizing a computer-driven system (Neuron Tracing System, NTS, Eutectic Inc). This reconstruction does not include the dendritic spines that account for more than 50% of the total dendritic area of these cells (see the following). The dendritic tree of the cell in Fig. 2 consists of 473 terminals, the area of the dendrites without the spines is 53,617 μm^2, and the total dendritic length is 10,839 μm. Interestingly, approximately 85% of this length is composed of thin ($<2.2\mu m$ diameter) dendritic branchlets bearing spines.

As in most neurons, the electrical properties of PC membranes are a function of voltage. In addition to several types of voltage-dependent regenerative responses, which are threshold phenomena (Llinás and Sugimori, 1980a), a prominent voltage- and time-dependent conductance governs the membrane resistance near the resting potential. As shown in Fig. 3a, these conductances, expressed as a partial repolarization of the membrane potential during prolonged hyperpolarizing current pulses, are already noticeable at low-current intensity. As a result of the fast development of these conductances (relative to the resting time

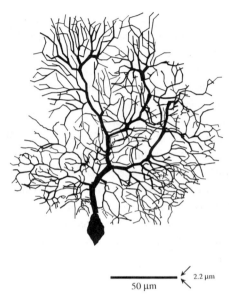

50 μm

2.2 μm

FIGURE 2.   Reconstruction of HRP-labeled Purkinje cell from guinea pig cerebellum. A semiautomatic computer tracing system (Eutectic, Inc.) was utilized for the reconstruction. The 3D structure of the cell was represented by 2,100 points that were used to generate either a cable (continuous) model of the cell or a compartmental model of it. The total membrane area (without the spines) is 53,617 μm^2; the total length of the spiny dendrites is 9,588 μm (88% of the total dendritic length). Assuming a density of 10 spines per 1 μm dendritic length and a membrane area of 1μm^2 for each spine implies a total dendritic area of 149,497 μm^2. Thus, almost 2/3 of the total membrane area of the dendrite is in the spines.

constant), even the input resistance at the resting potential can not be accurately measured. Clearly, linear models are not applicable to neurons with such properties. The use of experimental procedures to linearize the membrane properties, therefore, should be the first step toward the construction of a model for PCs.

Following the study of Crepel and Penit-Soria (1986) on anomalous rectification in PCs, we used cesium ions to effectively block these voltage-dependent conductances. Figure 3b shows the responses of the same cell as in Fig. 3a following substitution of the normal Ringer solution by cesium Ringer (2.5 mM CsCl). Three points are evident from comparing Figs. 3a and b:

1. In the presence of Cs^+ ions, the steady-state input resistance is much larger than that measured in normal conditions.
2. Cs^+ ions give rise to an increase in the membrane time constant.
3. Treatment with Cs^+ ions results in a linear I–V relation at the range of hyperpolarizing voltages used.

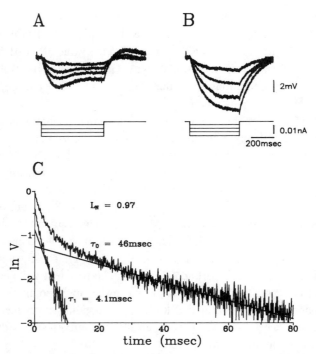

FIGURE 3.   Extracellular Cs^+ ions linearize the membrane properties of Purkinje cells. In a and b are superimposed traces of the membrane responses to rectangular current pulses of different intensities. In a, under normal conditions the membrane responses are governed by a prominent anomalous rectification. In b, the presence of 2.5 mM Cs^+ ions in the bathing solution gives rise to a voltage response that seems to behave passively. Now the input resistance is three times larger than in normal conditions. (c) Estimating the system time constant (τ_0), the first equalizing time constant (τ_1), and the electrotonic length (L_N) of the cell by peeling the transient voltage response of the neuron to a brief current pulse. The decay phase of the averaged voltage response to 100 brief (0.6 msec, 2 nA) hyperpolarizing current pulses was plotted on a semi-logarithmic graph. The final decay phase followed a single exponential with a time constant τ_0 of 46 msec; $\tau_1 = 4.06$ msec and $L_N = \pi/\{(\tau_0/\tau_1) - 1\}^{1/2} = 0.97$ (Rall, 1969).

These results suggest that under these conditions, the PC's membrane behaves passively. Consequently, the response of the cells to a brief hyperpolarizing current pulse was measured in the presence of Cs^+ ions. From this measurement, utilizing the *peeling* method of Rall (1969), the system time constant (τ_0) and the first equalizing time constant (τ_1) were recovered and the cable length (L_N) of the neuron was estimated. The example shown in Fig. 3c was obtained from the same cell depicted in Fig. 2. This cell has a relatively long time constant (46 msec) and, although the dendritic tree is large, the cell is electrically rather compact ($L_N = 0.97$). The voltage response for the short current pulse can also be used for estimating the input resistance (Durand *et al.*, 1983). It was found to be 13.21 MΩ for that cell, in agreement with the value of 13.01 MΩ obtained from the steady-state response (not shown). Our results from 20 Purkinje cells in cesium Ringer conditions show that the average τ_0 is 81 msec, the average R_N is 28 MΩ, and the average L_N is 0.59.

Based on the morphological data, cable and compartmental models were constructed for the cell shown in Fig. 2. Spines were incorporated into the models as described previously. We assumed that each spiny branchlet bears 10 spines per 1 μm length, and that the area of each spine is 1 μm^2 (Harris and Stevens, 1988 a,b). Initially, a cable model was used for computing the value of R_m that matches the experimental value of R_N, assuming that R_m is uniform over the soma-dendritic surface and that $R_i = 70$ $\Omega \cdot$ cm. The resultant R_m value (17,000 $\Omega \cdot$ cm^2) and the experimentally measured τ_0 (46 msec) were used to calculate the specific membrane capacity ($C_m = \tau_0/R_m = 2.71$). These values were then employed to build the compartmental model. The transient voltage response of this model to a small (3 nA) and brief (0.6 msec) hyperpolarizing pulse (Fig. 4a, continuous line) was compared to the corresponding experimental response (dotted curve). The discrepancy between the two curves shows that some, or all, of our assumptions were incorrect.

The assumption that was examined is that the membrane properties are uniform. As shown in Fig. 4b, even if the membrane resistance in the dendrites ($R_{m,dend}$) is 250 times larger than that of the soma ($R_{m,soma}$), the theoretical voltage transient still deviates considerably from the experimental transient. Other possibilities of nonuniform R_m distributions, such as different ratios of $R_{m,soma}$: $R_{m,dend}$ (from 1:250 to 50:1) and spine membrane resistance different from that of the dendrites ($R_{m,spine}$: $R_{m,dend}$, from 1:10 to 10:1), failed to produce voltage transients that match the experimental results satisfactorily.

The second assumption that was examined is the value of the specific axial resistivity (R_i). As shown in Fig. 4c, increasing R_i from 70 $\Omega \cdot$ cm to 250 $\Omega \cdot$ cm results in much better agreement between model prediction and experimental results. Increasing R_i, however, entails a decrease in R_m (to maintain the experimental R_N) and, therefore, C_m has to be increased to about 4 μF/cm^2 (to maintain the measured τ_0). The best fit was obtained only when both nonuniform R_m ($R_{m,soma}$: $R_{m,dend} = 1:250$) and a high R_i of 250 $\Omega \cdot$ cm were measured (Fig. 4d). With these assumptions, unlike Fig. 4c, even at the early phase of decay the theoretical transient (continuous line) fits the experimental results (dotted line). In addition, these assumptions enforce a C_m of 1.64 μF/cm^2, which is in much better agreement with the commonly used value of 1 μF/cm^2.

Once the model parameters have been estimated, the model can be used to

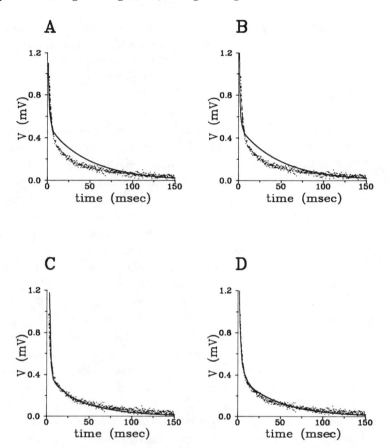

FIGURE 4. The model transients could match the experimental transients only when nonuniform R_m properties and high R_i value were assumed. (a–d) Comparison between the model predictions (continuous line) and the experimental voltage transient (dotted line) obtained from the cell shown in Fig. 2. In a, a uniform R_m was assumed, with $R_m = 17,000 \ \Omega \cdot cm^2$; $R_i = 70 \ \Omega \cdot cm$, and $C_m = 2.71 \ \mu F/cm^2$. In b, R_m of the dendritic membrane was 250 times larger than R_m at the soma. Parameters used: $R_i = 70 \ \Omega \cdot cm$, $C_m = 2.2 \ \mu F/cm^2$, $R_{m,dend} = 90,000 \ \Omega \cdot cm^2$, $R_{m,soma} = 360 \ \Omega \cdot cm^2$. In c, as in a, a uniform R_m was assumed, with $R_i = 250 \ \Omega \cdot cm$. Parameters used: $R_m = 11,000 \ \Omega \cdot cm^2$, $C_m = 4.18 \ \mu F/cm^2$. In d, the best fit (*faithful model*) was obtained assuming nonuniform R_m and high R_i value. Parameters used: $R_i = 250 \ \Omega \cdot cm$, $C_m = 1.64 \ \mu F/cm^2$, $R_{m,dend} = 110,000 \ \Omega \cdot cm^2$, $R_{m,soma} = 440 \ \Omega \cdot cm^2$. In all four cases, the model input resistance and time constant matched the corresponding experimental values. The compartmental modeling was performed using SPICE.

explore the behavior of transient membrane potentials along the cell structure. For example, the following phenomena can be theoretically studied: the passive spread of synaptic potentials, generated at different dendritic sites; the loss in synaptic efficacy due to dendritic spines; and the passive spread of action potentials from the soma into the dendritic tree.

The computed behavior of spatially distributed synaptic inputs into PCs is shown in Fig. 5. Three synaptic inputs at different dendritic sites were simulated. All three inputs were placed on spine heads, modelling the parallel fiber synapses. The locations of these inputs are marked by the three top arrows in Fig. 6. Synaptic input was simulated by a brief conductance change with a time-course described by an α-function (Fig. 5a). Figure 5b shows that the synaptic potentials at the spine head membranes reach an amplitude of about 20 mV, with only a moderate dependence on the spine location. At the spine base, however, a significant difference between the three synaptic potentials is predicted (Fig. 5c). The amplitude of the synaptic potential at the distal location is more than six times larger than that of the proximal synapse. This results from differences in the input impedance at the three spine base locations; it decreases as the distance from the soma decreases. At the somatic level (Fig. 5d), the reverse order is found. The largest synaptic potential, with an amplitude of only 0.11 mV, was generated at the proximal site. The large voltage attenuation (by a factor of about 200) from the spine head to the soma and the resultant small somatic response imply that hundreds of parallel-fiber inputs are needed to generate a somatic action potential.

It is interesting to note that although the proximal synaptic input reaches the largest maximum, after 15 msec the *distal* synaptic potential is larger than the proximal potential (Fig. 5d). This phenomenon, which results form the asymmetry in the dendritic structure and the nonuniformity in the membrane properties, may suggest that distal synaptic inputs differ from proximal inputs in their capability to temporally integrate low-frequency inputs. Another point worth mentioning is that, with the spine and synaptic parameters used, the spine morphology causes less than a 20% loss in synaptic efficacy (as compared to the same input impinging directly at the spine base).

The passive spread of the somatic Na^+-dependent action potential into the dendritic tree of PCs was investigated by Llinás and Sugimori (1980b). According to their description, somatic spikes appear at the dendritic level as small, fast potentials whose amplitude depends strongly on the distance from the cell body. The present model was utilized to investigate the passive antidromic invasion into the dendritic tree of the PC. For this purpose, the passive soma compartment was replaced by an excitable compartment with Hodgkin and Huxley membrane

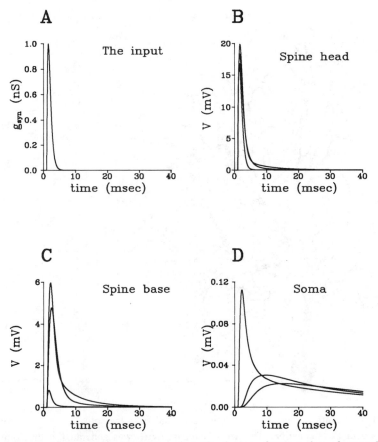

FIGURE 5.   Dendritic morphology determines the degree of attenuation of synaptic potentials along the tree. Synaptic inputs into a spine head located at three different dendritic sites (top three arrows in Fig. 6) were simulated. (a) The time course of the simulated synaptic conductance change (an α-function). The maximal conductance change was 1 nS at 0.5 msec, with a driving force of 100 mV. Similar EPSPs were generated at the spine head membrane, regardless of spine location (b). At the spine base, however, a significant difference in EPSPs was observed (c). The larger EPSP is at the distal location, whereas the smallest one is at the most proximal site. Note the marked attenuation from the spine head into the spine base. At the cell body (d), the largest synaptic potential (from the proximal input) is only 0.12 mV (almost 200-fold attenuation). The simulations were performed on the model of the cell shown in Fig. 2 with the parameters of Fig. 4d. Each spine was modeled by two compartments, one for the head membrane and one for the spine neck. Spine neck resistance, 190 MΩ; spine membrane area, 1 μm^2; $R_{m,spine}$ = $R_{m,dend}$.

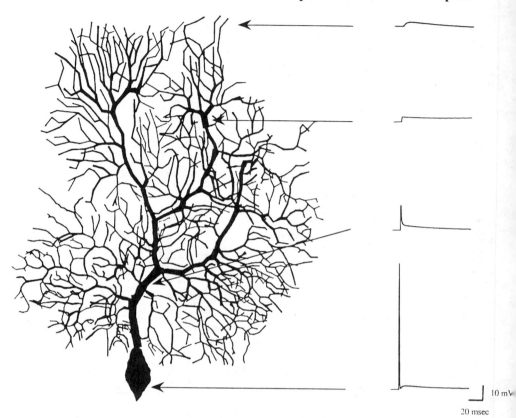

10 mV

20 msec

FIGURE 6.   Antidromic spread of action potential from the soma into the dendritic tree
shows a mark attenuation. The action potential (AP) was generated at the soma compartment utilizing Hodgkin and Huxley (Hodgkin *et al.*, 1952) equations. At the cell
body, the AP (generated by a brief current pulse) reaches an amplitude of 90 mV at 0.5
msec. At the terminal tip, however, the small (3 mV) depolarization reaches its peak
after 8 msec. This delayed depolarization is reflected at the cell body in a small after-
depolarizing potential. Similar potentials were recorded experimentally along the dendritic
tree of Purkinje cells (Llinás and Sugimori, 1980b). Simulations were performed utilizing
the model of the cell shown in Fig. 2 with the parameters of Fig. 4d, replacing the passive
soma compartment with an H&H compartment.

kinetics. As shown in Fig. 6 (lower trace), such a compartment can generate a
fast (0.4 msec rise-time) action potential with an amplitude of 90 mV. In agreement with the physiological results, this action potential appears at dendritic
locations as a small fast depolarization. As the distance from the soma increases,
the passive reflection of the somatic action potential becomes smaller in amplitude
and slower in rise time. At the dendritic terminals, the relatively fast somatic

action potential appears as a small (3 mV), prolonged depolarization (Fig. 6, top trace). When this potential reaches its peak value, the entire tree (including the soma) is almost equipotential. This is reflected as a prolonged after-depolarization at the soma (Fig. 6, lower trace). It is not unlikely that during high-frequency firing, which is so characteristic of PC activity, this prolonged depolarization of the whole tree will add up to a significant voltage displacement that might led to an increase in the excitability of the soma and the dendrites.

B. Modeling Morphologically Characterized Axonal Trees

The axon we chose to model for the present demonstration was adapted from the work of Schwark and Jones (1989, Fig. 5B). This axon innervates areas 3a, 3b, and 4 of the cat somatosensory cortex; the origin of this axon is not known (Fig. 7a). No details about its diameter or the degree of myelination were given. Therefore, we assumed that the axon is not myelinated and that the diameter of the stem (primary) branch is 2.5 μm, that of the secondary collaterals is 1 μm, whereas the diameter of high-order collaterals is 0.4 μm (Florence and Casagrande,1987; Peters, 1987; Schuz and Munster, 1985). We also assumed that the thin (0.4 μm) collaterals bear varicosities (release sites) at a spatial frequency of 4 μm (Florence and Casagrande, 1987; Kisvarday et al., 1987) and that the diameter, as well as the length, of each varicosity is 1.6 μm (Peters, 1987; Rockland, 1989). A total of 977 varicosities, each represented by a single compartment, were simulated. The Hodgkin and Huxley (Hodgkin et al., 1952) PDEs at 20°C were used to model the excitable properties of the axon.

An interesting question that can be explored using this model is the effect of morphological inhomogeneities (branch points, axonal varicosities) on the temporal disparity along the modeled axon. Figure 7b shows the distribution of the arrival times of the action potential peak for the 977 release sites of the modeled axon. The first peak corresponds to the varicosities at areas 3a and 3b, and the proximal varicosities of area 4. The second peak is mostly due to the distal varicosities in area 4. Figure 7b demonstrates that, under the assumption of the present study, the output sites from this axon are not activated simultaneously. Hence, a single cortical axon may serve as a differential timing device. (Areas 3a and 3b are activated 2–3 msec prior to the activation of area 4.)

Figure 7c shows the arrival times for the 977 release sites, assuming that the branch points per se do not affect the propagation velocity. This calculation was performed by computing the electrical distance (i.e., in units of λ) of each varicosity from the origin. The propagation delay to each of these varicosities was calculated directly from the constant propagation velocity when the distance

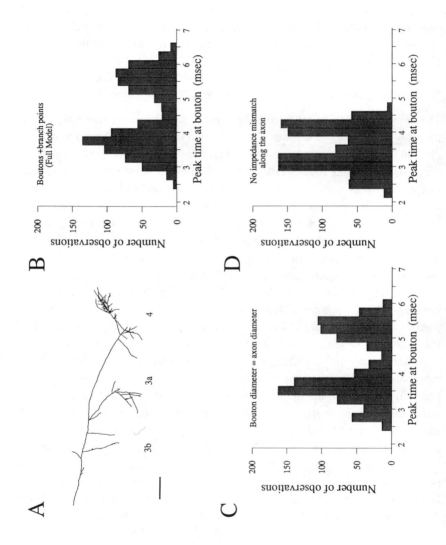

is scaled in units of λ. For the H&H action potential at 20°C this velocity is 2.843 λ/msec). Comparing Fig. 7b to Fig. 7c, one sees that the presence of branch points along the axon increases the delay of the action potential propagation by about 16–26%. As shown in Fig. 7d, the presence of axonal varicosities, however, has only a small effect on this delay.

IV.   Discussion

The underlying assumption of the present study is that the single nerve cell is computationally a rather powerful unit, and that its input–output properties are determined by complex interactions between the cell's morphology, physiology and synaptic architecture. Exploration of a neuron model that encompasses the anatomical and physiological detail will help in understanding the principles that underlie the computational capabilities of the cell. These principles then can be formalized and used in biologically realistic models of neural networks.

This chapter highlights the different stages in the construction of realistic models of morphologically and physiologically characterized dendritic trees, utilizing as an example a Purkinje cell from the guinea pig cerebellum (Fig. 2). Theoretical techniques for incorporating the large number of dendritic spines, an important morphological feature in many neurons types, into the neuron model were also introduced (Fig. 1). We also demonstrated the procedure of recovering

---

FIGURE 7.   The morphology of the reconstructed axon implies a temporal disparity between the different release sites. In a, an HRP-filled axon from the primary somatosensory cortex of the cat is shown. This axon was digitized from Fig. 5b of Schwark and Jones (1989). Scale bar = 300 μm. Axonal varicosities (synaptic boutons). In b–d, the action potential peak time at the different synaptic boutons along the modeled axon is shown. In b, the diameter of each bouton was 1.6 μm, whereas in c, the bouton diameter is identical to the diameter of the axon (no varicosities). In d, the distribution of action potential peak times at the boutons was calculated directly from the *electrotonic* distance of the bouton from the axon origin, assuming a constant velocity of 2.841 λ/msec. In this way, the delay due to geometrical irregularities (branch points or varicosities) was neglected. Comparing b and c to d shows that the presence of an impedance mismatch at the branch points significantly contributes to the propagation delay along the axon. Model parameters: The diameter of the primary axonal branch is 2.5 μm; that of the secondary collaterals is 1 μm, whereas the diameter of higher-order collaterals is 0.4 μm. The length of each bouton is 1.6 μm. The axon was assumed to be unmyelinated, with H&H kinetics at 20°C. The axon was modeled by 2,366 compartments, utilizing AXONTREE (Manor *et al.*, 1991a,b).

the biophysical parameters of the neuron by matching the theoretical predictions with the experimental results (Figs. 3 and 4). The complexity of signal processing in dendritic neurons is demonstrated in Figs. 5 and 6, utilizing the (faithful) model that was constructed for the Purkinje neuron under study. Finally, Fig. 7 shows that the details of the axonal tree may also play a role in the information-processing function of the cell.

The neuron models that were demonstrated in this work take into account only part of the available information on neuron physiology. Dendritic and somatic nonlinearities, axon myelinization, the distribution of different channels and their kinetics, etc. were not considered. Thus, these models should be viewed as an essential basis for more comprehensive models. Nonetheless, although limited, the present study highlights several general principles:

(1) *The single nerve cell, by itself, can function as a network of (almost) independent subunits.* In Fig. 5, it was shown that the receptive area of the neuron, the dendritic tree, is an electrically distributed system (not isopotential). Consequently, following the activation of localized synaptic inputs, a significant potential difference exists between different regions of the tree. Because of this partial electrical decoupling between different parts of the same dendritic tree, each subregion of the tree can perform a specific input–output operation, enabling the neuron to process information in parallel (Koch *et al.*, 1982; Rall, 1970; Rall and Segev, 1987).

(2) Voltage attenuation resulting from dendritic structure implies that the *site of the synaptic input is functionally significant.* In the present study, an attenuation factor as large as 200 was predicted for the PC dendritic tree. Consequently, hundreds of synaptic inputs are required for activating the neuron. Furthermore, the difference in attenuation factor for distal versus proximal inputs implies that the efficacy of the input depends on its location. Indeed, several studies have shown that synaptic inputs from different sources are distributed *nonrandomly* along the dendritic tree. (See Shepherd, 1990.) This specificity of mapping of synaptic information onto the dendritic tree may have important implications for information processing (e.g., Koch *et al.*, 1982; Rall and Segev, 1987).

(3) The dendritic tree behaves as a substantial *delay line* for its synaptic inputs. Furthermore, distal inputs are subject to significantly longer delays than proximal inputs. The present work shows that somatic potentials from distal inputs may peak as much as 10 msec later than potentials from corresponding proximal inputs (Fig. 5; also Segev *et al.*, 1990; Stratford *et al.*, 1989). In principle, this can be used by the neuron to *compute temporal correlations between inputs impinging at different distances from the soma.* (See Koch *et al.*, 1982; Rall, 1964, 1989.)

(4) *The axon can function as a differential timing device.* The propagation velocity in thin axonal processes may be very slow (less than 1 m/sec), leading to a significant propagation delay. Thus, different output sites of the same axon may be activated at different times (Fig. 7). Experimental and theoretical studies have also shown that information may be selectively channeled at axonal branch points and other regions of low safety-factor for action potential propagation (Grossman *et al.*, 1979; Khodorov and Timin, 1975; Moore *et al.*, 1983; Parnas and Segev, 1979).

Two additional principles that were not determined in the present study should also be considered:

(5) *The postsynaptic potentials do not sum linearly.* Because synaptic input is a localized *conductance change* to specific ions, the synaptic input is not a linear current source. This inherent *nonlinearity* could be used by the neuron to perform rather interesting computations (Torre and Poggio, 1978; Rall, 1964).

(6) *Dendritic excitability can fraction the tree into independent logical units.* The dendritic tree of many neuron types is endowed, in addition to the chemically activated channels, with a variety of excitable (voltage- and time-dependent) channels. These channels, which were shown to cluster *selectively* along the soma-dendritic surface (Tank *et al.*, 1988; Ross *et al.*, 1990; Westenbroek *et al.*, 1990) can generate a dendritic spike (Llinás and Nicholson, 1971), which serves as an output for a local summing point. Furthermore, interactions between these summing points give rise to complex electrical behavior (Llinás and Sugimori, 1980b; Llinás and Yarom, 1981). (See also theoretical implication of dendritic excitability in Rall and Segev, 1987; Shepherd and Brayton, 1987.)

Taken together, points 1–6 imply that the output from the axonal terminals is *temporally separated from the input events.* Because of the delay induced by the dendrites as well as by the axon, the neuron may perform a specific computation at its dendrites while the axon carries (and processes) other information. Consequently, the real neuron should not be modeled by a simple isopotential unit, which perform a *sum and fire* operation, as presently used in connectionist models. A network of such simple units should, at the least, be used to simulate the operation of a single neuron. Even then, it is quite unlikely that a network of *identical units* will suffice.

By no means do we imply that the understanding of the function of the nervous system requires a complete reconstruction of every neuron in the system. Indeed, we found that similar model parameters were adequate for three PCs that were completely reconstructed and physiologically characterized. To a large extent, the three corresponding models behaved similarly. We conclude, therefore, that our model of the PC can, in principle, represent all PCs. The message is, therefore, that the first step in analyzing the information-processing function of a specific neuronal system is to construct a detailed model for each type of neuron that comprises the system. Following such a detailed modeling study, a more simplified, yet realistic, unit will be built for each neuron type. Then the behavior of the whole network can be theoretically explored and compared to that of the actual neural network under study.

Acknowledgment

We thank our colleagues, R. Werman and S. Hochstein, for their comments on the manuscript. This work was supported by grants from the Office of Naval Research, #N000014-91-J-1350, and from the Israeli National Council for Research and Development, #2659-1-86.

References

BAER, S. M., and RINZEL J. (1991). "Propagation of Dendritic Spikes Mediated by Excitable Spines: A Continuum Theory," *J. Neurophys.* **65,** 874–890.

BARRETT, J. N., and CRILL, W. E. (1974. "Specific Membrane Properties of Cat Motoneurones," *J. Physiol. (London)* **293,** 301–324.

BUTZ, E. G., and COWAN, J. D. (1974). "Transient Potentials in Dendritic Systems of Arbitrary Geometry," *Biophys. J.* **14,** 661–689.

COOLEY, J. W., and DODGE, F. A. (1966). "Digital Computer Simulation for Excitation and Propagation of the Nerve Impulse," *Biophys. J.* **6,** 583–599.

CREPEL, F., and PENIT-SORIA, J. (1986). "Inward Rectification and Low Threshold Calcium Conductance in Rat Cerebellar Purkinje Cells," *J. Physiol. (London)* **372,** 1–23.

DURAND, D., CARLEN, P. L., GUREVICH, N., HO, A., and KUNOV, H. (1983). "Electrotonic Parameters of Rat Dentate Granule Cells Measured Using Short Current Pulses and HRP Staining," *J. Neurophysiol.* **50,** 1080–1096.

FLESHMAN, J. W., SEGEV, I., and BURKE, R. E. (1988). "Electrotonic Architecture of Type-Identified α Motoneurons in the Cat Spinal Cord." *J. Neurophysiol.* **60,** 60–85.

FLORENCE, S. L., and CASAGRANDE, V. A. (1987). "Organization of Individual Afferent Axons in Layer IV of Striate Cortex in a Primate," *J. Neurosci.* **7**(12), 3850–3868.

GILBERT, C. D., and WIESEL, T. N. (1979). "Morphology and Intercortical Projections of Functionally Characterized Neurones in the Cat Visual Cortex," *Nature* **280,** 120–125.

GRAY, E. G. (1959). " Axo-Somatic and Axo-Dendritic Synapses of the Cerebral Cortex: An Electron Microscope Study," *J. Anat.* **93**, 420–433.

GROSSMAN, Y., PARNAS, I., and SPIRA, M. E. (1979). "Differential Conduction Block in Branches of a Bifurcating Axon," *J. Physiol. (London)* **205**, 283–305.

HARRIS, K. M., and STEVENS, J. K. (1988a). "Dendritic Spines of Rat Cerebellar Purkinje Cells: Serial Electron Microscopy with Reference to Their Biophysical Characteristics," *J. Neurosci.* **12**, 4455–4469.

HARRIS, K. M., and STEVENS, J. K. (1988b). "Study of Dendritic Spines by Serial Electron Microscopy and Three-Dimensional Reconstruction." in *Neurobiology and Neurology, Vol. 37, Intrinsic Determinants of Neuronal Form and Function.* R. J. Lasek, M. M. Black (eds.), Alan Liss, New York.

HINES, M. (1984). "Efficient Computation of Branched Nerve Equations, *Int. J. Bio-Med Comp.* **15**, 69–76.

HODGKIN, A. L., HUXLEY, A. F., and KATZ, B. (1952). "Measurement of Current-Voltage Relations in the Membrane of the Giant Axon of *Loligo,*" *J. Physiol. (London)* **116**, 424–448.

HOLMES, R. W. (1986). "A Continuous Cable Method for Determining the Transient Potential in Passive Dendritic Trees of Known Geometry, *Biol. Cybern,* **55**, 115–124.

HOLMES, R. W. (1989). "The Role of Dendritic Diameter in Maximizing the Effectiveness of Synaptic Inputs," *Brain Res.* **478**, 127–137.

HOLMES, R. W., SEGEV, I., and RALL, W. (1991). "Interpretation of Equalizing Time Constant and Equivalent Cylinder Formulae Electrotonic Length Estimates in Multi-Cylinder of Branched Neural Structures (submitted to *J. Neurophysiol.*).

HORWITZ, B. (1981). "An Analytical Method for Investigating Transient Potentials in Neurons with Branching Dendritic Trees," *Biohys. J.* **36**, 155–192.

HUMPHREY, A. L., SUR, M., ULRICH, D. J., and SHERMAN, S. M. (1985). "Projection Patterns of Individual X- and Y-Cell Axons from the Lateral Geniculate Nucleus to Cortical Area 17 in the Cat," *J. Comp. Neurol.* **223**, 159–189.

JACK, J. J. B., NOBLE, D., and TSIEN, R. W. (1975). *Electrical Current Flow in Excitable Cells.* Oxford University Press, Oxford.

KHODOROV, B. I., and TIMIN, YE. N. (1975). "Nerve Impulse Propagation Along Non-uniform Fibres (Investigations Using Mathematical Models)," *Prog. Biophys. Molec. Biol.* **30**, 145–184.

KISVARDAY, Z. F., MARTIN, K. A. C., FRIEDLANDER, M. J., and SOMOGYI, P. (1987). "Evidence for Interlaminar Inhibitory Circuits in the Striate Cortex of the Cat," *J. Comp. Neurol.* **260**, 1–19.

KOCH, C., and POGGIO, T. (1985). "A Simple Algorithm for Solving the Cable Equation in Dendritic Trees of Arbitrary Geometry," *J. Neurosci. Meth.* **12**, 303–315.

KOCH, C., POGGIO, T., and TORRE, V. (1982). "Retinal Ganglion Cells: A Functional Interpretation of Dendritic Morphology," *Philos. Trans. R. Soc. Lond.* B **298**, 227–263.

LLINÁS, R., and NICHOLSON, C. (1971). "Electrophysiological Properties of Dendrites and Somata in Alligator Purkinje Cells," *J. Neurophysiol.* **34**, 532–551.

LLINÁS, R., and SUGIMORI, M. (1980a). "Electrophysiological Properties of *in vitro* Purkinje Cell Somata in Mammalian Cerebellar Slices," *J. Physiol. (London)* **305,** 171–195.

LLINÁS, R., and SUGIMORI, M. (1980b). "Electrophysiological Properties of *in vitro* Purkinje Cell Dendrites in Mammalian Cerebellar Slices," *J. Physiol. (London)* **305,** 197–213.

LLINÁS, R., and YAROM, Y. (1981). "Properties and Distribution of Ionic Conductances Generating Electroresponsiveness of Mammalian Inferior Olivary Neurones *in vitro,*" *J. Physiol. (London)* **315,** 569–584.

MANOR, Y., GONCZAROWSKI, J., and SEGEV, I. (1991a). "Propagation of Action Potentials along Complex Axonal Trees: Model and Implementation," *Biophys. J.* (in press).

MANOR, Y., KOCH, C., and SEGEV, I. (1991b). "The Effect of Geometrical Irregularities on Propagation Delay in Axonal Trees," *Biophys. J.* (in press).

MASCAGNI, M. V. (1989). "Numerical Methods for Neuronal Modeling," C. Koch and I. Segev (eds.), Methods in Neuronal Modeling (pp. 439–484). MIT Press, Cambridge, Massachusetts.

MOORE, J. W., STOCKBRIDGE, N., and WESTERFIELD, M. (1983). "On the Site of Impulse Generation in a Neuron," *J. Physiol. (London)* **336,** 301–311.

PARNAS, I., and SEGEV, I. (1979). "A Mathematical Model for the Conduction of Action Potentials along Bifurcating Axons," *J. Physiol (London)* **295,** 323–343.

PERKEL, D. H., MULLONEY, B., and BUDELLI, R. W. (1981). "Quantitative Methods for Predicting Neuronal Behavior," *Neuroscience* **6,** 823–837.

PETERS, A. (1987). "Synaptic Specificity in the Cerebral Cortex," in G. M. Edelman, W. E. Gall, and W. M. Cowan (eds.), *Synaptic Function,* (pp. 373–397), John Wiley, New York.

POGGIO, T., and TORRE, V. (1977). "A New Approach to Synaptic Interaction," in H. Heim and G. Palm (eds.), *Lecture Notes in Biomathematics:* Theoretical Approaches to Computer Systems, *Vol. 21* (pp. 88–115). Springer, Berlin, Heidelberg.

RALL, W. (1959). "Branching Dendritic Trees and Motoneuron Membrane Resistivity," *Exptl. Neurol.* **1,** 491–527.

RALL, W. (1964). "Theoretical Significance of Dendritic Trees for Neuronal Input–Output Relations," in R. Reiss (ed.), *Neural Theory and Modeling* (pp. 73–97). Stanford University Press, Stanford, California.

RALL, W. (1967). "Distinguishing Theoretical Synaptic Potentials Computed for Different Soma-Dendritic Distributions of Synaptic Input, *J. Neurophysiol.* **30,**1138–1168.

RALL, W. (1969). "Time Constants and Electrotonic Length of Membrane Cylinders in Neurons," *Biophys. J.* **9,** 1483–1508.

RALL, W. (1970). "Cable Properties of Dendrites and Effects of Synaptic Location," in P. Anderson and J. Jansen (eds.), *Excitatory Synaptic Mechanism* (pp. 175–187). Universitetsforlaget, Oslo, Norway.

RALL, W. (1974). "Dendritic Spines, Synaptic Potency and Neuronal Plasticity," in C. D. Woody, K. A. Brown, and J. D. Knispel (eds.), *Cellular Mechanisms Subserving Changes in Neuronal Activity, Vol. 3* (13–21). Brain Information Service Research Report, Los Angeles.

RALL, W. (1977). "Core Conductor Theory and Cable Properties of Neurons," in E. R. Kandel (ed.), *Handbook of Physiology, Vol. 1, Part. 1: The Nervous System* (pp. 39–97). American Physiology Society, Bethesda, Maryland.

RALL, W. (1989). "Cable Theory for Dendritic Neurons," in C. Koch and I. Segev (eds.), *Methods in Neuronal Modeling* (pp. 9–62). MIT Press, Cambridge, Massachusetts.

RALL, W., and RINZEL, J. (1973). "Branch Input Resistance and Steady Attenuation for Input to One Branch of a Dendritic Neuron Model," *Biophys. J.* **13,** 648–688.

RALL, W., and SEGEV, I. (1987). "Functional Possibilities for Synapses on Dendrites and on Dendritic Spines," in G. M. Edelman, E. E. Gall, and W. M. Cowan (eds.), *Synaptic Function* (pp. 605–636). John Wiley, New York.

RINZEL, J., and RALL, W. (1974). "Transient Response in a Dendritic Neuron Model for Current Injection at One Branch," *Biophys. J.* **14,** 759–790.

ROCKLAND, K. (1989). "Bistratified Distribution of Terminal Arbors of Individual Axons Projecting from Area V1 to Middle Temporal Area (MT) in the Macaque Monkey," *Visual. Neurosci.* **3,** 155–170.

ROSS, W. N., LASSER-ROSS, N., and WERMAN, R. (1990). "Spatial and Temporal Analysis of Calcium-Dependent Electrical Activity in Guinea Pig Purkinje Cell Dendrites," *Proc. R. Soc. Lond.* **240,** 173–185.

SCHUZ, A., and MUNSTER, A. (1985). "Synaptic Density on the Axonal Tree of a Pyramidal Cell in the Cortex of the Mouse," *Neurosci.* **15**(1), 33–39.

SCHWARK, H. D., and JONES, E. G. (1989). "The Distribution of Intrinsic Cortical Axons in Area 3b of Cat Primary Somatosensory Cortex," *Expt. Brain Res.* **78,** 501–513.

SEGEV I., and RALL, W. (1988). "Computational Study of an Excitable Dendritic Spine," *J. Neurophysiol.* **2,** 499–523.

SEGEV, I., FLESHMAN, J. W., and BURKE, E. (1989). "Compartmental Models of Complex Neurons," in C. Koch and I. Segev (eds.), *Methods in Neuronal Modeling* (pp. 63–96) MIT Press, Cambridge, Massachusetts.

SEGEV, I., FLESHMAN, J. W., and BURKE, R. E. (1990). "Computer Simulation of Group Ia EPSPs Using Morphologically Realistic Models of Cat Alpha-Motoneurons," *J. Neurophysiol.* **6,** 648–660.

SEGEV, I., FLESHMAN, J. W., MILLER, J. P., and BUNOW, B. (1985). "Modeling the Electrical Behavior of Anatomically Complex Neurons Using a Network Analysis Program: Passive Membrane," *Biol. Cybern.* **53,** 27–40.

SEGEV, I., WHITE, E. L., and GUTNICK, M. J. (1989). Detailed "Compartmental Model of an EM Reconstructed Spiny Stellate Cell in the Mouse Neocortex," *Soc. Neurosci. Abstr.* **15,** 256.

SERENO, M. I., and ULINSKI, P.S. (1987). "Caudal Topographic Nucleus Isthmi in the Turtle, *Pseudemys scripta*," *J. Comp. Neurol.* **261,** 319–346.

SHELTON, D. P. (1985). "Membrane Resistivity Estimated for the Purkinje Neurons by Means of a Passive Computer Model," *Neuroscience* **14,** 111–131.

SHEPHERD, G. M. (1990). *The Synaptic Organization of the Brain.* Oxford University Press, Oxford.

SHEPHERD, G. M., and BRAYTON, R. K. (1987). "Logic Operations Are Properties of

Computer-Simulated Interactions between Excitable Dendritic Spines," *Neuroscience* **12,** 151–166.

STOCKBRIDGE, N. (1989). "Theoretical Response of a Bifurcating Axon with a Locally Altered Axial Resistivity," *J. Theor. Biol.* **137,** 339–354.

STRATFORD, A. U., MASON, A., LARKMAN, A. U., MAJOR, G., and JACK, J. J. B. (1989). "The Modelling of Pyramidal Neurons in the Visual Cortex," in R. Durbin, C. Miall, and G. Mitchison, (eds.), *The Computing Neuron* (Chapter 16). Addison-Wesley, Wokingham, England.

TANK, D. W., SUGIMORI, M., CONNOR, J. A., and LLINÁS, R. R. (1988). "Spatially Resolved Calcium Dynamics of Mammalian Purkinje Cells in Cerebellar Slice," *Science,* **242,** 773–777.

TORRE, V., and POGGIO, T. (1978). "A Synaptic Mechanism Possibly Underlying Directional Selectivity to Motion," *Proc. R. Soc. Lond. B.* **202,** 409–416.

TRAUB, R. (1977). "Motoneurons of Different Geometry and the Size Principle," *Biol. Cybernetics* **25,** 163–176.

TURNER, D. A. (1984). "Segmental Cable Evaluation of Somatic Transients in Hippocampal Neurons (CA1, CA3 and Dentate)," *Biophys. J.* **46,** 73–84.

WESTENBROEK, R. E., AHLIJANIAN, M. K., and CATTERALL, W. A. (1990). "Clustering of L-Type Ca^{2+} Channels at the Base of Major Dendrites in Hippocampal Pyramidal Neurons," *Nature* **347,** 281–284.

Chapter 8

Functions of Very Distal Dendrites: Experimental and Computational Studies of Layer I Synapses on Neocortical Pyramidal Cells

LARRY J. CAULLER and BARRY W. CONNORS

Section of Neurobiology
Division of Biology and Medicine
Brown University, Providence, Rhode Island

I. The Significance of Cortical Layer I

Layer I of the neocortex presents an interesting problem in neural design. It is a dense neuropil about 0.1–0.15 mm thick in rats, consisting largely of axons and synapses from many sources, and dendrites arising from neuron somata in other layers (Marin-Padilla, 1984; Vogt, 1991). Layer I itself is almost devoid of cell bodies. The full thickness of neocortex varies from about 1.5 to 3 mm, and the somata of its principal neurons, the pyramidal cells, are scattered through-out layers II through VI. Synapses within layer I are a major source of input to many pyramidal cells, and these contacts are made with the most distal branches of the pyramidal cells' apical dendrites. Pyramidal cells with somata in layer V present the most extreme case. Their apical dendrites often stretch into layer I, and thus the cells receive synaptic contacts that are more than 1 mm from their cell bodies. The importance of these layer I inputs can be inferred from the structure of the apical dendrite; many ascend as a thick trunk through layers IV and III with relatively few oblique branches, and only upon reaching layers II and I do they branch profusely to form a distal apical tuft (Fig. 1).

The sources of afferents to layer I are quite diverse. Perhaps the most widely acknowledged input to layer I originates in the midline and intralaminar nuclei of the thalamus (Herkenham, 1986). Emphasis on the thalamic projections may be traced to the well-known recruitment experiments of Dempsey and Morrison (1942), in which the neocortex was cumulatively excited by repetitive stimulation of the midline thalamus. Such stimulation synaptically activates the most superficial layers of cortex (Foster, 1980) via widespread projections to layer I (Friedman *et al.*, 1987). The slow and diffuse nature of recruitment activation promotes the popular view that layer I inputs mediate a modulatory influence that biases the cortical response to the more direct inputs that terminate in middle layers. This view has been reinforced by observations that projections from cholinergic (Bear *et al.*, 1985; DeLima and Singer, 1986) and monoaminergic (Emson and Lindvall, 1979) modulatory systems also end densely in the superficial layers of the cortex.

It has now been well-established that the neocortex itself is another major source of input to layers I and II. (See reviews by Pandya and Yeterian, 1985; Zeki and Shipp, 1988.) We will use Pandya's term and call these cortico-cortical connections *backward* projections, because they transmit information from higher-order cortical areas back to earlier stages in the sensory hierarchies. For example, in monkeys, individual fibers of the backward projection from visual areas V2 and V3 course through subcortical white matter to the primary visual area V1. There they ascend directly to layer I and travel horizontally within it as far as 4 mm, forming numerous terminal arbors in layers I and II (Rockland and Virga, 1989). We have shown that backward projections to the rat somatosensory system also involve long horizontal fibers in layer I (Cauller and Connors, 1990). The potential significance of these higher-order cortico-cortical connections has prompted models of their role in memory and sensory processes (Edelman, 1987; Rolls, 1989). These models are supported by evidence that backward cortical projections to layers I and II mediate behaviorally relevant cortical potentials and signal conscious touch sensation in primary somatosensory areas of monkeys (Cauller and Kulics, 1991).

Interest in backward cortico-cortical influences has raised questions about the nature of distal synaptic inputs in layer I. The few studies that have addressed this question *in vivo* recorded the response of single units in the primary visual area V1 during general anesthesia. Cooling secondary visual areas has mixed modulatory effects upon the V1 response to visual stimuli (Sandell and Schiller, 1982), and stimulation of V2 or V3 evokes moderate responses in those V1 units that projected to the stimulated sites (Bullier, *et al.*, 1988). Most recently,

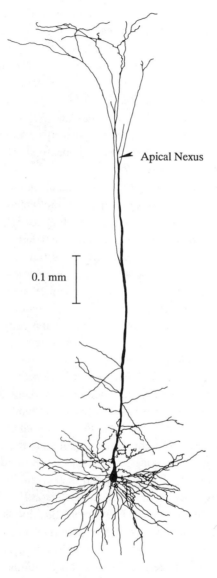

FIGURE 1.   Camera-lucida drawing of biocytin-filled neocortical pyramidal neuron. The cell body was in layer Vb and the most superficial dendrites at the top of the drawing, the distal apical tuft, were in layer I. Intracellular recordings shown in Figs. 3 and 4 were obtained from this cell. The arrow points to the *apical nexus*, defined as the most distal section of the apical trunk where it meets the apical tuft. Measurements described in the text and Fig. 10 were obtained from this point on the simulated cell.

Mignard and Malpeli (1991) reported that the backward projection from V2 to V1 is sufficient to generate complex visual responses in layers II, III, and V in the absence of thalamocortical responses in layers IV and VI. However, experimental conditions may have confounded these studies. The behaviorally relevant component of the somatosensory-evoked cortical potential generated by excitation of layers I and II in awake monkeys is selectively abolished during unconscious states of slow-wave sleep (Cauller and Kulics, 1988) and general anesthesia (Arezzo et al., 1981). Therefore, experiments in anesthetized animals may better represent a state in which the influence of backward projections to layers I and II is relatively suppressed.

The influence of dendritic structure and function on synaptic inputs is a classic problem in neuroscience. Dendrites of the spinal motoneuron (Fleshman et al., 1988; Jack and Redman, 1971) and cerebellar Purkinje cell (Llinás and Sugimori, 1981) have been well-studied; however, the obviously large differences between neurons make it dangerous to generalize. Dendritic functions in cerebral cortex are not well-characterized. The best-studied apical dendrites are those of pyramidal cells in hippocampus. In the CAI region, synapses are distributed relatively uniformly across apical dendrites for distances of about 0.5–0.7 mm. Some measurements imply that these dendrites are electrotonically compact, and synaptic inputs to the distal segments are about as effective as proximal inputs (Andersen et al., 1980). However, analysis of the problem is complicated by the strong likelihood that CA1 dendrites do not have passive membranes; intradendritic recordings reveal strong nonlinearities in their electrical properties (Benardo et al., 1982; Wong et al., 1979). The large pyramidal cells of neocortex present a problem somewhat different from the pyramidal cells of hippocampus. Many layer V cells have two major dendritic domains that are spatially segregated: one close to the soma consisting of the basal dendrites and proximal (oblique) branches of the apical dendrite, and the second distant from the soma comprised of the many apical tufts within layers I and II (Fig. 1).

Thus, layer I inputs to neocortex have interesting biophysical, anatomical, and behavioral relevance, but have been surprisingly neglected by physiologists. Our goal here is to introduce an approach to the study of layer I synapses onto pyramidal cells, using both experimental and computational methods. The general questions we address are: What are the physiological characteristics of layer I synapses onto pyramidal cells, and how does the long dendritic distance between a layer V cell body and a layer I synapse influence the function of those synapses?

II.   The Synaptic Response to Activation of Horizontal Layer I Afferents

This chapter is essentially a case study, illustrating an approach that combines quantitative morphology, physiology, and computational analysis to understand

the functions of a complex synaptic–neuronal interaction in the cortex. The data
to be presented here, unless noted otherwise, were collected from one neuron,
the layer Vb pyramidal cell reconstructed in Fig. 1. We have used a variation
of the *in vitro* slice of rat somatosensory neocortex (Connors *et al.*, 1982) to
examine the effectiveness of layer I inputs to pyramidal cells whose bodies lie
0.5–1 mm deeper, in layers III or V (Cauller and Connors, 1989, 1990). The
horizontal fibers in layer I (HLI) were isolated by disconnecting all deeper
horizontal fibers with a cut perpendicular to the surface, extending from just
below layer I downward through subcortical white matter (Fig. 2; also Cauller
and Connors, 1989). Layer I was stimulated on one side of the cut and the
response mediated by HLI fibers passing to the other side was recorded extracellularly and intracellularly. Impaled cells were filled with dyes (usually
biocytin) to relate dendritic structure to synaptic response and to permit quantitative morphological reconstruction for computational modeling. In addition to

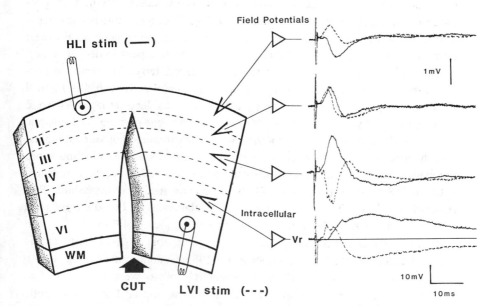

FIGURE 2.   Schematic drawing of cortical slice preparation and representative extracellular and intracellular recordings from different cortical layers. The horizontal path in
layer I (HLI) was isolated by the cut extending from white matter to just below layer I.
HLI was stimulated at least 0.3 mm from the cut and the response was recorded from
the other side. Extracellular field potentials in layers I, II, and III evoked by stimulation
of HLI (solid traces) are superimposed upon responses to stimulation of layer VI (broken
traces). The bottom traces are responses to each input recorded intracellularly from a
layer V pyramidal neuron.

providing greater access to cortical circuitry and stability for extended intra-cellular impalements, the cortical slice preparation avoids anesthesia and permits control over the neurochemical milieu.

The HLI pathway can be reasonably well-isolated in cortical slices. Although many local axon collaterals of underlying pyramidal cells may ascend to layer I, local collaterals do not extend horizontally within layer I as far as extrinsic backward inputs. Horseradish peroxidase (HRP) injected into layer I more than 0.3 mm on one side of cut slices was not transported to cells on the other side (Cauller and Connors, 1990). Therefore, HLI was stimulated at least 0.3 mm from the cut to avoid antidromic activation of the pyramidal cells on the other side. The usual method of activating the cortical slice, by stimulation of vertical fibers ascending through layer VI, evokes a complex multiphasic distribution of synaptic and antidromic activity throughout the overlying cortex. In contrast, current source-density analysis indicates that the HLI-evoked response is acti-vated by synaptic excitation restricted to layers I and II with no evidence of synchronous activation in deeper layers (Cauller and Connors, 1990). This spec-ificity is illustrated by the field potentials in Fig. 2. Large, short-latency nega-tivities evoked by HLI stimulation are restricted to the most superficial aspect of the slice, while layer VI-evoked negativities come to a peak in deeper layers.

HLI-evoked responses were also recorded intracellularly. All identified neu-rons that responded to HLI stimulation were pyramidal cells with distal apical tufts extending into layer I. Some pyramidal cells as deep as layer Vb were strongly excited by HLI inputs; our illustrative cell was one of these. The synaptic response to HLI inputs differs from that evoked by stimulation of vertical inputs. The characteristic response to strong vertical activation, via layer VI stimulation, involves a short EPSP that is abruptly truncated by a $GABA_A$ receptor-mediated chloride IPSP and a long-lasting $GABA_B$ receptor-mediated potassium IPSP (Connors et al., 1982, 1988). In contrast, stimulation of HLI inputs activates a long EPSP that lasts at least 50 ms with little or no sign of IPSPs (Figs. 3 and 4). The pyramidal cell discussed here was a type capable of generating intrinsic bursts of action potentials (McCormick et al., 1985; Connors and Gutnick, 1990); when the layer I-evoked EPSP was large enough, bursts were generated in an all-or-none fashion (Fig. 3, top trace). Both vertical and HLI-evoked synaptic responses were completely blocked by DNQX (10 μM), an antagonist of non-NMDA-type glutamate receptors.

The first indication that HLI synapses might be electrotonically distant from the cell body was that manipulation of the membrane potential by current injection at the somatic penetration site had little effect upon the amplitude of the HLI-evoked EPSP (Cauller and Connors, 1990). Remarkably, the HLI-evoked EPSP amplitude was nearly constant across a wide range of membrane potentials (Fig.

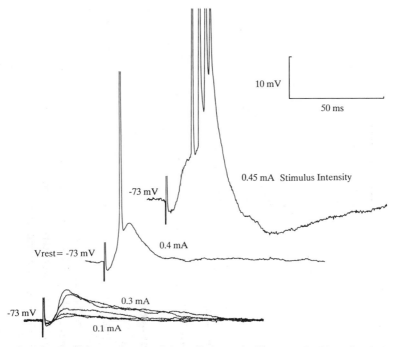

FIGURE 3.   Intracellular responses of the cell shown in Fig. 1 evoked by stimulation of the isolated HLI. Stimulus intensity was increased from 0.1 to 0.45 mA (0.2 ms duration) in steps of 0.05 mA. The strongest intensity evoked a burst of action potentials (peaks clipped). All other traces are the average of five responses. Each response was evoked with the cell at resting potential (-73 mV).

4c). A complicating factor is that the neuron exhibited strong rectification in the range from resting potential to spike threshold (Fig. 4b). Such rectification is a common characteristic of neocortical pyramidal cells (Connors *et al.*, 1982; Stafstrom *et al.*, 1985) and may be due to noninactivating, voltage-sensitive conductances near the cell body. The HLI-evoked EPSP that reaches the cell body from the distal apical dendrites may be distorted by the actions of the same voltage-sensitive conductances responsible for somatic rectification. There is also some evidence for fast voltage-sensitive conductances capable of generating spike-like responses along the apical trunk of neocortical pyramidal neurons (Amitai *et al.*, 1990; Huguenard *et al.*, 1989; Deschênes, 1981; Pockberger, 1991; Purpura, 1967). Preliminary results using whole-cell patch clamping in our laboratory (Cauller and Connors, unpublished) indicate that distal HLI inputs may activate regenerative inward currents in nonvoltage-clamped dendritic regions. For these reasons, it seems likely that the HLI-evoked EPSP that reaches the cell body has been significantly distorted by active dendritic conductances.

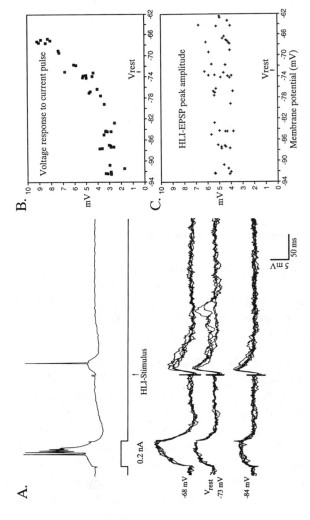

FIGURE 4. Intracellular responses to equal HLI stimuli (0.25 mA) and equal intracellular current pulses (0.2 nA) while manipulating the membrane potential by constant current injection. Membrane rectification is indicated by the change in amplitude of the voltage response to the constant current pulse. The mean peak amplitude of the HLI-EPSP remains constant across the same range of membrane potentials that show rectification.

In many cells, the initial HLI-evoked EPSP was followed by a late, voltage-sensitive depolarization that increased the duration of excitation for up to 200 ms (Fig. 4a). This late depolarization was selectively abolished by the NMDA-type glutamate receptor antagonist APV (50 μM) without affecting the initial HLI-evoked EPSP. The late depolarization resembles the APV-sensitive poly-synaptic activity described in infant (Luhmann and Prince, 1990) and mature cortex (Sutor and Hablitz, 1989). However, it is probably not mediated by polysynaptic activity because it was abolished by injecting enough somatic cur-rent to hyperpolarize the membrane potential by 10–20 mV from rest; no residual, underlying synaptic activity was then observed. This late depolarization appears to be generated by an active dendritic conductance that is triggered by the initial HLI-evoked EPSP and modulated by NMDA receptors. The finding that this late activity was more sensitive to somatic potential manipulations than the initial HLI-evoked EPSP indicates the active dendritic conductance may be electroton-ically closer to the cell body than the HLI synapses. Indeed, it seems likely that the propagation of HLI-evoked EPSPs down the long apical trunk to the cell bodies of layer V pyramids is especially vulnerable to active modulation from numerous sources.

III.   Computational Model of a Layer V Cell: Determination of Parameters

We undertook a series of simulations to evaluate the electrotonic properties of cortical cells. Intracellular staining with biocytin (Horikawa and Armstrong, 1988) permits quantitative morphometric reconstruction following physiological characterization of the neurons' response to synaptic inputs. The isolated HLI preparation offers a unique advantage because the location of layer I synaptic inputs is restricted to a specific portion of the dendritic tree, namely, the distal apical tufts. The central aim of our simulations was to predict the characteristics of inputs to the distal apical tufts assuming passive membrane properties. These characteristics could then be compared to the actual data.

We have most thoroughly simulated the passive properties of the layer Vb pyramidal cell shown in Fig. 1, using the program NEURON (Hines, 1989). In NEURON, the complex, branched structures of cells are represented by a series of linked compartments, each of which is modeled as a set of simple linear electrical elements. (For review of the theory and strategies of the compartmental modeling of neurons, see Segev *et al.*, 1985, 1989.) The somadendritic structure of our layer V neuron was drawn and measured using camera lucida and a 100 × oil immersion lens. Measurements were used to build a representation of the cell in NEURON that was composed of 360 separate compartments. Simulations were run on a Sun Sparcstation under Unix with XWindows, and it took about 50 ms of run-time per simulated time step.

The first problem encountered in modeling neurons is the plethora of unknown biophysical parameters; neuronal structure can be directly determined with some confidence, but electrical properties must usually be inferred from measurements made at a single site, namely, the soma. Methods have been well-described for estimating parameters (e.g., Nitzan *et al.*, 1990; Rall, 1959; Segev *et al.*, 1989), but a common and frustrating conclusion is that there is no unique set of parameters that fits the limited data (Stratford *et al.*, 1989). Nevertheless, by exploring a reasonable range of parameters, it is often possible to set narrow limits on likely neuronal behaviors, and that is what we have attempted here.

The first step toward characterizing the passive computational properties of the reconstructed neuron was to establish the parameter values that yield simulated somatic input impedances ($_{sim}R_{in}$) and time constants ($_{sim}\tau_m$) equal to the physiologically observed values ($_{obs}R_{in}$ and $_{obs}\tau_m$). The observed input impedance for this layer V cell measured by dV/dI just negative to rest was 14.8 MΩ. This relatively low input impedance is characteristic of layer V pyramids (Connors *et al.*, 1982), which are among the largest in rat cortex. The resting membrane potential remained stable at -73 mV for more than an hour. The $_{obs}\tau_m$ for the cell was 7.6 ms, measured as -1/slope along the linear portion of the log-linear charging curve (Fig. 5).

A fundamental unknown in mammalian neurons is the intracellular axial resistivity (R_a), and this parameter was a primary independent variable in our simulations. It is very likely that the mammalian value of R_a is greater than the 30 $\Omega \cdot$ cm value measured in squid giant axon (Cole, 1972), and it may be greater than the commonly used value of 70–100 $\Omega \cdot$ cm (Shelton, 1985), which is based upon rough estimates in spinal motoneurons (Barrett and Crill, 1974). One approach to finding a suitable value has been to adjust R_a until the early, fast transients of the simulated charging curve fit the early portion of the observed curve. The early transients result from equalization of the potential between the multiple dendritic compartments of complex neurons and are enhanced, therefore, by increased intracellular resistance. Such an approach has yielded R_a estimates exceeding 300 $\Omega \cdot$ cm in neocortical pyramidal cells (Stratford *et al.*, 1989). However, although we also observe early charging transients (Fig. 5), they typically last no more than 1–2 ms and are subject to significant errors arising from electrode capacitance and adjustment of the capacitance compensation of the electrometer bridge circuit. (See Wilson and Park, 1989.) Thus, we chose to study the effect of R_a by explicitly varying its value across a wide, but perhaps physiologically reasonable, range of 50–400 $\Omega \cdot$ cm.

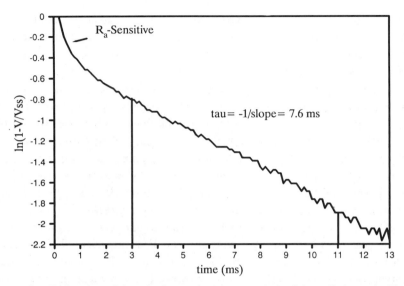

FIGURE 5.   Estimation of the membrane time constant (τ_m) of the cell shown in Fig. 1. τ_m (7.6 ms) was measured from the linear portion of the natural log-linear charging curve, between 3 and 11 ms. Ten intracellular responses to a weak hyperpolarizing current pulse (-0.2 nA) recorded from resting potential were averaged, normalized with respect to the steady-state potential generated during an extended current pulse, and plotted as the natural log of the difference as the normalized potential approached steady state. The early transient (<2 ms) corresponds to the portion of the simulated charging curve that is augmented by increasing R_a. The same procedure was employed to measure the time constant of the simulated charging curve. (See Fig. 10.)

For each R_a value, R_m was adjusted to yield the $_{obs}R_{in}$. In agreement with simulations of vagal motor neurons (Nitzan *et al.*, 1990), the required R_m fell by less than 20–50% across the eight-fold range of R_a values, and for R_a values above 300 $\Omega \cdot$ cm, the required R_m was nearly constant. However, across the entire range of R_a–R_m values that yielded the $_{obs}R_{in}$, the $_{sim}\tau_m$ (assuming a specific membrane capacitance C_m of 1 μF/cm^2, as reviewed by Cole, 1972) was much lower than the $_{obs}\tau_m$. It was necessary to increase the simulated C_m to 4 μF/cm^2, far above generally accepted physiological values, to yield the $_{obs}\tau_m$. However, leaving C_m at 1 μF/cm^2, the $_{sim}\tau_m$ could also be increased to the observed value by increasing the membrane area of the simulated dendritic sections, which, in turn, required greater R_m values to yield the $_{obs}R_{in}$.

Increasing the membrane area of the reconstructed neurons is justified for two reasons. First, it is likely that the overall structural dimensions, dendritic diameters and lengths, were underestimated due either to tissue shrinkage or optical distortions. Thus, we have explored the effects of assuming that the cell had shrunk uniformly by 10% during processing. Secondly, the dendrites of pyramidal neurons are covered with spiny evaginations that greatly increase their membrane surface area. Compensation for either spines or shrinkage effectively increased $_{sim}\tau_m$, and both were simulated across reasonable ranges.

The dendritic spines on pyramidal neurons greatly increase the cells' membrane area, and provide a large surface for synaptic contacts. At the same time, the thin necks of the spines are believed to isolate the synapses both electrotonically (Rall, 1978) and biochemically (Harris and Stevens, 1989; Holmes and Levy 1990; Zador et al., 1990). The relatively long, thin necks of the spines (1–5 μm long, 0.1–0.5 μm diameters) and the low conductance represented by the very small area of the spine head result in very high simulated input impedances (approaching 1 GΩ) encountered by current injected directly into the head of the spine. Accordingly, small synaptic currents would generate a much greater local depolarization at a spine head than if applied directly to shafts of thick dendrites, which have much lower input impedances. Due to the very small membrane area of individual spines (approximately 1 μm^2) and the correspondingly small capacitance they represent, spines charge very easily; dendritic potentials reach the spine heads with relatively little attenuation despite the high neck impedance. Therefore, spines may be simulated without explicitly attaching spine elements, by directly increasing the effective area of dendritic segments as if the surface were convoluted without changing the length or diameter of the dendrite (Holmes, 1986). Directly inserting area in this way effectively increased the time constant of our simulated neuron to the observed values.

To evaluate the suitability of simply adding surface area to simulate spines, we have compared this direct insertion method with simulations of a dendrite explicitly covered with discrete spines of varying densities and dimensions. In all cases, for a range of R_a values (50–200 Ω · cm), R_m and the amount of area inserted per unit length of dendrite were adjusted to yield a fixed input impedance and time constant measured at the middle of the dendrite. In the case of explicit spines, the total area inserted was equal to the area of each spine times the spine density. Direct area insertion was accomplished by increasing the specific membrane conductance ($1/R_m$) and capacitance (C_m) in proportion to the increase in area. Borrowing an example from Holmes and Levy (1990), to insert 1,000 μm^2 into a dendritic segment 1,000 μm long (i.e., 1 μm^2/um) and $1/\pi$ μm in diameter (i.e., cylinder area = 1000 μm^2), then C_m for that segment was doubled and R_m was halved.

Table I presents the results of our explicit spine simulations for comparison with the direct area insertion technique. In very good agreement with Holmes (1986), the amount of direct area insertion required to yield a given R_{in} and time constant was nearly equal to the area added by explicit spine insertions for all

TABLE I.   Comparison of the Direct Area Insertion Technique with Explicit
Spines in a Simulated Dendrite.

	Area Insertion ($\mu m^2/\mu m$)					Voltage Attenuation (% Reduction)				
		small spine		large spine			small spine		large spine	
R_a ($\Omega \cdot cm$)	direct	thick neck	thin neck	thick neck	thin neck	direct	thick neck	thin neck	thick neck	thin neck
50	1.02	1.07	1.07	1.05	1.05	89.7	89.6	89.6	89.6	89.6
100	1.34	1.37	1.37	1.37	1.38	79.7	79.6	79.6	79.7	79.7
150	1.69	1.73	1.74	1.74	1.74	70.3	70.1	70.1	70.1	70.1
200	2.16	2.14	2.14	2.16	2.16	61.0	61.1	61.1	61.4	6.13

A dendrite 400 μm long and 1 μm in diameter was simulated. For each R_a, the R_m and the amount
of inserted membrane area (direct-insertion or explicit spines) per μm length was adjusted to yield
R_{in} = 300 MΩ and τ_m = 5 ms measured at the center of the dendrite. Small spines (head diameter
= head length = 0.4 μm; area = 0.97 μm^2) and large spines (head diameter = head length =
0.8 μm; area = 2.48 μm^2) were modeled with either long, thin spine necks (length = 1.5 μm;
diameter = 0.1 μm) or short, thick spine necks (thick neck diameter = head diameter; length =
0.15/diameter), so as to keep the total spine areas constant. Area insertions are in units of μm^2
added area per μm length of dendrite. Voltage attenuations are the percent of the potential generated
at the (central) site of current injection that reaches the dendrite 200 μm away.

cases. Interestingly, neither the area per spine nor the spine neck diameter
significantly affected the amount of area required to meet the charging conditions.
Furthermore, attenuation of steady-state potential along the simulated dendrite
was the same regardless of whether spines were explicitly modeled or simulated
with the direct area insertion technique.

Table II presents the R_m's and direct area insertions required to make our
simulated neuron yield the $_{obs}R_{in}$ and $_{obs}\tau_m$ as a function of R_a across a range of
shrinkage up to 10%. Figure 6 plots the range of direct area insertion densities
required as a function of R_a. Since lower R_m values were required at greater R_a
values, more area was required to control $_{sim}\tau_m$ by increasing the membrane
capacitance. However, more area yielded greater conductance, which required
further R_m adjustments such that the R_a–R_m relation became non-monotonic
complex with area manipulations (Table II).

The effect of shrinkage-compensation on $_{sim}\tau_m$ was the same as direct area
insertion, and reduced the amount of required area insertion almost equally across
the R_a range. In fact, large shrinkage factors exceeded the requirement for direct
area insertion especially at low R_a values (Table II). In our simulations, area
was inserted uniformly along all dendritic compartments (excluding somatic
sections) regardless of dendrite diameter. This means that the area added was
proportionally greater in thinner dendrites than thick. This assumption of

TABLE II.   Effects of Area Insertion and Somatic Leak on Estimation of Membrane Parameters for Layer V Pyramidal Cell.

| | shrink = 0% | | | | shrink = 10% | | | |
| | no leak | | G_{sl} = 10 nS | | no leak | | G_{sl} = 10 nS | |
R_a ($\Omega \cdot$ cm)	R_m ($\Omega \cdot$ cm²)	area/l (μm²/μm)	R_m ($\Omega \cdot$ cm²)	area/l (μm²/μm)	R_m ($\Omega \cdot$ cm²)	area/l (μm²/μm)	R_m ($\Omega \cdot$ cm²)	area/l (μm²/μm)
50	7935	0.97	9155	0.78	—	—	—	—
75	8053	1.35	9283	1.11	8021	.27	9247	0.07
100	8097	1.65	9339	1.39	8097	0.54	9339	0.32
125	8120	1.93	9369	1.63	8136	0.78	9384	0.53
150	8128	2.18	9357	1.84	8156	1.00	9380	0.71
175	8129	2.43	9348	2.04	8167	1.21	9383	0.88
200	8140	2.68	9326	2.22	8160	1.40	9375	1.04
225	8145	2.92	9304	2.41	8165	1.60	9359	1.19
250	8151	3.18	9274	2.59	8167	1.79	9343	1.34
275	8160	3.43	9265	2.78	8169	1.99	9322	1.49
300	8163	3.69	9257	2.98	8166	2.18	9317	1.64

Dimensions of Layer V Pyramidal cell without Added Area Insertions.

	dendritic length (μm)	membrane area (μm²)	area/length (μm²/μm)
total	13101	50591	3.86
basal	6688	23056	3.45
trunk	732	8523	11.64
oblique	2928	10035	3.43
tuft	2736	8366	3.05

Direct area insertion (added area in μm² per μm length of dendrite) and R_m required by the modeled pyramidal cell to yield $_{sim}R_{in}$ = 14.8 MΩ and $_{sim}\tau_m$ = 7.6 ms. For either 0% or 10% shrink compensation, either 0 or 10 nS somatic leak conductance (G_{sl}) was assumed. Blank cells at R_a = 50 Ω · cm reflect the finding that 10% shrink compensation exceeded the requirement for area insertion. The lower table shows the observed dendritic area densities (without spines or additional insertions) for comparison with the area insertions (i.e., 1.25 μm²/μm = 32% of the total dendritic area density).

homogeneous spine area is obviously an oversimplification. We made some limited counts of actual spine densities on the layer V cell. Because an observer's view is partially occluded by the dendritic trunk itself, simple counts of observed spine density are necessarily underestimates; our values were corrected as described by Feldman and Peters (1979). Spines were nearly absent on the proximal 50–75 μm of the apical and basal trunks and relatively low (1 spine/μm dendritic

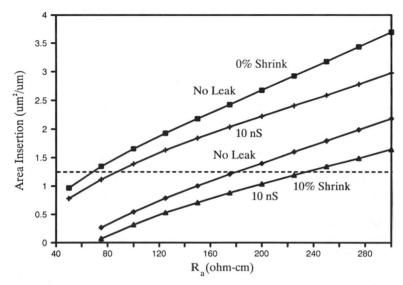

FIGURE 6.   Direct area insertion (μm^2 of membrane area per μm length of dendrite) across a range of axial resistivity (R_a) required by the simulation to match the observed input impedance and time constant. The top two curves assumed 0% shrinkage and the bottom two assumed 10% general shrinkage. The top curve of each pair assumed no somatic leak conductance and the bottom of each pair assumed 10 nS leak.

length) on the thinnest terminal dendrites. The highest spine density was on the middle sections of the trunks (1.6 spines/μm). The average, 1.3 spines/μm, falls within the range of spine densities observed in Golgi preparations of cortical pyramidal cells (Feldman, 1984). As a standard parameter, in the following simulations we directly inserted 1.25 μm^2 of membrane area per μm of dendritic length (i.e., average spine density × average spine area = 1.3 × 0.96), which corresponds to increasing the total dendritic area by approximately 32% (Table II).

For the sake of simplicity and lack of evidence, computational models of neurons generally assume that R_m and R_a are homogeneously distributed across all parts of the cell (but see Holmes and Woody, 1989). A notable exception is the widely recognized possibility that the cell body compartment is more conductive than the rest of the cell, due either to intrinsic membrane properties or to a leak conductance resulting from penetration of the cell with sharp microelectrodes. The latter possibility is supported by findings that input resistances measured by the *gigaseal* patch electrode technique are usually an order of magnitude greater than R_{in}'s measured with sharp microelectrodes (Pongracz *et al.*, 1991). We included a somatic leak conductance (G_{sl}) in our simulations. As shown in Table II and Fig. 6, adding a leak conductance increased the R_a

TABLE III. Best-Fit Membrane parameters for Layer V Pyramidal Cell.

	R_m ($\Omega \cdot$ cm)	R_m ($\Omega \cdot$ cm^2)	G_{sl} (nS)	λ_i		R_a ($\Omega \cdot$ cm)	R_m ($\Omega \cdot$ cm^2)	G_{sl} (nS)	λ_i
		0% shrink					10% shrink		
A→	69	8014	0.00	5.39 ←A		—	—	—	—
	75	8495	4.00	5.32		—	—	—	—
	100	10155	15.12	5.04		—	—	—	—
	125	11528	21.97	4.80		—	—	—	—
	150	12668	26.58	4.59		—	—	—	—
	175	13638	29.94	4.41	C→	181	8159	0.00	3.36 ←C
	200	14464	32.50	4.25		200	8617	4.12	3.28
	225	15108	34.40	4.10		225	9165	8.57	3.19
	250	15633	35.91	3.95		250	9643	12.11	3.11
	275	16079	37.16	3.82		275	10073	15.06	3.03
B→	300	16435	38.17	3.70 ←B	D→	300	10473	17.55	2.95 ←D
	325	16724	39.03	3.59		325	10798	19.59	2.88
	350	16945	39.74	3.48		350	11131	21.48	2.82
	375	17145	40.39	3.38		375	11399	23.04	2.76
	400	17277	40.93	3.29		400	11660	24.45	2.70

R_a, R_m, and G_{sl} values were adjusted, assuming either 10% or no shrinkage, to make the input impedance and time constant of the simulated cell equal those of the observed measures. In all cases, 1.25 μm^2/μm was directly inserted to compensate for spine area. The length constant index (λ_i) corresponding to each set of R_a–R_m values is shown.[3] The four sets of parameters used in the text to demonstrate the range of effects are indicated by A through D.

and R_m required to yield $_{sim}R_{in}$ = $_{obs}R_{in}$, thereby reducing the area insertion required to fit the $_{obs}\tau_m$.

To obtain a physiologically reasonable spectrum of simulation parameters, we fixed the amount of direct area insertion at 1.25 μm^2/μm according to the observed spine density, and adjusted the simulated R_m and G_{sl} across a range of R_a values to fit the $_{obs}R_{in}$ and τ_m. The results, assuming 0% or 10% shrinkage, are shown in Table III. The minimum values at G_{sl} = 0 correspond to the points in Fig. 6 that intersect the observed level of spine area density (dashed line). As R_a was increased, an increase in G_{sl} was necessary to make $_{sim}R_{in}$ = $_{obs}R_{in}$, which, in turn, required an increase in R_m to yield $_{sim}\tau_m$ = $_{obs}\tau_m$, requiring further adjustments of G_{sl}, and so on. Since at a given G_{sl}, the $_{sim}\tau_m$ increased monotonically with R_m, the parameter sets presented in Table III are unique (i.e., one pair of R_m and G_{sl} values for each R_a at given values of shrink and area insertion).

In the following simulations, we employ the four sets of parameters highlighted in Table III. These sets were chosen because they represent the extremes of what we consider the likely range for the reconstructed layer V cell. All include area insertions to model spine effects. Two sets of parameters include minimal R_a

values obtained while assuming either 0% or 10% shrink compensation, with no somatic leak conductance. The other two sets of parameters also compensate for either 0% or 10% shrinkage, while employing the high value of $R_a = 300$ $\Omega \cdot$ cm that is associated with relatively large R_m and G_{sl} values. Despite the wide range of parameter values, as shown in the following, all four parameter sets place strong constraints on possible mechanisms of the layer I-evoked synaptic response.

IV.   Steady-State and Transient Responses of the Modeled Pyramidal Neuron

With all the basic parameters of the model in hand, it is possible to evaluate its characteristics and behavior in a variety of useful ways. Here, we will describe the responses of the model cell to both steady-state and transient inputs. One of the most striking features of a large pyramidal cell is its complex dendritic geometry. To the eye, it seems immediately obvious that the distal parts of the apical dendrite will behave differently compared to the more proximal basal dendrites. To examine this explicitly, we first simulated the effects of applying focal, prolonged current steps to various sites on the cell. Figure 7 shows results comparing inputs to the distal branches of the apical dendritic tuft (A), the more proximal oblique branches of the apical dendrite (0), and the basal dendrites (B) and the soma (S).

Current was injected into each terminal dendritic segment, halfway between the last branch point and the dendritic tip, or into the soma. First, it is worth pointing out a generally useful principle of symmetry concerning the spread of steady-state potential from a site of current injection:

The steady-state potential generated at a site x *by a unit current injection at a site* y *{i.e.,* $V_{ss}(y \rightarrow x)$*} is equal to the steady-state potential generated at site* y *by the same amount of current injected at site* x *{i.e.,* $V_{ss}(x \rightarrow y) = V_{ss}(y \rightarrow x)$*}.*

This is illustrated in Fig. 7a, which shows that the average steady-state potential in the soma generated by current injected into the dendritic tips (shaded bars) is equal to the average steady-state potential generated in the tips by equal somatic current injection (solid bars).

Comparisons between the different classes of dendrites is very revealing. For all four parameter sets, the apical tuft dendrites were much more isolated from the soma than either the basal or apical oblique dendrites. This is reflected in the steady-state data, in two ways. First, the spread of potential from the soma to the apical tuft, or from the apical tuft to the soma, is much more attenuated

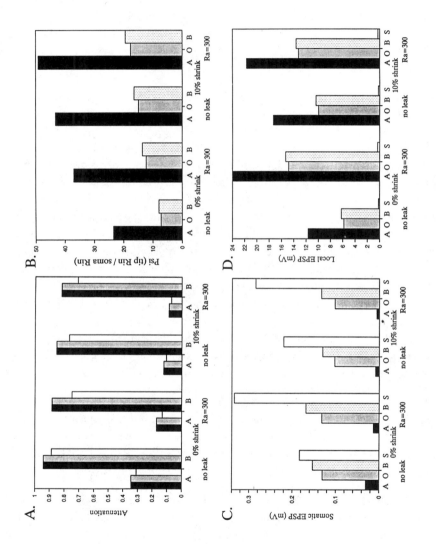

than the spread between the soma and the other basal dendrites (Fig. 7a). Second, the average input impedance measured at the apical dendrites is much greater than at the basal or oblique dendrites (Fig. 7b). This latter result has several ramifications. Applying the symmetry principle stated above, the relative input impedance equals the *asymmetry index*, Ψ, developed by Nitzan *et al.*, (1990) to describe the isolation of distal dendrites on vagal motor neurons.[1] This index

[1]Ψ is defined as the ratio of attenuation factors (Nitzan *et al.*, 1990):

$$\Psi = \frac{AF(d \to s)}{AF(s \to d)},$$

where $AF(d \to s)$ equals the steady-state potential at the soma generated by current injection at the dendrite relative to the local dendritic potential generated by that current:

$$AF(d \to s) = \frac{V_{ss}(d \to d)}{V_{ss}(d \to s)}, \quad \text{and} \quad AF(s \to d) = \frac{V_{ss}(s \to s)}{V_{ss}(s \to d)}.$$

Therefore, since $V_{ss}(d \to d)$ equals the input impedance at the dendrite times the injection current, and $V_{ss}(s \to s)$ equals the somatic input impedance times the injection current, Ψ equals the ratio of dendritic to somatic input impedances because, according to the preceding symmetry principle, $V_{ss}(d \to s)$ equals $V_{ss}(s \to d)$.

FIGURE 7. Averaged measures showing the isolation of the distal apical (A), basal (B), and oblique (O) dendrites relative to the soma (S) across the range of four parameter sets of 0% and 10% shrinkage, with and without somatic leak compensation (parameter sets A, B, C, and D in Table III). Measurements or manipulations were made at the middle of the terminal sections (25–250 μm) on each part of the tree (A, $n = 17$; B, $n = 60$; O, $n = 24$). (a) The steady-state membrane potential at the distal apical and basal dendrites during current injection into the soma (solid bar) is equal to the potential at the soma during the same current injected into the distal dendrites (shaded bar). Attenuation factors are the ratio of these steady-state potentials relative to the somatic potential generated by the same current injected directly into the soma. The open bars compare the relative efficacy of the unitary synaptic charge reaching the soma from the distal dendrites. Synaptic efficacy is defined as the area under the EPSP at the soma from a synapse on the terminal dendrites relative to area under EPSP of an equivalent synapse directly on the soma.[2] (b) The steady-state input impedance of the distal dendrites relative to the somatic input impedance. This value is equivalent to the asymmetry index, Ψ.[1] (c) The peak amplitude of the unitary EPSP (1 nS peak conductance, alpha function time constant = 0.5 ms) measured in the soma from synapses applied to the soma or to the distal dendrites. In cases where somatic leak conductances were assumed (and thus, $R_a = 300$), the leak was sealed to make the somatic EPSP measurements. While the somatic EPSP was significantly less than that shown when the leak was included, the EPSPs from the distal dendrites, especially the apical tuft, were only slightly less. (d) The peak amplitude of the local unitary EPSP, at the site of the synapse. The somatic EPSP amplitudes in (c) equal those in (d).

equals the ratio of steady-state voltage attenuation from the site of current injection in the dendrite to the soma, relative to the attenuation from the soma to the dendrite. Like vagal motor neurons, the asymmetry index of the basal and oblique dendrites ranged from 7 to 15. In contrast, the apical asymmetry index ranged from 25 to 50. An important implication of the relatively high input impedance of the apical tuft dendrites is that a given amount of current will depolarize the apical tuft dendrites much more than it will the proximal dendrites.

The uniqueness of the apical tufts was even more apparent when transient potentials were simulated. Synaptic potentials were generated by alpha function conductance changes (Jack *et al.*, 1975) modeled upon recent measures of unitary EPSPs in hippocampus (1 nS peak conductance, 0.5 ms time constant, 0 mV reversal potential as seen in Bekkers and Stevens, 1989). Thus,

$$g_{syn}(t) = \frac{g_{max} \cdot t}{\tau \cdot e^{-(t-\tau)/\tau}},$$

where $g_{syn}(t)$ is the synaptic conductance as a function of time, and the peak value, g_{max}, is reached at $t = \tau$. Unitary synaptic conductances for layer I synapses have never been measured, but synapse ultrastructure is unremarkable and they are similar in size to synapses elsewhere in the cortex (Vaughan and Peters, 1973). Synapses were applied directly to the soma, or to the same points

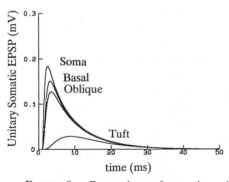
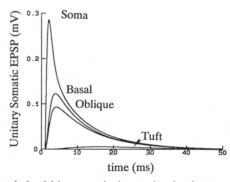

FIGURE 8. Comparison of somatic and distal dendritic synaptic inputs in simulated neuron. Simulated unitary synaptic conductances were applied directly to the soma or to the distal dendritic compartments, and EPSPs were measured in the soma. Two sets of parameters were employed to show the effect of compensation for shrinkage and somatic leak conductance, including the implication that $R_a = 300 \ \Omega \cdot cm$. In the case where shrinkage and leak were compensated, the somatic leak was sealed to measure the EPSPs in the soma. While the somatic EPSP was significantly less than that shown when the leak was included, the EPSPs from the distal dendrites, especially the apical tuft, were only slightly smaller than that shown.

at the middle of the terminal dendritic sections where the input impedance was measured, with no spine assumed. Figures 7c and 8 show that the unitary EPSPs reaching the soma from the apical tuft dendrites were much more attenuated than EPSPs originating on basal or oblique dendrites. The peak synaptic potential was more attenuated than the steady-state potential; while the steady-state potential from the apical tuft reached 10–35% of the direct somatic potential, the peak synaptic potential reaching the soma from the apical tuft was just 5–15% of the peak somatic EPSP. In addition to the peak amplitude, another commonly used index of synaptic efficacy is the relative time integral of the somatic potential generated by distant synapses. This index represents the total charge that reaches the soma from a dendritic synapse relative to the charge injected by an equal synapse located directly on the soma.[2] As shown in Fig. 7a (open bars), the synaptic-integral index very nearly equaled the steady-state attenuation.

Dendritic synapses were most effective, and the steady-state potential least attenuated, with the parameter set that assumed no shrinkage and no somatic leak conductance. The cell was most compact under these conditions because the R_a was lowest relative to R_m, which yielded the greatest length constant index, and the actual physical length was not stretched to compensate for shrinkage. However, these conditions should be considered an unlikely extreme. All attempts to approximate more realistic conditions by compensating for shrinkage or somatic leak conductance increased attenuation and synaptic distance to the apical tuft. The effect of two parameters sets on averaged, unitary inputs to the apical tuft, basal, and oblique dendrites are compared to the direct somatic EPSP in Fig. 8. Our findings are in general agreement with the findings of Stratford et al. (1989, cf. their Fig. 16.5; also Holmes and Woody, 1989). However, Stratford et al. simulated and reconstructed a very different type of pyramidal cell; it had a prominent apical trunk, but no distal apical tuft. They found that the tip of the apical trunk was much more synaptically distant than the terminal sections of the basal and oblique dendrites.

Interestingly, steady-state attenuation and synaptic efficacy were well-correlated with λ_i, the length constant index, $\sqrt{(R_m/4R_a)}$ (Table III).[3] The finding

---

[2]The time integral is defined as:

$$\int V(t)\ dt\ =\ \int R_{in} \cdot I(t)\ dt\ =\ \int R_{in} \cdot [dQ(t)/dt]\ dt\ =\ R_{in} \cdot Q_{total}.$$

In calculating the ratio of integrals, R_{in} is a constant, so the ratio reflects the relative total charge, Q.

[3]Since the length constant (λ) of a cylindrical cable is defined as:

$$\lambda\ =\ \sqrt{(dR_m/4R_a)},$$

the length constant index (λ_i), $\sqrt{(R_m/4R_a)}$, equals the length constant of an equivalent cylinder relative to the root of the diameter.

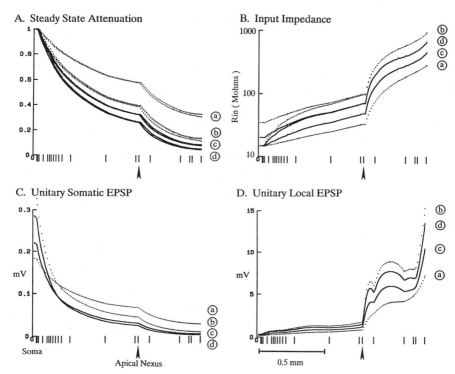

A. Steady State Attenuation

B. Input Impedance

C. Unitary Somatic EPSP

D. Unitary Local EPSP

FIGURE 9.   Input impedance and synaptic efficacy as a function of dendritic location. Measures from the simulated cell were obtained at 10 μm intervals from the soma to the tip of a terminal tuft dendrite (dendrite with arrowhead in Fig. 1). The location of each branch point along the dendrite is indicated by hash marks, with the soma at the origin and the apical nexus indicated by the arrow. Each measure was obtained under the four sets of parameters indicated in Table III: a and b, 0% shrink (broken traces); c and d, 10% shrink (solid traces); a and c, no somatic leak assumed; b and d, somatic leak assumed such that $R_a = 300 \ \Omega \cdot$ cm. (a) The steady-state potential at each point generated by somatic current injection was normalized with respect to the somatic potential generated by direct somatic current injection as in Fig. 7a. The somatic potential generated by peripheral current injection is superimposed to show it is exactly equal to the peripheral potential generated by equal somatic current (symmetry principle). The trace slightly below each steady-state attenuation trace is the relative efficacy (see legend for Fig. 7a) for a unitary synapse applied at each point along the dendrite. (b) The input impedance for steady-state current injection at each point along the dendrite (log-linear scale). Curves connecting no leak and $R_a = 300 \ \Omega \cdot$ cm conditions are input impedances with the somatic leak included showing the effect of the leak does not extend beyond the nexus. (c) The peak amplitude of the unitary EPSP measured in the soma. (d) The peak amplitude of the unitary EPSP measured at the site of the synapse.

that the increased R_a and R_m values required by shrinkage and leak compensation yielded decreased length constant indices indicates that R_a increased more sharply than R_m. Accordingly, axial resistivity was apparently the dominant factor in increasing the isolation of the distal dendrites when shrinkage and somatic leak were compensated.

An important consequence of the high input impedance at the apical tuft dendrites was the large amplitude of local synaptic events measured there (Fig. 7d). At the apical tuft synapses, the average peak depolarization generated by a unitary event was 12–24 mV (60–75 times the corresponding somatic event) compared to average peaks of 6–16 mV generated by unitary synapses on the terminal basal or oblique dendrites. Figure 9 tracks the increase of input impedance and local synaptic potential at 10 μm intervals, from the soma to the tip of an apical tuft dendrite. At the point where the apical trunk connects with the tuft, the impedance and local synaptic potential suddenly increase because of an abrupt decrease in dendritic diameter; thus, the steady-state potential and synaptic integral suddenly decrease. To emphasize the importance of this boundary, we have chosen the name *apical nexus* to identify the connection point between the apical trunk and the terminal apical tuft (Figs. 1 and 9; arrowheads). Figure 9 also demonstrates how the long apical trunk isolates the apical tuft from the rest of the cell (*cf.* functions to either side of the apical nexus).

The experimental observation that the HLI-evoked EPSP was relatively insensitive to somatic manipulations of the membrane potential was cited in Section II as evidence for the remote location of the layer I synapses. Likewise, simulated changes of the membrane potential by somatic current injection had little effect upon the apical tuft synaptic potential because the steady-state potential was greatly attenuated from soma to tuft. To mimic our experimental HLI activation, 100 unitary synapses (1 nS peak conductances) were scattered among the compartments of the apical tufts (<1 synapse per 10 μm) and activated simultaneously. Figure 10 illustrates that even the most electrotonically compact parameters isolated the apical tufts beyond the effective reach of somatic manipulations. However, although a very small EPSP reached the soma, the synaptic potential at the apical nexus was very large and summated almost linearly with the potential generated by somatic current injection.

V.   Efficacy and Mechanisms of Synaptic Inputs to Layer I

In our simulations of the passive layer V pyramidal cell, the distal apical tuft was so electronically distant that even massive synaptic conductances there had relatively small somatic effects. When the apical tufts were subjected to

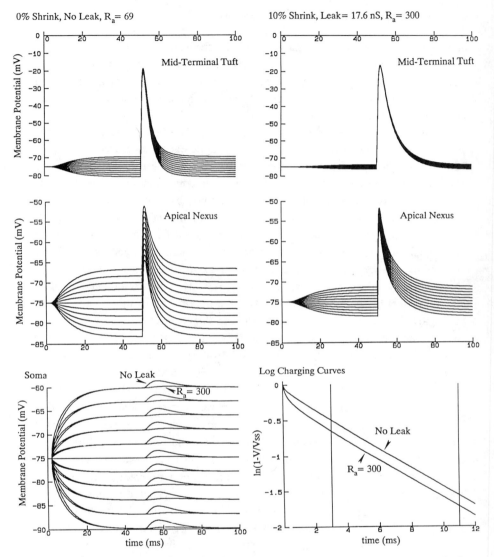

FIGURE 10. Simulation of EPSPs generated by layer I synapses. Responses were evoked by 100 unitary synapses distributed across the distal apical tuft dendrites during somatic current injection (-1 nA to $+1$ nA in 0.2 nA increments). The amplitude of the somatic EPSP is nearly constant across the range of somatic membrane potentials generated by somatic current injection because the steady-state potential from the soma is greatly attenuated at the site of the synapses in the distal apical tuft. The amplitude of the somatic EPSP is greatly attenuated under the more realistic conditions, assuming 10% shrinkage and a somatic leak conductance such that $R_a = 300 \ \Omega \cdot$ cm. The simulated charging curve is plotted the same as the observed curve in Fig. 5 to show that the relatively high R_a implied by the somatic leak conductance yields an early transient like the observed curve.

simultaneous activation of 10 distributed synapses, each with 1 μA peak conductance (i.e., 1000 × unitary), using the most compact cell parameters assuming no leak and no shrinkage, a peak EPSP of just 10.5 mV was generated in the soma. Experimentally, the half-widths of HLI-evoked EPSPs *in vitro*, measured as the duration of the EPSP at one-half peak amplitude, increased with stimulus intensity to greater than 50 ms at just subthreshold. In contrast, the half-widths of simulated somatic EPSPs generated by tuft synapses decreased with increasing strength of activation, such that the half-width of the maximally activated EPSP was less than 14 ms. It appears, therefore, that assumptions of purely passive membrane in these pyramidal cells cannot account for the observed effectiveness of the layer I inputs.

However, the passive model does provide significant insights into the mechanism of the HLI-evoked response. Assuming the most compact conditions (i.e., no shrinkage, no leak), activation of 100 unitary, 1 nS synaptic conductances widely distributed across the apical tuft dendrites depolarized the terminal sections greater than 50 mV and depolarized the apical nexus more than 17 mV, while the somatic EPSP was just 1.6 mV (Fig. 10). On the other hand, although leak and shrinkage compensation decreased the somatic EPSP to less than 0.3 mV (0.35 mV when the leak was sealed), these more realistic conditions increased the local depolarization generated by 100 unitary synapses to more than 55 mV and the nexus potential to more than 20 mV. To achieve somatic EPSPs greater than 1 mV under these more realistic conditions, it was necessary to increase the peak conductances of the 100 synapses to 10 nS, which depolarized the nexus by 55 mV while depolarizing the terminal tuft dendrites almost to the EPSP reversal potential.

It seems very likely, given the evidence for voltage-sensitive conductances in the apical trunk (Section II), that the very large dendritic potentials that would be generated by even a few unitary synaptic inputs to the distal apical tuft could trigger dendritic spikes. Thus, layer I synaptic events would be actively amplified and perhaps regeneratively conducted to the soma. This possibility implies that the remote isolation of the apical tuft may actually improve the effectiveness of layer I inputs. Rather than thinking of layer I inputs as a source of weak modulatory influences, it may be more appropriate to treat the distal apical tuft as an independent integrative unit that actively communicates with the cell body and more proximal dendrites via the long apical trunk.

The apical nexus, at the base of the apical tuft, represents an important summing node for the distal tree. Due to the thickness of the apical trunk (mean diameter ≈ 3.7 μm), the somatic potential reaches the base of the tuft with moderate attenuation, while further spread into the thin tuft dendrites is severely attenuated (Fig. 9a). The apical nexus sums large synaptic potentials generated in the tuft with the somatic potential. The tuft and somatic activity are otherwise nearly independent. The presence of voltage-sensitive calcium channels at the base of the tuft would be ideally suited to trigger an active response. The relatively greater volume of the trunk would also tend to keep intracellular calcium concentrations low, thereby avoiding auto-inactivation or calcium-activated potassium conductance before threshold is reached.

These conclusions regarding the effect of isolation for the operation of the distal apical tuft are analogous to the isolation of synapses on the heads of dendritic spines. Several investigators have recently concluded that the thin spine neck does not significantly affect synaptic efficacy (Harris and Stevens, 1989; Holmes and Levy, 1990; Zador et al., 1990). Similarly, it is now widely believed that the spread of potential from the parent dendrite to the spine head is only slightly attenuated by the spine neck. However, synaptic depolarization of the spine head would be amplified by its high input impedance. The slow, voltage-sensitive NMDA channels in the spine head would be especially affected by repetitive synaptic events. This reasoning would predict that voltage-sensitive channels in the spine head would be more readily activated by synapses on longer, thinner spines with high input impedances than by synapses on short, fat spines (Miller et al., 1985). Accordingly, to the extent that NMDA channels mediate long-term potentiation (Nicholl et al., 1989), long, thin spines might be more plastic. Furthermore, since the high input impedance of the distal tuft dendrites would augment the spine input impedance, such distal spine synapses should be most readily potentiated. However, we have found (Cauller and Connors, unpublished experiments) that the HLI-evoked response in vitro cannot be potentiated by repetitive stimulation. This may indicate that the distal inputs were already potentiated to an asymptotic maximal level of efficacy in the slice.

Further explorations of distal inputs to pyramidal cells should focus on the presence and effects of voltage-sensitive and calcium-activated conductances in the apical dendrites. In addition, it seems likely that cholinergic, GABAergic or perhaps ambient NMDA receptor-mediated inputs strongly modulate the effectiveness of distal tuft synapses in layer I, by changing the electrotonic properties of the apical trunk.

VI. Summary

Recent work has suggested that *backward* cortico-cortical projections, which end largely on distal apical dendrites in layer I, are important for higher cortical functions. By isolating horizontal afferents to layer I in an *in vitro* neocortical slice, we have found that layer I synapses can strongly excite pyramidal cells as deep as layer Vb. To examine the computational properties of these distal inputs, we recorded the layer I response of a layer V pyramidal cell, stained it with biocytin, morphometrically reconstructed it, and simulated its passive electrotonic structure. A physiologically reasonable range of cellular parameters was determined, such that the input impedance and time constant of the simulated cell matched the intracellularly measured values. Assuming the dendrites were passive, it was not possible to generate a simulated somatic EPSP in response to synaptic inputs to the distal apical dendrites that matched the observed intracellular response. However, the passive simulation did reveal that, due to the very high input impedance of the distal apical dendrites, the effect of just a few unitary synapses would be more likely to activate voltage-sensitive apical dendritic conductances than would synapses on other dendrites closer to the cell body. We conclude, therefore, that layer I inputs to the distal apical dendrites of layer V pyramidal cells are probably amplified by active conductances along the apical dendrite. We suggest that the distal apical tuft in layer I operates as an isolated, nonlinear integrator whose effectiveness may be modulated by synaptic inputs along the apical trunk.

Acknowledgments

We thank Michael Hines for his generous and invaluable assistance in the use of NEURON, and Isabelle Bulthoff for providing some of the morphological data. This study was supported by the Office of Naval Research (N00014-90-J-1701), a postdoctoral fellowship (NS08376) from the NIH to L.J.C., and a Research Career Development Award (NS01271) from the NIH to B.W.C.

References

AMITAI, Y., FRIEDMAN, A., GUTNICK, M. J., and CONNORS, B. W. (1990). *European Neuroscience Meeting,* abstract.

ANDERSEN, P., SILVENIUS, H., SUNBERG, S. H., and SVEEN, O. (1980). "A Comparison of Distal and Proximal Dendritic Synapses on CA1 Pyramids in Guinea Pig Hippocampal Slices *in vitro,*" *Journal of Physiology (London)* **307,** 273–299.

AREZZO, J. C., VAUGHN, H. G., and LEGATT, A. D. (1981). "Topography and Intracranial Sources of Somatosensory Evoked Potentials in the Monkey: II. Cortical Components," *Electroencephalography, Clinical Neurophysiology* **28,** 1–18.

BARRETT, J. N., and CRILL, W. E. (1974). "Specific Membrane Properties of Cat Motoneurones," *Journal of Physiology (London)* **239,** 310–324.

BEAR, M. F., CARNES, K. M., and EBNER, F. E. (1985). "An Investigation of Cholinergic Circuitry in Cat Striate Cortex Using Acetylcholinesterase Histochemistry," *Journal of Comparative Neurology* **234,** 411–430.

BEKKERS, J. M., and STEVENS, C. F. (1989). "NMDA and Non-NMDA Receptors Are Co-Localized at Individual Excitatory Synapses in Cultured Rat Hippocampus," *Nature* **341**, 230–233.

BENARDO, L. S., MASUKAWA, L. M., and PRINCE, D. A. (1982). "Electrophysiology of Isolated Hippocampal Pyramidal Dendrites," *Journal of Neuroscience* **2**, 1614–1622.

BULLIER, J., McCOURT, M. E., and HENRY, G. H. (1988). "Physiological Studies on the Feedback Connection to the Striate Cortex from Cortical Areas 18 and 19 of the Cat," *Experimental Brain Research* **70**, 90–98.

CAULLER, L. J., and CONNORS, B. W. (1989). "Origin and Function of Horizontal Layer I Afferents to Rat SI Neocortex," *Society for Neuroscience Abstracts* **15**, 281.

CAULLER, L. J., and CONNORS, B. W. (1990). "Horizontal Layer I Inputs to Primary Somatosensory (Barrelfield) Neocortex of Rat," *Society for Neuroscience Abstracts* **16**, 242.

CAULLER, L. J., and KULICS, A. T. (1988). "A Comparison of Awake and Sleeping Cortical States by Analysis of the Somatosensory-Evoked Response of Postcentral Area 1 in Rhesus Monkey," *Experimental Brain Research* **72**, 584–592.

CAULLER, L. J., and KULICS, A. T. (1991). "The Neural Basis of the Behaviorally Relevant N1 Component of the Somatosensory-Evoked Potential in Awake Monkeys: Evidence that Backward Cortical Projections Signal Conscious Touch Sensation," *Experimental Brain Research,* **84**, 607–619.

COLE, K. S. (1972). *Membranes, Ions and Impulses.* University of California Press, Berkeley, California.

CONNORS, B. W., and GUTNICK, M. J. (1990). "Intrinsic Firing Patterns of Diverse Neocortical Neurons," *Trends in Neuroscience* **13**, 99–104.

CONNORS, B. W., GUTNICK, M. J., and PRINCE, D. A. (1982). "Electrophysiological Properties of Neocortical Neurons *in vitro,*" *Journal of Neurophysiology* **48**, 1302–1320.

CONNORS, B. W., MALENKA, R. C., and SILVA, L. R. (1988). "Two Inhibitory Postsynaptic Potentials, and GABA_A and GABA_B Receptor-Mediated Responses in Neocortex of Rat and Cat," *Journal of Physiology (London)* **406**, 443–468.

DeLIMA, A. D., and SINGER, W. (1986). "Cholinergic Innervation of the Cat Striate Cortex: A Choline Acetyltransferase Immunocytochemical Analysis," *Journal of Comparative Neurology* **250**, 324–338.

DEMPSEY, E. W., and MORRISON, R. S. (1942). "The Production of Rhythmically Recurrent Cortical Potentials after Localized Thalamic Simulation," *American Journal of Physiology* **135**, 293–300.

DESCHÊNES, M. (1981). "Dendritic Spikes Induced in Fast Pyramidal Tract Neurons by Thalamic Stimulation," *Experimental Brain Research* **43**, 304–308.

EDELMAN, G. M. (1987). *Neural Darwinism: The Theory of Neuronal Group Selection.* Basic Books, New York.

EMSON, P. C., and LINDVALL, P. (1979). "Distribution of Putative Neurotransmitters in the Cortex," *Neuroscience* **4**, 1–30.

FELDMAN, M. L. (1984). "Morphology of the Neocortical Pyramidal Neuron," in A. Peters and E. G. Jones (eds.), *Cerebral Cortex, Vol. 1* (pp. 123–200). Plenum, New York.

FELDMAN, M. L., and PETERS, A. (1979). "A Technique for Estimating Total Spine Numbers on Golgi-Impregnated Dendrites," *Journal of Comparative Neurology* **179**, 761–794.

FLESHMAN, J. W., SEGEV, I., and BURKE, R. E. (1988). "Electrotonic Architecture of Type-Identified α-Motoneurons in the Cat Spinal Cord," *Journal of Neurophysiology* **60**, 60–85.

FOSTER, J. A. (1980). "Intracortical Origin of Recruiting Responses in the Cat Cortex," *Electroencephalography & Clinical Neurophysiology* **48**, 639–653.

FRIEDMAN, D P., BACKEVALIER, J., UNGERLEIDER, L. G., and MISHKIN, M. (1987). "Widespread Thalamic Projection to Layer I of Primate Cortex," *Society for Neuroscience Abstracts* **13**, 251.

HARRIS, K. M., and STEVENS, J. K. (1989). "Dendritic Spines of CA1 Pyramidal Cells in the Rat Hippocampus: Serial Electron Microscopy with Reference to Their Biophysical Characteristics," *Journal of Neuroscience* **9**, 2982–2997.

HERKENHAM, M. (1986). "New Perspectives on the Organization and Evolution of Nonspecific Thalamocortical Projections" in E. G. Jones and A. Peters (eds.), *Cerebral Cortex, Vol. 5* (pp. 403–445). Plenum, New York.

HINES, M. (1989). "A Program for Simulation of Nerve Equations with Branching Geometries," *International Journal of Biomedical Computing* **24**, 55–68.

HOLMES, W. R. (1986). *Cable Theory Modelling of the Effectiveness of Synaptic Inputs to Cortical Pyramidal Cells*. Ph.D. Dissertation, University of California, Los Angeles.

HOLMES, W. R., and LEVY, W. B. (1990). "Insight into Associative Long-Term Potentiation from Computational Models of NMDA Receptor-Mediated Calcium Influx and Intracellular Calcium Concentration Changes," *Journal of Neurophysiology* **63**, 1148–1168.

HOLMES, W. R., and WOODY, C. D. (1989). "Effects of Uniform and Non-Uniform Synaptic Activation Distributions' on the Cable Properties of Modeled Cortical Pyramidal Cells," *Brain Research* **505**, 12–22.

HORIKAWA, K., and ARMSTRONG, W. E. (1988). "A Versatile Means of Intracellular Labelling: Injection of Biocytin and Its Detection with Avidin Conjugates," *Journal of Neuroscience Methods* **25**, 1–11.

HUGUENARD, J. R., HAMILL, O. P., and PRINCE, D. A. (1989). "Sodium Channels in Dendrites of Rat Cortical Pyramidal Neurons," *Proceedings of the National Academy of Sciences, USA* **86**, 2473–2477.

JACK, J. J. B., NOBLE, D., and TSIEN, R. W. (1975). *Electrical Current Flow in Excitable Cells* (2nd ed.). Clarendon Press, Oxford.

JACK, J. J. B., and REDMAN, S. J. (1971). "An Electrical Description of the Motoneurone and Its Application to the Analysis of Synaptic Potentials," *Journal of Physiology (London)* **349**, 205–226.

LLINÁS, R., and SUGIMORI, M. (1981). "Electrophysiological Properties of *in vitro*

Purkinje Cell Dendrites in Mammalian Cerebellar Slices," *Journal of Physiology (London)* **305**, 197–213.

LUHMANN, H., and PRINCE, D. A. (1990). "Transient Expression of Polysynaptic NMDA Receptor-Mediated Activity during Neocortical Development," *Neuroscience Letters* **111**, 109–115.

MARIN-PADILLA, M. (1984). "Neurons of Layer I: A Developmental Analysis," in A. Peters and E. G. Jones (eds.), *Cerebral Cortex, Vol. 1*, (pp. 447–478). Plenum, New York.

McCORMICK, D. A., CONNORS, B. W., LIGHTHALL, J. W., and PRINCE, D. A. (1985). "Comparative Electrophysiology of Pyramidal and Sparsely Spiny Stellate Neurons of the Neocortex," *Journal of Neurophysiology* **54**, 782–806.

MIGNARD, M., and MALPELI, J. G. (1991). "Paths of Information Flow through Visual Cortex," *Science* **251**, 1249–1251.

MILLER, J. P., RALL, W., and RINZEL, J. (1985). "Synaptic Amplification by Active Membrane in Dendritic Spines," *Brain Research* **325**, 325–330.

NICOLL, R. A., KAUER, J. A., and MALENKA, R. C. (1989). "The Current Excitement in Long-Term Potentiation," *Neuron* **1**, 97–103.

NITZAN, R., SEGEV, I., and YAROM, Y. (1990). "Voltage Behavior along the Irregular Dendritic Structure of Morphologically and Physiologically Characterized Vagal Motoneurons in the Guinea Pig," *Journal of Neurophysiology* **63**, 333–346.

PANDYA, D. N., and YETERIAN, E. H. (1985). "Architecture and Connections of Cortical Association Areas," in A. Peters and E. G. Jones (eds.), *Cerebral Cortex, Vol. 4*, (pp. 3–61). Plenum, New York.

POCKBERGER, H. (1991). "Electrophysiological and Morphological Properties of Rat Motor Cortex Neuron *in vivo*," *Brain Research* **539**, 181–190.

PONGRACZ, F., FIRESTEIN, S., and SHEPHERD, G. M. (1991). "Electrotonic Structure of Olfactory Sensory Neurons Analyzed by Intracellular and Whole Cell Patch Clamp Techniques," *Journal of Neurophysiology* **65**, 747–758.

PURPURA, D. P. (1967). "Comparative Physiology of Dendrites," in G. C. Quarton, T. Melnechuk, and F. O. Schmitt (eds.), *The Neurosciences* (pp. 372–393). Rockefeller University Press, New York.

RALL, W. (1959). "Branching Dendritic Trees and Motoneuron Membrane Resistivity," *Experimental Neurology* **2**, 503–532.

RALL, W. (1978). "Dendritic Spines and Synaptic Potency," in R. Porter (ed.), *Studies in Neurophysiology* (pp. 203–209). Cambridge University Press, Cambridge, UK.

ROCKLAND, K. S., and VIRGA, A. (1989). "Terminal Arbors of Individual (feedback) Axons Projecting from V2 to V1 in the Macaque Monkey: A Study Using Immunohistochemistry of Anterogradely Transported *Phaseolus vulgarisleucoagglutinin*," *Journal of Comparative Neurology* **285**, 54–72.

ROLLS, E. T. (1989). "Functions of Neuronal Networks in the Hippocampus and Neocortex in Memory," in J. H. Byrne and W. O. Berry (eds.), *Neural Models of Plasticity* (pp. 240–265). Academic Press, San Diego.

SANDELL, J. H., and SCHILLER, P. H. (1982). "Effect of Cooling Area 18 on Striate Cortex Cells in Squirrel Monkey," *Journal of Neurophysiology* **48**, 38–48.

SEGEV, I., FLESHMAN, J. W., MILLER, J. P., and BUNOW, B. (1985). "Modeling the Electrical Behavior of Anatomically Complex Neurons Using a Network Analysis Program: Passive Membrane," *Biological Cybernetics* **53,** 27–40.

SEGEV, I., FLESHMAN, J. W., and BURKE, B. (1989). "Compartmental Models of Complex Neurons," in C. Koch and I. Segev (eds.), *Methods in Neuronal Modeling* (pp. 63–96). MIT Press, Cambridge, Massachusetts.

SHELTON, D. P. (1985). "Membrane Resistivity Estimated for the Purkinje Neuron by Means of a Passive Computer Model," *Neuroscience* **14,** 111–131.

STAFSTROM, C. E., SCHWINDT, P. C., CHUBB, M. C., and CRILL, W. E. (1985). "Properties of Persistent Sodium Conductance and Calcium Conductance of Layer V Neurons from Cat Sensorimotor Cortex *in vitro*," *Journal of Neurophysiology* **53,** 153–170.

STRATFORD, K., MASON, A., LARKMAN, A., MAJOR, G., and JACK, J. (1989). "The Modelling of Pyramidal Neurones in Visual Cortex," in R. Durbin, C. Miall, and G. Mitchison (eds.), *The Computing Neuron* (pp. 296–321). Addison-Wesley, Wokingham, UK.

SUTOR, B., and HABLITZ, J. J. (1989). "EPSPs in Rat Neocortical Neurons in *in vitro*: II. Involvement of N-methyl-D-aspartate Receptors in the Generation of EPSPs," *Journal of Neurophysiology* **61,** 621–634.

VAUGHAN, D. W., and PETERS, A. (1973). "A Three-Dimensional Study of Layer I of Rat Parietal Cortex," *Journal of Comparative Neurology* **149,** 355–370.

VOGT, B. A. (1991). "The Role of Layer I in Cortical Function," in A. Peters and E. G. Jones (eds.), *Cerebral Cortex, Vol. 9*. Plenum, New York.

WILSON, C. J., and PARK, M. R. (1989). "Capacitance Compensation and Bridge Balance Adjustment in Intracellular Recording from Dendritic Neurons," *Journal of Neuroscience Methods* **27,** 51–75.

WONG, R. K. S., PRINCE, D. A., and BASBAUM, A. (1979). "Intradendritic Recordings from Hippocampal Neurons,] *Proceedings of the National Academy of Sciences, USA* **76,** 986–990.

ZADOR, A., KOCH, C., and BROWN, T. H. (1990). "Biophysical Model of a Hebbian Synapse," *Proceedings of the National Academy of Sciences, USA* **87,** 6718–6722.

ZEKI, S., and SHIPP, S. (1988). "The Functional Logic of Cortical Connections," *Nature* **353,** 311–317.

PART II

ION CHANNELS AND
PATTERNED DISCHARGE,
SYNAPSES, AND NEURONAL
SELECTIVITY

This section considers the neuronal properties underlying patterned discharge in single neurons and the ability of neurons to respond selectively to temporal and spatial patterns of input.

Schwindt (Chapter 9) reviews the relation between specific ion channels and the input/output properties and discharge patterns of the large Betz cells of the motor cortex. Identified Na^+, Ca^{++}, and K^+ currents with different voltage and time dependence are considered as mechanisms for producing the different types of temporal modulation of firing rate of motor cortex neurons seen during motor tasks. Some of these ion channels are sensitive to the voltage history of the neuron, and others are responsive to modulator neurotransmitters. Hence, a given neuron will respond with a different type of discharge pattern for the same synaptic sensory input pattern, depending on the complement of ion channels present, their density, and the presence or absence of modulator transmitters.

McCormick, Huguenard, and Stowbridge (Chapter 10) point out that many of the patterns of electrical activity observed in the brain that had been previously attributed to network properties are now seen to have counterparts in the intrinsic properties of individual neurons. In the case of relay neurons in the thalamus, the neurons exhibit different modes of discharge (burst or tonic) in different behavioral states, and these discharge patterns are due to specific ion channels whose conductances are functions of membrane voltage and the presence of neurotransmitters. They report on new simulations that confirm and extend the contribution of low-threshold calcium currents and a hyperpolarization-activated cation current in the generation of rhythmic bursting.

Antón, Granger, and Lynch (Chapter 11) consider various functions that have been used to model postsynaptic potentials (PSPs). They introduce a computationally efficient closed-form interactive solution for lumped circuits. This interactive difference-of-exponentials function, which can account for nonlinear interactions of synapses due to synaptically induced changes in membrane time constants and ionic driving force, allows them to explore the temporal properties of multiple synaptic inputs, or *input trains*. They use this function in models of olfactory tufted/mitral cells. In addition, they outline a hierarchical approach, using physiologically detailed models at the synaptic level, but using greatly simplified neurons in their network models combining large populations of neurons.

Koch and Poggio (Chapter 12) address the status of an approach they call the biophysics of computation, with a focus on the simplest nonlinearity in neuron computation: multiplication. They compare the computational ability of multiplicative or polynomial neurons with thresholded linear units. Polynomial units are more powerful than perceptron units but less powerful than hidden-layer

233

neural nets. Networks of polynomial units are shown to approximate smooth functions arbitrarily well. The experimental evidence for multiplication-like interactions in neurons and in simple neural systems is reviewed, and candidate synaptic mechanisms are outlined. A mechanism for multiplication among two neuronal populations is presented based on the distribution of threshold values within the neuron populations.

Borg-Graham and Grzywacz (Chapter 13) present a model retinal circuit that accounts for the emergence of neurons that selectively respond to stimulus motion in a particular direction in the retina. They identify three critical elements for directionality of neuron response: an asymmetric input-to-output distribution on the directionally selective neuron dendrite; intracellular resistivity of the dendrite, which allows on-path interactions; and inhibitory shunting of sufficient duration.

Chapter 9   Ionic Currents Governing Input–Output Relations of Betz Cells

PETER C. SCHWINDT

Department of Physiology and Biophysics
University of Washington School of Medicine
Seattle, Washington

I. Introduction

Most messages between neurons are transmitted by a sequence of action potentials constituting a frequency-coded signal. The sequence of action potentials results from the transduction of postsynaptic currents. Traditionally, we have assumed that the postsynaptic currents simply sum to evoke greater or fewer action potentials per unit time, but we now know that the transduction process is much more complicated and interesting. This process depends on several factors, such as the properties of the postsynaptic receptors, the electrotonic structure of the neuron, and the site(s) of spike initiation. A primary factor, however, is the system of ionic conductances in the postsynaptic membrane. This chapter will focus on several ionic conductances and their role in shaping and modifying the repetitive firing of action potentials in large pyramidal neurons (*Betz cells*) from layer V of area 4γ of cat sensorimotor cortex.

Only three ionic currents are required to generate repetitive firing—a fast inward current to cause the spike upstroke, a fast outward current to reset the spike, and another outward current to adequately space the spikes during low-rate firing. In certain non-bursting neurons of nudibranch molluscs, spike spacing is provided solely by a transient potassium current, the *A-current* (Connor and Stevens, 1971). Central mammalian neurons seem to rely predominantly on a slow calcium-dependent potassium current to space the spikes; but Betz cells, like many neurons investigated recently, possess many types of ion channels. Some of these *excess* channels help to fine-tune the basic response. The repetitive response of a Betz cell is quite plastic, however. A stereotyped stimulus can evoke different firing patterns depending on factors such as resting potential or the presence of a neuromodulator. As we will see in this chapter, some of these excess channels are responsible for this plasticity.

There is evidence for differences in both firing properties and membrane conductance systems among cell types (reviewed by Llinás, 1988), so the cell type chosen for study is significant. The pyramidal neurons from layer V of area 4γ provide the output of cat primary sensorimotor cortex (Hassler and Muhs-Clement, 1964). In the cat, there is no distinct anatomical separation between the primary motor and sensory areas of cortex, but large layer V cells in area 4γ are known to send their axons to the brain stem and spinal cord to influence motor activity. Many of these axons descend to lower centers in the pyramidal tract (Biedenbach *et al.*, 1986).

Pyramidal tract neurons fire repetitively when an animal performs specific behavioral tasks, and their firing correlates with various parameters of force or movement (Cheney and Fetz, 1980; Evarts, 1968). The recent studies of Lemon and Mantel (1989) suggest that the firing rate of even a single corticomotoneuronal neuron (i.e., a pyramidal tract neuron that makes a monosynaptic connection with a spinal motoneuron) can significantly modulate EMG activity. Such experiments show that the firing rate of single cortical neurons has physiological significance, and they provide the primary impetus for discovering the rules by which synaptic input is converted to a spike train output in these neurons.

Repetitive firing properties of Betz cells have been examined by injecting depolarizing current through the intracellular recording microelectrode (Fig. 1a; also Koike *et al.*, 1970; Stafstrom *et al.*, 1984a). This is meant to mimic the net excitatory synaptic current that would normally reach the soma from synapses all over the neuron. Figure 1b shows that the firing rate is linearly related to injected current. The linear relation holds for any interspike interval, though the slope of the relation is steeper for early interspike intervals. In this respect, the input–output relations of Betz cells are similar qualitatively to those of spinal motoneurons (Kernell, 1965), which they influence.

The response of a Betz cell is simpler, therefore, than certain other neurons. Hippocampal CA3 pyramidal neurons (Wong and Prince, 1978), cerebellar Purkinje cells (Llinás and Sugimori, 1980), thalamic neurons (Llinás and Jahnsen, 1982), and even certain neurons from rodent neocortex (Connors and Gutnick, 1990) are capable of firing single or repetitive bursts of action potentials to a prolonged stimulus in addition to (or instead of) the rhythmic firing displayed

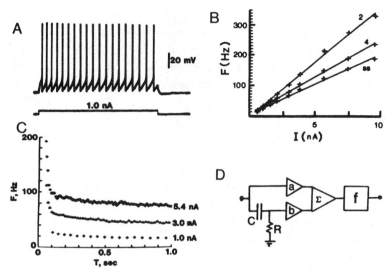

FIGURE 1.   Repetitive firing properties of Betz cells. (a) Repetitive firing (upper) evoked
by 1 nA, 1 s injected current pulse (lower). (b) Plot of instantaneous firing rate (*F*,
reciprocal interspike interval) versus injected current (*I*) for the second (2), fourth (4),
and last (ss) interspike interval during 1 s of evoked firing. (c) Plot of instantaneous firing
rate versus time (*T*) during three injected current pulses of indicated amplitude. (d) Circuit
whose response to voltage input mimics response of a Betz cell to current input. See text
for details.

by the Betz cells. On the other hand, it is still rather mysterious how a *simple*
linear input–output relation results from the combined activation of several nonlinear conductances. After 19 years, there remains only one study in which a
neuron's repetitive firing behavior has been successfully reconstructed from an
empirical model based solely on direct measurements of ionic currents (Connor
and Stevens, 1971). This study showed that a large dynamic range of repetitive
firing requires a specific mechanism for low-rate firing, but even this study
provided no intuitive feeling for why the resulting input–output relation is linear.

An important aspect of the repetitive response of Betz cells is its variation
with time. When a Betz cell is depolarized from resting potential (typically,
−70 mV) by injecting a constant current pulse, firing rate declines from an
initial high rate to a slower tonic rate. This decline (spike frequency adaptation)
usually consists of two phases in Betz cells, an initial decline lasting about 100
ms (fast adaptation) and a subsequent decline (slow adaptation) of longer but
more variable duration (Fig. 1c).

The presence of adaptation suggests that the cell may be sensitive to stimulus
rate of change (which is infinitely fast for a current step) as well as stimulus

amplitude. Such rate sensitivity would be important when the underlying synaptic drive varies with time. Remarkably, only a handful of studies have investigated the repetitive response of neurons to a time-varying stimulus, in spite of the likelihood that time-varying stimuli are common during normal function. The response of Betz cells to a prolonged ramp of injected current has been investigated (Stafstrom *et al.*, 1984a). The response could not be described by a simple sum of terms proportional to the amplitude and the rate of change of the stimulus. The response of a Betz cell to a current step or ramp is, however, qualitatively similar to the response of the simple circuit shown in Fig. 1d to a step or ramp of voltage. In this circuit, the signals from each branch are weighted (by a or b), summed in Σ and converted to a pulse train by f, a linear voltage-to-frequency converter. The RC term in Fig. 1d is much longer than the passive membrane time constant. This term represents the activation of three slow potassium conductances that will be described shortly. The fastest of these three conductances has time constants of about 40 ms, and the slowest has time constants of several seconds. Activation of the slowest conductance requires repetitive firing at a minimum rate for a minimum time. Because of these slow conductances, firing rate may not reach steady state even during long-lasting ramps, and it exhibits adaptation after a long-lasting current ramp, just as occurs after a current step (Stafstrom *et al.*, 1984a).

The ionic mechanisms underlying these and other responses have been investigated using intracellular recording and a single microelectrode voltage clamp (SEVC) of the soma of Betz cells in an *in vitro* brain slice preparation. The brain slice allows control of the extracellular environment. Of course, the ultimate destination of a cell's axon cannot be determined in the slice, but recorded cells have electrical and anatomical features in common with fast-conducting pyramidal tract neurons (Stafstrom *et al.*, 1984b; Takahashi, 1965).

The voltage clamp allows direct visualization of whole-cell ionic currents (as opposed to inferring their nature from membrane potential responses). A detailed biophysical study of the ionic currents is difficult in this preparation, however. The recorded currents may be distorted because the membrane where the channels are located may not be isopotential with the soma. Nevertheless, this technique is useful for identifying the types of ionic conductances that the neuron possesses and certain of their properties, such as gating mode, kinetics, activation potential, etc. It has been possible to infer roles for the different conductances from these properties and, in some cases, to identify a pharmacological agent that reduces only one conductance. The effect of that agent on cell firing can be observed, and at least one specific function of that conductance can be deduced.

II. Persistent Sodium Current

Depolarizing a Betz cell between resting potential and spike threshold evokes two ionic currents that are unusual but very prominent in Betz cells. One of these is a persistent sodium current; the other is a very slow potassium current that appears to be activated by sodium influx.

The persistent sodium current, I_{NaP}, activates rapidly during depolarizations positive to -60 mV (Stafstrom *et al.*, 1982, 1985). Figure 2d shows the inward current evoked by a long-lasting voltage step and its elimination by tetrodotoxin (TTX). Blockade by TTX identifies the current as a sodium current, because

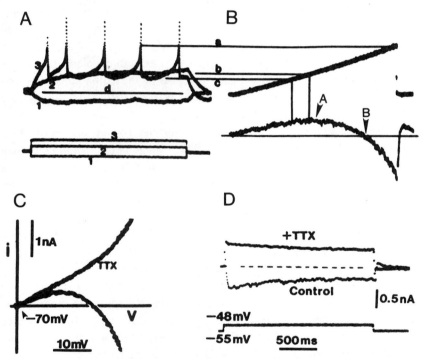

FIGURE 2. Persistent sodium current. (a) Superimposed responses (upper) of a Betz cell to the three like-numbered injected current pulses (lower). Line *d* marks resting potential. Repetitive spikes are truncated. (b) Slow ramp depolarization in voltage clamp (upper) and corresponding ionic current (lower). Lines *a–c* allow correlation of ionic current with corresponding voltages in a and b. (c) Current-voltage (i-V) relation from same cell before and after application of tetrodotoxin (TTX). (d) Currents (upper) evoked by step depolarization (lower) in another cell before and after TTX application. Panels a–c adapted from Fig. 1 in Stafstrom *et al.* (1982).

TTX is a specific blocker of voltage-gated sodium channels. Whether I_{NaP} flows through non-inactivating sodium channels that are distinct from the usual transient sodium channels is still controversial. Nevertheless, I_{NaP} constitutes a functionally distinct current that has been found in several neurons of the mammalian central nervous system (Llinás, 1988).

Some ideas of I_{NaP} function are gained by comparing the voltage range where it is activated to the excursions of membrane potential during responses evoked by injected current. Figure 2a shows two subthreshold responses and one suprathreshold response evoked in a Betz cell. Figure 2b shows a slow ramp voltage command applied to the same cell and the corresponding inwardly rectifying ionic current. Slightly larger depolarization evokes a large transient inward current (resulting from activation of the axon initial segment or transient somatic sodium channels), which cannot be controlled by the SEVC. The difference between the current-voltage (I–V) relations before and after TTX application in Fig. 2c gives the I–V relation for I_{NaP}. I_{NaP} increases with depolarization and is known to be actively present at voltages traversed by the action potential (Stafstrom *et al.*, 1985), but technical limitations of the SEVC have precluded a determination of its full range of activation.

One clear function of I_{NaP} is to set the threshold voltage for regenerative depolarization. Over much of the subthreshold voltage range, net ionic current is outward because I_{NaP} is superimposed on a larger outward current (mainly, a linear *leakage* current). During injection of a steady or slowly varying current, the threshold for regenerative depolarization occurs at the potential where the net I–V relation has zero slope (point A in Fig. 2b). During depolarizations past point B in Fig. 2b, the net current becomes inward. Regenerative depolarization would proceed from this point even if injected current were reduced to zero. Thus, *spike threshold* is set by I_{NaP} in Betz cells rather than by the transient sodium current. Since I_{NaP} is persistent, threshold voltage is not expected to be influenced significantly by the rate of depolarization, and little accommodation (rise in threshold voltage) is seen during slow depolarization of Betz cells (Koike *et al.*, 1968; Stafstrom *et al.*, 1984b). In contrast, neural membrane with no persistent inward current (e.g., an axon) may fail to generate an action potential if membrane potential rises too slowly (Schlue *et al.*, 1974) because the transient sodium channels can inactivate completely during the slow depolarization.

Activation of I_{NaP} causes the inward curvature of the net I–V curve. As a consequence of this curvature, the response of Betz cells to a subthreshold depolarizing current pulse is about twice as large as to a hyperpolarizing pulse of the same size (*cf.* traces 1 and 2 in Fig. 2a); that is, the effective input resistance becomes larger as the cell becomes depolarized, and synaptic current that reaches the soma produces a larger postsynaptic potential than it would if the membrane were at rest. If the persistent sodium channels extended into the dendrites, depolarizing synaptic current would be transmitted along the dendritic tree much more effectively than hyperpolarizing synaptic current, but the spatial distribution of these channels is unknown at present.

I_{NaP} is on at potentials traversed in the interspike interval during repetitive firing (e.g., between lines a–c in Figs. 4a,b). Based on results from spinal motoneurons (Schwindt and Crill, 1982), I_{NaP} may play a fundamental role in shaping the Betz cell's basic input–output relation. Normally, spinal motoneurons also exhibit both a linear input–output relation and a persistent inward current (carried predominantly by calcium ions). In motoneurons lacking this current, firing rate tends to increase sublinearly and saturates as injected current is increased. Apparently, the persistent inward current is needed to insure that firing rate increases with injected current. In its absence, the injected current is insufficient to maintain depolarization in the face of the increased slow potassium currents that accompany faster firing rates. Presumably, I_{NaP} performs a similar function in the Betz cells, but this is difficult to test experimentally, as all pharmacological agents tried so far block both I_{NaP} and the transient sodium current.

III. Sodium-Dependent Potassium Current

I_{NaP} is the dominant ionic current activated between resting potential and spike threshold during depolarizations maintained for several seconds. During long depolarizations, the inward current decreases, and the net current may even become outward. Even if net current does not become outward during the depolarization, a slowly decaying *tail* of outward current appears when membrane potential is repolarized to resting potential (Fig. 3a). This outward current is caused by sodium influx because it is abolished along with I_{NaP} by TTX (Fig. 3b), and neither current is affected by preventing calcium influx. The outward current can result in a long-lasting afterhyperpolarization (AHP), which may be observed when the recording amplifier is switched out of the voltage clamp mode at the end of a voltage step (Fig. 3c). The outward current is present after inhibition of the sodium-potassium pump; it is associated with increased membrane conductance, and its reversal potential varies with extracellular potassium concentration as expected of a potassium current (Schwindt *et al.*, 1989). Apparently, it is a sodium-dependent potassium current analogous to the better-known calcium-dependent potassium current.

Single sodium-dependent (calcium-independent) potassium channels have been identified in other preparations. In isolated membrane patches from cultured

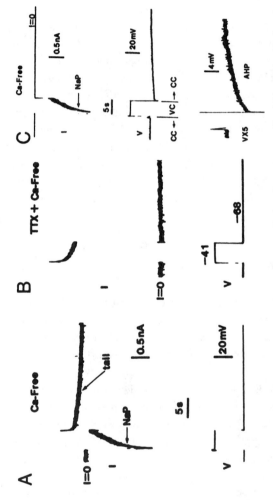

FIGURE 3. Sodium-dependent potassium current. All data from same cell with calcium replaced by cobalt in perfusing solution. (a) Persistent sodium current (NaP, upper trace) evoked by a voltage step (V, lower trace). Slow decay (tail) of outward current follows repolarization of membrane potential. (b) Persistent sodium current and outward tail are both abolished by TTX. (c) Step depolarization (middle trace) caused by switching into voltage clamp (VC) from constant current (CC) recording mode evokes the persistent sodium current (upper). A long-lasting afterhyperpolarization (AHP) follows the depolarization (lower) when voltage clamp is turned off. Membrane potential in mV, indicated by numbers on voltage trace in b, lower, is the same in each panel. Time base in a applies to all panels. Adapted from Figs. 2, 3 in Schwindt et al. (1989).

avian neurons, these channels are activated when intracellular sodium concentration rises above 12 mM (Haimann *et al.*, 1990). Presumably, Betz cells have similar channels that are activated by the rise of intracellular sodium concentration accompanying I_{NaP} activation. The apparently slow kinetics of $I_{K(Na)}$ in Betz cells may simply reflect the time required for intracellular sodium concentration first to build up during activity and then to decay to resting levels.

One would predict that adequate sodium influx during spike activity also would activate these channels, and, indeed, a similar slowly decaying potassium current is seen when calcium influx is prevented and a voltage clamp is imposed following a train of evoked spikes. One manifestation of this current in the absence of voltage clamp is an AHP that may take up to 30 s to decay following 1 s of repetitive firing (Figs. 4a,b; also Schwindt *et al.*, 1988b). This slow, calcium-insensitive AHP is not seen after a single spike; firing must occur at about 20 Hz for 200 ms (or 200 Hz for 20 ms) before it appears. The duration and amplitude of this AHP increase as firing is prolonged or firing rate is made faster. This slow AHP is enhanced when episodes of repetitive firing occur within several seconds of each other (Fig. 4c). During each successive firing episode, the same injected current pulse evokes fewer spikes, and slow adaptation is reduced (Fig. 4d).

Activation of $I_{K(Na)}$ appears to be the primary cause of slow adaptation of firing rate in Betz cells. In most Betz cells, this slow decline of firing rate is still prominent when calcium influx is blocked (Fig. 4d, trace 1; also Schwindt *et al.*, 1989). Under this experimental condition, $I_{K(Na)}$ is the only outward current present with activation kinetics slow enough to gradually reduce excitability during depolarization. More direct evidence of the importance of $I_{K(Na)}$ in slow adaptation is provided by the observations that both $I_{K(Na)}$ and slow adaptation are reduced or abolished by low doses of muscarinic or adrenergic agonists (Figs. 5c,d; also Schwindt *et al.*, 1989; Foehring *et al.*, 1989).

In summary, $I_{K(Na)}$ provides a repolarizing mechanism that has slow onset but a long-lasting effect. Its activation results in a slow decline of firing rate during prolonged depolarization and a long period of subnormal excitability afterward. Its effect on excitability can be modulated by at least two classes of neurotransmitters.

IV. Calcium-Dependent Potassium Currents

Calcium-dependent potassium currents, $I_{K(Ca)}$, have been identified during voltage clamp of Betz cells, but their general features may be deduced from the AHP that follows a 1 s train of spikes. Comparing the AHP before and after

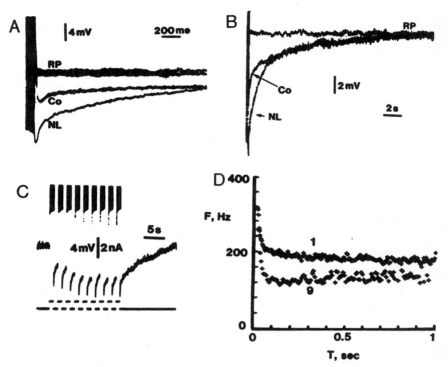

FIGURE 4.   Calcium-insensitive afterhyperpolarizations (AHPs) and spike frequency adaptation. (a,b) Superimposed traces of AHPs following a train of 100 spikes evoked at 100 Hz by individual brief injected current pulses. The same spike train was evoked in normal perfusate (NL) and after substitution of cobalt for calcium (Co). Panel b is same data as in panel a shown at slower sweep speed. The difference between the AHP in each perfusate gives the calcium-sensitive AHP. Traces labeled RP indicate holding potential (-65 mV) in each perfusate. (c) High-gain record of firing episodes (upper) evoked by a train of 1 s injected current pulses repeated each 2 s (lower). Cobalt was substituted for calcium in the perfusate. Downward deflections on voltage record (upper) are slow, calcium-insensitive AHPs that follow each firing episode and summate during repeated episodes. (d) Plot of instantaneous firing rate (F) versus time (T) during the first (1) and ninth (9) episodes of c. Note lower mean firing rate and absence of slow adaptation during the ninth episode. Spikes in panels a–c are truncated. Panels c,d from Fig. 1 in Schwindt *et al.* (1989).

blockade of calcium influx reveals that $I_{K(Ca)}$ contributes only to the first few seconds of the AHP in Betz cells (Figs. 4a,b; also Schwindt *et al.*, 1988b). The late portion of the AHP is insensitive to calcium blockade and corresponds to the slow decay of $I_{K(Na)}$, as described previously. The calcium-sensitive portion of the AHP consists of two parts. Most prominent is an early component of about 150 ms maximum duration (labeled mAHP in Fig. 5a). This component has been called the medium-duration AHP to distinguish it from a fast AHP of a few ms duration that occurs at the foot of every spike. The medium-duration AHP is also present after a single spike, though it grows in amplitude and duration as up to 10 spikes are evoked at 100 Hz (Schwindt *et al.*, 1988c). A slow calcium-dependent component follows the medium-duration AHP (labeled sAHP in Fig. 5a). This component appears only when multiple spikes are evoked at a fast rate, much like the sodium-dependent AHP.

These two calcium-dependent AHPs differ in their sensitivity to pharmacological agents as well as time course. The medium AHP—and the corresponding current, $I_{K(Ca)M}$—is abolished by apamin, a peptide isolated from bee venom (Fig. 5a; also Schwindt *et al.*, 1988b). In studies of single channels, apamin has been shown to specifically block a small conductance (SK) calcium-dependent potassium channel (Blatz and Magleby, 1987). The SK channel is very sensitive to intracellular calcium concentration, but displays little voltage dependence.

The slower calcium-dependent AHP—and the corresponding current, $I_{K(Ca)S}$— is unaffected by apamin (Fig. 5a); instead, it is reduced (together with $I_{K(Na)}$) by muscarinic and adrenergic agonists, which have no effect on $I_{K(Ca)M}$ (Figs. 5c,e; also Foehring *et al.*, 1989; Schwindt *et al.*, 1988b). Though no corresponding single channel has been identified yet, the whole-cell $I_{K(Ca)S}$ of Betz cells also appears to have little or no voltage dependence (unpublished observations). It is likely, therefore, that the time courses of both $I_{K(Ca)M}$ and $I_{K(Ca)S}$ simply reflect the time course over which intracellular calcium concentration remains above its resting level. If this interpretation is correct, $I_{K(Ca)S}$ must be even more sensitive to intracellular calcium concentration than $I_{K(Ca)M}$, as it persists after the apamin-sensitive component has decayed completely.

The apamin-sensitive $I_{K(Ca)M}$ is the dominant calcium-dependent potassium current in Betz cells and provides the basic mechanism for spacing spikes during repetitive firing. Application of apamin results in much faster firing in response to a given injected current pulse (Fig. 5b; also Schwindt *et al.*, 1988c). The slope of the input–output relation is also steepened, but slow adaptation of firing rate is unaltered.

It is harder to test the function of $I_{K(Ca)S}$ because it is blocked by the same neurotransmitter agonists that block $I_{K(Na)}$. Blockade of calcium influx does sometimes reduce slow adaptation of firing rate in cells where $I_{K(Na)}$ is small and $I_{K(Ca)S}$ is large (Schwindt *et al.*, 1988b). On this basis, it appears that the function of $I_{K(Ca)S}$ is similar to $I_{K(Na)}$, but $I_{K(Ca)S}$ normally appears to play a secondary role.

At this point, it is interesting to compare the slow potassium currents of Betz cells to those of hippocampal pyramidal cells. What I have called $I_{K(Ca)S}$ in Betz

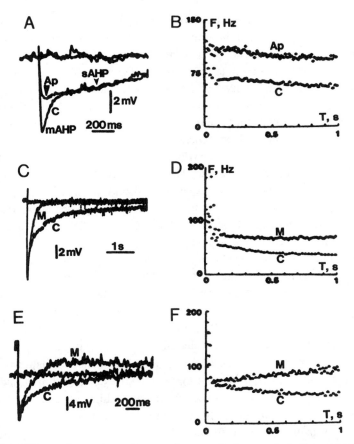

FIGURE 5.   Effects of apamin and muscarine on afterhyperpolarizations (AHPs) and repetitive firing. (a) Superimposed records of AHPs following a train of 50 spikes evoked at 100 Hz by individual brief injected current pulses in control perfusate (C) and after the addition of apamin (Ap). Only the medium-duration AHP (mAHP) is blocked by apamin; the slow AHP (sAHP) is unaffected. Topmost horizontal traces in panels a,c,e indicate holding potentials were identical before and after drug application. (b) Plot of instantaneous firing rate (F) versus time (T) evoked by injection of a 2 nA current pulse before and after drug application. (c) Superimposed AHPs following 1 s of firing evoked at 143 Hz before (C) and after (M) application of 5 μM muscarine. (d) Response of same cell to 1 nA current pulse in each condition. (e) Superimposed AHPs in another cell following 1 s of repetitive firing evoked by a 4 nA current pulse before (C) and after (M) application of 50 μM muscarine. (f) Response of same cell to 1 nA current pulse in each condition.

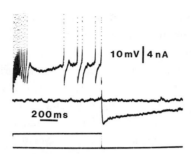

FIGURE 6.   Altered firing pattern caused by low intracellular concentration of BAPTA. Response (upper) of a Betz cell to an injected current pulse (lower) after impalement with a microelectrode containing 2.7 M KCl plus 2 mM dimethyl-BAPTA. Spikes are truncated.

cells is similar to the slow calcium-dependent potassium current called I_{AHP} in the hippocampal neurons (Lancaster and Adams, 1986). In these cells, I_{AHP} underlies the slow AHP, controls spike spacing, and causes adaptation of firing rate (Lancaster and Adams, 1986; Madison and Nicoll, 1984). Both $I_{K(Ca)S}$ and I_{AHP} decay slowly, seem to have little voltage-dependence, are sensitive to adrenergic and muscarinic agonists, and are insensitive to apamin or low doses of tetraethylammonium (TEA) (Lancaster and Adams, 1986; Lancaster and Nicoll, 1987). The role of I_{AHP} in both spike frequency adaptation and long-lasting reduction of excitability in hippocampal neurons seems to be supplanted by $I_{K(Na)}$ in Betz cells, which also employ the separate calcium-dependent potassium current, $I_{K(Ca)M}$, to control basic spike spacing. The reason for these differences in mechanisms is unclear, but is probably related to normal firing patterns and functions.

The firing properties of Betz cells can be made to resemble those of hippocampal neurons by applying low intracellular concentrations of the calcium chelator BAPTA (Schwindt *et al.*, 1990). In this situation (Fig. 6), firing rate declines to zero during the first 200 ms of depolarization by a constant current pulse; membrane potential repolarizes for a longer period and rises subsequently to trigger a few more spikes. A large, long-lasting AHP follows the depolarization. This effect seems to be caused by a large, selective enhancement of $I_{K(Ca)S}$, which dominates the cell's firing properties. A large calcium-dependent AHP, which can last over 10 s, follows even a brief spike train. The corresponding ionic current is identified as $I_{K(Ca)S}$ because it is insensitive to apamin or TEA but is abolished by muscarinic and adrenergic agonists. The mechanism by which low BAPTA concentrations produce this effect in Betz cells is unclear, but theoretical considerations suggest that low concentrations of a fast, mobile calcium buffer can temporarily result in a larger-than-normal calcium concentration beneath the membrane after a large calcium influx (Sala and Hernandez-Cruz, 1990). Whatever the mechanism, the significance of this result is that a small change of intracellular calcium concentration during neural activity can cause a normally minor current to dominate and radically alter firing properties. It would be interesting to know, on the one hand, if this effect can be provoked by an endogenous substance and, on the other hand, if differences in firing properties among neurons depends on intrinsic calcium buffering as well as ion channel expression.

V. Calcium-Dependent Cation Current

As mentioned before, a low dose of muscarine can, because of its reduction of $I_{K(Ca)S}$ and $I_{K(Na)}$, reduce or abolish the slow decline of firing rate that normally occurs during injection of a constant current pulse. When repetitive firing is evoked in the presence of muscarine concentrations greater than 10 μM, firing rate actually accelerates during the current pulse (Fig. 5f; also Schwindt, 1988b). The cause of this acceleration is the spike-related activation of a slow inward current during the firing. The presence of this inward current is signaled by the appearance of a slow afterdepolarization (ADP), which replaces the usual slow AHP following the firing (Fig. 5e). The ionic mechanism of this slow ADP has not been investigated in detail, but it probably is caused by a calcium-dependent cation current. The ADP is unaffected by TTX but is eliminated by blocking calcium influx. It does not appear to be a calcium-dependent chloride current, as it is seen when the microelectrode contains impermeant anions, in which case chloride equilibrium potential is expected to be near -70 mV (Scharfman and Sarvey, 1987).

This current has not been seen in the absence of sufficient concentrations of muscarinic agonists, but it is mentioned here to round out the description of how the temporal variation of firing rate can be controlled by a neuromodulator. In the absence of muscarine, firing rate declines during prolonged depolarization. Low concentrations of muscarine cause firing to become tonic; higher concentrations cause firing to accelerate during the same current pulse. These effects were demonstrated using muscarine, but these cells are known to receive cholinergic innervation, and the acetylcholine acts via muscarinic cholinergic receptors (Krnjevic and Phillis, 1963). Presumably, similar effects can be produced *in vivo* by firing of presynaptic cholinergic fibers.

VI.   Slow Inward Cation Current

This voltage-gated inward current, termed I_h, is carried by both potassium and sodium ions; it is turned on by hyperpolarization and turned off by depolarization (Figs. 7a,b). The conductance turns on near -55 mV and reaches a maximum near -100 mV in Betz cells, and its kinetics are relatively slow and independent of membrane potential (Spain *et al.*, 1987). Both its activation and inactivation consist of a fast phase (time constant ≈ 40 ms) and a slow phase (time constant ≈ 300 ms). It is blocked in a voltage-dependent manner (more block with more hyperpolarization) by extracellular cesium, but it is insensitive to barium, which distinguishes it from a fast inward rectifier current carried exclusively by potassium ions in many cell types.

I_h is active at resting potential (-70 mV), and its activation increases sharply with hyperpolarization (half-activation occurs at about -82 mV). The presence of this inward (depolarizing) current ensures that resting potential will remain well positive of potassium equilibrium potential (about -100 mV in 3 mM extracellular potassium concentration). An empirical model of this current, based on voltage clamp data, suggests that it may play several other roles (Spain *et al.*, 1987). Its activation results in a membrane potential response that resembles the medium-duration AHP, and modeling studies have suggested that it influences the shape and apparent voltage dependence of spike-activated AHPs (Schwindt *et al.*, 1988a).

The modeling studies also suggested that I_h can influence the temporal variation

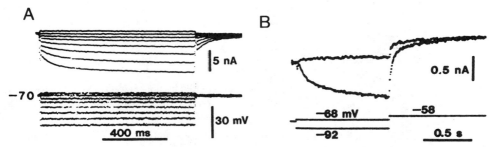

FIGURE 7.   Slow inward cation current. (a) Slow inward cation current (upper) evoked by hyperpolarizing voltage steps (lower). Also note slow decay of the current following the voltage steps. (b) Superimposed traces of currents (upper) evoked by different combinations of voltage steps (lower). Note that current during the step to -58 mV is more inward when preceded by the step (to -92 mV) that evokes a large inward cation current. See text for further explanation.

of firing rate. When a Betz cell is depolarized, I_h slowly turns off; but while it is turning off, it constitutes an *extra* depolarizing current (Fig. 7b). One would then predict that the cell would fire faster during the several hundred ms it takes I_h to decay; that is, I_h turnoff should contribute to the initial adaptation of firing rate. Furthermore, one would predict that if the cell were hyperpolarized (activating more I_h) before it were depolarized, the initial firing rate would be even

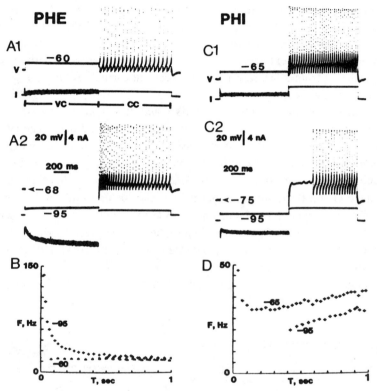

FIGURE 8.   Two classes of Betz cells can be distinguished by their response to prepulses. In panels a and c, prepulse potential is maintained for 1 s by voltage clamp (VC in a1); amplifier is then switched in constant current mode (CC in a1), and a constant depolarizing current pulse is injected for 1 s. Numbers on voltage traces indicate membrane potential in mV. Cell of panels a,b exhibits post-hyperpolarization excitation (PHE): Initial firing rate during the current pulse is faster following the more negative prepulse. Panel b shows instantaneous firing rate versus time following each prepulse. Note absence of spike frequency adaptation following prepulse to −60 mV. Cell of panels c,d exhibits post-hyperpolarization inhibition (PHI): Initial firing rate is slower following the more negative prepulse, and the onset of firing is delayed following the prepulse to −95 mV. Panel d shows instantaneous firing rate versus time following each prepulse in this cell.

faster (because the decay of I_h starts from a larger initial value). A more general expression of this idea is that the temporal variation of firing rate during depolarization will depend on the preceding membrane potential.

As shown in Figs. 8a,b, one class of Betz cells fulfills these predictions (Spain *et al.*, 1991a). When membrane potential is held at -60 mV before a current pulse is injected, no adaptation of firing rate occurs during the pulse (Figs. 8a1,b, trace -60); but if membrane potential is held hyperpolarized before applying the same pulse, the initial firing rate is much higher than the final rate, to which it decays after several hundred ms (Figs. 8a2,b, trace -95). The initial firing rate is, in fact, a graded function of prepulse potential and duration. Negative prepulses (up to a limit at -100 mV) cause a faster initial firing rate, as do longer prepulses to a given potential (up to a limit of about 1 s). This class of Betz cells is said to exhibit posthyperpolarization excitation (PHE). A low-threshold calcium conductance could cause the PHE response, but in Betz cells the response persists when calcium influx is blocked. The PHE response is reduced instead by extracellular cesium, and both I_h and the initial firing rate are reduced to the same degree. Intracellular staining of physiologically identified PHE cells places them among the largest pyramidal cells of layer V. They have anatomical features characteristic of fast-conducting corticofugal neurons, brief action potentials (<0.4 ms duration), and relatively low input resistances (9 MΩ). Hyperpolarizing prepulses have a quite different effect on firing rate in another group of Betz cells that are more influenced by one of the voltage-gated potassium currents described next.

VII. Voltage-Gated Potassium Currents

Betz cells possess at least two voltage-gated potassium currents (Spain *et al.*, 1991b). Both currents may be evoked by adequate depolarization in the presence of TTX (Figs. 9a,b), and both are present when calcium influx is blocked or when a calcium chelator is injected intracellularly. Both currents are transient, but are distinguished by their different time courses of inactivation. During a step depolarization, the fast transient potassium current, $I_{K(V)F}$, inactivates within 20 ms (Fig. 9a), whereas the slow transient current, $I_{k(V)S}$, takes longer than 10 s to decay completely (Figs. 9b,d). Both currents activate rapidly (Figs. 9a,c), suggesting that both can contribute to spike repolarization. This idea can be tested because each current responds differently to pharmacological agents. The fast transient conductance is reduced about half by 1 mM TEA, a dose with little effect on the slow transient current. The slow transient conductance is

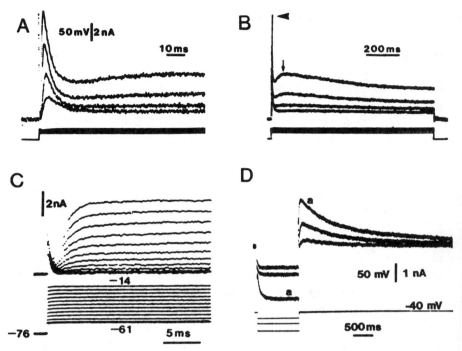

FIGURE 9. Voltage-gated potassium currents. (a,b) Same records at different sweep speeds obtained in presence of TTX. Panel a shows time course of fast transient current (upper) evoked by voltage steps (lower) between −50 and −37 mV from a holding potential of −70 mV. Slow transient current is most prominent in panel b. (Note 20 times slower time base.) Peak of slow transient current is marked by downward arrow. Fast transient current is brief upward deflection marked by horizontal arrowhead. (c) Rapid onset of slow transient current is apparent after pharmacological blockade of fast transient current. Numbers by voltage traces (lower) indicate membrane potential in mV. (d) 1 s hyperpolarizing prepulses remove inactivation from slow transient current. Note slow time course of subsequent inactivation. Current traces labeled *a* correspond. Adapted from Figs. 1, 4 in Spain *et al.* (1991a).

reduced about half by 200 μM 4-aminopyridine (4-AP), a dose having no effect on the fast transient current.

    Effects on spike repolarization were tested after blocking calcium influx, because low doses of TEA are known to block the large-conductance (BK) calcium- and voltage-dependent potassium channels (Blatz and Magleby, 1987) that have been implicated as a mechanism of spike repolarization in other neurons; but calcium blockade has little or no effect on spike repolarization in Betz cells (Figs. 10a,b), suggesting that the $I_{K(Ca)}$ of Betz cells activates too slowly or BK channels are too sparse in Betz cells to have much effect. With calcium blocked,

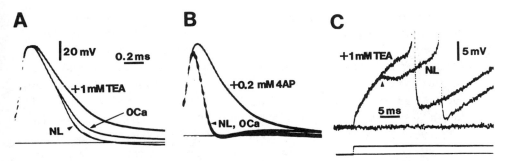

FIGURE 10.   Effects of blocking agents on spike repolarization and spike latency. Spikes in panels a and b were evoked by 3 ms injected current pulses (not shown). Changing from normal perfusate to one containing EGTA, raised magnesium and no added calcium (0Ca) had little or no effect on spike duration, whereas the subsequent addition of tetraethylammonium (TEA; panel a) or 4-aminopyridine (4AP; panel b) caused large spike broadening. (c) The interruption of the rise of membrane potential to spike threshold (marked by upward arrowhead on trace NL) during injection of a current pulse (lower) also was abolished by TEA. See text for further explanation. Adapted from Fig. 6 in Spain et al. (1991a).

both TEA and 4-AP significantly widen the spike (Figs. 10a,b), suggesting each of the two voltage-dependent potassium currents is needed for rapid spike repolarization.

Rapid repolarization of the spike is particularly important in Betz cell function because of the presence of I_{NaP} and a slowly inactivating calcium current. When potassium currents are reduced, these persistent inward currents tend to keep the membrane depolarized, resulting in epileptiform bursts of spikes or prolonged, cardiac-like action potentials (Stafstrom et al., 1985; Schwindt et al., 1988c).

The voltage-dependent properties of the transient potassium currents allow them to play a wider role in Betz cell function. Both currents are first activated at potentials below spike threshold, and both are only partially inactivated at normal resting potential (Spain et al., 1991b). When a Betz cell is depolarized from resting potential by an injected current pulse, it usually displays an interruption of the rise of membrane potential to the first spike (Fig. 10c, trace NL). This delay of firing appears to result from activation of $I_{K(V)F}$ because the interruption of depolarization occurs at the potential where it is first activated, and the interruption is abolished when $I_{K(V)F}$ is blocked pharmacologically (Fig. 10c).

It was described earlier how hyperpolarizing prepulses cause the initial firing rate to become faster in one group of Betz cells. In another group of Betz cells, a hyperpolarizing prepulse causes the initial firing rate to become slower than if the membrane were depolarized from resting potential (Figs. 8c,d; also Spain et al., 1991b). When depolarization is preceded by a hyperpolarization in these cells, the minimum firing rate occurs several hundred ms after the onset of the depolarization, and larger hyperpolarizations result in slower minimum rates (Fig. 8d). After reaching a minimum, firing rate gradually accelerates up to the final tonic rate (which is not influenced by the prepulse), but it may take the cell up to 5 s to achieve this rate. This effect is also graded with prepulse duration such that 1 s prepulses are maximally effective. This group of Betz cells is said to exhibit post-hyperpolarization inhibition (PHI). Intracellular staining of PHI cells reveals they are also large layer V pyramidal neurons, but somewhat smaller than PHE cells. They have anatomical features characteristic of slowly conducting corticofungal neurons as well as wider spikes (0.6–1.2 ms duration) and higher input resistance than PHE cells.

The slow transient potassium current, $I_{K(V)S}$, appears to be responsible for the PHI response, as this response is eliminated selectively by low doses of 4-AP. The explanation of the PHI response is as follows. A hyperpolarizing prepulse removes some of the partial inactivation of $I_{K(V)S}$ that exists at resting potential. $I_{K(V)S}$ is then activated more fully during the subsequent depolarization (*cf.* Fig. 9d) and thereby constitutes an *extra* repolarizing current that remains present until $I_{K(V)S}$ inactivates. The inactivation of $I_{K(V)S}$ is so slow, however, that it can maintain this repolarizing influence for many interspike intervals.

PHI cells have a much smaller I_h than PHE cells. This appears to result from a lower channel density and not just from fewer channels, as would naturally result from the smaller size of these cells (Spain *et al.*, 1991b). Other Betz cells show a mixed response consisting of an early enhancement and a late reduction of firing rate following a hyperpolarizing prepulse. Apparently, the mixed response results from the combined effects of I_h and $I_{K(V)S}$. Altogether, these observations demonstrate that the temporal variation of firing rate during a maintained stimulus can be altered simply by altering the preceding membrane potential. In addition, different groups of Betz cells have different voltage-dependent firing properties and may be divided into different functional classes on this basis.

VIII. Conclusions

In this chapter, we have seen that the repetitive firing properties of Betz cells depend on several types of ionic conductances. Some conductances, like $I_{K(Ca)M}$ and I_{NaP}, seem to provide the basic mechanisms for repetitive firing. Several other conductances govern the temporal variation of firing rate. The temporal pattern of firing is subject to alteration because at least two of the ionic currents involved, $I_{K(Na)}$ and $I_{K(Ca)S}$, can be altered by neurotransmitters, and the influence

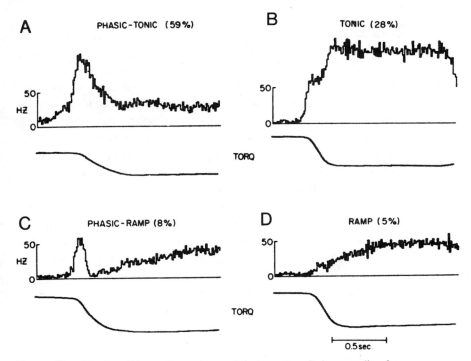

FIGURE 11.   The four firing patterns observed during extracellular recording from neurons in monkey motor cortex. See text for further explanation. Adapted from Fig. 1 in Cheney and Fetz (1980).

of two other currents, I_h and $I_{K(V)S}$, varies with the time-history of membrane potential before the stimulus.

Because it possesses a variety of conductances, an individual Betz cell is able to compute several different output functions for a given input. The firing patterns of cortical output neurons during a behavioral task illustrate this idea.

Cheney and Fetz (1980) observed the four types of firing patterns shown in Fig. 11 while recording extracellularly from corticomotoneuronal neurons in monkeys trained to maintain isometric torsion about their wrists after a visual cue. One interpretation of the four firing patterns would be that they result from four corresponding patterns of synaptic input. This would require four different presynaptic circuits, one to provide each temporal pattern of synaptic input. It is equally possible, however, that all four firing patterns result from the activation of a single presynaptic circuit that simply generates a step of synaptic current.

The most commonly observed response, the phasic–tonic pattern of Fig. 11a, closely resembles the adaptation of firing rate seen in most Betz cells during depolarization from resting potential by injection of a current step. The tonic pattern of Fig. 11b could result from application of the same step input to cells simultaneously receiving input from adrenergic or cholinergic afferents. We have seen that purely tonic firing can also be evoked in those Betz cells that exhibit the PHE response if they are relatively depolarized before the command step arrives. The least common responses, the ramp patterns of Figs. 11c,d, could result from application of the command step to cells simultaneously receiving strong cholinergic input. We have seen that an acceleration of firing rate also can be evoked in those Betz cells that exhibit the mixed or the PHI responses if the command step arrives when these cells are relatively hyperpolarized.

These examples point to the important role that the membrane conductances of individual neurons may play in the functioning of a neural network. The multiple ionic mechanisms that govern repetitive firing imbue each neuron of the network with its own computational power. This computational power may be as important as the temporal pattern of synaptic input in determining a neuron's output.

References

BIEDENBACH, M. A., DEVITO, J. L., and BROWN, A. C. (1986). "Pyramidal Tract of the Cat: Soma Distribution and Morphology," *Experimental Brain Research* **61**, 303–310.

BLATZ, A. L., and MAGLEBY, K. L. (1987). "Calcium-Activated Potassium Channels," *Trends in Neuroscience* **10**, 463–467.

CHENEY, P. D., and FETZ, E. E. (1980). "Functional Classes of Primate Corticomotoneuronal Cells and Their Relation to Active Force," *Journal of Neurophysiology* **44**, 773–791.

CONNOR, J. A., and STEVENS, C. F. (1971). "Prediction of Repetitive Firing Behaviour from Voltage Clamp Data on an Isolated Neurone Soma," *Journal of Physiology (London)* **213**, 31–53.

CONNORS, B. W., and GUTNICK, M. J. (1990). "Intrinsic Firing Patterns of Diverse Neocortical Neurons," *Trends in Neuroscience* **13**, 99–104.

EVARTS, E. V. (1968). "Relation of Pyramidal Tract Activity to Force Exerted during Voluntary Movement," *Journal of Neurophysiology* **31**, 14–27.

FOEHRING, R. C., SCHWINDT, P. C., and CRILL, W. E. (1989). "Norepinephrine Selectively Reduces Slow Ca^{2+}- and Na^{+}-Mediated K^{+} Currents in Cat Neocortical Neurons," *Journal of Neurophysiology* **61**, 245–256.

HAIMANN, C., BERNHEIM, L., BERTRAND, D., and BADER, C. R. (1990). "Potassium Current Activated by Intracellular Sodium in Quail Trigeminal Ganglion Neurons," *Journal of General Physiology* **95**, 961–979.

HASSLER, R., and MUHS-CLEMENT, K. (1964). "Architektonischer Aufbau des Sensorimotorischen und Parietalen Cortex der Katze," *Hirnforschung* **6**, 377–420.

KERNELL, D. (1965). "The Adaptation and the Relation between Discharge Frequency and Current Strength of Cat Lumbosacral Motoneurones Stimulated by Long-Lasting Injected Currents," *Acta Physiologica Scandinavia* **65**, 65–73.

KOIKE, H., MANO, N., OKADA, Y., and OSHIMA, T. (1970). "Repetitive Impulses Generated in Fast and Slow Pyramidal Tract Cells by Intracellularly Applied Current Steps," *Brain Research* **11**, 263–281.

KOIKE, H., OKADA, Y., and OSHIMA, T. (1968). "Accommodative Properties of Fast and Slow Pyramidal Tract Cells and Their Modification by Different Levels of Their Membrane Potential," *Experimental Brain Research* **5**, 189–204.

KRNJEVIC, K., and PHILLIS, J. W. (1963). "Pharmacological Properties of Acetylcholine Sensitive Cells in the Cerebral Cortex," *Journal of Physiology (London)* **166**, 328–350.

LANCASTER, B., and ADAMS, P. R. (1986). "Calcium-Dependent Current Generating the Afterhyperpolarization of Hippocampal Neurons," *Journal of Neurophysiology* **55**, 1268–1282.

LANCASTER, B., and NICOLL, R. A. (1987). "Properties of Two Calcium-Activated Hyperpolarizations in Rat Hippocampal Neurones," *Journal of Physiology (London)* **389**, 187–203.

LEMON, R. N., and MANTEL, G. W. H. (1989). "The Influence of Changes in Discharge Frequency of Corticospinal Neurones on Hand Muscles in the Monkey," *Journal of Physiology (London)* **413**, 351–378.

LLINÁS, R. (1988). "The Intrinsic Electrophysiological Properties of Mammalian Neurons: Insights into Central Nervous System Function," *Science* **242**, 1654–1666.

LLINÁS, R., and JAHNSEN, H. (1982). "Electrophysiology of Mammalian Thalamic Neurons *in vitro*," *Nature* **297**, 406–408.

LLINÁS, R., and SUGIMORI, M. (1980). "Electrophysiological Properties of *in vitro* Purkinje Cell Somata in Mammalian Cerebellar Slices," *Journal of Physiology (London)* **305**, 171–195.

MADISON, D. V., and NICOLL, R. A. (1984). "Control of the Repetitive Discharge of Rat CA1 Pyramidal Neurones *in vitro*," *Journal of Physiology (London)* **354**, 319–331.

SALA, F., and HERNANDEZ-CRUZ, A. (1990). "Calcium Diffusion Modeling in a Spherical Neuron: Relevance of Buffering Properties," *Biophysics Journal* **57**, 313–324.

SCHARFMAN, H. E., and SARVEY, J. M. (1987). "Responses to GABA recorded from Identified Rat Visual Cortical Neurons," *Neuroscience* **23**, 407–422.

SCHLUE, W. R., RICHTER, D. W., MAURITZ, K.-H., and NACIMIENTO, A. C. (1974). "Responses of Cat Spinal Motoneuron Somata and Axons to Linearly Rising Currents," *Journal of Neurophysiology* **37**, 303–309.

SCHWINDT, P. C., and CRILL, W. E. (1982). "Factors Influencing Motoneuron Rhythmic Firing: Results from a Voltage Clamp Study," *Journal of Neurophysiology* **48**, 875–890.

SCHWINDT, P. C., SPAIN, W., and CRILL, W. E. (1988a). "Influence of Anomalous Rectifier Activation on Afterhyperpolarizations of Neurons from Cat Sensorimotor Cortex *in vitro*," *Journal of Neurophysiology* **59**, 468–481.

SCHWINDT, P. C., SPAIN, W. J., and CRILL, W. E. (1989). "Long-Lasting Reduction of Excitability by a Sodium-Dependent Potassium Current in Cat Neocortical Neurons," *Journal of Neurophysiology* **61**, 233–244.

SCHWINDT, P. C., SPAIN, W. J., and CRILL, W. E. (1990). "Anomalous Effects of Intracellular Ca^{2+} Chelation in Cat Neocortical Neurons," *Society for Neuroscience Abstracts* **16**, 355.

SCHWINDT, P. C., SPAIN, W. J., FOEHRING, R. C., CHUBB, M. C., and CRILL, W. E. (1988b). "Slow Conductances in Neurons from Cat Sensorimotor Cortex *in vitro* and Their Role in Slow Excitability Changes," *Journal of Neurophysiology* **59**, 450–467.

SCHWINDT, P. C., SPAIN, W. J., FOEHRING, R. C., STAFSTROM, C. E., CHUBB, M. C., and CRILL, W. E. (1988c). "Multiple Potassium Conductances and Their Functions in Neurons from Cat Sensorimotor Cortex *in vitro*," *Journal of Neurophysiology* **59**, 424n449.

SPAIN, W. J., SCHWINDT, P. C., and CRILL, W. E. (1987). "Anomalous Rectification in Neurons from Cat Sensorimotor Cortex *in vitro*," *Journal of Neurophysiology* **57**, 1555–1576.

SPAIN, W. J., SCHWINDT, P. C., and CRILL, W. E. (1991a). "Two Transient Potassium Currents in Pyramidal Neurones from Cat Sensorimotor Cortex," *Journal of Physiology (London)* **434**, 591–607.

SPAIN, W. J., SCHWINDT, P. C., and CRILL, W. E. (1991b). "Post-Inhibitory Excitation and Inhibition in Layer V Pyramidal Neurones from Cat Sensorimotor Cortex," *Journal of Physiology (London)* **434**, 609–626.

STAFSTROM, C. E., SCHWINDT, P. C., CHUBB, M. C., and CRILL, W. E. (1985). "Properties of Persistent Sodium Conductance and Calcium Conductance of Layer V Neurons from Cat Sensorimotor Cortex *in vitro*," *Journal of Neurophysiology* **53**, 153–170.

STAFSTROM, C. E., SCHWINDT, P. C., and CRILL, W. E. (1982). "Negative Slope Conductance due to a Persistent Subthreshold Sodium Current in Cat Neocortical Neurons *in vitro*," *Brain Research* **236**, 221–226.

STAFSTROM, C. E., SCHWINDT, P. C., and CRILL, W. E. (1984a). "Repetitive Firing in Layer V Neurons from Cat Neocortex in vitro," *Journal of Neurophysiology* **52**, 264–277.

STAFSTROM, C. E., SCHWINDT, P. C., FLATMAN, J. A., and CRILL, W. E. (1984b). "Properties of the Subthreshold Response and Action Potential Recorded in Layer V Neurons from Cat Sensorimotor Cortex *in vitro*," *Journal of Neurophysiology* **52**, 244–263.

TAKAHASHI, K. (1965). "Slow and Fast Groups of Pyramidal Tract Cells and Their Respective Membrane Properties," *Journal of Neurophysiology* **28**, 908–924.

WONG, R. K. S., and PRINCE, D. A. (1978). "Participation of Calcium Spikes during Intrinsic Burst Firing in Hippocampal Neurons," *Brain Research* **159**, 385–390.

Chapter 10   Determination of State-Dependent Processing in Thalamus by Single Neuron Properties and Neuromodulators

DAVID A. MCCORMICK,[1] JOHN HUGUENARD,[2]
and BEN W. STROWBRIDGE[1]

[1]*Yale University School of Medicine*
Section of Neurobiology
New Haven, Connecticut
[2]*Stanford University School of Medicine*
Department of Neurology
Stanford, California

I.   Introduction

One of the most prominent features of neurons in the nervous system is their ability to generate action potentials in varying patterns. Since the discovery of the electroencephalogram (EEG) in the late 1800s it has been known that electrical activity in the brain occurs in multiple patterns depending in large part upon the behavioral state of the animal (Caton, 1875; Magoun, 1958; Brazier, 1980). During periods of slow-wave sleep, for example, the electrical activity in the forebrain is marked by the presence of synchronous slow waves in the frequency range of 1–12 Hz, while during periods of arousal and attentiveness, these slow waves disappear and are replaced by a more desynchronous pattern of activity with a much broader representation of frequencies. Extracellular and intracellular recordings *in vivo* have revealed that these various patterns of activity in the EEG arise from the combined electrical activity of large numbers of forebrain neurons. (See, e.g., reviews by Andersen and Andersson, 1968; Steriade and Deschênes, 1984.) While these investigations of gross electrical activity in the mammalian forebrain were revealing complex and recurring patterns of neuronal firing, detailed intracellular recordings from reduced preparations such as the squid giant axon or motoneurons of the spinal cord were detailing the basic properties of action potential generation and synaptic transmission (Hodgkin and Huxley, 1952; Eccles, 1964). In contrast to the seemingly complex nature of gross electrical activity in the forebrain, electrical activity at the single neuronal level appeared to be quite simple: Action potentials were generated by only two currents, depolarization through the entry of sodium and repolarization through the exit of potassium, and synaptic potentials were either brief and excitatory or brief and inhibitory. The apparent simple nature of neurons and neuronal interactions through action potential generation and synaptic transmission suggested that the complex patterns of electrical activity generated in the forebrain was an emergent property arising from the complicated interconnections of large numbers of neurons and not so much due to the properties of the neurons themselves (e.g., Andersen and Andersson, 1968). However, as techniques became available to perform detailed analysis on single neurons and their elements, a plethora of neuronal currents and neurotransmitter actions were revealed (reviewed by Llinás, 1988; Storm, 1990; McCormick, 1990b), indicating that even single neurons are capable of generating quite complicated patterns of electrical activity due to the properties of the currents intrinsic to each neuron. In this manner, single neurons have the ability to generate what was once thought to arise from a large network of neurons; in other words, single neurons can behave as a network. However, not all neurons behave similarly. Just as different classes of cells may have strikingly different morphologies, such as the flattened dendritic tree of the cerebellar Purkinje cell versus the more radially oriented dendritic tree of many thalamic relay neurons, different classes of neurons also have unique intrinsic electrophysiological properties (e.g., Llinás, 1988; McCormick, 1990b). Thus, the electrical properties of neurons in the nervous system are, in general, consistent within a group, (e.g., relay neurons), but markedly distinct between groups (e.g., relay neurons versus Purkinje cells). Presumably, these differences in the electrophysiological properties of different types of neurons is necessary to fulfill their proper role in information processing in the CNS and may allow markedly different postsynaptic responses to identical presynaptic inputs. The presence of unique electrophysiological properties in different types of neurons in each neuronal circuit has this practical implication: An understanding of the cellular basis of neuronal processing in the nervous system not only requires a thorough understanding of how neurons are anatomically interconnected, but also how each constituent neuron and neuronal element behaves and how these elements interact with each other to give rise to the properties of the circuit as a whole.

During the past few years, we have been investigating in detail the electrophysiological properties of neurons in the cerebral cortex and thalamus and how these properties are altered by neuromodulatory neurotransmitters. This research is aimed at understanding the cellular basis of the different modes of activity, information processing and excitability exhibited by forebrain neurons during various states of sleep, arousal, and attentiveness. Of the various cell types in the forebrain, the electrophysiological properties of thalamic neurons are a particularly useful and poignant example, for these cells have the ability, either singularly or as a group, to generate the same basic pattern of activity *in vitro* that the forebrain generates *in vivo:* slow rhythmic oscillation during the relative lack of ascending modulatory input, and single-spike activity (also known as tonic activity) during the activation of modulatory inputs (e.g., McCormick, 1989; Steriade *et al.,* 1990). Understanding these different modes of neuronal activity in thalamic relay neurons will not only lead to a better understanding of the functional role of the thalamus, but will also enhance our understanding of neurons throughout the nervous system, since electrophysiological investigations have revealed that cells at all levels of the neuraxis possess the ability to exhibit different modes of rhythmic oscillation (e.g., Llinás, 1988). In addition, thalamic neurons strongly express two important currents, the low-threshold Ca^{++} current and the hyperpolarization-activated cation current, both of which are found in most central neurons and have been implicated in the generation of rhythmic activity.

In this chapter, we will briefly review some of the relevant experimental data obtained from thalamic neurons, describe a computational model of this data, and discuss some of the functional consequences of different modes of action potential generation in central neurons.

II.   Electrophysiological Properties of Thalamic Neurons

Thalamic relay neurons occupy a unique and important position in the nervous system, for nearly all sensory/motor information that reaches the neocortex must first pass through the thalamus by forming synaptic connections with relay neurons. Likewise, the cerebral cortex heavily innervates thalamic relay neurons in a reciprocal fashion (Jones, 1985). Thus, the intrinsic properties of thalamic relay cells can potentially have a strong influence upon the pattern and course of electrical activity in the forebrain. Extracellular recordings from thalamic relay neurons during different behavioral states *in vivo* have revealed that these cells display two basic patterns of electrical activity:

1. Rhythmic burst firing in which clusters of 2–8 action potentials arrive at high frequency (300–500 Hz) within each burst, with interburst frequencies of 1–12 Hz.
2. Single-spike activity in which action potentials are generated in trains of single spikes.

(See Lamarre *et al.*, 1971; McCarley *et al.*, 1983; Steriade and Deschênes, 1984; Fourment *et al.*, 1984.)

Intracellular recordings both *in vivo* and *in vitro* by Steriade and Deschênes (1984) and Jahnsen and Llinás (1984a,b) revealed that these two states of neuronal activity were due in large part to the intrinsic electrophysiological properties of these neurons. Of particular importance to the ability of these cells to exhibit multiple states of neuronal activity was the presence of a strong low-threshold Ca^{++} current, also known as the transient Ca^{++} current, or T-current, I_T (Jahnsen and Llinás, 1984a,b; Coulter *et al.*, 1989a; Crunelli *et al.*, 1989; Hernandez-Cruz and Pape, 1989). This Ca^{++} current is unique in that it is activated by

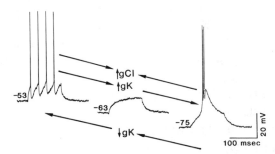

FIGURE 1.   Different firing modes of thalamic relay neurons. Intracellular injection of a 120 msec constant current pulse into a guinea pig LGNd relay neuron at a membrane potential of -53 mV results in a train of four single spikes. Injection of the same current pulse while the cell is at -63 mV, in contrast, results in a largely passive membrane response. Hyperpolarization of the neuron to -75 mV results in the removal of inactivation of the low-threshold Ca^{++} current and thus the appearance of a low-threshold Ca^{++} spike and the corresponding burst discharge. *In vivo,* these changes in membrane potential are brought about in response to changes in resting ionic conductances. Decreases in K^+ conductance (gK), for example through the stimulation of α_1, H_1, or muscarinic receptors, results in depolarization of the relay neuron and a switch in firing mode. Increases in gK (due, for example, to the removal of stimulation of α_1, H_1, or muscarinic receptors) have the opposite effect. Increases in gCl tend to inhibit all types of neuronal activity.

depolarization at relatively negative membrane potentials (starting at approximately -75 mV as seen in Fig. 4), and inactivates moderately rapidly. (See, e.g., Fig. 4.) Thus, depolarization of thalamic neurons from a relatively hyperpolarized level results in the generation of a low-threshold Ca^{++} spike that subsequently generates a high-frequency burst of 3–8 fast Na^+- and K^+-mediated action potentials (Fig. 1, -75 mV). Tonic depolarization of the neuron, in contrast, can result in complete inactivation of I_T. During this state, depolarizing influences either cause only passive depolarization of the membrane (Fig. 1, -63 mV) or the generation of action potentials (Fig. 1, -53 mV), depending upon the amplitude of the input and the membrane potential of the cell. Thus, thalamic neurons possess at least three states of neuronal activity:

1. Burst firing.
2. Single-spike firing.
3. Silent and in between either burst firing or single-spike activity.

Which state is exhibited depends upon the membrane potential of the cell (Jahnsen and Llinás, 1984a,b), which in turn depends upon the influence of a variety of neuromodulatory transmitters (reviewed in McCormick, 1989).

Detailed investigation of the electrophysiological properties of thalamic neurons has revealed no less than 12 distinct ionic currents, including:

1. A fast and transient Na^+ current, I_{Nat}, underlying the generation of fast action potentials.
2. A persistent Na^+ current, I_{Nap}, which is activated below spike threshold and is non-inactivating.
3. A low-threshold Ca^{++} current, I_T.
4. A high-threshold Ca^{++} current, I_L.
5. A hyperpolarization-activated cation current I_h, and a number of different K^+ currents.

(See Jahnsen and Llinás, 1984a,b; Coulter et al., 1989a.) Potassium currents that are present in thalamic neurons include:

6. A leak K^+ current, which helps determine resting membrane potential and which is modulated by neurotransmitters (McCormick and Prince, 1988).
7. A Ca^{++}-activated K^+ current, which helps to repolarize action potentials and may be similar to I_C (Jahnsen and Llinás, 1984b) and at least three different K^+ currents activated by depolarization.

These include:

8. A rapidly inactivating K^+ current, I_A (Huguenard, unpublished observations), and two slowly inactivating K^+ currents:

9. One that activates rapidly at relatively negative membrane potentials and is known as I_D or I_{As} (McCormick, 1990a), and

10. The other, I_K, that activates more slowly and at more positive membrane potentials (Huguenard and Coulter, unpublished observation).

These distinct ionic currents interact in a manner which allows an individual thalamic relay neurons to display two completely different and distinct modes of action potential generation: the rhythmic burst firing mode and the single-spike firing mode. (See Figs. 1–3.) The ionic basis of single-spike activity has been reviewed and modeled elsewhere (Yamada *et al.*, 1989; Storm, 1990). By contrast, the ionic basis by which central neurons rhythmically generate bursts of action potentials, or by which they exhibit multiple modes of action potentials, has not received widespread attention, although notable exceptions include the mathematical models of thalamic neuronal activity by Rose and colleagues (Rose and Hindmarsh, 1989). Therefore, we would like to review what is known about the physiological properties of thalamic relay neurons that allow them to rhythmically generate bursts of action potentials.

A.   *Ionic Basis for Rhythmic Oscillation in Single Thalamic Neurons*

Intracellular and extracellular recordings from thalamic relay neurons *in vivo* have revealed that these cells can generate at least two different types of rhythmic burst firing:

1. Spindle waves, which appear as clusters of 7–12 Hz oscillation that waxes and wanes over a period of 1–2 seconds (Steriade and Deschênes, 1984).

2. Slow 1–4 Hz rhythmic burst firing, which is highly regular and appears during the removal of all or most afferent inputs or activity (Lamarre *et al.*, 1971; Fourment *et al.*, 1984; McCormick and Pape, 1990a).

The first type of rhythmic oscillation, spindle waves, appears to result from both the intrinsic properties of thalamic neurons and from the interactions of different types of thalamic neurons, namely, the relay cells and the inhibitory GABAergic neurons of the nucleus reticularis (Steriade and Deschênes, 1984). In contrast, the 1–4 Hz rhythmic oscillation appears to be an intrinsic property of thalamic relay cells resulting from the interaction of known ionic currents (McCormick and Pape, 1990a). Therefore, in the present model, we have simulated the slow 1–4 Hz oscillatory mode while keeping in mind that thalamic neurons are also capable of other frequencies of oscillation.

The ability of thalamic relay neurons to generate slow 1–4 Hz rhythmic burst firing has been proposed to result from the interaction of the low-threshold Ca^{++} current, I_T, and the hyperpolarization-activated cation current, I_h (McCormick and Pape, 1990a). In thalamic relay cells, as in almost all cells of the CNS studied thus far, hyperpolarization of the membrane results in the activation of a time- and voltage-sensitive inward current known as I_h. Removal of various ions in the medium bathing thalamic cells *in vitro* has demonstrated that I_h is carried by both Na^+ and K^+ in approximately equal proportions and, therefore, has an equilibrium potential of about -40 mV. Hyperpolarization of the membrane potential, negative to approximately -60 mV, results in the activation of I_h, which subsequently causes a depolarizing *sag* in the membrane potential back towards rest over the next few seconds. Indeed, if the membrane potential is held depolarized to, say, -55 mV, with intracellular injection of current, and this current is suddenly removed, the membrane potential will hyperpolarize to approximately -85 mV and then steadily depolarize back towards a resting value of approximately -65 mV. This depolarization can result in the generation of a Ca^{++} spike through activation of the low threshold Ca^{++} current, I_T. In addition, in neurons that possess the proper balance of these two currents, the generation of a burst of action potentials can result in sufficient deactivation (turning off) of I_h so as to allow another hyperpolarization to -85 mV as I_T inactivates and the Ca^{++} spike ends. Thus, we have the following scenario: Hyperpolarization of a thalamic relay neuron activates I_h. Activation of I_h slowly depolarizes the neuron (Fig. 2b, I_h activate). The slow depolarization activates I_T, which results in the generation of a Ca^{++} spike (Fig. 2b, I_T activate). During the generation of the Ca^{++} spike, I_h deactivates and I_T inactivates. As the Ca^{++} spike repolarizes, I_h is once again activated, thereby again causing a slow depolarization of the neuron, which can again activate another Ca^{++} spike (Fig. 2). The time period between Ca^{++} spikes is important because during this period the inactivation of I_T is removed. This de-inactivation of I_T determines the amplitude and duration of the next Ca^{++} spike (Fig. 2b, removal I_T inactivation). Depolarization of thalamic neurons through the intracellular injection of current (Figs. 2a, 3), or through the actions of neuromodulatory transmitters (as seen shortly), results in a complete block of rhythmic burst firing as a result of I_T inactivation and the lack of I_h activation.

Hyperpolarization of thalamic neurons yielding membrane potentials negative to approximately -85 mV also results in abolition of rhythmic burst firing due to lack of activation of I_T, even though I_h is strongly activated (Fig. 3). In this manner, thalamic neurons can only generate rhythmic 1–4 Hz burst firing at membrane potentials between approximately -75 and -85 mV (Fig. 3).

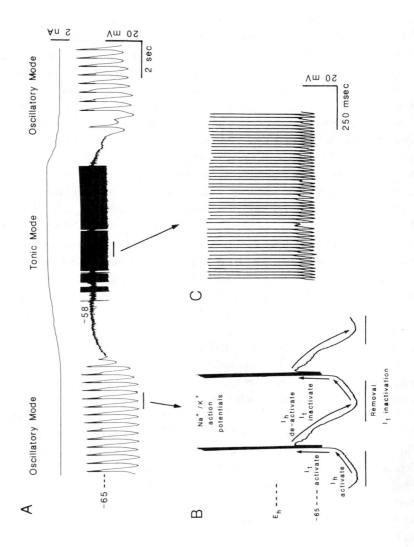

A   Oscillatory Mode   Tonic Mode   Oscillatory Mode

−65 ‑‑‑‑   −58   2 nA
                                                                                 20 mV

                                                                                 2 sec

B

Eₕ ‑‑‑‑‑

                                              Na⁺/K⁺
                                              action
                                              potentials

                                              I_h
                                              de‑activate
−65 ‑‑‑ activate I_t
I_t inactivate
I_h
activate
                                              Removal
                                              I_t inactivation

C   20 mV

                                                                                 250 msec

266

III.   Neuromodulation of Thalamic Neuronal Activity

Depolarization of thalamic relay neurons both abolishes the ability of these cells to rhythmically oscillate and shifts their neuronal activity to the single-spike, or tonic, mode of action potential generation (Figs. 2a,c). Extracellular and intracellular recordings *in vivo* have revealed that the single-spike mode of action potential generation is prevalent during periods of arousal and attentiveness and is associated with an increase in transmission of synaptic inputs from, for example, the retina (Hirsch *et al.*, 1983; McCarley *et al.*, 1983). Intracellular recordings *in vivo* in awake behaving cats have revealed that this shift in firing mode is associated with a tonic depolarization of the membrane potential, an associated abolition of rhythmic burst firing, and the production of single-spike activity (Hirsch *et al.*, 1983).

The cellular mechanisms by which thalamic relay neurons are depolarized out of the oscillatory mode and into the single-spike mode of action potential generation have recently been investigated with intracellular recordings *in vitro*. Application of the neurotransmitters that are released by the neuromodulatory systems arising in the brainstem and hypothalamus—namely, acetylcholine, norepinephrine, and histamine (acting through muscarinic, α_1, and H_1 receptors, respectively)—all give rise to a strong depolarization of thalamic relay cells through reduction of a relatively linear K^+ conductance that is active at rest (McCormick and Prince, 1987; McCormick and Prince, 1988; McCormick,

---

FIGURE 2.   Intrinsic rhythmic oscillatory properties of cat LGNd relay neurons. In the absence of current injection and influence of modulatory neurotransmitters, cat LGNd relay neurons rhythmically generate low-threshold Ca^{++} spike-mediated bursts of action potentials at a rate of 1–4 Hz (a,b). Experimental results *in vitro* revealed that this oscillatory state arose from the interaction of the low-threshold Ca^{++} current I_T and the hyperpolarization activated cation current I_h (b). In this manner, activation of the low-threshold Ca^{++} current, I_T, depolarizes the membrane towards threshold for a burst of Na^+- and K^+-dependent fast action potentials. The depolarization deactivates a portion of I_h that was activated immediately before the Ca^{++} spike. Repolarization of the membrane due to I_T inactivation is followed by a hyperpolarizing overshoot, due to the reduced depolarizing effect of I_h. The hyperpolarization in turn de-inactivates I_T and activates I_h, which depolarizes the membrane towards threshold for another Ca^{++} spike. Depolarization of the relay neuron with the intracellular injection of current (top trace in a) results in an abolition of rhythmic burst firing and a switch to the tonic mode of action potential generation (expanded in c for detail). Removal of this depolarizing influence results in a shift back to the oscillatory mode of action potential generation. From McCormick and Pape, (1990a).

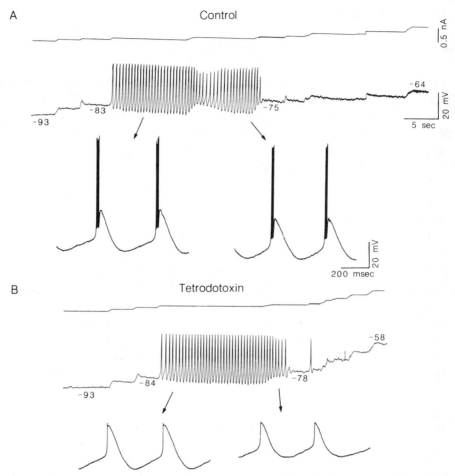

FIGURE 3.   Generation of rhythmic low-threshold Ca^{++} spikes is an intrinsic property to thalamic relay neurons. Changing the membrane potential from -93 to -64 reveals that rhythmic oscillations only occur at potentials between -83 and -75 mV. Further depolarization of the neuron positive to -55 mV resulted in the complete switch to the single-spike firing mode of action potential generation. (See Fig. 2.) Local application of the Na^+ channel poison tetrodotoxin results in an abolition of fast action potentials, but does not alter the generation of low-threshold Ca^{++} spikes (expanded in b for detail), indicating that this oscillation is intrinsic to thalamic relay neurons and does not require the generation of fast action potentials. From McCormick and Pape, (1990a).

unpublished observations). This depolarization is typically 10–20 mV at its maximal extent (e.g., after maximal stimulation of the neurotransmitter receptors) and can completely switch the relay neuron out of the rhythmic burst firing mode into the single-spike mode of action potential generation. Thus, it has been suggested that the tonic depolarization that is associated with increases in arousal, attentiveness, and the waking state is directly the result of increased release of norepinephrine, acetylcholine, and histamine as a result of increased activity in the neurons that contain these neurotransmitters (reviewed in McCormick, 1989).

A second major neurotransmitter response that has been revealed in the LGNd can be activated by norepinephrine (through β receptors), histamine (through H_2 receptors), and serotonin and appears to be mediated by an increase in intracellular concentrations of cyclic AMP (McCormick and Pape, 1990b). Activation of these receptors results in a change in the voltage dependence of the hyperpolarization-activated cation current I_h such that this current is activated at more depolarized levels (McCormick and Pape, 1990b). This rightward shift in the activation curve for I_h is associated with a number of secondary consequences, including an increase in apparent input conductance at membrane potentials negative to -55 mV and a reduced propensity to oscillate (McCormick and Pape, 1990b). These secondary effects have been proposed to result directly from the shift in voltage sensitivity of I_h (McCormick and Pape, 1990b).

In an effort to confirm these hypotheses and to test out new ones in an ideal preparation, we have recently developed a computational model of neuronal activity in single thalamic neurons. In addition to acting as a *yard stick* of our understanding of thalamic neurons, this model has served very useful in developing and testing new experiments, which can then be applied to thalamic neurons *in vitro*. In addition, computational models have allowed us to readily perform *thought* experiments that are not easily achieved in living neurons. Next, we present the equations used for the thalamic model and present and discuss some of the results that we have obtained with this model.

IV.   Computational Simulation of Thalamic Neuronal Activity

The two modes of action potential generation in thalamic neurons were successfully modeled using Hodgkin–Huxley-style equations based upon a single compartmental model containing six different currents: the transient Na^+ current, I_{Nat}, which underlies action potential generation; a delayed rectifier K^+ current I_K; a transient, voltage-activated K^+ current, I_A; the transient Ca^{++} current, I_T; the hyperpolarization-activated cation current, I_h; and a leak K^+ current, I_{Kleak}

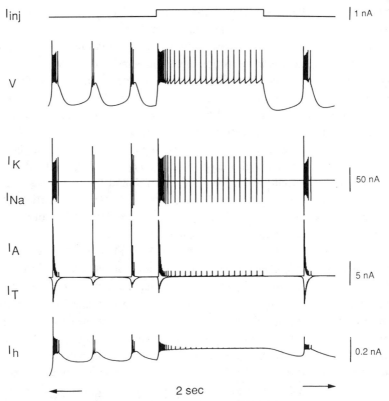

FIGURE 4.   Two different modes of action potential generation in simulation of neuronal activity in thalamic relay neurons. The simulation began at a membrane potential of -85 mV. At this membrane potential, I_h is strongly activated. The activation of I_h depolarizes the cell and activates I_T, thereby generating a Ca^{++} spike. This Ca^{++} spike activates a high-frequency burst of short duration action potentials mediated by the voltage-sensitive Na current, I_{Na}, and the delayed rectifier K current I_K. The transient K current, I_A, also contributes to the repolarization of action potentials during the burst. As I_T inactivates, the low-threshold Ca^{++} spike ends and the cell repolarizes, thus activating I_h. Activation of I_h once again depolarizes the cell and activates another I_T-mediated burst of two fast action potentials. Injection of depolarizing current (I_{inj}) results in an initial burst of action potentials followed by a prolonged train of spikes. Return of I_{inj} to zero is associated with reinstatement of rhythmic burst firing.

(Fig. 4). Removal of the transient Na^+ current (I_{Nat}) and the transient K^+ current (I_A) revealed the generation of rhythmic Ca^{++} spikes that underlie the generation of rhythmic burst discharges *in vivo* (Fig. 3b) and in the simulation (Figs. 7– 10). Previous authors have performed computational simulations of the contri- butions of transient Na^+ and K^+ currents on neuronal firing (Hodgkin and

Huxley, 1952; Connor and Stevens, 1971; McMullen and Ly, 1988; Yamada *et al.*, 1989). In contrast, the ability of central neurons to rhythmically generate low-threshold Ca^{++} spikes has not received detailed mathematical study, although the model of thalamic burst firing by Rose and Hindmarsh (1989) is a notable exception. Therefore, we focus our attention here upon the simulation of rhythmic Ca^{++} spikes in a thalamic neuron using four different currents: I_T, I_K, I_h, and I_{Kleak}.

A.   Modeling of Ionic Currents in Thalamic Neurons

In vitro and *in vivo* analysis of the different types of currents in thalamic neurons has yielded substantial information concerning the voltage and time-dependent properties of a subset of these currents (e. g., I_T and I_h), but little or no information on the spatial distribution of these currents. Therefore, the present simulation is based upon a *single compartmental model* in which the entire membrane acts in unison. Although this is an obvious gross simplification of the complex and presumably important morphological features of thalamic relay neurons, our ability to obtain results with a single compartmental model that are very similar to real thalamic neurons attests to the usefulness of this morphologically simplistic approach. Future versions of the present model will incorporate multiple compartmental models as information concerning the spatial distribution of these currents becomes available.

The intracellular recordings of rhythmic burst generation in thalamic relay neurons represented in Figs. 2 and 3 were obtained at a temperature of 35° C (McCormick and Pape, 1990a) and, therefore, all equations used in the present computational model were normalized to this temperature. The amplitude of each of the four different currents, I_T, I_K, I_h, and I_{Kleak} at each time step, Δt, was calculated from the value of that current's conductance at that time step and the voltage difference between the membrane potential (V) at that time and the ion's equilibrium potential (E), according to Ohm's law:

$$I_{(t)} = g_{(t)} \cdot (V - E). \tag{1}$$

Each of the voltage-sensitive conductances (gT, gK, and gh) were calculated at each time step according to Hodgkin–Huxley theory (Hodgkin and Huxley, 1952), in which

$$g_t = g_{max} \cdot m^i \cdot h^j, \tag{2}$$

where g_{max} is equal to the single-channel conductance times the total number of channels, m and h are independent activation and inactivation variables or gates, and i and j denote the apparent number of each type of gate per channel. For a channel to be conducting, all of its voltage-dependent gates must be open, i.e., the i activation gates (m) and the j inactivation (h) gates. The hyperpolarization-activated cation current I_h does not inactivate (McCormick and Pape, 1990a) and, therefore, is described without an inactivation variable. Similarly, I_K inactivates, although very slowly over a period of seconds (Rudy, 1988) and, therefore, we have chosen to ignore inactivation of I_K for simplification. In accord with the methods of Hodgkin and Huxley, the shifting of the channel gates between the permissive state (e.g. m) and the non-permissive state (e.g $1 - m$) was modeled with first-order reaction kinetics of the form,

$$1 - m \underset{\beta_n}{\overset{\alpha_n}{\rightleftharpoons}} m. \tag{3}$$

Thus, the rate of change of the activation variable m with respect to the time (dm/dt) follows the differential equation:

$$\frac{dm}{dt} = \alpha_m \cdot (1 - m) - \beta_m \cdot m. \tag{4}$$

If we voltage-clamp the neuron at a particular membrane potential for an infinite period of time such that m comes to steady state (i.e., $dm/dt = 0$) and rearrange Eq. (4), we find

$$m_\infty = \frac{\alpha_m}{\alpha_m + \beta_m}. \tag{5}$$

This relationship between the rate variables α_m and β_m and the steady-state activation variable m was utilized in the simulation of I_T and I_K. The steady-state activation variable for I_h, in contrast, was calculated from a second-order Boltzman equation. (See shortly.) In the non-steady-state situation, the activation variable m or inactivation variable h can be calculated from the general solution of Eq. (4):

$$m = m_0 + (m_\infty - m_0) \cdot [1 - e^{(-t/\tau_m)}], \tag{6}$$

where the time constant τ_m is described by

$$\tau_m = \frac{1}{\alpha_m + \beta_m}. \tag{7}$$

Equation (6) was utilized to calculate the amplitude and time course of I_T and I_h under voltage-clamp conditions, as illustrated in Figs. 5 and 6. In this manner, once equations that describe α_m and β_m (and α_h and β_h) are obtained, then the precise amplitude and time course of the current can be calculated.

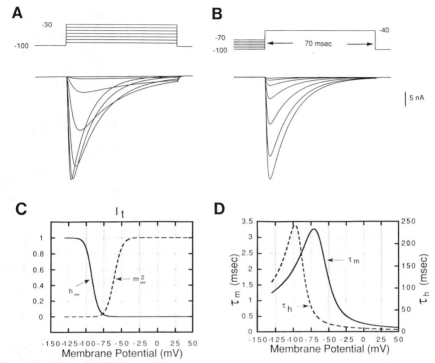

FIGURE 5.   Properties of low-threshold Ca^{++} current used in the computational simulations. Voltage steps from a holding potential of -100 mV in 10 mV increments to -30 mV illustrate some of the activation and inactivation characteristics of I_T. Changing the level of the holding potential prior to stepping to -40 mV illustrates the voltage dependence of steady-state inactivation. The voltage dependence and kinetics of the activation (m_∞) and inactivation (h_∞) parameters is illustrated in c and d.

B.   Stepwise Current Clamp Simulations of Thalamic Neurons

Simulation of neuronal activity requires a starting point for all of the activation and inactivation variables for the different currents involved. Usually, a convenient starting point is to assume that the neuron is at resting membrane potential so that all of the different variables are at steady state. However, thalamic neurons rhythmically oscillate at rest and, therefore, do not have a true *resting* membrane potential. Therefore, we chose the most hyperpolarized membrane potential as a starting point (approximately -85 mV) and calculated the steady-state values of the activation and inactivation variables for each of the three voltage-dependent currents (I_T, I_K, I_h). This would be achieved in a *real* neuron if the cell were voltage-clamped to -85 mV until each of these currents came to steady state (i.e., for many seconds). The simulation then began by *releasing* the neuron

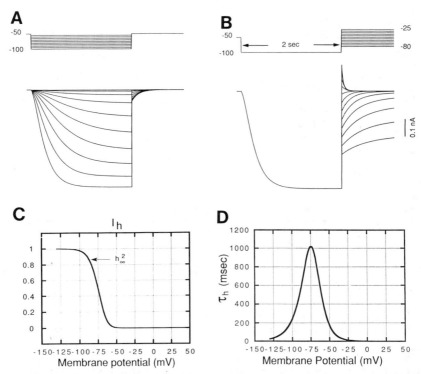

FIGURE 6.   Properties of the hyperpolarization-activated cation current I_h used in the computational simulation. (a) Voltage steps from a holding potential of -50 mV to -100 mV illustrate the voltage dependence and kinetics of activation of I_h. (Note lack of inactivation.) (b) Examination of the voltage dependence of the kinetics of deactivation of I_h tail currents also reveals the reversal potential (-45 mV). The steady-state activation curve and kinetics for I_h are illustrated in c and d.

from this starting point. Next, we calculated the change in membrane potential resulting from the flow of ionic currents, the subsequent change in values using Eq. (4) due to the change in membrane potential, and the subsequent change in each ionic current resulting from the change in activation and inactivation variables, using Eqs. 1 and 2. (See the preceding.) This calculation occurred in a time-step manner in which each step was < or = 25 μsec and used the following relationship:

$$\frac{dV}{dt} = \frac{I_{inj} + I_T + I_K + I_h + I_{Kleak}}{C_i}, \tag{8}$$

where C_i is the input capacitance of the cell. Measurement of the apparent input capacitance of the cat LGNd relay neuron in Figs. 2 and 3 revealed a value of

0.29 nF, which was similar to the average of other cat LGNd relay neurons (0.44 nF). The differential equation relating rate of change in voltage with respect to time was numerically solved in discrete time steps using a two-step Euler integration method.

1.   Transient and Low-threshold Ca^{++} Current, I_T.   The transient and low-threshold Ca^{++} current I_T was described (Fig. 5) by the equation:

$$I_T = gT_{max} \cdot m^2 \cdot h \cdot (V - E_{Ca}). \tag{9}$$

The exact value of the total T-conductance, gT_{max}, in cat LGNd relay neurons is difficult to obtain. In an effort to estimate this conductance, we measured the maximal rate of change of the membrane potential during the generation of Ca^{++} spikes in guinea pig and cat thalamic neurons after blocking I_{Nat} with tetrodotoxin, I_h with local application of Cs^+, and transient K^+ currents with 4-aminopyridine (McCormick, unpublished observations). Intracellular injection of a depolarizing current pulse was used to trigger a low-threshold Ca^{++} spike while the cell was hyperpolarized to approximately -100 mV. In this manner, dV/dt during the Ca^{++} spike peaked at about 30 mV/msec, corresponding to a peak current of approximately 9 nA, according to Eq. (8). Simulation of these conditions in the present model indicated that similar results can be obtained with a maximal T-conductance of around 0.2–0.4 μS. This value of the gT_{max} is also similar to that estimated from the T-conductance obtained from dissociated thalamic neurons *in vitro*. Rat relay neurons exhibit a peak T-conductance of about 1.18 nS per pF of capacitance, after correcting to a temperature of 35° C (Huguenard, unpublished observations). Given that the apparent input capacitance used in the present model is 290 pF, this yields a value of 0.34 μS for gT_{max}. In the present simulations, we utilized a value of 0.4 μS for the maximal T-conductance.

Voltage clamp recordings from rat thalamic relay neurons, enzymatically dissociated and recorded with the whole-cell patch clamp technique *in vitro*, have recently been utilized to analyze the voltage-dependent and kinetic properties of the low-threshold Ca^{++} current. (See Fig. 5; also Coulter *et al.*, 1989a; Huguenard, unpublished observations). These voltage clamp data can be well described by Eq. (9) and the following set of definitions for activation rate variables, α_{mt}, β_{mt}, and the inactivation rate variables, α_{ht}, β_{ht}:

$$\alpha_{mt} = \frac{-0.075 \cdot (V + 50)}{e^{((V+50)/-7.5)} - 1}, \tag{10}$$

$$\beta_{mt} = \frac{0.0101 \cdot (V + 51)}{e^{((V+51)/4.4)} - 1}, \tag{11}$$

$$\alpha_{ht} = \frac{3.44 \times 10^{-4} \cdot (V + 79)}{e^{((V + 79)/4.45)} - 1}, \tag{12}$$

$$\beta_{ht} = \frac{-3.093 \times 10^{-4} \cdot (V + 85)}{e^{((V + 85)/-4)} - 1}. \tag{13}$$

Interestingly, these equations predict low-threshold Ca^{++} spikes that are very similar to those that have been recorded from rodent thalamic neurons maintained *in vitro* (not shown). However, this is somewhat problematic, since rodent thalamic neurons do not generate the slow oscillatory 1–2 Hz rhythm nearly as well as cat LGNd neurons. Similarly, our computational model using the aforementioned parameters for I_T failed to maintain oscillatory Ca^{++} spikes (not shown). Closer examination of the low-threshold Ca^{++} spikes from rat and cat thalamic relay neurons indicate that those in cat are substantially longer in duration than those in rat (Crunelli *et al.*, 1989; McCormick, unpublished observations), a finding that can be mimicked by assuming a difference in inactivation kinetics of I_T between the two species. Therefore, we have included an additional variable in our model by which α_{ht} and β_{ht} are multiplied, slowing the rate of inactivation. We have found that a two-fold slowing of α_{ht} and β_{ht} is sufficient to mimic the generation of rhythmic Ca^{++} spikes in cat LGNd neurons, although the individual simulated Ca^{++} spikes are still shorter in duration than those recorded from live cat LGNd neurons. (Compare Figs. 3b and 7.)

The equilibrium potential for Ca^{++} was calculated according to the Nernst equation assuming a baseline level of Ca^{++} in the intracellular milieu of 50 nM and an extracellular concentration of 1.2 mM. During neuronal activity, the intracellular concentration of Ca^{++} will change in a continuous manner, and, therefore, it is necessary to calculate the subsequent changes in the equilibrium potential for Ca^{++}. However, at present, no information is available concerning the dynamics of Ca^{++} buffering in thalamic neurons. In the present simulation, we adopted the simple proportional model of Ca^{++} diffusion used by Traub (1982) in the modeling of hippocampal pyramidal cells in which the diffusion of $[Ca^{++}]$ out of the shell of cytoplasm just beneath the membrane is given by:

$$\frac{d[Ca^{++}]}{dt} = \beta \cdot [Ca^{++}]. \tag{14}$$

Using Faraday's constant for relating current flow to fractions of a mole of ions, we then calculated the concentration of Ca^{++} in the 100 nm of cytoplasm just below the inner face of the membrane at each time step according to the following equation:

$$[Ca^{++}]_t = [Ca^{++}]_{t-1} + \Delta t \cdot (-5.18 \times 10^{-3} \cdot I_T/ \tag{15}$$
$$(\text{Area} \cdot \text{depth}) - \beta \cdot [Ca^{++}]_{t-1}),$$

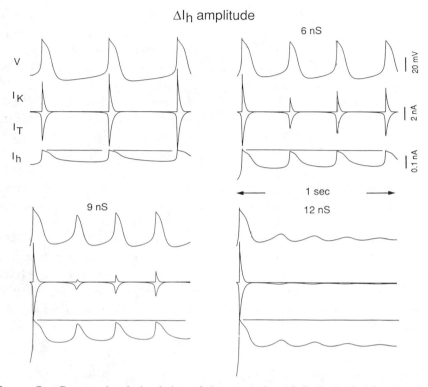

FIGURE 7. Computational simulation of the generation of rhythmic Ca^{++} spikes in thalamic neurons and the effect of changing the amplitude of I_h. Releasing the model with 3 nS maximal h-conductance at a membrane potential of -85 mV results in depolarization of the membrane potential due to activation of I_h. This depolarization activates I_T, which generates a low-threshold Ca^{++} spike. The peak of the spike is truncated by the activation of the K$^+$ current, I_K. The low-threshold Ca^{++} spike repolarizes due to inactivation of I_T. During the Ca^{++} spike, I_h falls nearly to zero (line). Upon termination of the Ca^{++} spike, the membrane potential repolarizes and activates I_h, which subsequently depolarizes the membrane and thus starts a new cycle. Changing the maximal h-conductance from 3 nS to 9 nS has the effect of increasing the frequency of oscillation and decreasing the amplitude of each Ca^{++} spike. Increasing the maximal h-conductance to 12 nS results in only a damped oscillation, which terminates after approximately 1 second.

where concentration is in moles per liter, time is in msec, current is in nA, and area is in μm^2. The area of membrane to use is not clear, since the location of the low-threshold Ca^{++} current in thalamic neurons has not yet been determined. Although it has been suggested that this current is largely somatic in origin (Steriade and Llinás, 1988), *in vitro* patch clamp studies suggest that dendrites may also express I_T (Coulter *et al.*, 1989a). In the present model, we have used the average total membrane area of relay neurons in the cat LGNd, which is about 38,000 μm^2 (Bloomfield *et al.*, 1987). The rate of diffusion and buffering of Ca^{++} in thalamic neurons has not yet been experimentally investigated. Therefore, we used a wide range of values for the diffusion rate variable β and have found that values near or greater than 1 yield satisfactory rhythmic Ca^{++} spikes. At values substantially less than 1, the concentration of Ca^{++} increases dramatically and remains in the 100 nm below the inner surface of the membrane for prolonged periods of time, thereby reducing the ability of the model to rhythmically oscillate by reducing the amplitude of I_T.

2. Delayed Rectifier K Current, I_K.   At present, no voltage clamp data has been published on the different types of voltage-activated K currents in thalamic relay neurons. Although we found that voltage-sensitive K currents were not necessary to replicate rhythmic Ca^{++} spikes in the present simulation, we did find that these currents modulated to a significant degree the amplitude-time course of these Ca^{++} spikes. Therefore, we have elected to include a delayed rectifier K current by adjusting the equations developed by Yamada *et al.*, (1989) for bullfrog sympathetic ganglion cells to 35°C. In this manner, the following relations for I_K were obtained:

$$I_K = gK_{max} \cdot n^4 \cdot (V + 105), \tag{16}$$

where the equilibrium potential for K^+ was taken as -105 mV (McCormick and Prince, 1988). The Hodgkin–Huxley kinetic equations for the activation variable n were:

$$\alpha_{mk} = \frac{0.096 \cdot (-45 - V)}{e^{((-45-V)/5.0)} - 1.0}, \tag{17}$$

$$\beta_{mk} = 1.5 \cdot e^{((-50-V)/40)}, \tag{18}$$

I_K in most neuronal systems is known to slowly inactivate (Rudy, 1988). However, since this inactivation occurs over a very slow time course (seconds), we elected not to include it here.

3. Hyperpolarization-Activated Cation Current, I_h.   Hyperpolarization of thalamic relay neurons results in the activation of an inward current carried by Na^+ and K^+ ions and denoted I_h for *hyperpolarization-activated* (Fig. 6a). This current has an equilibrium potential of -40 mV (Fig. 6b), which is approximately halfway between E_{Na} and E_K, is relatively small with a total membrane conductance of between 5 and 25 nS, and does not inactivate (Fig. 6a; also McCormick and Pape, 1990a). The rate of activation of I_h varies from a time constant of greater than 1 second at threshold (approximately -60 mV) to around 100 msec at membrane potentials, where the current is fully activated (e.g., -100 mV). (See Fig. 6d.) The rate of deactivation of I_h varies in just the opposite manner, becoming much faster with depolarization (McCormick and Pape, 1990a; also Fig. 6d). In the present model, we have modeled I_h with the following set of equations:

$$I_h = gh_{max} \cdot m_h^2 \cdot (V + 40), \tag{19}$$

where the steady-state value of the activation variable m_h is described by:

$$m_{h\infty} = \frac{1}{1 + e^{((V+69)/7.1)}}, \tag{20}$$

and the time constant of m_h is given by:

$$\tau_h = \frac{1,000}{e^{(-6.26-0.0767*V)} + e^{(7.13+0.1079*V)}}. \tag{21}$$

The rate of change of m_h was then calculated in the simulation according to the following relation:

$$\frac{dm_h}{dt} = \frac{m_{h\infty} - m_h}{\tau_h}. \tag{22}$$

4.   Leak K Current I_{Kleak}.   Leak K^+ currents are important, for they strongly influence the resting membrane potential of the cell and the ability of the cell to demonstrate rhythmic oscillations. In addition, at least one type of leak K^+ current is under the influence of a number of modulatory neurotransmitter systems in the LGNd (McCormick and Prince, 1987, 1988). Here, we modeled the leak K^+ current by:

$$I_{Kleak} = gKleak_{max} \cdot (V + 105), \tag{23}$$

where g_{Kleak} was typically 6 nS in accordance with the apparent input resistance of the neuron in Figs. 2 and 3 at rest and in the absence of I_h (approximately 165 MΩ).

C.   Simulation of Two-State Firing of Thalamic Relay Neurons

Cat thalamic relay neurons *in vitro* generate rhythmic bursts of action potentials due to the presence of the rhythmic occurrence of low-threshold Ca^{++} spikes (Figs. 2, 3). We have previously proposed that the rhythmic Ca^{++} spikes result from the interaction between the low-threshold Ca^{++} current I_T and the hyperpolarization-activated cation current I_h, while the fast action potentials are mediated by the traditional mechanisms of a fast inward sodium current and various voltage-sensitive K^+ currents (McCormick and Pape, 1990a). The most negative membrane potential achieved between the rhythmic Ca^{++} spikes is approximately -85 mV. Therefore, we simulated the neuronal activity that resulted from releasing the neuron from a holding potential of -85 mV and included, in addition to I_T, I_K, I_h and I_{Kleak}, two other voltage-activated currents, I_{NA} and I_A, to mediate fast action potentials. Under these circumstances, the simulation successfully replicated the ability of thalamic neurons to rhythmically generate Ca^{++} spike-mediated bursts of action potentials at a frequency of approximately 2–3 Hz (Fig. 4). Intracellular injection of a depolarizing current pulse (I_{inj}) resulted initially in a high-frequency burst of action potential resulting from the activation of I_T, followed by a continuous train of action potentials signaling the switch of the neuron to the tonic or single-spike mode of action potential generation (Fig. 4). Removal of the depolarizing current injection allowed the cell to return to rhythmic oscillation (Fig. 4). In this manner, the simulation was able to replicate with a fair degree of accuracy the two firing modes of thalamic relay neurons. In accordance with results obtained *in vitro,* block of fast Na^+ current I_{Na} resulted in the appearance of rhythmic Ca^{++} spikes only (Figs. 7–10). In addition, the transient K^+ current I_A was also found not to be necessary for the generation of rhythmic Ca^{++} spikes, although it did modulate the shape of each Ca^{++} spike (Huguenard, unpublished results). Therefore, to simplify the present model, we included only the four currents: I_T, I_K, I_h, and I_{Kleak}.

1. Simulation of Alteration in Amplitude and Voltage Sensitivity of I_h. By removing the transient Na^+ and K^+ currents I_{Nat} and I_A, we were able to generate rhythmic Ca^{++} spikes with only three active currents, I_T, I_K, and I_h, and one passive current I_{Kleak} (Figs. 7–10). Using the model, we have investigated a number of factors that influence the ability of thalamic relay neurons to rhythmically generate Ca^{++} spikes and how these factors influence this oscillation. Among the different variables that will influence the ability of neurons to oscillate, perhaps the most interesting are the amplitude and voltage dependency of I_T, I_h, and I_{Kleak}, since these appear to be under the influence of neuromodulatory neurotransmitters (McCormick and Prince, 1987, 1988; McCormick and Pape, 1990a). However, investigating the influence of subtle changes in these three currents on rhythmic oscillations is difficult to perform *in vivo* or *in vitro*. Therefore, in the present simulation, we have investigated the role of each of these different currents in the generation of rhythmic Ca^{++} spikes in thalamic neurons.

To investigate the influence of I_h on rhythmic Ca^{++} spike generation, gT_{max}

was set at 0.4 μS, $g\mathrm{K}_{max}$ at 2 μS, $g\mathrm{Kleak}_{max}$ at 6 nS, and $g\mathrm{h}_{max}$ was varied from 0 to 20 nS. In this manner, rhythmic Ca^{++} generation was found to occur only with values of $g\mathrm{h}_{max}$ in the range of 3 and 12 nS (Fig. 7). With $g\mathrm{h}_{max}$ less than 3 nS, rhythmic oscillation failed due to an inability of I_h to overcome the influence of I_{Kleak} and depolarize the neuron to activate a Ca^{++} spike (not shown). In contrast, values of $g\mathrm{h}_{max}$ above approximately 10 nS resulted in an abolition of oscillatory activity due to the reduction in the extent to which the membrane potential hyperpolarized after each Ca^{++} spike (due to the persistence of I_h during this time period) and, subsequently, the lack of sufficient removal of inactivation of I_T to support the generation of rhythmic Ca^{++}; spikes (Fig. 7, 12 nS). Increasing $g\mathrm{h}_{max}$ from 3 nS to 12 nS gradually increased the rate at which Ca^{++} spikes are generated from a frequency of approximately 2 Hz to one of about 4 Hz and decreased the amplitude of I_T during each Ca^{++} spike (Fig. 7). This result replicates intracellular recordings from cat thalamic neurons *in vitro* in which reduction of I_h with local application of Cs^+ is found to result in an increased propensity of the neurons to oscillate, an increase in the amplitude of the Ca^{++} spike during these oscillations, and a slowing of the frequency of oscillations (McCormick and Pape, 1990a). Also in similarity with the model, strong block of I_h *in vitro* is found to result in an abolition of rhythmic oscillatory burst firing (McCormick and Pape, 1990a).

Activation of a number of neuromodulatory receptors on thalamic neurons, including β-adrenergic, serotoninergic, and H_2 histaminergic, can result in shifts in the activation curve for I_h (McCormick and Pape, 1990b). The maximal shift in activation curve that could be obtained from these agonists was about 5–10 mV positive and was proposed to underlie the subsequent abolition of rhythmic oscillation. The effects of smaller changes in the voltage dependency of I_h have not yet been investigated (McCormick and Pape, 1990b). In the present simulation, we found that, like *in vitro*, positive shifts of the voltage dependency of I_h by 5 mV resulted in an abolition of the ability of the neuron to generate rhythmic Ca^{++} spikes (*cf.* Fig. 8, 0 mV and +5 mV). In contrast, a shift of I_h voltage dependence to more negative membrane potentials by 5 mV resulted in an enhancement of the rhythmic Ca^{++} spikes and a slowing of the frequency of oscillation (Fig. 8, −5 mV). Investigations of smaller changes in voltage sensitivity of I_h revealed changes in the amplitude of Ca^{++} spikes and frequency of Ca^{++} spike generation that were in between these values. These results suggest that weak activation or removal of activation of postsynaptic receptors coupled to adenylyl cyclase may be capable of controlling the frequency of rhythmic burst firing *in vivo* during different behavioral states. This suggestion remains to be adequately tested in thalamic relay neurons *in vitro* or *in vivo* with small doses of norepinephrine, serotonin or histamine.

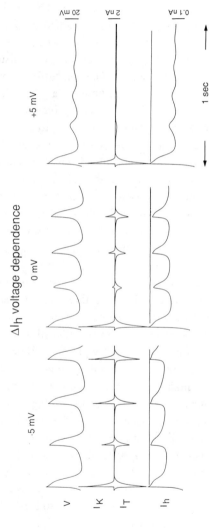

FIGURE 8. Effects of changing voltage dependence of I_h on rhythmic oscillation in thalamic neurons. Shifting the voltage dependence and kinetics of activation of I_h by 5 mV more positive results in the ability of thalamic neurons to generate rhythmic Ca^{++} spikes to be dampened, as has been demonstrated *in vitro* through the stimulation of β-adrenergic receptors or serotoninergic receptors (McCormick and Pape, 1990b). Shifting the voltage dependence by 5 mV to more negative membrane potentials results, in contrast, results in an enhancement of the rhythmic Ca^{++} spikes and a slowing of the rhythm.

282

2.   Simulation of Alteration in Amplitude of I_T.   Alteration of the maximal T-conductance revealed that in the present simulations (with gh_{max} set at 9 nS), rhythmic Ca^{++} spikes could be generated with any value of gT_{max} greater than approximately 0.3 μS (Fig. 9). Increasing gT_{max} was found to strongly influence the rate of rhythmic Ca^{++} spike generation, such that increases in gT_{max} resulted in an increase in the frequency of oscillation (Fig. 9). Although similar experiments have not yet been performed *in vitro* or *in vivo*, block of the low-threshold Ca^{++} current is known to block the generation of low-threshold Ca^{++} spikes whether they occur rhythmically or not (Jahnsen and Llinás, 1984a,b; McCormick, unpublished observations). These results indicate that not only is the amplitude of each Ca^{++} spike-mediated burst *in vivo* related to the density of T-conductance in the neuron, but the rate at which these Ca^{++} spikes are generated is also influenced.

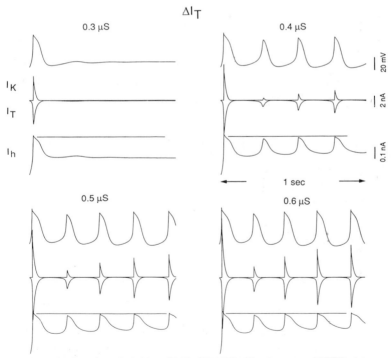

FIGURE 9.   Effects of alteration in the amplitude of the low-threshold Ca^{++} current I_T on the generation of rhythmic Ca^{++} spikes. Reducing the maximal conductance of I_T below approximately 0.32 μS abolished the ability of the simulated thalamic neuron to oscillate. Increasing the maximal conductance of I_T increased the amplitude of the Ca^{++} current during each Ca^{++} spike and increased the rate of rhythmic Ca^{++} spike generation.

3. Simulation of Alteration in Amplitude of I_{Kleak}.

Alterations in the amplitude of gKleak$_{max}$ affects three properties of thalamic neurons:

1. The course of the membrane potential between Ca^{++} spikes when I_{Kleak} forms a substantial portion of the total ionic current.
2. The input resistance of the neuron at these membrane potentials.
3. The membrane time constant of the neuron at these membrane potentials.

When gKleak$_{max}$ is reduced to values less than approximately 4 nS (with gh$_{max}$ at 9 nS and gT$_{max}$ at 0.4 μs), rhythmic oscillation fails due to a reduction in the hyperpolarization of the neuron after each Ca^{++} spike and a subsequent decrease in removal of inactivation of I_T (Fig. 10, 2 nS). Similarly, when gKleak$_{max}$ is too large (e.g., 12 nS), rhythmic oscillation also fails due to an inability of I_h

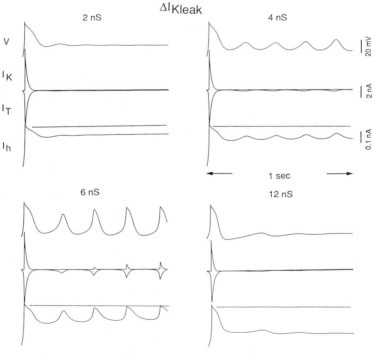

FIGURE 10. Effect of changing the amplitude of I_{Kleak} on rhythmic Ca^{++} spike generation. With too small a value for leak K conductance (2 nS), rhythmic oscillation fails due to a decrease in hyperpolarization after each Ca^{++} spike (which is necessary to remove inactivation of I_T). In contrast, when the leak K conductance is made too large, rhythmic oscillation also fails due to a failure of I_h to depolarize the neuron sufficiently to activate sufficient I_T to generate a rhythmic Ca^{++} spike. Values of gKleak$_{max}$ between 4 and 10 nS supported the generation of rhythmic Ca^{++} spikes.

to depolarize the neuron sufficiently to activate a Ca^{++} spike (Fig. 10, 12 nS). The generation of rhythmic Ca^{++} spikes was supported by values of $g\text{Kleak}_{max}$ between 4 and 10 nS (Fig. 10). Increasing the amplitude of $g\text{Kleak}_{max}$ decreased the rate of rhythmic oscillation, although these changes in rate were not as robust as those seen with changes in gh_{max} or gT_{max} (cf. Figs. 7–10). These results indicate that the passive, as well as the active, properties of thalamic neurons contribute to the ability of these cells to generate different types of neuronal activity.

D.   Summary of Thalamic Simulation

The present simulation confirms and extends the important contribution of the low-threshold Ca^{++} current I_T and the hyperpolarization-activated cation current I_h in the generation of rhythmic burst firing in thalamic relay neurons. We have shown that simulation of currents that have voltage dependence and kinetics similar to those recorded in actual thalamic neurons can replicate some of the basic electrophysiological properties of these neurons. These results suggest that the generation of rhythmic Ca^{++} spikes can be achieved with only two voltage-gated currents, I_T and I_h, while depolarization-activated K^+ currents are important in modulating the shape of the individual Ca^{++} spikes (not shown).

V.   Functional Implications of Multistate Neuronal Activity

The presence of multiple states of activity in forebrain neurons has a number of important consequences for the pattern and fate of synaptic and neuronal activity generated. Since the earliest electrophysiological investigations, it has been recognized that different behavioral states are associated with different types of neuronal activity in the thalamus and neocortex (reviewed in Magoun, 1958; Brazier, 1980). Changes in these different states of the nervous system can be viewed as changes in the pattern of activity of neurons and neural circuits from the propensity to generate one subset of oscillatory activities to that of another. For example, periods of drowsiness or EEG-synchronized sleep are associated with the appearance of cooperative rhythmic burst firing in thalamocortical circuits, generally in the frequency range of 1–12 Hz (for example, Steriade and Deschênes, 1984), while during periods of arousal and cognition, these slower frequencies are abolished and neuronal activity is characterized by either a less synchronous nature or the appearance of oscillations in the range of 40–60 Hz. In this manner, neurons and neuronal circuits appear to have intrinsic harmonic frequencies that are brought out under different circumstances in accordance with the status of modulatory neurotransmitter agents and fast synaptic potentials. This is functionally important, since many different neurological phenomena also have intrinsic frequencies that arise from a perversion of the natural oscillatory phenomenon of the forebrain. The example, absence seizures are expressed as a marked 3 Hz synchronous oscillation appearing throughout the forebrain and especially in the neocortex and thalamus. (See review by Buzsaki *et al.*, 1990). This abnormal hypersynchronous 3 Hz rhythm appears to be a perversion of the neural circuits responsible for the generation of the normal 7–12 Hz spindle waves: namely, thalamic neurons and intra-thalamic interactions. Indeed, recent results have shown that drugs of choice for the treatment of absence seizures, such as ethosuximide, reduce the T-current by approximately 30% in clinically relevant concentrations (Coulter *et al.*, 1989b). Our present simulations indicate that such a reduction in I_T may very well block the ability of thalamic neurons to generate rhythmic bursts of action potentials (Fig. 9), thereby removing the pacemaker for absence seizures. In addition to absence seizures, certain frequencies of Parkinsonian tremor may also depend upon rhythmic burst firing in thalamic neurons (Buzsaki *et al.*, 1990) and, therefore, the ability of these cells to generate this form of activity may have important implications for the treatment of this disorder.

In normal animals, the presence of the different states of neuronal activity in thalamic neurons has the functional consequence of dramatically changing the manner in which sensory stimuli are transmitted to the neocortex. Let us consider each of the different possible states that a thalamic neuron may exhibit:

1. Rhythmic burst firing.
2. In the burst firing mode but not rhythmically bursting.
3. In between the burst firing mode and the single-spike firing mode.
4. In the single-spike firing mode.

During periods in which rhythmic burst firing is prevalent, thalamic relay neurons appear to respond best to inputs that are within the frequency range of this intrinsic oscillation (e.g., 1–12 Hz), and in this manner act as a *band-pass* filter. When synaptic inputs arrive at frequencies higher than just 12 Hz, rhythmic burst firing can be disrupted due to a decrease in the hyperpolarizing phase between each burst, and, therefore, a decrease in the removal of inactivation of I_T, which is critical for generating each Ca^{++} spike. At frequencies between 1 and 12 Hz, those synaptic potentials that are in phase with the rhythmic oscillatory activity may add or facilitate each burst, while those that are not will not be transmitted to the neocortex and will antagonize the oscillatory activity (McCormick and Feeser, 1990). Thus, during active periods of rhythmic burst firing, the transmission of sensory information to the neocortex would appear to be markedly degraded or limited to the frequency domain of the oscillation. This observation may help to explain the finding that the responsiveness of thalamic neurons to sensory stimuli is significantly reduced during periods of rhythmic burst firing (Livingston and Hubel, 1981).

In thalamic neurons that are in the burst firing mode but not actively engaged in oscillation, the burst firing mechanism acts to greatly emphasize events that occur at low (<12 Hz) frequencies such that single events that arrive at frequencies less than once per second can give rise to a high-frequency burst of 3–6 action potentials. As the frequency of excitatory input increases, the number of spikes per burst decreases until it completely fails at frequencies in excess of approximately 12 Hz (McCormick and Feeser, 1990). This property of the burst firing mode may allow thalamic neurons to markedly enhance synaptic potentials relating to some functionally significant event, such as the detection of a moving object in a dimly lit visual scene. The arrival of the synaptic potentials in the thalamus corresponding to this event may trigger a single Ca^{++} spike, or a rhythmic barrage of Ca^{++} spikes, which may then generate and transmit a high-frequency (300–500 Hz) barrage of fast Na^{+}- and K^{+}-mediated action potentials to the neocortex. An additional consequence of the burst firing mode is that sustained responses would be less prevalent. Thus, the thalamic relay neuron may respond to a sensory-evoked barrage of excitatory postsynaptic potentials with an initial burst response and then either switch to the single-spike mode or fall silent (Jahnsen and Llinás, 1984a; McCormick and Feeser, 1990).

If thalamic neurons are depolarized positive to approximately -65 mV, the transient low-threshold Ca^{++} current is inactivated and burst firing in response to excitatory inputs is abolished (Jahnsen and Llinás, 1984a). Under these conditions, the neuron neither is in the burst firing mode nor has it reached firing threshold for generation of trains of action potentials (-55 mV). Therefore, only the largest excitatory postsynaptic potentials (>10 mV), or barrages of EPSPs, will be transmitted to the neocortex. Further depolarization of the neuron as a result of the influence of neuromodulatory neurotransmitters, such as acetylcholine and norepinephrine (McCormick and Prince, 1987, 1988), will result in an increase in sensitivity of the neuron to depolarizing inputs such that smaller and smaller sensory-evoked EPSPs will be transmitted to the cortex. A further consequence of the single-spike firing mode is that the *bandwidth* of the neuron is extended to range from zero to at least 100 Hz, and thus gives rise to a more faithful representation of the excitatory synaptic input (McCormick and Feeser, 1990). This increase in the dynamic range of thalamic neuronal activity with depolarization into the single-spike firing mode probably contributes to the well-known increase in responsiveness of thalamic neurons to sensory stimulation during periods of arousal and attentiveness (Livingstone and Hubel, 1981; McCormick, 1989).

VI. Conclusions

Thalamic neurons serve as an example of multistate neurons in the CNS. These cells are capable of at least two intrinsic modes of action potential generation: slow rhythmic burst firing and single-spike activity. The generation of rhythmic burst firing appears to result from the interaction of the low-threshold Ca^{++} current and a *pacemaker* current, the hyperpolarization-activated cation current known as I_h. Computational simulation of thalamic neurons indicates that alterations in either of these currents will not only affect the ability of thalamic neurons to oscillate, but also the frequency of oscillation. Through alterations of specific ionic currents, such as I_h, the ascending modulatory neurotransmitter inputs are able to dramatically influence the firing mode of thalamic neurons. These alterations in thalamic neuronal activity, in turn, have important consequences on the processing and transmittal of sensory information to the neocortex.

Acknowledgments

We thank Pratik Mukerjee for his work on earlier versions of this model. This work was supported by the National Institute of Neurological Disorders and Stroke, the John and Esther Klingenstein Fund, the Sloan Foundation, the Jane and Peter Pattison Fund, the Jacob Javits Center for Neuroscience, and by a gift from IBM.

References

ANDERSEN, P., and ANDERSSON, S. A. (1968). *Physiological Basis of the Alpha Rhythm.* Appleton, Century, Croft, New York.

BLOOMFIELD, S. A., HAMOS, J. E., and SHERMAN, S. M. (1987). "Passive Cable Properties of Neurones in the Lateral Geniculate Nucleus of Cat," *Journal of Physiology* **383,** 653–692.

BRAZIER, M. A. B. (1980). "The Historical Development of Neurophysiology," in J. A. Hobson and M. A. B. Brazier (eds.), *The Reticular Formation Revisited.* Raven Press, New York.

BUZSAKI, G., SMITH, A., BERGER, S., FISHER, L. J., and GAGE, F. H. (1990). "Petit mal Epilepsy and Parkinsonian Tremor: Hypothesis of a Common Pacemaker," *Neuroscience* **36,** 1–14.

CATON, R. (1875). "Researches on Electrical Phenomena of Cerebral Gray Mater," *Transactions of the Ninth International Medical Congress* **3**, 246–249.

CONNOR, J. A., and STEVENS, C. F. (1971). "Inward and Delayed Outward Membrane Currents in Isolated Neural Somata under Voltage Clamp," *Journal of Physiology* **213**, 1–19.

COULTER, D. A., HUGUENARD, J. R., and PRINCE, D. A. (1989a). "Calcium Currents in Rat Thalamocortical Relay Neurons: Kinetic Properties of the Transient, Low Threshold Current," *Journal of Physiology* **414**, 587–604.

COULTER, D. A., HUGUENARD, J. R., and PRINCE, D. A. (1989b). "Specific Petit mal Anticonvulsants Reduce Calcium Currents in Thalamic Neurons," *Neuroscience Letters* **98**, 74–78.

CRUNELLI, V., LIGHTOWLER, S., and POLLARD, C. E. (1989). "A T-Type Ca^{++} Current Underlies Low-Threshold Ca^{++} Potentials in Cells of the Cat and Rat Lateral Geniculate Nucleus," *Journal of Physiology* **413**, 543–561.

ECCLES, J. C. (1964). *The Physiology of Synapses*. Academic Press, New York.

FOURMENT, A., HIRSCH, J. C., MARC, M. E., and GUIDET, C. (1984). "Modulation of Postsynaptic Activities of Thalamic Lateral Geniculate Neurons by Spontaneous Changes in Number of Retinal Inputs in Chronic Cats: 1. Input–Output Relations," *Neuroscience* **12**, 453–464.

HERNANDEZ-CRUZ, A., and PAPE, H.-C. (1989). "Identification of Two Calcium Currents in Acutely Dissociated Neurons from the Rat Lateral Geniculate Nucleus," *Journal of Neurophysiology* **61**, 1270–1283.

HIRSCH, J. C., FOURMENT, A., and MARC, M. E. (1983). "Sleep-Related Variations of Membrane Potential in the Lateral Geniculate Body Relay Neurons of the Cat," *Brain Research* **259**, 308–312.

HODGKIN, A. L., and HUXLEY, A. F. (1952). "A Quantitative Description of Membrane Current and Its Application to Conduction and Excitation of Nerve," *Journal of Physiology* **117**, 500–544.

JAHNSEN, H., and LLINÁS, R. (1984a). "Electrophysiological Properties of Guinea-Pig Thalamic Neurons: an *in vitro* Study," *Journal of Physiology* **349**, 105–226.

JAHNSEN. H., and LLINÁS, R. (1984b). "Ionic Basis for the Electroresponsiveness and Oscillatory Properties of Guinea-Pig Thalamic Neurons *in vitro*," *Journal of Physiology* **349**, 227–247.

JONES, E. G. (1985). *The Thalamus*. Plenum, New York.

LAMARRE, Y., FILION, M., and CORDEAU, J. P. (1971). "Neuronal Discharge of the Ventrolateral Nucleus of the Thalamus during Sleep and Wakefulness in the Cat: I. Spontaneous Activity," *Experimental Brain Research* **12**, 480–498.

LIVINGSTONE, M. S., and HUBEL, D. H (1981). "Effects of Sleep and Arousal on the Processing of Visual Information in the Cat," *Nature* **291**, 554–561.

LLINÁS, R. R. (1988). "The Intrinsic Electrophysiological Properties of Mammalian Neurons: Insights into Central Nervous System Function," *Science* **242**, 1654–1664.

MAGOUN, H. W. (1958). *The Waking Brain*. Charles C. Thomas, Springfield, Illinois.

McCARLEY, R. W., BENOIT, O., and BARRIONUEVO, G. (1983). "Lateral Geniculate

Nucleus Unitary Discharge in Sleep and Waking: State- and Rate-Specific Aspects," *Journal of Neurophysiology* **50,** 798–818.

McCORMICK, D. A. (1989). "Cholinergic and Noradrenergic Modulation of Thalamocortical Processing," *Trends in Neuroscience* **12,** 215–221.

McCORMICK, D. A. (1990a). "Possible Ionic Basis for Lagged Visual Responses in Cat LGNd Relay Neurones," *Society for Neuroscience Abstract* **16,** 159.

McCORMICK, D. A. (1990b). "Membrane Properties and Neurotransmitter Actions," in G.M. Shepherd (ed.), *The Synaptic Organization of the Brain*, 3rd ed. (pp. 32–66). Oxford University Press, New York.

McCORMICK, D. A., and FEESER, H. R. (1990). "Functional Implications of Burst Firing and Single Spike Activity in Lateral Geniculate Relay Neurons," *Neuroscience* **39,** 103–113.

McCORMICK, D. A., and PAPE, H.-C. (1990a). "Properties of a Hyperpolarization Activated Cation Current and Its Role in Rhythmic Oscillation in Thalamic Relay Neurons," *Journal of Physiology* **431,** 291–318.

McCORMICK, D. A., and PAPE, H.-C. (1990b). "Noradrenergic and Serotonergic Modulation of the Hyperpolarization Activated Cation Current I_h in Thalamic Relay Neurons," *Journal of Physiology* **431,** 319–342.

McCORMICK, D. A., and PRINCE, D. A. (1987). "Actions of Acetylcholine in the Guinea Pig and Cat Lateral and Medial Geniculate Nuclei," *Journal of Physiology* **392,** 147–165.

McCORMICK, D. A., and PRINCE, D. A. (1988). "Noradrenergic Modulation of Firing Pattern in Guinea Pig and Cat Thalamic Neurons, *in vitro*," *Journal of Neurophysiology* **59,** 978–996.

McMULLEN, T. A., and LY, N. (1988). "Model of Oscillatory Activity in Thalamic Neurons: Role of Voltage- and Calcium-Dependent Ionic Conductances," *Biological Cybernetics* **58,** 243–259.

ROSE, R. M., and HINDMARSH, J. L. (1989). "The Assembly of Ionic Currents in a Thalamic Neuron: I. The Three -Dimensional Model," *Proceedings of the Royal Society of London Series B* **237,** 267–288.

RUDY, B. (1988). "Diversity and Ubiquity of K Channels," *Neuroscience* **25,** 729–749.

STERIADE, M., and DESCHÊNES, M. (1984). "The Thalamus as a Neuronal Oscillator," *Brain Research Reviews* **37,** 1093–1113.

STERIADE, M., and LLINÁS, R. R. (1988). "The Functional States of the Thalamus and the Associated Neuronal Interplay," *Phsyiological Reviews* **68,** 649–742.

STERIADE, M., JONES, E. G., and LLINÁS, R. R. (1990). *Thalamic Oscillations and Signalling*. John Wiley, New York.

STORM, J. F. (1990). "Potassium Current in Hippocampal Pyramidal Cells," *Progress in Brain Research* **83,** 161–187.

TRAUB, R. D. (1982). "Simulation of Intrinsic Bursting in CA3 Hippocampal Neurons," *Neuroscience* **7,** 1233–1242.

YAMADA, W. M., KOCH, C., and ADAMS, P. R. (1989). "Multiple Channels and Calcium Dynamics," in C. Koch and I. Segev (eds.), *Methods in Neuronal Modeling*. MIT Press, Cambridge, Massachusetts.

Chapter 11

Temporal Information Processing in Synapses, Cells, and Circuits

Philip S. Antón, Richard Granger, and Gary Lynch

Bonney Center for the Neurobiology of Learning and Memory
University of California,
Irvine, California

I. Introduction

Biological systems face information-processing problem domains significantly more complex than those that have been successfully addressed by artificial systems. A primary question about information processing in the brain is how disparate specialized peripheral systems can communicate with central telencephalic systems; at the core of the mystery is the role of time. From the time constants of particular ion channel openings to the stylized rhythmic activity patterns prevalent across neocortex, time is a dominant factor in flow and transformation of information in the brain. This chapter addresses issues of temporal processing at multiple levels of brain organization:

CHANNELS. Two primary glutamate receptor channel subtypes, NMDA and AMPA, have distinctly different time courses, yet are co-localized at telencephalic synapses (Bekkers and Stevens, 1989) and are often coactive, especially during induction of synaptic long-term potentiation (LTP). The interactions of these time courses lead to sensitivity of cells to temporal sequences (Granger *et al.*, 1991).

Single Neuron Computation
Neural Nets: Foundations to Applications

SYNAPSES.   The optimal activity pattern for synapse-specific LTP induction is the rhythmic *theta-burst* stimulus (TBS) (Larson and Lynch, 1989); this also is the only pattern that occurs in both experimental induction paradigms and in natural physiological activity. This activity pattern has been shown to lead to the network-level ability for hierarchical clustering in cortical networks (Ambros-Ingerson *et al.*, 1990).

CELLS.   If the biophysics underlying all cellular potentials were well understood, then biophysical models of temporal interactions among temporally patterned inputs would be preferable to physiological models on grounds of precision. However, in many situations, cellular potentials are well characterized in terms of their shape and amplitude, whereas the precise mechanisms underlying these responses remain uncertain. In such cases, it is desirable to model temporal cellular behavior by combining physiological PSP curves juxtaposed in time. This requires improvement on alpha functions, which cannot accurately model complex PSP shapes (Antón, 1991).

LOCAL CIRCUITS.   Many cortical areas contain *local circuits* including numerous excitatory cells and a smaller number of inhibitory interneurons. Often these interneurons arborize extensively within a local region, innervating and being innervated by most of the excitatory cells in that region. Complex temporal interactions among excitatory and inhibitory elements in local circuits can give rise to simple *competitive* effects: Only the most strongly activated cell in such a *patch* responds; the rest are inhibited (Coultrip *et al.*, 1991).

NETWORKS.   Temporal processing in networks of different anatomical architectures can differ considerably; each such network suggests different computational functionality. For instance, the olfactory bulb includes the unusual features of glomeruli and graded dendrodendritic synapses, and it has markedly more inhibitory (granule) cells than excitatory (mitral/tufted) cells. In contrast, olfactory (piriform) cortex has no glomerular structure, few or no dendrodendritic synapses, and many more excitatory than inhibitory cells. These distinct architectural features have given rise to extremely different computational hypotheses (Granger *et al.*, 1989; Antón, 1991; Antón *et al.*, 1991).

SYSTEMS.   The olfactory bulb and olfactory cortex are strongly connected by both feedforward (bulb–cortex) and feedback (cortex–bulb) pathways; the systems interaction among these two distinct networks gives rise to computational abilities not apparent in either one in isolation (Granger *et al.*, 1990b; Ambros-Ingerson *et al.*, 1990).

This chapter begins by deriving a novel strategy for accurately modeling temporal interactions among synaptic inputs, and then proceeds to describe a series of modeling results, ranging from synaptic to network interactions, based on complex temporal processing of time-varying inputs.

II.   Physiological Models of Cellular PSPs

A.   Introduction

Biophysical single-cell models are relatively computationally expensive, but true to known cell membrane and electrical properties; physiological-level single-cell models are less expensive but do not necessarily accurately reflect membrane and channel properties. Yet the latter have the advantage that physiological EPSPs can be curve-fit with great accuracy, and thus the detailed temporal characteristics of unit responses can be simulated appropriately. The same would be true of biophysical models only if the biophysics of cellular potentials were sufficiently well-understood, which still is not the case.

Extant biophysical models of single neurons—based on the electrical properties of cell membranes and channels—include cable theory (Rall, 1960, 1962, 1969, 1989; Jack et al., 1975; Koch et al., 1982) and compartmental models (Rall, 1964; Segev et al., 1985, 1989; Bunow et al., 1985). Models at a physiological level of description have also been used when the computations under study are believed to arise from temporal combinations of membrane potential transients or patterns of input combined with rigid or plastic synaptic weights (Perkel, 1964; Walløe et al., 1969; Jack et al., 1975; Shamma, 1989; Antón, 1991; Antón et al., 1991). These physiological-level models are motivated by the simple linear summation of membrane transients observable in cells at low levels of input (Shepherd, 1979). Their efficiency permits large studies of the temporal interplay of excitatory and inhibitory postsynaptic potentials (EPSPs and IPSPs) as well as investigations into the maps between cellular inputs and responses via spiking thresholds or firing-rate functions.

If the biophysics underlying all cellular potentials were well understood, then biophysical models would be preferable to physiological models on the grounds of precision. However, in many situations, cellular potentials are well characterized in terms of their shape and amplitude, whereas the precise mechanisms and sources underlying these responses remain uncertain. In such cases, physiological models can attempt to model cellular behavior by combining copies of PSP measurements juxtaposed in time. Unfortunately, the simplest physiological models do not accurately reflect the nonlinearities due to time-constant and driving-force interactions between multiple PSPs. In addition, the functions typically employed for the representation of PSPs often do not contain enough parameters to allow accurate curve-fitting of actual PSPs measured from neurons. These shortcomings can become extremely relevant if, for instance, the size of PSPs begins to be reduced as a result of nonlinear interactions that change the effective membrane time constant and synaptic driving force. Also, particular shapes of PSPs cannot be modeled using intuitively realistic alpha functions, yet precise PSP shapes may be crucial for studies involving patterns of input, including temporal overlap. In addition, silent inhibition cannot be modeled in the simplest physiological representations, since their inhibitory effects are due to changes in the effective membrane time constant rather than an IPSP summed with membrane potential.

We present a set of novel PSP functions—derived from single lumped-circuit cell representations—which do accurately model these interactive effects and fit precise physiological PSP data (Antón, 1991; Antón *et al.*, 1991). These new PSP functions provide computationally efficient, closed-form iterative solutions to the lumped circuits, permitting iterative maintenance of interactive effects in an efficient manner during simulation while retaining the simplicity of the physiological modeling approach. These new findings partially bridge the gap between biophysical and physiological models without greatly increasing the computational complexity of the physiological simulations. The result is a novel physiological model for temporal integration of synaptic inputs to nerve cells (Fig. 1), including the functions for interactive effects among PSPs through time-constant and driving-force effects.

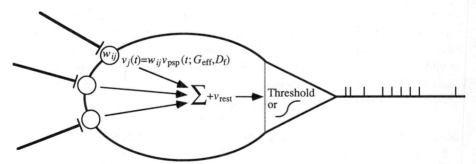

FIGURE 1.   Components in a typical physiological model of a cell. Synaptic strength variables w_{ij} (or G_{syn_0}) are multiplied by the PSP shape function v_{psp} to yield the synaptic potential contribution. These contributions are then summed linearly with the resting potential v_{rest} to obtain the membrane potential for the cell. This potential is then compared to a firing threshold or firing-rate function to determine cell output.

B.   PSP Functions

The PSP temporal aspects included in a physiological simulation are modeled using PSP functions. Let us represent PSP transients by a weight variable w_{ij} between the ith afferent cell and the jth cell multiplied by the shape function v_{psp}:

$$v_j = w_{ij}v_{psp}. \tag{1}$$

Of course, if weights are not an issue, then we can set $w_{ij} = 1$ so the PSP will be represented by v_{psp} alone.

Commonly used functions and the properties they model include:

- Delta functions $v_j = w_{ij}\delta(t - t_0)$
  (modeling a simple time-delayed synaptic weight)
- Step functions $v_j = w_{ij}s(t_1,t_2)$
  (modeling a non-instantaneous synaptic weight)
- Decaying exponentials $v_j = w_{ij}e^{-at}$
  (modeling a decaying temporal value)
- Alpha functions $v_j = ew_{ij}\alpha te^{-\alpha t}$
  (modeling nonzero rise and fall times)
- Simple difference-of-exponentials functions
  $v_j = (e^{-b_jt} - e^{-G_m t/C_m})D_{f_j}(0)G_{syn0_j}/(G_m - b_jC_m)$
  derived later (modeling realistic rise and fall times)
- Interactive difference-of-exponentials functions (Eq. (6)) derived later (modeling time-constant and driving-force effects as well as realistic rise and fall times)

(See Fig. 2.) While alpha and the difference-of-exponentials functions are visually similar to a typical PSP transient, quantitative measurements of the rise and fall times of actual PSPs often yield shapes that a single-parameter alpha function cannot match. Thus, the difference-of-exponentials functions (derived later, and similar to the cable theory solution by Rall for a current step pulse (Rall, 1962, 1969, 1989)) are the most complex yet also the most realistic.

As pointed out by Shamma, the use of simple delta functions will concentrate the simulator's emphasis on the synaptic weights rather than the temporal properties of the PSPs, thus allowing clear study of the importance of these weights on the network computation without interference from temporal effects (Shamma, 1989). On the other hand, difference-of-exponentials have enough parameters to allow curve fitting of actual PSP measurements from the soma; these curves implicitly reflect the electrotonic changes imposed by the cell's dendritic structure

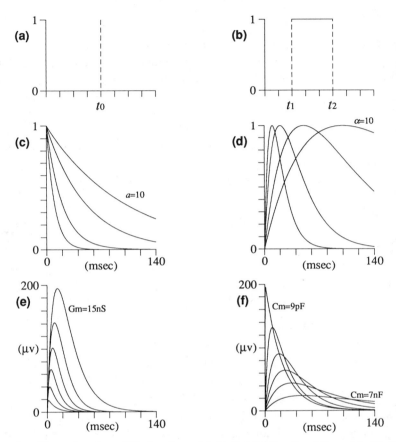

FIGURE 2. Plots of various PSP functions: (a) Unit delta function $\delta(t - t_0)$. (b) Unit step function $s(t_1,t_2)$. (c) Unit exponential decay e^{-at}. $a = $ 10,20,50, and 100. (d) Unit alpha functions $eate^{-at}$. $\alpha = $ 10,20,50, and 100. (e) Difference-of-exponentials $(e^{-b_jt} - e^{-G_mt/C_m})$ $D_{f_j}(0)G_{syn0_j}/(G_m - b_jC_m)$. $C_m = $ 270 pF and $b = $ 100, while $G_m = $ 15 nS, 30 nS, 54 nS, 100 nS, 200 nS, and 500 nS. (f) Same function as in e, but with $G_m = $ 54 nS, $b = $ 45, and $C_m = $ 9 pF, 270 pF, 800 pF, 1600 pF, 3000 pF, and 7000 pF.

on a PSP when traveling from the dendritic input site to the soma, since most intracellular physiological data presently involves somatic measurements. Curve fitting, therefore, can allow the physiological simulation to reflect some of the influences imposed by the spatial geometry that is not directly simulated. This benefit is most relevant in simulations where the primary output is axonal, since cell firing is determined by somatic membrane potential relative to the firing threshold.

1. A Simple Iterative PSP Function for Interactive Effects. Nonlinear interactions between PSPs begin to take effect as the number of inputs to a cell increases. Increased synaptic conductances increase the overall effective membrane conductance (G_{eff}) of the cell, thus speeding the membrane decay constant G_{eff}/C_m due to faster discharging of the membrane capacitance (C_m) (Rall, 1962; Shepherd, 1979). Such time-constant changes will therefore change the amplitude as well as the rise and fall times of PSPs. Changes in membrane potential also cause changes in subsequent synaptic currents, since the driving forces at the synapses are different from the resting driving forces.

In addition to summation of PSPs, silent inhibitions might be possible for a cell type under study if the membrane potential at input time equals the reversal potential for the synapse (Coombs et al., 1955; Rall, 1962, 1964; Jack et al., 1975). In this case, the synaptic activations result in the opening of synaptic channels but lead to no visible PSPs due to the absence of a driving force. Nevertheless, the channel openings change the overall effective membrane constant and thus effect the shape and amplitude of other inputs to the cell.

To obtain a PSP function that permits consideration of these interactive effects in a physiological simulation, an interactive PSP function was derived from a lower-level lumped-circuit representation of a cell (Antón, 1991). This approach uses the superposition theorem on the *RC* circuit representation (Fig. 3a) to obtain the voltage contribution due to each individual synaptic activation. In superposition, all the synaptic batteries but the one for the synapse in question are short-circuited. This leaves the other synaptic conductances in parallel with the membrane conductance, yielding an effective conductance of

$$G_{eff_j}(t) = G_m + \sum_{k \neq j} G_{syn_k}(t),\qquad(2)$$

as seen by the *j*th synapse. For an exponential synaptic conductance transient (Colquhoun, 1981), the current injected into the equivalent *RC* circuit of Fig. 3b is

$$i_j(t) = (E_{syn_j} - v_{m_j}(t))\, G_{syn0_j} e^{-b_j t},\qquad(3)$$

where b is the channel-closing time constant, E_{syn_j} is the synaptic reversal potential, $v_{m_j}(t) = v_j(t) + v_{rest}$ is the membrane potential due to the *j*th synapse, and v_{rest} is the resting membrane potential.

Now let us consider a simulation time step Δt in which the changes in effective conductance $G_{eff_j}(t)$ and synaptic driving force $(E_{syn_j} - v_{m_j}(t))$ during the time step are small (or small enough to yield the desired simulation accuracy). The

(a)

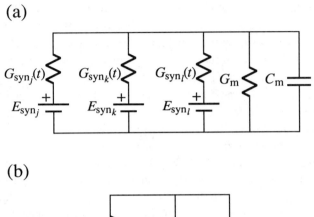

(b)

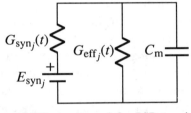

FIGURE 3.   Development of RC circuit used for PSP transient equation. (a) Typical lumped-circuit model of a cell's passive dendritic tree and soma with parallel paths representing multiple synapses (Rall, 1962; Tuckwell, 1988). (b) Circuit for response of E_{syn_j} that combines all other synaptic conductances with the membrane conductance.

differential equation for the jth synapse's membrane circuit during the interval $t \in (T, T + \Delta t]$ is then approximated by:

$$I_j e^{b_j t} = C_\text{m} \frac{dv_j}{dt} + G_{\text{eff}_j} v_j, \tag{4}$$

where $I_j = (E_{\text{syn}_j} - v_{\text{m}_j}(T))G_{\text{syn}0_j}$ and $G_{\text{eff}_j}\text{-}G_{\text{eff}_j}(T)$ are constants.
    The solution of this differential equation is given by:

$$v_j(T + \Delta t) = \frac{(E_{\text{syn}_j} - v_{\text{m}_j}(T))\, G_{\text{syn}0_j} e^{-b_j T}}{(G_{\text{eff}_j}(t) - b_j C_\text{m})} \, [e^{-b_j \Delta t} - e^{-G_{\text{eff}_j}(t)\Delta t/C_\text{m}}]$$
$$+ v_j(T)e^{-G_{\text{eff}_j}(t)\Delta t/C_\text{m}} \tag{5}$$

for $b_j \neq G_{\text{eff}_j}/C_\text{m}$, since the synaptic conductance has decayed down to the value $G_{\text{syn}0_j} e^{-b_j T}$ by time T.
    We now define variables for the driving force, active component, and passive component of $v_j(t)$. For PSPs with nonzero initial driving force, let:

$$D_{\text{f}_j}(T) = (E_{\text{syn}_j} - v_{\text{m}_j}(T)),$$

$$A_j(t) = \frac{D_{f_j}(0)G_{\text{syn}0_j}e^{-b_jT}}{G_{\text{eff}_j}(t) - b_jC_m} [e^{-b_j\Delta t} - e^{-G_{\text{eff}_j}(t)\Delta t/C_m}], \text{ and}$$

$$P_j(t) = v_j(T)e^{-G_{\text{eff}_j}(t)\Delta t/C_m}.$$

At this point, we may wish to update the driving force from $D_{f_j}(T)$ to $D_{f_j}(t)$ to obtain a better approximation of $v_j(t)$. Thus, the expression for $v_j(t)$ becomes:

$$v_j(t) = A_j(t)\frac{D_{f_j}(t)}{D_{f_j}(0)} + P_j(t).$$

Solving for $v_j(t)$ then yields the PSP function:

$$v_j(t) = \frac{A_j(t) + P_j(t)}{1 + A_j(t)/D_{f_j}(0)} \tag{6}$$

Note that in the case of a silent inhibitory input k, there is no driving force to cause a voltage contribution directly through $v_k(t)$. Thus, no PSP function for the silent input is needed. An inhibitory effect is reflected, however, through the maintenance of the effective membrane conductance $G_{\text{eff}_j}(t)$ for the other PSP inputs $j \neq k$. Thus, the PSP function expressed in Eq. (6) can model silent inhibitions as well as the other interactive effects between PSPs in a lumped-circuit model while remaining quite efficient and scaling linearly in the number of synapses modeled.

Interestingly, if we wish to neglect the interactive effects on membrane time-constant and driving force, we can use the superposition approach on the RC circuit representation to obtain a PSP function that is more flexible than alpha functions when curve-fitting actual data. In this case, D_{f_j} and G_{eff_j} are constants, and the continuous solution to the RC circuit can be shown to be:

$$v_j(t) = \frac{D_{f_j}(0)G_{\text{syn}0_j}}{(G_m - b_jC_m)} [e^{-b_jt} - e^{-G_mt/C_m}] \tag{7}$$

(Antón, 1991). This function can be used very efficiently by pre-calculating and storing the PSP voltage transient for each simulation time step in an array for later use during summation; each active synapse would then only require a simple addition during run time. Such an approach may be useful if the levels of activity are low and silent inhibitions are not present.

C. Summation

Once the voltage contributions have been determined by the PSP function, simple linear summation of these contributions together with the resting potential of the cell determines the simulated cell potential. Even the more sophisticated PSP functions using Eq. (6) can be summed linearly due to the use of the linear superposition theorem in the equation's derivation. The cell membrane potential is then given by:

$$v_m(t) = v_{rest} + \sum_j v_j(t) \tag{8}$$

for the PSP functions $v_j(t)$. Note that any long-term inputs that may elevate all the resting potentials of the cells under study can be modeled and studied by merely changing the resting potential summed together with the transient PSP inputs.

D.   Output Determination

After the inputs to the cell have been combined into a resultant potential value, the model must somehow make the decision whether to fire or not. Common approaches in physiological models include using a firing threshold, an instantaneous firing-rate function, or a direct implementation of the Hodgkin–Huxley or polynomial equations for voltage-dependent activations.

The most straightforward yet least efficient approach is to model directly the nonlinear dynamics of the excitable membrane of the axon via Hodgkin–Huxley or polynomial equations. Here, the simulation would merely input the summed voltage into the desired nonlinear equations to produce continuous action potential voltages (Jack *et al.*, 1975; Bunow *et al.*, 1985; Segev *et al.*, 1989; Rinzel and Ermentrout, 1989).

The action-potential waveforms could, however, be represented by a simplified instantaneous spike or linear approximations together with an output determination function. This simplified approach is especially appropriate if the primary function of the action potentials in the simulation is merely to activate efferent synapses or if the simulation time steps are of the same order as action potential dynamics.

One such approach simplifies the voltage dependence of axonal activation to a comparison of total membrane potential with an abstract firing threshold (Perkel, 1964; Walløe *et al.*, 1969; Jack *et al.*, 1975; Getting, 1989; Antón, 1991; Antón *et al.*, 1991). In this integrate-and-fire method, an axonal output is generated when the membrane potential crosses an activation threshold. This threshold may be fixed or variable to reflect absolute and relative refractory periods of the cell. Absolute refraction is modeled by an infinite threshold, while relative refraction is modeled by some type of elevated threshold behavior after firing. Hyperexcitability can even be included by following the relative refractory period with a period containing a lower-than-normal firing threshold.

A second approach is to use an instantaneous firing-rate function to transform the cell activity level into an output firing frequency (Shamma, 1989). This approach simplifies the threshold and refractory period behavior into a simple transformation that models the peak firing frequency (reflecting the absolute refractory period), scaling of firing frequency (reflecting a decaying relative-refractory period), and minimum activity level for output (reflecting the minimum firing threshold). Sigmoidal functions such as

$$z(t) = \frac{z_{\max}}{1 + e^{a(v_b - v)}} \tag{9}$$

(Shamma, 1989) are useful for this purpose since they suppress low-level inputs, have a nearly linear mid-range, and have a decaying peak to maximum. (See sketch in Fig. 1.) Here, $z(v_b) = z_{\max}/2$ and the slope of $z(t)$ at v_b is $a/4$.

III.   Physiological Modeling of Temporal Integrative Properties

Lumped-cell physiological modeling of single neurons has proven quite useful in studying the capabilities that temporal integration gives neuronal architectures. Here, we present a few examples of physiologically-based phenomena involving both small and large numbers of interacting neurons.

A.   *Physiological Modeling of Threshold Gradations in the Olfactory Bulb Reveal a Frequency-to-Spatial Transformation*

This simulation of the olfactory bulb (Fig. 4) illustrates the usefulness of physiological modeling, given the neural properties deemed relevant to the function under study (Antón, 1991; Antón *et al.*, 1991). In this application, the interactive PSP equations (Eq. (6)) were employed, since large numbers of excitatory inputs converge on each primary cell in the bulb (about 6,250 ON inputs in the rat (Shepherd, 1979; Mori, 1987)). The goal of the simulation was to test whether gradations by depth of the firing thresholds in the primary mitral/tufted (M/T) cells could lead to a transformation of the frequency-related inputs from the olfactory nerve (ON) into a spatial representation composed of the number of M/T cells firing. In this case, the primary feature of the bulb neurons that needed to be simulated was the temporal summation of the PSPs due to the excitatory nerve inputs. (See Fig. 5.) As expected, an increase of the input frequency led to more EPSPs overlapping in time and, therefore, a higher level of activity in the primary cells, causing additional cells with higher thresholds to fire. The result was a sigmoidal transformation curve for bulbar activity given the frequency-related concentration inputs. (See Fig. 6.)

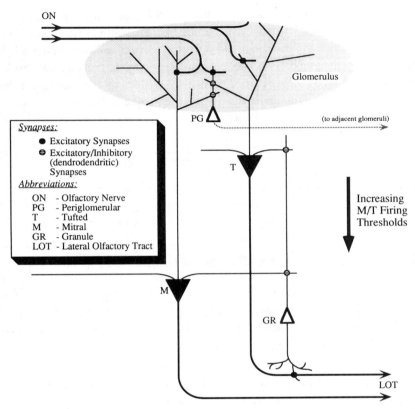

ON

Glomerulus

(to adjacent glomeruli)

PG

T

Increasing
M/T Firing
Thresholds

M

GR

LOT

Synapses:

● Excitatory Synapses
⊖ Excitatory/Inhibitory
   (dendrodendritic)
   Synapses

Abbreviations:

ON - Olfactory Nerve
PG - Periglomerular
T   - Tufted
M   - Mitral
GR - Granule
LOT - Lateral Olfactory Tract

FIGURE 4.   Schematic diagram of the simulated anatomy of an olfactory bulb glomerular section (Shepherd, 1979; Mori, 1987). Mitral/tufted cell axons (bottom lines in the figure) comprise the lateral olfactory tract (LOT), the primary output from bulb and input to layer I olfactory cortex.

Physiological modeling had a number of benefits in this case. First, representation of the temporal aspects of the PSPs provided testing of the transformation hypothesis. Second, the efficiency of the method allowed large numbers of cells to be simulated. Extensions of the simulation to include large numbers of cells, centrifugal inputs, or more interacting structures (e.g., multiple glomeruli) could be included without a large impact on the simulation time, since the computational complexity scales linearly with the number of synapses simulated. Third, the interactive PSP functions reduced concerns that possible saturation and other nonlinear effects would be relevant to the transformation and yet excluded from the simulation. Fourth, the efficiency allowed for rapid testing of simulation parameters when compared to simulation methods that consume large amounts of time. Fifth, the model produced realistic data (somatic potential

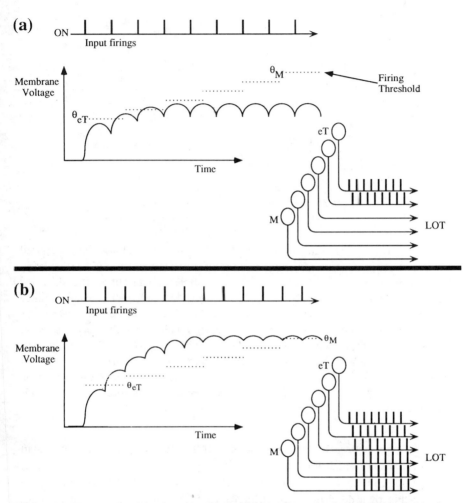

FIGURE 5.   Schematic of the frequency-to-spatial transformation based on threshold gradations. The dashed lines illustrate the increased firing threshold by depth of the mitral/tufted cells (Schneider and Scott, 1983; Mori, 1987). The lower section illustrates the increased number of mitral/tufted cells responding to the increased olfactory nerve (ON) firing frequency due to the gradation in firing thresholds (θ).

combined with linear approximations of action potential spikes) that permitted the use of one's experience with physiological data in understanding the results. More abstract methods that do not provide temporal PSP data would not permit the use of these visual intuitions.

While the physiological approach had many benefits, there were properties that could not be considered due to the nature of the physiological model. The

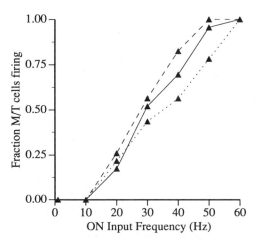

FIGURE 6.   The number of mitral/tufted (M/T) cells responding within a single simulated glomerulus in response to varying frequencies of afferent olfactory nerve (ON) stimulation in the observed range for olfactory nerve firings from 1Hz to 60Hz. The simulated glomerulus consisted of 625 ON inputs, 23 excitatory M/T cells, 90 inhibitory granule interneurons, and 15 periglomerular interneurons. The three curves shown illustrate the sigmoidal transformation with saturation levels varying with interneuronal synaptic strength. M/T inhibitory synaptic conductances G_{syn0} were 0.5 nS (long dashes), 1 nS (solid line), and 2 nS (short dashes). M/T firing thresholds were set at -52.9 mV $+$ $0.2\,i$ mV for the ith cell, where i increased with depth from the most external tufted cell ($i = 1$) to the deepest mitral cell ($i = 22$). The number of dendrodendritic connections with granule cells was equal for each M/T cell. (See Antón *et al.*, 1991.)

lumped-circuit approach eliminated effects on the PSPs due to local rather than global effects unless abstract approximations are made to compensate for them. For example, inputs to the granule cells from the mitral cells are made on spine heads rather than dendritic shafts (Shepherd, 1979; Mori, 1987). These spines tend to facilitate the reciprocal response back onto the same mitral cell (Antón, 1991). An approximation of this effect might be to lower the granule-to-mitral dendrodendritic activation threshold to ensure a proper reciprocal response (Antón *et al.*, 1991). Of course, this approximation would also facilitate lateral inhibitions to other mitral cells due to the lumped nature of the granule membrane potential. This problem might be corrected at the programming level by activating the lower granule-to-mitral threshold only if the same synapse received mitral-to-granule activation. Consideration of such abstract programming fixes must be weighed together with the increased effort and complexity the fixes impose, and the need to consider such problems at all must be weighed given the task presented to the simulation work.

B.   Temporal Integration Provides Winner-Take-All Circuits

Layer II olfactory ;ortex is composed of many times more excitatory cells than inhibitory cells. Th :se fewer inhil 'tory interneurons exhibit relatively fixed radial axonal aborizations, contacting a ıocal region or *patch* of cortex. (See, e.g., Van Hoesen and Pandya, 1975; Haberly, 1983.) The resulting local-circuit architecture can be modeled by several excitatory cells innervating and receiving feedback from a common inhibitory interneuron. Simulations of physiological interactions among cells in a local circuit of this kind were performed. (See Coultrip *et al.*, 1991). Excitatory cells in the simulation received input during a cycle of naturally occurring rhythmic activity (the olfactory-hippocampal theta rhythm). Since the neuron receiving the most input activation reached firing threshold first, it was able to activate the local inhibitory cell, which, in turn, prevented the rest of the cells from firing. (See Fig. 7.) This organization,

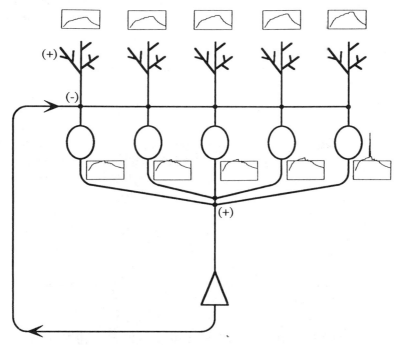

FIGURE 7.   A simple winner-take-all local circuit consisting of five excitatory cells (circles) and a single inhibitory cell (triangle). The top boxes show the dendritic input levels, while the middle boxes show the somatic potential and firing spikes. The cell on the far right received the greatest amount of input and was the first and only cell to fire. All other cells in this patch were subsequently inhibited by the resulting activation of the inhibitory cell at the bottom. (See Coultrip *et al.*, 1991.)

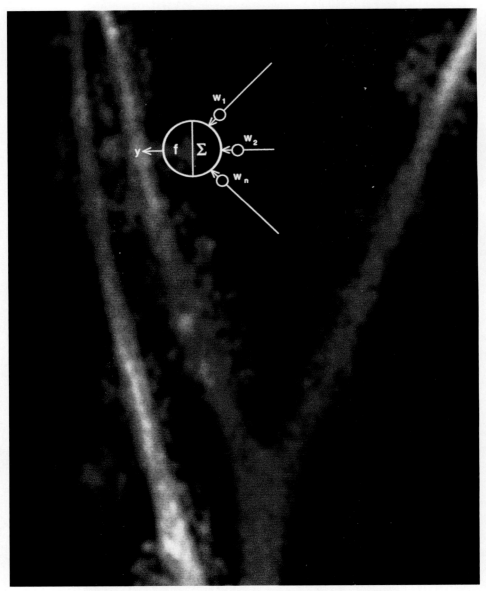

CHAPTER 4, FIGURE 2. Living hippocampal pyramidal neuron. This image of a portion of the dendritic tree was obtained using confocal scanning laser microscopy. Dendritic spines can be clearly resolved using this method. A typical PE is shown superimposed on one of the spines. The computations occurring in each of these spines (Fig. 11) appear to be considerably more complex than those in a typical PE (Fig. 1) and there are about 10^5 spines on each neuron. Rather than viewing PEs as neuron-like, it may be more appropriate to compare a PE to a small part of a neuron. In this case, the neuron could be compared to a multilayered neural network. It remains to be seen how easily a neural network can capture the essential computations that occur in a real neuron.

(a)

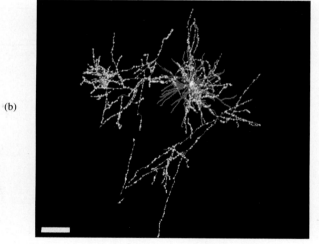

(b)

CHAPTER 14, FIGURE 1.   Detailed three-dimensional structure of a layer VI pyramidal neuron from the primary visual cortex (area V1) of the cat. The receptive field properties of this neuron were studied, and then the cell was labeled intracellularly with horseradish peroxidase. Later, the cell was processed histologically, and reconstructed using a computer-assisted method. The neuron is shown here as a wire figure, in the vertical plane (a) and horizontal plane (b). Green, dendrites; red, axon; white/blue, boutons. The boutons form two clear clusters in layer IV. One cluster coincides with the branching of the apical dendrite (detailed in c); the other is displaced horizontally by about 325 μm from the apical dendrite. Scale bar = 100 μm. *Figure continues.*

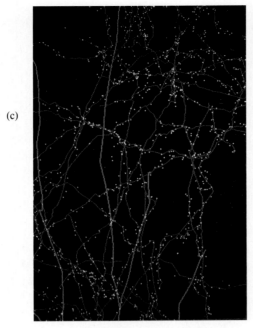

(c)

CHAPTER 14 FIGURE 1 *continued*

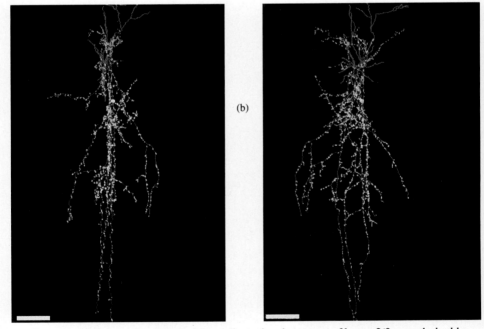

(a)

(b)

CHAPTER 14, FIGURE 2.   Detailed three-dimensional structure of layers 2/3 smooth *double bouquet* neuron in the primary visual cortex (area V1) of the cat. Methods and color codes are as for Fig. 1. The neuron is shown in the vertical plane in a front view (a) and side view (b). The axon arborizes below the somadendrite, and extends down to layer 6. The boutons cluster in layers 3/4 border, and also deeper in layer 4. Scale bar = 100 μm.

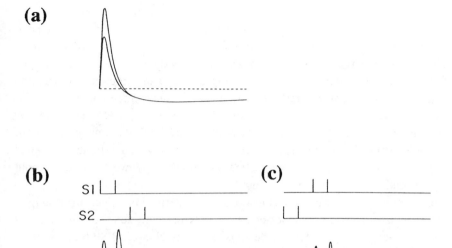

FIGURE 8.   SPICE simulations of single-cell responses to temporally patterned inputs. (a) Simulated postsynaptic potentials (measured at the soma) in response to single pulse input, before and after potentiation. Before potentiation, both afferent pathways ($S1$ and $S2$) had identical PSPs (peak conductance G_{max} = 2.5 nS; approximate peak synaptic current of 35 pA). After potentiation of pathway $S1$, its peak conductance was 5.0 nS; the $S2$ pathway remained unchanged. (b) Simulated somatic PSP in response to the temporal activation sequence $S1$–$S1$–$S2$–$S2$ (activated at 10 msec intervals, i.e., at 100 Hz), before and after potentiation of the $S1$ pathway. (c) Response to the reverse sequence $S2$–$S2$–$S1$–$S1$, before and after potentiation of the $S1$ pathway.

leading with $S1$ exhibited a response 35% larger than before potentiation, whereas leading with $S2$ led to a response only 29% larger than before potentiation; thus, potentiation of the $S1$ pathway caused a selective increase in the response to the sequence $S1$–$S1$–$S2$–$S2$ that is 21% larger than the increase to the reverse sequence. These simulation results predict differential LTP expression as a function of temporal order of activation of potentiated and unpotentiated pathways.

IV.   Discussion

Physiological models can be useful tools in our efforts to study brain function through the simulation of neural systems. Various PSP functions can be selected, depending on the emphasis of the simulation on synaptic weights, temporal summation, or interactive effects between synaptic inputs. The parameters of these functions—whether abstract or biophysical—adjust the approximation of PSP amplitudes, rise times, fall times, and channel dynamics, depending on the function chosen. Interactive effects can impose shape and amplitude modulation that may be important to the computations under study. Shape modulation, for example, may be important when testing the differences between simultaneous and slightly offset arrival times of activation and inhibitory inputs to a cell. Amplitude modulation may be important when incoming signal strength is critical to the cellular operation or when testing the difference between one large input and several smaller simultaneous inputs.

Even more sophisticated PSP functions can be derived to reflect, for example, more realistic synaptic conductance transients or more sophisticated channel models. (See Antón, 1991.) The benefits of a more complex PSP function must be weighed, however, against the increased computational overhead of their implementations.

Note that physiological models can simulate short- and long-term synaptic and cellular modifications. Changes in synaptic strength can be incorporated by changing either the synaptic weights w_{ij} in the simpler PSP functions or the peak synaptic conductance G_{syn0_j} in the difference-of-exponentials functions. Other changes in the effectiveness of receptor activation or transmitter removal rates can be reflected through changes in the synaptic time constant b_j in the difference-of-exponentials functions. Still other changes in resting membrane potential v_{rest} can be incorporated directly into the summation expression (Eq. (8)). The synaptic weights can be static or changed during a simulation by supervisory modulations of the preceding parameters while incurring negligible computational costs. Static weights are useful when studying temporal input patterns or during the performance operation of the system. Dynamic weights are useful in testing the effect of various learning rules on the system under study or to investigate the structure as a possible memory site.

Note also that some electrotonic effects are indirectly included in the more accurate PSP functions, since their parameters are adjusted to produce results that match actual PSP data that reflects the electrotonic effects present in the cell. Thus, if somatic potential is the primary measurement to be modeled (e.g., if cell firing is the primary concern), then the PSP shapes they match have already undergone some electronic degradation; fitting the PSP functions to actual data then includes these effects in the PSP shapes as long as data is obtained from representative input locations.

The physiological approach is more efficient than cable and compartmental models (Rall, 1962, 1964, 1989; Jack *et al.*, 1975; Segev *et al.*, 1989; Koch *et al.,* 1982), since simple, closed-form, non-recursive analytical solutions are obtained for each individual PSP expression and for the subsequent linear summation of these membrane transients. Cable theory provides many analytical expressions for voltage transients in abstract situations that unfortunately contain extremely complex expressions and functions that preclude their use in the simulation of large numbers of neurons. Also, compartmental methods are quite useful in representing complex dendritic trees, cable effects, and active membranes, but require more computational overhead. If these added capabilities are deemed irrelevant to the interactions or computations under study, then the more efficient physiological approach may be used. Furthermore, the membrane voltage expressions of Koch *et al.* (1982) involve either convolution integrals for voltage transients or multiplicative Volterra expressions for steady-state membrane potentials. While the steady-state equations are simple enough, they are not applicable if the transient case is important. The more general transient-case expressions involve more run-time computational overhead in calculating local responses than does the physiological model while excelling when cable effects are studied.

In addition, actual PSPs can be modeled using the physiological approach to reasonably reflect biophysical interactions without knowing the actual electrical properties of the cells in question. While multi-compartmental simulations may more accurately model the interactions of PSPs, since they model the spatial dynamics of a cell, the electrical properties of a cell's dendritic tree and soma must unfortunately be investigated in great detail. If approximations of the actual cell properties are used through curve-fitting of physiological data, then much of the foundation for the accuracy of the multi-compartmental approach is undermined; observed phenomena would then be based on approximations and would thus have similar validity to the physiological approach. Investigations of computations not requiring spatial input properties can then be investigated physiologically with lower overhead while maintaining time-constant and driving-force effects.

As discussed previously, however, the more detailed mathematical or biophysical approaches can be useful if the detail they provide is necessary for the entire study or if the relevance of substructures to the computation under study needs to be studied in greater detail. In the latter case, the physiological approach can then be used to simulate the relevant factors in a more efficient manner that permits the subsequent simulation of larger numbers of cells and cell structures, increasing the field-of-view of the simulation.

While the physiological methods presented before, then, use simple lumped-circuit approximations of cellular membrane potentials, they do provide a model that is of considerable value for testing the significance of detailed physiological variables on the computations performed by neuronal aggregates. Interactive effects can be approximated without resort to compartmental simulations by using difference-of-exponentials functions that reflect changes in PSP amplitudes and shapes due to effective changes in the membrane time constant and synaptic driving forces. The efficiency of this physiological approach allows very large numbers of interacting cells to be modeled, facilitating a larger view of neuronal operation and studies into the function of different interacting structures.

Compartmental simulations excel at simulating spatial influences in varied dendritic trees (Rall, 1964, 1967, 1989; Jack *et al.*, 1975; Segev *et al.*, 1989). While their computational overhead is much higher than that for physiological models, they can provide accurate simulation of complex structures, including both temporal and spatial properties. In addition, the existence of simulation environments—e.g., SPICE (Segev *et al.*, 1985; Bunow *et al.*, 1985), SABER (Carnevale *et al.*, 1990), and GENESIS (Wilson and Bower, 1989)—permits a trade-off of lower programming time for increased computer time.

A powerful approach to studying neuronal structures, then, is to use these more detailed models (e.g., compartmental simulations, mathematical models) to answer questions of relevance to the computation under study, and to follow with more abstract modeling (e.g., fewer-compartment simulations, physiological models) to study the relevant complexities while reducing the computational overhead to allow simulations of larger structures with greater extent. This approach will also help in the often formidable task of teasing apart the relevant parameters, parameter ranges, and anatomical details from those that may be irrelevant to some particular task. The multilevel approach was used, for example, to test the spine effects on the mitral/granule cell interaction in greater detail using SABER while keeping the larger olfactory bulb simulation at the more abstract physiological level (Antón, 1991; Antón *et al.*, 1991). Furthermore, the frequency-to-spatial transformation itself, tested at the physiological level, was subsequently employed in a more abstract simulation involving cortico–bulbar interactions in the olfactory system (Ambros-Ingerson, 1990; Ambros-Ingerson *et al.*, 1990; Granger *et al.*, 1990a). In these studies, a sigmoidal function matching the physiological simulation data was used to relate each abstract glomeruli input level to the fraction of primary M/T cells activated in each respective glomerular slice. This sigmoid reflects, therefore, the frequency-to-spatial transformation effect on a population of cells without direct simulation of the individual M/T cell potentials and the firing threshold gradation.

A model is useful precisely to the extent that it is smaller than the phenomenon it purports to describe. Yet being smaller, it omits detail present in the object of study, raising the question of which details are to be omitted and which retained. As in the physical sciences, neurobiological models at different levels of description retain different amounts of detail. We have given examples of modeling at the synaptic, cellular, and circuit levels, and shown that, on occasion, complex emergent phenomena at one level may be incorporated in somewhat simplified form into higher-level models. We have focused throughout on the processing of signals across time, since this has seemed to us to be a central theme of information processing in the brain. Interestingly, temporal processes that might seem low-level have been found in our models to have network-level effects; for instance, our modeling efforts have indicated that the differences in frequency of firing of afferents from the olfactory nerve are processed in the olfactory bulb so as to change the frequency coding to spatial coding, and this in turn enables a form of encoding that is used by the olfactory cortex model to recognize families of odors. Moreover, slight asynchronisms in afferent activation of single cells have been shown to cause differential amounts of long-term potentiation (Larson and Lynch, 1989); unexpected was the possibility of differential LTP *expression,* but that is what is predicted by our single-cell modeling efforts. Further modeling at a more abstract level suggests that this differential expression can be used for the computationally useful task of temporal sequence recognition (Granger *et al.,* 1991). This is part of a growing pattern of results in which emergent results from a low level of modeling give rise to important features of a simplified higher-level model.

Acknowledgments

P. S. Antón was supported by a Ford Foundation Pre-Doctoral Fellowship and a University of California President's Dissertation Year Fellowship. Further funding was provided by ONR Grant #N00014-89-J-3179 (R. Granger and G. Lynch, PI's).

References

AMBROS-INGERSON, J. (1990). "Computational Properties and Behavioral Expression of Cortical–Peripheral Interactions Suggested by a Model of Olfactory Bulb and Piriform Cortex." Ph.D. Thesis, University of California, Irvine, California.

AMBROS-INGERSON, J., GRANGER, R., and LYNCH, G. (1990). "Simulation of Paleocortex Performs Hierarchical Clustering," *Science* **247**, 1344–1348.

ANTÓN, P. S. (1991). "Simulations of Information Processing, Control, and Plasticity Effects in the Olfactory Bulb." Ph.D. Thesis, University of California, Irvine, California.

ANTÓN, P. S., LYNCH, G., and GRANGER, R. (1991). "Computation of Frequency-to-Spatial Transform by Olfactory Bulb Glomeruli," *Biol. Cybern.* (in press).

BEKKERS, J. M., and STEVENS, C. F. (1989). "NMDA and Non-NMDA Receptors Are Co-Localized at Individual Excitatory Synapses in Cultured Rat Hippocampus," *Nature* **341**, 230–233.

BUNOW, B., SEGEV, I., and FLESHMAN, W. (1985). "Modeling the Electrical Behavior of Anatomically Complex Neurons Using a Network Analysis Program: Excitable Membrane," *Biol. Cybern.* **53,** 41–56.

CARNEVALE, N. T., WOOLF, T. B., and SHEPHERD, G. M. (1990). "Neuron Simulations with SABER," *J. Neurosci. Meth.* **33,** 135–148.

COLQUHOUN, D. (1981). "How Fast Do Drugs Work?" *Trends Pharmacol. Sci.* **2**(8), 212–217.

COOMBS, J. S., ECCLES, J. C., and FATT, P. (1955). "The Inhibitory Suppression of Reflex Discharges from Motoneurons," *J. Physiol.* **130,** 396–413.

COULTRIP, R., GRANGER, R., and LYNCH, G. (1991). "A Cortical Model of Winner-Take-All Competition via Lateral Inhibition," *Neural Networks* (in press).

GETTING, P. A. (1989). "Reconstruction of Small Neural Networks," in C. Koch and I. Segev (eds.), *Methods in Neuronal Modeling.* MIT Press, Cambridge, Massachusetts.

GRANGER, R., AMBROS-INGERSON, J., and LYNCH, G. (1989). "Derivation of Encoding Characteristics of Layer II Cerebral Cortex," *J. Cog. Neurosci.* **1**(1), 61–87.

GRANGER, R., AMBROS-INGERSON, J., ANTÓN, P., and LYNCH, G. (1990a). "Unsupervised Perceptual Learning: A Paleocortical Model," in S. J. Hanson and C. R. Olson (eds.), *Connectionist Modeling and Brain Function: The Developing Interface.* MIT Press, Cambridge, Massachusetts.

GRANGER, R., AMBROS-INGERSON, J., STAUBLI, U., and LYNCH, G. (1990b). "Memorial Operation of Multiple, Interacting Simulated Brain Structures," in M. Gluck and D. Rumelhart (eds.), *Neuroscience and Connectionist Theory* (pp. 95–129). Lawrence Erlbaum, Hillsdale, New Jersey.

GRANGER, R., WHITSON, J., LARSON, J., and LYNCH, G. (1991). Unpublished data.

GROSSBERG, S. (1976). "Adaptive Pattern Classification and Universal Recording: I. Parallel Development and Coding of Neural Feature Detectors," *Biol. Cybern.* **23,** 121–134.

HABERLY, L. B. (1983). "Structure of the Piriform Cortex of the Opossum: I. Description of the Neuron Types with Golgi Methods," *J. Comp. Neurol.* **213,** 163–187.

JACK, J. J. B., NOBLE, D., and TSIEN, R. W. (1975). *Electric Current Flow in Excitable Cells.* Clarendon Press, Oxford.

KOCH, C., POGGIO, T., and TORRE, V. (1982). "Retinal Ganglion Cells: A Functional Interpretation of Dendritic Morphology," *Phil. Trans. R. Soc. Lond.* **B298,** 227–264.

LARSON, J., and LYNCH, G. (1989). "Theta Pattern Stimulation and the Induction of LTP: The Sequence in which Synapses are Stimulated Determines the Degree to which They Potentiate," *Brain Res.* **489,** 49–58.

MORI, K. (1987). "Membrane and Synaptic Properties of Identified Neurons in the Olfactory Bulb," *Prog. Neurobiol.* **29,** 275–320.

PERKEL, D. H. (1964). "A Digital-Computer Model of Nerve Cell Functioning," Memorandum RM-4132-NIH. The Rand Corporation, Santa Monica, California.

RALL, W. (1960). "Membrane Potential Transients and Membrane Time Constant of Motoneurons," *Exp. Neurol.* **2,** 503–532.

RALL, W. (1962). "Theory of Physiological Properties of Dendrites," *Ann. N.Y. Acad. Sci.* **96**(4), 1071–1092.

RALL, W. (1964). "Theoretical Significance of Dendritic Trees for Neuronal Input–Output Relations," in R. F. Reiss (ed.), *Neural Theory and Modeling*. Stanford University Press, Stanford, California.

RALL, W. (1967). "Distinguishing Theoretical Synaptic Potentials Computed for Different Soma-Dendritic Distributions of Synaptic Inputs," *J. Neurophysiol.* **30,** 1138–1168.

RALL, W. (1969). "Time Constants and Electrotonic Length of Membrane Cylinders and Neurons," *Biophy. J.* **9,** 1483–1508.

RALL, W. (1989). "Cable Theory for Dendritic Neurons," in C. Koch and I. Segev (eds.), *Methods in Neuronal Modeling*. MIT Press, Cambridge, Massachusetts.

RINZEL, J., and ERMENTROUT, G. B. (1989). "Analysis of Neural Excitability and Oscillations," in C. Koch and I. Segev (eds.), *Methods in Neuronal Modeling*. MIT Press, Cambridge, Massachusetts.

SCHNEIDER, S. P., and SCOTT, J. W. (1983). "Orthodromic Response Properties of Rat Olfactory Bulb Mitral and Tufted Cells Correlate with Their Projection Patterns," *J. Neurophysiol.* **50**(2), 358–378.

SEGEV, I., FLESHMAN, J. W., MILLER, J. P., and BUNOW, B. (1985). "Modeling the Electrical Behavior of Anatomically Complex Neurons Using a Network Analysis Program: Passive Membrane," *Biol. Cybern.* **53,** 27–40.

SEGEV, I., FLESHMAN, J. W., and BURKE, R. E. (1989). "Compartmental Models of Complex Neurons," in C. Koch and I. Segev (eds.), *Methods in Neuronal Modeling*. MIT Press, Cambridge, Masschusetts.

SHAMMA, S. (1989). "Spatial and Temporal Processing in Central Auditory Networks," in C. Koch and I. Segev (eds.), *Methods in Neuronal Modeling*. MIT Press, Cambridge, Massachusetts.

SHEPHERD, G. M. (1979). *The Synaptic Organization of the Brain, 2nd edition*. Oxford University Press, New York.

SHOEMAKER, P., and HUTCHENS, C. (1991). "Synchronous Analog Implementation of Olfactory Model," Technical Report. Mid-Year Report 1, Naval Ocean Systems Center, San Diego.

TUCKWELL, H. C. (1988). *Introduction to Theoretical Neurobiology, Vol. 1*. Cambridge University Press, Cambridge U.K.

VAN HOESEN, G., and PANDYA, D. (1975). "Some Connections of the Entorhinal (Area 28) and Perirhinal (Area 35) Cortices of the Rhesus Monkey: I. Temporal Lobe Afferents," *Brain Res.* **95,** 1–24.

WALLØE, L., JANSEN, J. K. S., and NYGAARD, K. (1969). "A Computer Simulated Model of a Second Order Sensory Neuron," *Kybernetik* **6,** 130–140.

WILSON, M. A., and BOWER, J. M. (1989). "The Simulation of Large-Scale Neural Networks," in C. Koch and I. Segev (eds.), *Methods in Neuronal Modeling*. MIT Press, Cambridge, Massachusetts.

YUILLE, A. L., and GRZYWACZ, N. M. (1989). "A Winner-Take-All Mechanism Based on Presynaptic Inhibition Feedback," *Neural Comp.* **1**(3), 334–347.

Chapter 12   Multiplying with Synapses and Neurons

CHRISTOF KOCH

Computation and Neural Systems Program,
Division of Biology,
and Division of Engineering and Applied Science
California Institute of Technology,
Pasadena, California

TOMASO POGGIO

Center for Biological Information Processing,
Department of Brain and Cognitive Sciences
and Artificial Intelligence Laboratory,
Massachusetts Institute of Technology,
Cambridge, Massachusetts

I. Introduction

Properties and limitations of neurons and synapses are crucial in determining the algorithms used by the brain to perform specific computational tasks. In this chapter, we focus on the simplest type of nonlinearity, *multiplication*. We discuss (a) the role that multiplication has in the computations underlying motion perception and learning and (b) the computational power of multiplicative or polynomial neurons. We then review a number of different biophysical mechanisms that give rise to multiplicative interactions. The specificity of these interactions ranges from pairs of individual synapses to small sets of neurons. Furthermore, some of the mechanisms exploit random variations in neuronal properties. As we will demonstrate, both algorithmic and hardware constraints argue for polynomial networks and for the multiplication-like interactions that represent their core.

Single Neuron Computation
Neural Nets: Foundations to Applications

II.   Why Multiplications?

The standard model of a neuron that is used in most theories of the brain is based on the assumption that the primary computational mechanism used by neurons is the threshold mechanism associated with the generation of action potentials. Although, strictly speaking, it is not a threshold (Sabah and Leibovic, 1969), in the sense that there exists a continuous, if very steep, voltage response to arbitrary currents, it can be well-described—given the limited bandwidth of the intracellular potential—by a nonstationary threshold (Noble and Stein, 1966). Almost all network models approximate this time-dependent nonlinearity by a stationary nonlinearity. Thus, the classical McCulloch and Pitts (1943) network uses the following caricature of a spike mechanism: A neuron is active if the sum of its inputs, excitatory and inhibitory, exceeds a certain threshold, and is otherwise silent. The majority of work carried out in the neural network field assumes this basic model of a neuron, sometimes with the modification of a sigmoid-like, smooth nonlinearity, rather than a binary threshold (Wilson and Cowan, 1972; Hopfield, 1984).

However, this threshold mechanism is not the only nonlinearity that plays an important role in information processing in the brain. Over the years, a substantial body of evidence has grown to support the presence of multiplicative-like operations. Early advocates of the idea that graded potentials, dendro-dendritic synapses, and very specific synaptic arrangements can be used to process information are Bullock (1959) and Shepherd (1972). In the following, we will discuss some of this evidence, in particular in relation to motion perception. Given that multiplication—or its digital cousin, the AND gate—is the simplest possible nonlinearity, neuronal elements implementing multiplicative interactions can process information. A body of literature has developed that investigates the computational abilities of such *polynomial* or sigma-pi units. We will highlight some of the key results and ask whether or not such interactions can be learned from examples. A number of different biophysical mechanisms exist that mimic a polynomial interaction over at least some range of their input variables. We will describe some of these mechanisms and discuss their possible involvement in the neuronal operations underlying the computation of the direction of motion.

A.   How Powerful are Multiplications?

In the analog domain, linear operations alone cannot provide a sufficiently powerful set of operations for information processing. Nonlinear operations are needed. The traditional candidate for neural networks is the threshold operation, corresponding to the biophysical mechanism of spike generation. There is, however, increasing evidence that much processing takes place without action potentials, just in terms of graded signals. From this point of view, the obvious linear operation to be considered, in addition to threshold, is multiplication. Computationally, multiplication is probably the *simplest* and most fundamental nonlinear operation. One of the authors (Poggio and Reichardt, 1980, and Poggio, 1983) has argued that an interesting and sufficiently powerful class of computational machines is represented by *polynomial algorithms*.

Polynomial algorithms, in this context, are nothing other than approximation schemes based on multivariate polynomials. Figure 1b shows in network notation a polynomial algorithm. The *hidden* units are computing the monomials, whereas the weights w are just the coefficients of the polynomial.

Because of Weierstrass's approximation theorem, which roughly states that every real continuous function can be approximated by polynominals, polynomial algorithms can approximate under weak conditions all *smooth* input–output transductions. They are equivalent to forward networks with one hidden layer, where the basic operation is multiplication (and powers). They are very similar to what has been called sigma-pi networks (Feldman and Ballard, 1982; Rumelhart *et al.*, 1986; Volper and Hampson, 1987; Mel, 1990a).

Unlike the classic thresholded linear unit, the output of a polynomial is computed as a sum of contributions from a set of monominals:

$$P(\mathbf{x}) = a_1 + b_1 x_1 + b_2 x_2 + c_1 x_1^2 + c_2 x_1 x_2 + \ldots . \qquad (1)$$

Thus, the output is given by the linear sum of monominals, where each monominal is the product of k inputs x_i. (See Fig. 1b.) Monominals have been rediscovered today as sigma-pi units, often also described as *higher-order* units, since the multiplicative operation can be used to implement second- and higher-order polynomial relations among a set of inputs (Giles and Maxwell, 1987). Recent work, especially by Mel (1990a,b) and Durbin and Rumelhart (1989), suggests that networks based on sigma-pi units may be more powerful and have other advantages with respect to the more traditional threshold-based networks (Poggio, 1975a).

1.   Relations to Boolean Functions.   It is easy to see the power of polynomial units in the case of boolean functions and circuits. Let us define a *boolean linear threshold function* as

$$y = \text{Thres}\left(\sum_{i=1}^{n} w_i x_i \right), \qquad (2)$$

where $\text{Thres}(x) = 1$ if $x \geq 0$, and 0 (or -1 in some implementations) if

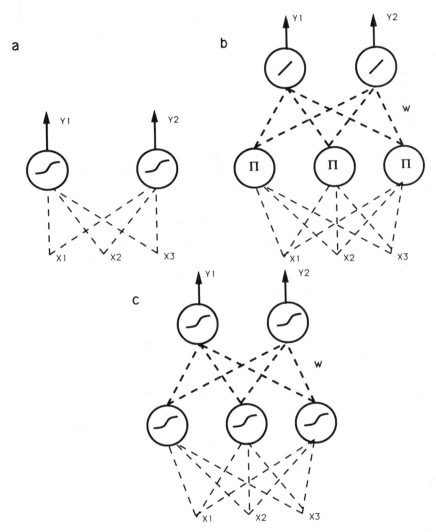

FIGURE 1.   Various neural network architectures with different computational powers.
(a) A standard perceptron with linear threshold unit (denotated with a sigmoidal threshold).
Its computational power is less than that of a network built out of multiplicative or sigma-
pi units shown in b, computing the linear sum of monominals in the input (Eq. (1)). The
schematic in c illustrates a linear threshold neural network with one hidden layer.

$x < 0$ and $x_i \in \{0,1\}$ or $\{-1,1\}$. We can now generalize to a boolean polynomial threshold function as:

$$y = \text{Thres}(P(\mathbf{x})), \tag{3}$$

where the polynominal $P(\mathbf{x})$ is as defined in Eq. (1). It is obvious that threshold units allow the synthesis of any boolean function by using networks with an input layer, a hidden layer with a threshold nonlinearity, and an output layer with threshold nonlinearities, i.e., a classical one-layer perceptron (Rosenblatt, 1962; also Fig. 1a). We also know that problems such as parity or exclusive-or, because they are non-separable, cannot be computed by a perceptron. However, they can trivially be implemented by a polynomial unit, even without having recourse to the threshold (e.g., $x_1 \oplus x_2 = x_1 x_2$ in the $\{-1,1\}$ representation). Thus, polynomial units are obviously more powerful than the standard perceptron, but less powerful than the standard hidden-layer neural network. (For details, see Bruck, 1989.)

B.   Relations to Smooth Functions

The situation regarding approximation of smooth functions in terms of networks based on sigmoidal units versus networks based on sigma-pi units is similar and quite clear. Both networks can approximate arbitrarily well any continuous function of n variables on a finite interval. In the case of sigmoid units, one hidden layer is sufficient for approximating any continuous function. (This does not even require a sigmoid at the output.) In the case of polynomial networks such as that illustrated in Fig. 1b, Weierstrass's theorem (Poggio, 1975a) ensures that a network composed of one input layer and one *hidden* layer of product units can represent arbitrarily well any continuous function on a finite interval. Depending on the accuracy of the required approximation, the polynomial may have a very large number of terms (and the networks of Fig. 1 a very large number of hidden units). Moore and Poggio (1988) proved a result that is related, and that also established a connection between multilayer perceptrons with units that take a nonlinear function (such as the sigmoid) of the sum of their inputs and polynomial networks. This result can be stated as follows:

Given a fixed polynominal of one variable, $f(x)$, for any polynominal of n variables, $g(y_1, y_2, \ldots, y_n)$, there exists a feedforward network of nodes, each of which computes $f(x)$ (where x is the weighted input to the node), that exactly computes g.

Networks of this type, again because of the Stone–Weierstrass theorem, can approximate arbitrarily well continuous functions on a bounded interval. This may give some insight into the case of sigmoidal nonlinearities, since the sigmoid function can be approximated by a polynomial (on a compact domain).

C. Experimental Evidence for Multiplicative-Like Interactions in Motion Perception

Some of the best evidence that multiplications are carried out in the nervous systems comes from an analysis of motion perception in insects and man. Data going back to the 1950s (Hassenstein and Reichardt 1956; reviewed in Reichardt, 1987) suggests that the key mechanism underlying the optomotor response of insects to moving stimuli is mediated by a correlation-like operation. In this scheme, the linearly filtered output from one receptor is multiplied by the temporally delayed and linearly filtered output from a neighboring receptor and then temporally averaged (Fig. 2a). As shown by Poggio and Reichardt (1973), any system able to signal the direction of motion in its temporal averaged response must contain a nonlinearity. This is because the time-averaged output of a linear interaction is identical to the result of the interaction of the time-averaged input signals. Furthermore, a multiplication, or second-order interaction, is the simplest possible interaction giving rise to directional selectivity. The strongest behavioral support for multiplication-like interactions in flies are as follows:

- The average optomotor response depends quadratically on the pattern contrast for small values of contrast (Buchner, 1976).
- The average optomotor response does not depend on the relative phases of the spatial Fourier component of a stimulus pattern moved at constant speed (*phase invariance*) (Hassenstein and Reichardt, 1956; Götz, 1972).
- The time-dependent optomotor response to a single sine-wave contains at most a second harmonic (*frequency-doubling*) (Geiger and Poggio, 1975; Pick, 1976).

It is important to emphasize that these properties usually only hold for sufficiently small values of contrast. For instance, while the average optomotor response of the fly is quadratic for small contrasts, it saturates for high-contrast values. Furthermore, all three properties are destroyed if the system contains higher-order interactions (e.g., cubic interactions) (Poggio and Reichardt, 1976). More recently, physiological data supporting the multiplication operation were obtained by intracellular recordings from identified direction-selective interneurons in the third optic ganglion of the blowfly *Calliphora* (Egelhaaf and Borst, 1989; Egel-

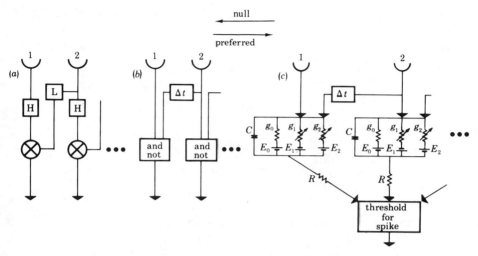

FIGURE 2.   Different models for detecting the direction of motion of the pattern $I(x,t)$ = $I(x - vt)$. One half the correlation detector (Hassenstein and Reichardt, 1956) is illustrated in a. The two inputs are multiplied after one input is low-pass filtered. If an average operation is performed on the output, the overall operation is equivalent to an autocorrelation on a linearly filtered version of $I(x - vt)$. Barlow and Levick (1965) proposed a related AND-NOT scheme to account for direction selectivity in the rabbit retina (b). Torre and Poggio (1978) proposed a biophysical implementation of this scheme (c), using the nonlinear interaction among an excitatory and a shunting type of inhibitory synapse on neighboring patches of dendrites. The underlying nonlinear interaction, approximated under certain conditions by Eq. (11), can mimic a multiplication. From Koch and Poggio (1987).

haaf et al., 1989). Remarkably, even for contrast values of 30%, no significant nonlinearity (beyond a multiplication) seems to distort the time course of the signal.

Psychophysical work in humans from a number of independent groups strongly supports models of the correlation types, albeit with spatio-temporal filter functions different from those in insects. From a mathematical point of view, all these models are equivalent to second-order multiplicative models. (For an overview, see Buchner, 1984; Hildreth and Koch, 1987; and Poggio et al., 1989). The general form for a second-order system with n discrete inputs, $I_i(t) = I(x_i,t)$, is:

$$y(t) = K_0 + \sum_{i=1}^{n} \int K_1(t - \tau)I_i(\tau)d\tau +$$

$$\sum_{i,j=1}^{n} \iint K_{i,j}(t - \tau_1, t - \tau_2)I_i(\tau_1)I_j(\tau_2)d\tau_1 d\tau_2,$$

(4)

where K_0 is a constant and the $K_{i,j}$ contain the spatio-temporal filtering and the nonlinear operations. Equation (4) contains n linear *kernels* and n^2 quadratic *kernels*. It is the quadratic cross-kernels of the sort $K_{i,j}$ with $i \neq j$ that mediate the direction-selective interaction. In the case of two photoreceptors spaced Δx apart and being stimulated by a constantly moving image, we have $I_1(t) = I(x_1 - vt)$ and $I_2(t) = I(x_2 - vt) = I(x_1 + \Delta x - vt)$ for $x_2 = x_1 + \Delta x$. It is easy to prove that:

Any model that consists of a finite number of inputs (corresponding to the photoreceptors) followed by linear operations and quadratic no-memory non-linearities (in any combination and order) is a special case of Eq. (4).

From this definition, it follows that there are several versions of the correlation model, for instance:

- The F model (low-pass filters instead of delays) (Thorson, 1966).
- The F-H model (low-pass filters and high-pass filters) (Poggio and Reichardt, 1976).
- The *elaborated* Reichardt model by van Santen and Sperling (1984).
- The *spatio-temporal* energy model of Adelson and Bergen (1985) and Heeger (1987).
- The model of Watson and Ahumada (1985), if provided with a quadratic nonlinearity at its output (a nonlinearity being needed in any true motion director—Poggio and Reichardt, 1973).

All the standard definitions of optical flow—which correspond to specific models for how to measure motion—can be represented by a functional power series expansions of the Volterra type. Equation (4) is a truncated Volterra series and, as such, may be considered an approximation of those definitions. Under general circumstances, terms of order higher than the second may not be negligible and the approximation will correspondingly be quantitatively poor.

D.   Relation to Recent Theories of Learning in Cortex

Mel (1990a) has suggested that cortical circuits may implement sigma-pi networks, with the function of memorizing input–output relations. (See also Mel and Koch, 1990.) Poggio (1990) has proposed a related theory of how part of the brain might work. The theory's main point is that the brain has modules, perhaps in cortical areas, dedicated to learning input–output mappings from sets of *examples*. These modules are realized according to the theory in terms of a

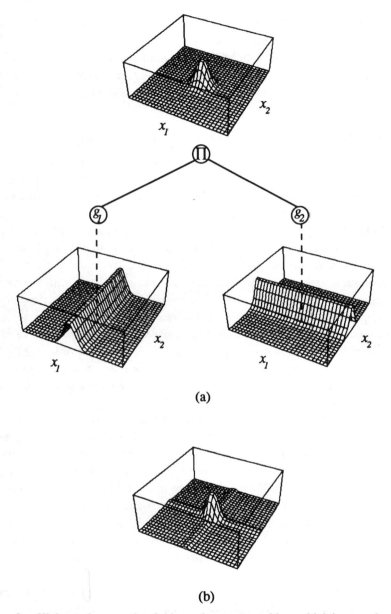

(a)

(b)

FIGURE 3.   Higher-order receptive fields can be constructed by multiplying together the outputs of lower-order receptive fields. Multidimensional neural response selectivities can be synthesized in this manner as demonstrated here for two 1D variables. This is the crucial operation required for some recent theories of learning in the brain (Mel, 1990a; Poggio and Girosi, 1990). the multiplication operation is approximated by a sum followed by a rectification in b. From Mel (1990a) with permission.

special class of layered networks, where the key operations are multiplications between signals that implement the product of Gaussian functions of the input (Poggio and Girosi, 1990). In both cases, multiplications play a critical role. The two theories rest on the assumption that neurons and/or dendrites perform multiplication-like operations. (See Mel, 1990a,b; also Fig. 3).

E.　Learning to Synthesize Algorithms from a Set of Multipliers

Suppose that biological multipliers are available of one or more of the types described earlier. Can one synthesize useful algorithms by learning from a set of examples using these primitives? The answer is positive, as shown a number of years ago by one of the authors, who considered exactly this problem (Poggio 1975a,b, 1983). Polynomial algorithms can be synthesized in terms of linear regression on the terms of the polynomial (that is, the outputs of the hidden units of Fig. 1b). A variety of algorithms for performing the regression can be used: pseudoinverse techniques, gradient descent, and even the iterative techniques suggested by Poggio (1975a), which are probably worth exploring especially for fast implementations.[1] The main (and almost trivial) result from the point of view of learning is that *any nonlinear mapping between two finite sets of vectors can be written as a polynomial.* (See Poggio, 1975a, 1983.) Invariance properties of the mapping can be exploited in elegant ways, together with prior information on the range and domain of the function, to reduce the complexity of the learning task.

Two recent demonstrations of the capabilities of polynomial algorithms for learning can be found in Kalaba *et al.*, (1989) and Drumheller (1989). The first paper shows that polynomial algorithms can be very effective. It demonstrates excellent performance of the pseudoinverse scheme of Poggio (1975a) used on a polynomial estimator of the mapping to be learned. Their results on the specific problem they considered (to estimate motion parameters from image data) were substantially better (by three orders of magnitude) using a second-order polynomial estimator than using the optimal linear scheme. A higher-order, i.e., cubic, estimator did little to improve upon the quadratic estimator for this specific problem.

Drumheller (1989) showed that direction-selective motion detectors can be synthesized from a set of examples, starting from a black-box with several spatially separated photoreceptor inputs and a single direction-selective output.

---

[1]The best linear approximation is first found, then the optimal second-order connections, up to the nth order; then the best linear correction is computed, and so on. A theorem (Poggio, 1975a) shows that this procedure converges to the optimal nth-order polynomial.

The structure of the black-box is assumed to be a kth-order polynomial in its inputs. The examples are pairs of motion sequences and their corresponding correct direction of the motion. Drumheller's (1989) analysis illustrates that second-order terms are most important for direction selectivity but higher-order terms can play a significant role.

One may also ask whether it is possible to learn to multiply by using some of the network techniques that are available. Unpublished experiments by C. Atkeson show that a single-hidden-layer sigmoidal network with four hidden units (as in Fig. 1c) can be made to approximate $x_1 x_2$ with very good precision using the standard back-propagation learning technique. The four sigmoidal units turn out to approximate $\frac{1}{4}((x_1 + x_2)^2 - (x_1 - x_2)^2)$ with remarkable precision. Other approximation techniques, such as flexible Fourier series, gamma poly-nominals, and Radial Basis Functions, also approximate multiplication very well (Marnyama *et al.*, personal communication, 1991). Each of them corresponds to a different class of networks with one hidden layer.

III.   Multiplication: Biophysical Mechanisms

Let us now survey and briefly discuss a number of biophysical mechanisms that can, *in principle,* give rise to a multiplicative-like interaction. The specificity of these mechanisms ranges from multiplicative interaction among pairs of cells to mechanisms involving populations of cells.

A.   *Nonlinear Synaptic Interaction via Silent Inhibition*

When two neighboring regions of a dendritic tree receive simultaneous synaptic input, the resulting postsynaptic potential is, in general, *not* the sum of the potentials generated by each synapse alone. (For a review, see Koch and Poggio, 1987). The interaction among synapses would be linear—as positioned by the vast majority of neural network models—if the synaptic input corresponds to a current injection. However, the binding of the neurotransmitter released by de-polarization of the presynaptic terminal opens—or, more rarely, closes—chan-nels specific to certain ions. Biophysically, this corresponds to a change in the postsynaptic membrane conductance, $g_{syn}(t)$, in series with a battery, the synaptic reversal potential E_{syn}. The synaptic current is given by the product of the conductance change times the effective driving potential, that is, the difference between the potential across the membrane $V(t)$ and the synaptic battery. Ac-cordingly, we have:

$$I_{syn}(t) = g_{syn}(t) \cdot (E_{syn} - V(t)). \qquad (5)$$

For simplicity, we here consider the case in which the synaptic input changes slowly compared to the time constant of the cell; in other words, we are only considering the stationary current and voltage in response to a constant synaptic input g_{syn}. Using Ohm's law, $V = K \cdot I_{syn}$, we can then write the following expression for the postsynaptic voltage V relative to the resting potential of the cell at the location of the synapse,

$$V = \frac{g_{syn}KE_{syn}}{1 + Kg_{syn}}, \tag{6}$$

where K denotes the input resistance at the location of the synapse. For an excitatory synapse, $E_{syn} > 0$, the cell is depolarized, i.e., $V > 0$ (assuming that $g_{syn} > 0$). In the case of a hyperpolarizing synapse, such as the GABA$_B$ receptor linked to a potassium channel with $E_{syn} < 0$, the membrane is hyperpolarized. A synapse with its reversal potential at or near the cell's resting potential, i.e., $E_{syn} = 0$, termed a *shunting* inhibition. Because the action of such a synapse on the membrane potential remains invisible if by itself, it is frequently also termed *silent* inhibition. The GABA$_A$ receptor, coupled to a chloride channel, appears to act like silent inhibition.

If the product of the synaptic induced conductance change and the synaptic input resistance is small, i.e., $Kg_{syn} \ll 1$, then Eq. (6) can be approximated by:

$$V \approx g_{syn} KE_{syn}. \tag{7}$$

In other words, the synaptic input can be treated as a current input, $g_{syn} E_{syn}$. In the other extreme case, that is, when the product of the synaptic conductance change and the synaptic input resistance is very large, i.e., $Kg_{syn} \gg 1$, we have:

$$V \approx E_{syn}, \tag{8}$$

that is, the synapse is saturated.

Let us now consider what happens if an excitatory synapse, with an associated conductance increase $g_e > 0$ and battery $E_e > 0$, is activated together with an inhibitory synapse of the shunting type with $g_i > 0$ and $E_i = 0$. We can write down the following equation for the stationary membrane potential (Fig. 2c):

$$V = \frac{g_e E_e}{g_e + g_i + 1/K}. \tag{9}$$

In other words, the effect of a shunting or silent synapse is a divisive inhibition of the cell's response. If the amplitudes of the excitatory and inhibitory conductance changes are small relative to $1/K$, we can use Taylor series expansion and expand the membrane potential as (Torre and Poggio, 1978):

$$V \approx E_e K(g_e - g_e^2 K - g_e g_i K + \ldots). \tag{10}$$

Thus, the membrane potential is given by a linear and a quadratic contribution from the excitatory synaptic input and a term dependent on the product of the excitatory and the inhibitory inputs. This *cross-term* expresses the interaction among two inputs and *is the first term in this expansion that can be used to process information* inside the passive dendritic tree. The conditions under which this nonlinear interaction are maximized are thin long dendrites or dendritic spines (with their corresponding high-input resistances) and a large inhibitory conductance change g_i. This argument can be generalized to the general case of an arbitrary bounded conductance change $g_{syn}(t)$ (Torre and Poggio, 1978).

Note that Eq. (10) does not correspond to a clean multiplication because of the presence of contaminating terms such as g_e^2 and higher-order terms that we neglected in our series expansion. It can be converted into a clean multiplication by subtracting the self-terms associated with the excitatory input g_e (Poggio and Torre, 1981).

1.   Relationship to Standard Neural Networks.   We only considered the nonlinear interaction among a pair of excitatory and inhibitory synapses. In general, however, the potential induced by any synapse will change the potential throughout the dendritic tree, thereby reducing the driving potential seen by a second synaptic input. In other words, the nonlinear interaction is always sublinear. What effect does this have on the standard linear threshold neural network model?

Let us write down the update equation associated with a *unit i* in a network of n such units,

$$C\frac{dV_i}{dt} = \frac{V_i}{K} + I_i + \sum_{j=1}^{n} T_{ij}U_j, \tag{11}$$

where C is the membrane capacity associated with unit i, I_i is an external current, T_{ij} the synaptic connectivity matrix, and U_j the output associated with the unit j. This output $U_j(t)$ is related to its *membrane* potential $V_j(t)$ via a stationary nonlinearity, usually a sigmoid, $U_j(t) = f(V_j(t))$. (For more details, see Wilson and Cowan, 1972, or Hopfield, 1984.)

This equation assumes that the synaptic input to a cell, $T_{ij}U_j$, adds linearly without any synaptic interaction occurring. Thus, standard neural network theory treats synaptic input not as conductance changes but as current inputs. As we have seen, a biophysically more realistic model of a unit corresponds to a polynomial unit:

$$C\frac{dV_i}{dt} = \frac{V_i}{K} + I_i + \sum_{j=1}^{n} T_{ij}U_j + \sum_{j,k=1}^{n} T'_{ijk}U_jU_k$$
$$+ \sum_{j,k,l=1}^{n} T''_{ijkl} U_jU_kU_l \ldots . \tag{12}$$

Note that the *connectivity* matrices T, T', T'', etc. are not independent of each other, but are functions of the cable properties and of the specific synaptic architecture chosen. In general, they are also functions of time t. To what extent these higher-order terms are important or not depends on the product of the synaptic input conductance g_{syn} and the synaptic input resistance K. As we discussed in the previous part of this chapter, polynomial units are computationally more powerful than the linear threshold unit underlying the standard neural network paradigm.

Note that our analysis is only correct as long as the dendritic membrane does not contain significant nonlinearities. These could, of course, further enhance the computational power of *real* neurons over their feebleminded *linear threshold* counterparts.

2.   Retinal Directional Selectivity.   One computation for which it is thought that such a mechanism is relevant is direction selectivity in ganglion cells in the turtle or rabbit retina. The principal working hypothesis on how retinal direction selectivity arises is that it is due to the interaction of an excitatory and an inhibitory pathway, implementing an AND-NOT or veto-like operation (Barlow and Levick, 1965; also Fig. 2b). In the preferred direction of motion, inhibition is delayed with respect to excitation, while the time course of excitation and inhibition overlap in the null direction. Pharmacological evidence argues for acetylcholine as being the excitatory and GABA$_A$ the inhibitory neurotransmitter regulating direction selectivity (Caldwell *et al.*, 1978; Ariel and Daw, 1982).

Torre and Poggio (1978) argued that the nonlinear interaction specified in Eq. (10) underlies this nonlinear operation (Fig. 2c). Thus, inhibition at the level of the ganglion cell vetoes the excitation provided by the acetylcholinergic starburst amacrine cell, approximating a multiplicative type of interaction. Direction selectivity arises from the spatial separation of the receptive fields of the excitatory and inhibitory pathways and from their different time courses. Particularly, some data suggest that the inhibitory pathway is slower than the excitatory pathway (Barlow and Levick, 1965; Wyatt and Daw, 1975). Intracellular electrophysiological evidence for shunting inhibition in direction-selective retinal ganglion cells stimulated with moving stimuli has been provided in the turtle and frog (Marchiafava, 1979; Watanabe and Murakami, 1984; also Amthor and Grzywacz, 1990). However, more recent evidence also supports the notion that the excitatory input to the ganglion cells, starburst amacrine cells, may show already asymmetric behavior in response to movement in the null and preferred directions (DeVoe *et al.*, 1985; also Chapter 13 by Borg-Graham and Grzywacz in this book).

Another important property of direction-selective ganglion cells is that small regions of their receptive field are directionally selective. This suggests the existence of a dozen or so direction-selective subunits, scattered throughout the receptive field of the cell (Barlow and Levick, 1965). Torre and Poggio (1978) pointed out that the subunits impose a theoretical problem if directional selectivity is to arise in the ganglion cells. Due to the locality of shunting or silent inhibition, its action would be limited to specific branches of the dendritic tree. In fact, it is possible to prove (Koch *et al.*, 1982, 1983), that silent inhibition obeys an *on-the-path* condition, such that silent inhibition located beyond the location of the excitatory synapse or not on the direct path between excitation and the soma has very little effect on the excitatory synapse. (See also Rall, 1964, 1967.) The dendritic tree of direction-selective retinal ganglion cells shows numerous fine and heavily branched processes, very suitable for implementing spatially selective subunits (Amthor *et al.*, 1984). More recent extracellular evidence for the involvement of silent inhibition in direction selectivity is given by Amthor and Grzywacz (1990).

An unresolved issue is to what extent shunting inhibition truly approximates a multiplication and can underly the second-order proper discussed in Section II.D (e.g., phase-invariance or frequency-doubling). Indeed, Grzywacz and Koch (1987) have shown that Eq. (10), describing the action of synaptic conductances on the membrane potential, only approximates second-order behavior over a relatively small range of voltage values if realistic nonlinearities, such as the retinal ON-OFF rectification, are taken into account.

B.   Facilitation through Silent Disinhibition

An interesting variant of the idea discussed in the previous section involves a *decrease* in synaptic inhibition of the silent or shunting type. In other words, the existence of a tonically active chloride conductance is assumed, a membrane conductance that is then closed or reduced upon synaptic activation. Our previous equations, (5)–(10), still hold, except that $g_i < 0$. In other words, inhibition would be disinhibited, thereby facilitating the cell's response. Thus, in the preferred direction, the time course of excitation coincides with the time course of disinhibition, while in the null direction, disinhibition arrives too late to facilitate the depolarizing response sufficient for it to reach threshold. As made explicit in Eq. (10), this sort of interaction can, in principle, approximate a multiplication, provided that the conductance decrease associated with disinhibition is larger than the excitatory conductance increase and is large compared to the synaptic input conductance $1/K$. From a logical point of view, this interaction implements an analog version of an AND-like gate.

Electrophysiological recordings from tangential interneurons in the second

optic ganglion of the crayfish have provided evidence for two different neurotransmitters, ACh and GABA, targeting the same synapse associated with a
chloride reversal potential (Pfeiffer-Linn and Glantz, 1989). Application of ACh
leads to a conventional *shunting inhibition,* while application of GABA reduces
the chloride conductance, effectively acting like a disinhibition. The involvement
of tangential cells in direction selectivity is at present unclear. However, it may
well be that this—or a similar synaptic interaction—is at the basis of direction
selectivity in the fly (Franceschini *et al.,* 1989).

   Finally, disinhibition of shunting inhibition may possibly explain the small
but significant facilitation reported already by Barlow and Levick (1965) for
motion in the preferred direction. (See also Grzywacz and Poggio, 1990.)

C. NMDA Receptors

A substantial proportion of excitatory postsynaptic potentials in the brain, in
particular in the cortex and hippocampus, are triggered by the release of glutamate
or aspartate at the presynaptic terminal. However, while these synapses may all
use the same neurotransmitter, the receptors in the postsynaptic membrane fall
into two different classes—those that can be activated by the exogenous substances, kainate and quisqualate, and those that are activated by an agonist termed
N-methyl-D-aspartate (NMDA). Activation of the non-NMDA receptors cause
relatively fast EPSPs, while NMDA receptor activation leads to much more
prolonged (100 msec or longer) EPSPs. In the cat visual cortex, at least half of
the visually evoked neuronal activity can blocked by pharmacological blockade
of the NMDA receptors (Miller *et al.,* 1989; Fox *et al.,* 1990).

   The current flowing through the NMDA receptor, mainly a mixture of Na^+
and K^+ with a small fraction of Ca^{2+} ions, depends in a unique manner on the
postsynaptic potential (Fig. 4). While the conductance increase associated with
the non-NMDA receptor and, indeed, with other types of receptors, such as
those using ACh or GABA, does not depend on the postsynaptic potential, the
conductance increase following NMDA receptor activation does. At levels near
the resting potential, the conductance increases only a little following synaptic
input, while it is much larger at more depolarizing values. The reason for this
behavior is the fact that at voltage values around the resting potential, external
Mg^{2+} ions enter the synaptic channel, blocking it effectively for a short duration.
This Mg^{2+} block is relieved at more depolarized levels (Nowak *et al.,* 1984;
Mayer *et al.,* 1984). Using data from cultured hippocampus cells, Jahr and
Stevens (1990) derived the following equation for the NMDA receptor associated
current:

$$I_{NMDA}(t) = (E_{NMDA} - V)g_{peak}\frac{e^{-t/\tau_1} - e^{-t/\tau_2}}{1 + \eta[Mg]e^{-\gamma V}}, \tag{13}$$

where $E_{NMDA} = 0\ mV, g_{peak} = 0.2\ nS, \tau_1 = 80$ msec, and $\tau_2 = 0.06$ msec, η $= 0.33$ per mM, $[Mg] = 2$ mM usually, and $\gamma = 0.66$ per mV; thus, the steepness of the voltage dependency is about 16. 7 mV. (For more details, see Zador et al., 1990). This voltage dependency (plotted in Fig. 4) is apparent in the negative slope conductance region in the current-voltage relationship of the NMDA response. Thus, this synapse acts like a molecular AND gate, only carrying significant current in the presence of both a presynaptic input and postsynaptic depolarization (Koch, 1987). It is, therefore, ideally suited to implement Hebb's rule for the associative form of long-term potentiation (Kelso et al., 1986). Likewise, given this negative slope conductance region of approximately constant slope, extending between -70 and -35 mV, multiplication could occur in the sense that the postsynaptic current is given by the product of the postsynaptic potential at the time the NMDA receptor was activated times the presynaptic input. Doubling either the membrane potential or the presynaptic input will approximately double I_{NMDA} (over a limited range). Driving the potential outside this narrow voltage window will cause a reduction of I_{NMDA}. Mel (1990b) has investigated in more detail to what extent NMDA synapses can act in a multiplicative manner.

In an elegant study, Fox et al. (1990) assessed recently the contribution of NMDA currents in cat visual cortex. By iontophoretic application of NMDA

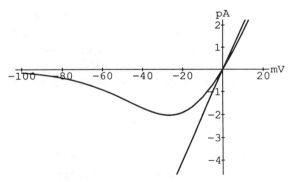

FIGURE 4.   Peak current through an NMDA synapse as a function of the postsynaptic voltage. We plotted the time-independent part of Eq. (14) with [Mg] = 1 mM. For the maximal amount of current to flow, the postsynaptic potential must be depolarized to approximately -30 mV. If no magnesium is present, the current is linear in the voltage, as for the non-NMDA synapse. A multiplication-like operation between presynaptic input and postsynaptic potential can be achieved in the -80 to -30 mV range, where the slope conductance is negative.

and its antagonist, APV, they could show that NMDA increases the gain of the visual driven response in superficial cells, i.e., a multiplicative effect, while activation of the non-NMDA receptor increases the overall level of activity, independent of the level of visual input, i.e., an additive effect. Cells in deep layers (IV, V, and VI) showed little effect of APV on visual-evoked responses.

Note that the spatial specificity of this mechanism is less than that of the previously discussed synaptic interaction, since the multiplicative interaction occurs between the NMDA associated conductance change and the postsynaptic voltage and not between two synaptic conductance changes. Thus, the spatial locus for multiplicative interaction to occur is some significant fraction of the dendritic tree.

D. Product of Firing Rates

The following mechanism allows one neuron to multiply by detecting coincidences in two independent trains of action potentials. This mechanism, proposed by Srinivasan and Bernard (1976), assumes that this neuron can be described by a passive membrane, with an associated membrane time constant τ, and a stationary threshold. Each presynaptic input gives rise to a postsynaptic EPSP decaying exponential in time: $V(t) = V_o e^{-t/\tau}$.

If a given voltage threshold V_T is reached, the neuron spikes and the membrane potential is reset to zero (*leaky integrate-and-fire neuron*) (Fig. 5). Our coincidence neuron C receives input from two presynaptic axons A and B, each firing action potentials at a given frequency f_A and f_B. We now make two auxiliary assumptions. The first one is $V_o < V_T < 2V_o$; in other words, while one presynaptic spike will not cause a large enough EPSP to discharge the cell, two EPSPs will. The second assumption prevents a single presynaptic axon from discharging the cell by itself by placing some restriction on the maximal discharge frequency f. The appropriate minimal separation time Δ is determined by the simple equation:

$$V_o e^{-\Delta/\tau} + V_o = V_T. \qquad (14)$$

If the inverse of the spiking frequency of one of the two inputs exceeds Δ, that input will be able, by itself, to trigger the postsynaptic cell. It is now clear that neuron C will only fire if an action potential from axon A arrives within Δ msec from an action potential along axon B and vice versa (Fig. 5). Therefore, assuming statistical independence of the two spike trains, it is straightforward to show (Srinivasan and Bernard, 1976) that the average frequency of the coincidence neuron C is given by:

$$f_C = 2\Delta f_A f_B. \qquad (15)$$

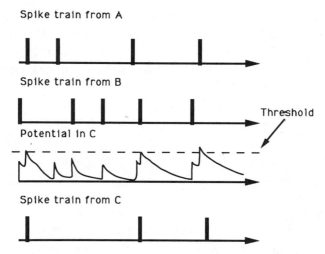

FIGURE 5.   A fire-and-integrate neuron C receives input from two independent input fibers A and B. Cell C only goes above threshold and generates an output spike if a spike on one input line arrives within Δ msec (Eq. (15)) of a spike in the second input. Thus, neuron C acts as a coincidence detector, adjusting its output firing frequency to be proportional to the product of the firing rates of A and B (Srinivasan and Bernard, 1976).

In other words, the output frequency of cell C is a random variable whose mean frequency is f_C. The principle underlying the multiplicative interaction is that the joint probability of two statistically independent events is equal to the product of their individual probabilities. Note that this mechanism relies on jitter in the input spike trains. If, for instance, both presynaptic inputs had identical frequencies and were perfectly regular, the output would be either zero or equal to the input frequency, depending on the relative phase of the two trains. Computer simulations of this mechanism indicate that increasing the jitter in the two input spike trains decreases the jitter in the output train (Srinivasan and Bernard, 1976).

It is at present unclear to what extent such a mechanism can approximate a multiplication if the integrate-and-fire cell is replaced by a more realistic neuron with a nonstationary threshold, such as the one in the Hodgkin and Huxley (1952) model, and many partially correlated synaptic inputs impinge on one cell instead of two. As a final point, note that the spatial locus of the multiplicative interaction is the entire neuron.

E.   Linear Summation of Noisy Linear Threshold Neurons

The following mechanism implements a multiplication among two neuronal populations via a squaring operation (Suarez and Koch, 1989a). Let us assume a population of n neurons whose output is zero if the *somatic potential* V is below a threshold V_T and whose output is linear for small V above this value: $f(V) = \alpha[V - V_T]$, where $[x] = x$ if $x > 0$, and 0 otherwise (Fig. 6a). We assume that a postsynaptic cell adds the contribution from n of these cells, such such that its response is given by:

$$R = \sum_{i=1}^{n} f(V). \tag{16}$$

If V_T is the same for all cells, $R = n\alpha(V - V_T)$ as long as $V > V_T$. However, we will now assume that the threshold varies randomly from cell to cell, let us say distributed uniformly between V_{t_1} and V_{t_1}. If V falls within these values, the function $\alpha(V - V_{T_1})$ is randomly sampled across this interval before being summed, and the system then simply computes the area below $f(V)$, similar to Monte Carlo integration methods (Fig. 6b). For a sufficiently large value of n, we then have that R is proportional to $(V - V_{T_1})^2$. (In general, if $f(V)$ is an mth-order polynomial, R will be proportional to $(V - V_{T_1})^{m+1}$.) Alternatively, if V_T were constant for all cells while the somatic potential of the neurons varied randomly between V_{T_1} and V for a given input, the same quadratic behavior in V would be obtained.

Thus, for the case of direction-selective cells in visual cortex, this random variation could be obtained by summing over a population of cells that are broadly tuned for the direction of motion with a certain distribution of preferred directions. In all cases, for values of V much higher than V_{T_2}, the output R will grow linearly, since the system will integrate only over a narrow range around V. Finally, more realistic neurons saturate at some output value, say, αV_m. R will then saturate when $V > V_{T_2} + V_m$ (Fig. 6b).

The output of this simple system, therefore, approximates a squaring operation (Fig. 7). To multiply two variables, A and B, we would, therefore, need two separate neuronal populations. One would have as input $A + B$ and feed into a cell whose output is proportional to $(A + B)^2$, while a second cell population computes $(A - B)^2$. Subtracting these two outputs gives the desired result, $4AB$.

FIGURE 6.   (a) Input–output relationship of a linear rectifying direction-selective cell. The cell does not fire if the *somatic potential* x is below a given threshold x_T. This threshold x_T is assumed to vary from one cell to the next. (b) The sum R of the responses of 50 such cells, with the threshold x_T being uniform distributed between $x = 1$ and $x = 3$. (See arrows in b.) R is quadratic for small values of x and saturates for large values. The dashed curve is $12.5(x - 1)^2$ and corresponds to the expected mean of R. From Suarez and Koch (1989a).

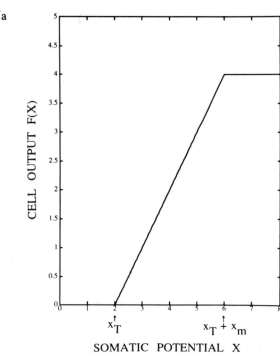

a

CELL OUTPUT F(X)

SOMATIC POTENTIAL X

x_T

$x_T + x_m$

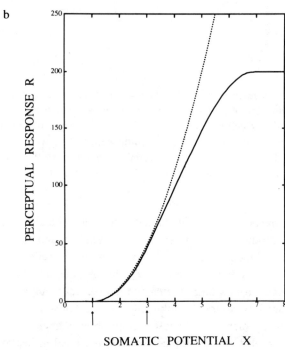

b

PERCEPTUAL RESPONSE R

SOMATIC POTENTIAL X

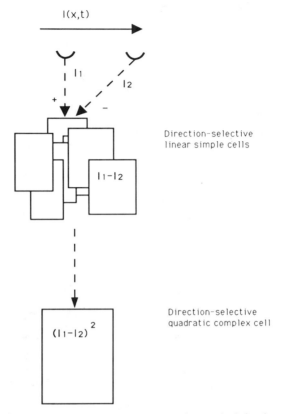

FIGURE 7.   Schematic indicating how direction-selective cortical simple cells with a linear response curve, as indicated in Fig. 6a, can synthesize quadratic behavior via a squaring operation. The complex cell receives input from a large number of simple cells and is expected to show second-order properties. Multiplication of I_1 with I_2 is achieved by squaring the sum, i.e., $(I_1 + I_2)^2$, and subtracting it from the difference, i.e., $(I_1 - I_2)^2$ (not shown here), as suggested as one possible implementation of Reichardt-type motion detectors.

Note that this corresponds to the functional schemata associated with the spatio-temporal energy model of motion computation (Adelson and Bergen, 1985). In our interpretation of their model, we assume that simple cells with spatio-temporal-oriented receptive fields followed by a threshold compute the direction of motion. The output from a large number of simple cells, approximately linear in the contrast (for low contrast) (Holub and Morton-Gibson, 1981), is then fed into complex cells, which should show second-order behavior (Fig. 7).

Like the spike-multiplication implementation discussed in Section III.D, this mechanism requires random variations, in this case in the threshold. It is the least specific of all the mechanisms we discussed in this chapter, requiring two distinct neuronal populations. Computer simulations indicate that $n = 50$ yields very satisfactory results (Suarez and Koch, 1989a; also Fig. 6b). Calculations carried out with more realistic neurons incorporating Hodgkin and Huxley dynamics (1952) with random variations in the densities of the sodium and potassium currents confirm these findings (Suarez and Koch, 1989b).

F.   Other Types of Multipliers

A number of other candidate mechanisms approximating a multiplication exist.

1.   Log-Exp Transform.   One can exploit a technique sometimes used in electronic circuits to multiply x with y:

$$x \cdot y = e^{\log(x) + \log(y)}. \tag{17}$$

In the case of visual input, the compressive logarithmic nonlinearity could be provided by the transduction between light and membrane potential in neurons in the outer retina. Exponential dependencies abound in the nervous system, for instance, between voltage-dependent membrane conductances and the membrane potential (for small values of the membrane potential before saturation sets in). In fact, Mead (1989) argues that the exponential nonlinearity is one of the most basic computational primitives in the nervous system and uses it in his attempt to build silicon versions of neural circuits.

2.   Synaptic Transfer Function.   Another mechanism exploits the nonlinear relationship between pre- and postsynaptic potential at a chemical synapse. In general, the relationship between V_{pre} and V_{post} is sigmoidal, saturating for high values of V_{pre} (Fig. 8). For values of V_{pre} above threshold, the synaptic transfer function approximates a higher-order polynomial, i.e., $V_{post} = \alpha V_{pre}^n$ with α some constant and $n > 1$. Such a mechanism was proposed to underlie gain control for wide-field motion in the visual system of the fly (Reichardt, *et al.*, 1983). A similar combination of sums and differences as outlined in Section III.E must then be used to multiply two signals, x and y.

Computing higher-order polynomials via the synaptic transfer function requires that the presynaptic potential be graded, i.e., can take on continuous values. At an axodendritic or axosomatic presynaptic terminal, the invasion of an action potential will always cause more or less the same depolarization given the constant amplitude of the action potential, ruling out the widespread use of this mechanism in cortex. Such a mechanism appears more likely in the retina, with its nonspiking horizontal, bipolar, and amacrine cells.

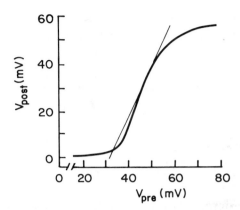

FIGURE 8.   The input–output relationship for a synapse, in this case the giant synapse in the squid stellate ganglion. In the appropriate range—for instance, around $V_{pre} \approx 35$ mV—the output V_{post} is a quadratic function of the input V_{pre}. The voltages are given with respect to the resting potential. Modified from Katz and Miledi (1967).

3.   Cooperative Binding.   In principle, the binding of one substance to another can implement a multiplication-like operation. For instance, let us assume that free intracellular calcium binds to calmodulin, the most common intracellular calcium buffer. High concentrations of this and similar buffers have been identified in postsynaptic structures and spines (Wood *et al.*, 1980). Calmodulin has four calcium binding sites; thus, the concentration of fully bound calmodulin, i.e., $[CM - Ca_4]$, can be described by four second-order differential equations of the type,

$$[Ca^{2+}] + [CM - Ca_n] \underset{b}{\overset{f}{\rightleftharpoons}} [CM - Ca_{n+1}],   (18)$$

where and f is the forward and b the reverse binding rate. If we assume that the binding is noncooperative and sequential, the steady-state concentration of the various calcium-bound calmodulin forms is given by:

$$[CM - Ca_n] = \frac{[total - CM] \cdot ([Ca^{2+}]/K_d)^n}{1 + \sum_{i=1}^{n} ([Ca^{2+}]/K_d)^i},   (19)$$

where $K_d = b/f$ is the dissociation rate of calmodulin and $[total - CM]$ the total amount of bound and free calmodulin. Thus, at low calcium concentration, the concentration of fully bound calmodulin is proportional to the fourth power of calcium. Each additional binding step to a buffer or enzyme will potentiate this association and sharpen the relationship between $[Ca^{2+}]$ and the bound buffer. Thus, Gamble and Koch (1987) showed numerically that a fast burst of action potentials (10 spikes at 333 Hz) to a synapse located on a spine elevates peak intracellular calcium in the spine head by about a factor of 5 in comparison to the peak calcium evoked after 10 spikes at 50 Hz. This difference becomes amplified 1,000-fold in the two situations if fully bound calmodulin is considered. This could have important functional consequences, given that a calcium–calmodulin-dependent protein kinase, CaMKII, may be crucial in the biochemical cascade underlying the establishment of long-term potentiation (Kennedy, 1989). At the moment, it is not known to what extent such computations, implemented by chemical rather than by electrical means, are used in the nervous system. (For a more detailed discussion of this, see Poggio and Koch, 1985.)

IV.   Conclusion

We have delineated a number of biophysical mechanisms in this chapter that can potentially implement multiplication-like operations. These mechanisms range from very specific ones operating at the level of individual synapses to mechanisms involving small populations of cells. Mechanisms in the first category, such as those involving shunting or silent inhibition, NMDA synapses, or the synaptic transfer function (Sections III.A,B,C, and 2.1, 2.2, 2.3 and F.2) are based on the assumption that the specificity of biological hardware makes it possible to implement nonlinear operations in the dendritic tree of neurons, prior to the spiking threshold at the cell body. The second category of mechanisms, implementing multiplication via detecting coincidences in the arrival of action potentials or via a squaring operation achieved by a Monte Carlo-like integration procedure (Sections III.D and E), does not require such specificity. In fact, both mechanisms in the latter category only lead to second-order behavior if random variations in either the arrival times of action potentials or the distribution of thresholds are present. It seems rather elegant that the nervous system may not only work *in spite of* random variations in neuronal properties but *because* of it.

Behavioral evidence in insects and psychophysical evidence in humans strongly support second-order models for motion perception formally equivalent to the Reichardt correlation model. However, so far, only a handful of studies have attempted to study the properties associated with second-order systems, such as phase-invariance, frequency-doubling, etc. (Section II.D) at the single cell level. Intracellular recordings from neurons in the third optic ganglion of the fly (Egelhaaf *et al.*, 1989) support second-order models (also Buchner, 1984), while extracellular recordings in the rabbit direction-selective retinal ganglion cells (Grzywacz *et al.*, 1990) do not support second-order models. In fact, a theoretical investigation (Grzywacz and Koch, 1987) has shown that quadratic properties are quite fragile at the single-cell level. Thus, it is at present unclear how the overall output of a system can show quadratic properties without these being expressed at the single-cell level.

Biological implementations of the operation of multiplication are critical to theories recently proposed of how the brain works. Both Poggio (1990) and Mel (1990a) have argued for a model of associative learning that views neurons as coding specific instances of mappings, e.g., memories. In this view, neurons correspond to multiplicative units, implementing an approximative multidimensional look-up table, with groups of synapses acting as individual table entries. The key requirement underlying these theories is the widespread availability of multiplication-like operations at the dendritic level.

We still know too little about the computational style used by the nervous system. As a consequence, there exists a large gap between computational theories of vision, motor control and learning, and their possible implementation in neural hardware. The model of computation provided by the digital computer is clearly unsatisfactory for the neurobiologist, given the evidence that neurons are very different devices from the simple digital switches suggested by the McCulloch and Pitts (1943) model. We have argued before (e.g., Koch and Poggio, 1987) that we need a *Biophysics of Computation,* linking specific biophysical mechanisms to specific instances of neuronal computations. It is obvious from this and other chapters in this book, that this program of mapping computations at the abstract level back into neuronal hardware is well under way.

Acknowledgment

CK wishes to acknowledge many useful discussions on multiplication in real neurons with Bartlett Mel. CK is supported by grants from the Office of Naval Research, a Presidential Young Investigator Award from the NSF, and funds from the James S. McDonnell Foundation. TP is supported by grants from the Office of Naval Research and the Incas and Ellen Whitaker chair.

References

ADELSON, E. H., and BERGEN, J. R. (1985). "Spatiotemporal Energy Models for the Perception of Motion," *J. Opt. Soc. Am. A* **2,** 284–299.

AMTHOR, F. R., and GRZYWACZ, N. M. (1991). "The Nonlinearity of the Inhibition Underlying Retinal Directional Selectivity," *Vis. Neurosci.* **6,** 197–206.

AMTHOR, F. R., OYSTER, C. W., and TAKAHASHI, E. S. (1984). "Morphology of ON–OFF Direction-Selective Ganglion Cells in the Rabbit Retina," *Brain Research* **298,** 187–190.

ARIEL, M., and DAW, N. W. (1982). "Pharmacological Analysis of Directionally Sensitive Rabbit Retinal Ganglion Cells," *J. Physiol.* **324,** 161–185.

BARLOW, H. B., and LEVICK, W. R. (1965). "The Mechanism of Directionally Selective Units in Rabbit's Retina," *J. Physiol.* **178,** 477–504.

BRUCK, J. (1989). "Harmonic Analysis of Polynomial Threshold Function," Ph.D. Thesis, Dept. of Electrical Engineering, Stanford University, Stanford, California.

BUCHNER, E. (1976). "Elementary Movement Detectors in an Insect Visual System," *Biol. Cybern.* **24,** 85–101.

BUCHNER, E. (1984). "Behavioral Analysis of Spatial Vision in Insects," in M. A. Ali (ed.), *Photoreception and Vision in Invertebrates* (pp. 561–621). Plenum, New York.

BULLOCK, T. H. (1959). "Neuron Doctrine and Electrophysiology," *Science* **129,** 997–1002.

CALDWELL, J. H., DAW, N. W., and WYATT, H. J. (1978). "Effects of Picrotoxin and Strychnine on Rabbit Retinal Ganglion Cells: Lateral Interactions for Cells with More Complex Receptive Fields," *J. Physiol.* **276,** 277–298.

DEVOE, R. D., GUY, R. G., and CRISWELL, M. H. (1985). "Directionally Selective Cells of the Inner Nuclear Layer of the Turtle Retina," *Invest. Opthalm. Vis. Sci.* **26,** 311.

DRUMHELLER, M. (1989). "Synthesizing a Motion Detector from Examples," Master's Thesis, Dept. of Brain and Cognitive Sciences, MIT, Cambridge, Massachusetts.

DURBIN, R., and RUMELHART, D. E. (1989). "Product Units: A Computationally Powerful and Biologically Plausible Extension to Backpropagation Networks," *Neural Comp.* **1,** 133–142.

EGELHAAF, M., and BORST, A. (1989). "Transient and Steady-State Response Properties of Movement Detectors," *J. Opt. Soc. Am. A* **6,** 116–127.

EGELHAAF, M., BORST, A., and REICHARDT, W. (1989). "Computational Structure of a Biological Motion-Detection System as Revealed by Local Detector Analysis in the Fly's Nervous System," *J. Opt. Soc. Am. A* **6,** 1070–1087.

FELDMAN, J. A., and BALLARD, D. H. (1982). "Connectionist Models and Their Properties," *Cognitive Sci.* **6,** 205–254.

FOX, K., SATO, H., and DAW, N. (1990). "The Effect of Varying Stimulus Intensity of NMDA-Receptor Activity in Cat Visual Cortex," *J. Neurophysiol.* **64,** 1413–1428.

FRANCESCHINI, N., RIEHLE, A., and LE NESTOUR, A. (1989). "Directionally Selective Motion Detection by Insect Neurons," in H. Stavenga and R. Hardie (eds.), *Facets of Vision* (pp. 99.360–390). Springer-Verlag, Berlin.

GAMBLE, E., and KOCH, C. (1987). "The Dynamics of Free Calcium in Dendritic Spines in Response to Repetitive Synaptic Input," *Science* **236,** 1311–1315.

GEIGER, G., and POGGIO, T. (1975). "The Orientation of Flies towards Visual Patterns: On the Search for the Underlying Functional Interactions," *Biol. Cybern.* **19,** 39–54.

GILES, L. C., and MAXWELL, T. (1987). "Learning, Invariance, and Generalization in High-Order Neural Networks," *Appl. Optics* **26,** 4972–4978.

GÖTZ, K. G. (1972). "Principles of Optomotor Reactions in Insects," *Bibl. Opthalmol.* **82,** 251–259.

GRZYWACZ, N. M., AMTHOR, F. R., and MISTLER, L. A. (1990). "Applicability of Quadratic and Threshold Models to Motion Discrimination in the Rabbit Retina," *Biol. Cybern.* (in press).

GRZYWACZ, N. M., and KOCH, C. (1987). "Functional Properties of Models for Direction Selectivity in the Retina," *Synapse* **1**, 417–434.

GRZYWACZ, N. M., and POGGIO, T. (1990). "Computation of Motion by Real Neurons," in S. F. Zornetzer, J. L. Davis, and C. Lau (eds.), *An Introduction to Neural and Electronic Networks*. Academic Press, Orlando, Florida.

HASSENSTEIN, B., and REICHARDT, W. E. (1956). "Systemtheoretische Analyse der Zeit-, Reihenfolgen- und Vorzeichenauswertung bei der Bewegungsperzeption des Rüsselkäfers *Chlorophanus*," *Z. Naturforsch.* **11b**, 513–524.

HEEGER, D. J. (1987). "A Model for Extraction of Image," *J. Opt. Soc. Amer. A* **4**, 1455–1471.

HILDRETH, E., and KOCH, C. (1987). "The Analysis of Visual Motion: From Computational Theory to Neuronal Mechanisms," *Ann. Rev. Neurosci.* **10**, 477–533.

HODGKIN, A. L., and HUXLEY, A. F. (1952). "A Quantitative Description of Membrane Current and Its Application to Conduction and Excitation in Nerve," *J. Physiol.* **117**, 500–544.

HOLUB, R. A., and MORTON-GIBSON, M. (1981). "Response of Visual Cortical Neurons of the Cat to Moving Sinusoidal Gratings: Response-Contrast Functions and Spatiotemporal Interactions," *J. Neurophysiol.* **46**, 1244–1259.

HOPFIELD, J. J. (1984). "Neurons with Graded Response Have Collective Computational Properties like Those of Two-State Neurons," *Proc. Natl. Acad. Sci. USA* **81**, 3088–3092.

JAHR, C. E., and STEVENS, C. F. (1990). "A Quantitative Description of NMDA Receptor Channel Kinetic Behavior," *J. Neurosci.* **10**, 1830–1837.

KALABA, LICHTENSTEIN, SIMCHOUY, and TESTFASTIAN (1989).

KATZ, B., and MILEDI, R. (1967). "A Study of Synaptic Transmission in the Absence of Nerve Impulses," *J. Physiol.* **192**, 407–436.

KELSO, S. R., GANONG, A. H., and BROWN, T. H. (1986). "Hebbian Synapses in Hippocampus," *Proc. Natl. Acad. Sci. USA* **83**, 5326–5330.

KENNEDY, M. B. (1989). "Regulation of Synaptic Transmission in the Central Nervous System," *Cell* **59**, 777–787.

KOCH, C. (1987). "The Action of the Corticofugal Pathway on Sensory Thalamic Nuclei: A Hypothesis," *Neurosci.* **23**, 399–406.

KOCH, C., and POGGIO, T. (1987). "Biophysics of Computation: Neurons, Synapses and Membranes," in G. M. Edelman, W. E. Gall, and W. M. Cowan (eds.) *Synaptic Function* (pp. 637–698). John Wiley, New York.

KOCH, C., POGGIO, T., and TORRE, V. (1982). "Retinal Ganglion Cells: A Functional Interpretation of Dendritic Morphology," *Phil. Trans. Roy. Soc. Lond. B* **298**, 227–264.

KOCH, C., POGGIO, T., and TORRE, V. (1983). "Nonlinear Interaction in a Dendritic Tree: Localization Timing and Role in Information Processing," *Proc. Natl. Acad. Sci. USA* **80**, 2799–2802.

MARCHIAFAVA, P. L. (1979). "The Responses of Retinal Ganglion Cells to Stationary and Moving Visual Stimuli," *Vision Res.* **19**, 1203–1211.

MAYER, M. L., WESTBROOK, G. L., and GUTHRIE, P. B. (1984). "Voltage-Dependent Block by Mg^{2+} of NMDA Responses in Spinal Cord Neurones," *Nature* **309**, 261–263.

McCULLOCH, W. S., and PITTS, W. (1943). "A Logical Calculus of Ideas Immanent in Neural Nets," *Bull. Math. Biophys.* **5**, 115–137.

MEAD, C. (1989). *Analog VLSI and Neural Systems.* Addison-Wesley, Reading, Massachusetts.

MEL, B. W. (1990a). "The Sigma-Pi Column: A Model of Associative Learning in Cerebral Neocortex," *CNS Memo* **6**, California Institute of Technology, Pasadena, California.

MEL, B. W. (1990b). "The Sigma-Pi Model Neuron: Roles of the Dendritic Tree in Associative Learning," *Soc. Neurosci. Abstr.* **16**, 205.4.

MEL, B. W., and KOCH, C. (1990). "Sigma-Pi Learning: On Radical Basis Functions and Cortical Associative Learning," in D. S. Touretzky (ed.), *Advances Neural Inf. Proc. Systems, Vol. 2,* (pp. 474–481). Morgan Kaufmann, San Mateo, California.

MILLER, K., CHAPMAN, B., and STRYKER, M. (1989). "Visual Responses in Adult Cat Visual Cortex Depend on N-methyl-D-aspartate Receptors," *Proc. Natl. Acad. Sci. USA* **86**, 5183–5187.

MOORE, B., and POGGIO, T. (1988). "Representation of Properties of Multilayer Networks," *International Neural Network Society Annual Meeting, Boston, September 6–10.*

NOBLE, D., and STEIN, R. B. (1966). "The Threshold Conditions for Initiation of Action Potentials by Excitable Cells," *J. Physiol.* **187**, 129–162.

NOWAK, L., BREGESTOVSKI, P., ASCHER, P., HERBET, A., and PROCHIANTZ, A. (1984). "Magnesium Gates Glutamate-Activated Channels in Mouse Central Neurones," *Nature* **307**, 462–465.

PFEIFFER-LINN, C., and GLANTZ, R. M. (1989). "Acetylcholine and GABA Mediate Opposing Actions on Neuronal Chloride Channels in Crayfish," *Science* **245**, 1249–1251.

PICK, B. (1976). "Visual Pattern Discrimination as an Element of the Fly's Orientation Behavior," *Biol. Cybern.* **23**, 171–180.

POGGIO, T. (1975a). "On Optimal Nonlinear Associative Recall," *Biol. Cybern,* **19**, 201–209.

POGGIO, T. (1975b). "On Optimal Discrete Estimation," in G. D. McCann and P. Z. Marmarelis (eds.), *Prof. 1. Symp. Testing and Identification of Nonlinear Systems* (pp. 30–37). California Institute of Technology, Pasadena, California.

POGGIO, T. (1983). "Visual Algorithms," in O. Braddick and A. C. Sleigh (eds.), Physical and Biological Processing of Images (pp. 128–153). Springer-Verlag, Berlin.

POGGIO, T. (1990). "A Theory of How the Brain Might Work," in *Cold Spring Harbor Symp. Quant. Biol.* **55**, 899–910.

POGGIO, T., and GIROSI, F. (1990). "Regularization Algorithms for Learning that are Equivalent to Multilayer Networks," *Science* **247**, 978–982.

POGGIO, T., and KOCH, C. (1985). "Ill-Posed Problems in Early Vision: From Computational Theory to Analogue Networks," *Proc. R. Soc. Lond. B* **226**, 303–323.

POGGIO, T., and REICHARDT, W. E. (1973). "Considerations on Models of Movement Detection," *Kybern.* **13**, 223–227.

POGGIO, T., and REICHARDT, W. E. (1976). "Visual Control of Orientation Behaviour in the Fly, Part II: Towards the Underlying Neural Interactions," *Quart. Rev. Biophys.* **9**, 377–438.

POGGIO, T., and REICHARDT, W. E. (1980). "On the Representation of Multi-Input Systems: Computational Properties of Polynomial Algorithms," *Biol. Cybern.* **37**, 167–186.

POGGIO, T., and TORRE, V. (1978). "A New Approach to Synaptic Interaction," in R. Heim and G. Palm (eds.), *Approaches to Complex Systems* (pp. 89–115). Springer-Verlag, Berlin.

POGGIO, T., and TORRE, V. (1981). "A Theory of Synaptic Interactions," in W. E. Reichardt and T. Poggio (eds.), *Theoretical approaches in Neurobiology* (pp. 28–38). MIT Press, Cambridge, Massachusetts.

POGGIO, T., YANG, W., and TORRE, V. (1989). "Optical-Flow: Computational Properties and Networks, Biological and Analog," in R. Durbin, C. Miall, and G. Mitchison (eds.), *The Computing Neuron* (pp. 355–370). Addison-Wesley, Reading, Massachusetts.

RALL, W. (1964). "Theoretical Significance of Dendritic Trees for Neuronal Input–Output Relations," in R. F. Reiss (ed.), *Neural Theory of Modelling* (pp. 73–97). Stanford University Press, Stanford, California.

RALL, W. (1967). "Distinguishing Theoretical Synaptic Potentials Computed for Different Soma-Dendritic Distributions of Synaptic Inputs," *J. Neurophysiol.* **30**, 1138–1168.

REICHARDT, W. (1987). "Evaluation of Optical Motion of Information by Movement Detectors," *J. Comp. Physiol. A* **161**, 533–547.

REICHARDT, W., POGGIO, T., and HAUSEN, K. (1983). "Figure-Ground Discrimination by Relative Movement in the Visual System of the Fly, Part II: Towards the Neural Circuitry," *Biol. Cybern.* **46S**, 1–30.

ROSENBLATT, F. (1962). *Principles of Neurodynamics.* Spartan, New York.

RUMELHART, D. E., HINTON, G., and McCLELLAND, J. L. (1986). "A General Framework for Parallel Distributed Processing," in D. E. Rumelhart and J. L. McClelland (eds.), *Parallel Distributed Processing: Explorations in the Microstructure of Cognition, Vol. 1,* (pp. 45–76). MIT Press, Cambridge, Massachusetts.

SABAH, N. H., and LEIBOVIC, K. N. (1969). "Subthreshold Oscillatory Responses of the H-H Cable Model for the Squid Giant Axon," *Biophys. J.* **9**, 1206–1222.

SHEPHERD, G. M. (1972). "The Neuron Doctrine: A Revision of Functional Concepts," *Yale J. Biol. Med.* **45**, 584–599.

SRINIVASAN, M. V., and BERNARD, G. D. (1976). "A Proposed Mechanism for Multiplication of Neural Signals," *Biol. Cybern.* **21**, 227–236.

SUAREZ, H., and KOCH, C. (1989a). "Linking Linear Threshold Units with Quadratic Models of Motion Perception," *Neural Comput.* **1**, 318–320.

SUAREZ, H., and KOCH, C. (1989b). "Linking Simple Cells with Quadratic Models of Motion Perception," *Neurosci. Abstr.* **16**, 419.3.

THORSON, J. (1966). "Small-Signal Analysis of a Visual Reflex in the Locust," *Kybern.* **3**, 52–66.

TORRE, V., and POGGIO, T. (1978). "A Synaptic Mechanism Possibly Underlying Directional Selectivity to Motion," *Proc. R. Soc. Lond. B* **202**, 409–416.

VAN SANTEN, J. P. H., and SPERLING, G. (1984). "Elaborated Reichardt Detectors," *J. Opt. Soc. Am. A.* **2**, 300–320.

VOLPER, D. J., and HAMPSON, S. E. (1987). "Learning Using Specific Instances," *Biol. Cybern.* **57**, 57–71.

WATANABE, S.-I., and MURAKAMI, M. (1984). "Synaptic Mechanisms of Directional Selectivity in Ganglion Cells of Frog Retina as Revealed by Intracellular Recordings," *Jpn. J. Physiol.* **34**, 497–511.

WATSON, A. B., and AHUMADA, A. J., Jr. (1985). "Model of Human Visual-Motion Sensing," *J. Opt. Soc. Am. A* **2**, 322–342.

WILSON, H. R., and COWAN, J. D. (1972). "Excitatory and Inhibitory Interactions in Localized Populations of Model Neurons," *Biophys. J.* **12**, 1–24.

WOOD, J. G., WALLACE, W., WHITAKER, J. N., and CHEUNG, W. Y. (1980). "Immunocytochemical Localization of Calmodulin and a Heat-Labile Calmodulin-Binding Protein (CaM–BP$_{80}$) in Basal Ganglia of Mouse Brain," *J. Cell Biol.* **84**, 66–76.

WYATT, H. J., and DAW, N. W. (1975). "Directionally Sensitive Ganglion Cells in the Rabbit Retina: Specificity for Stimulus Direction, Size and Speed," *J. Neurophysiol.* **38**, 613–626.

ZADOR, A., KOCH, C., and BROWN, T. H. (1990). "Biophysical Model of a Hebbian Synapse," *Proc. Natl. Acad. Sci. USA* **87**, 6718–6722.

Chapter 13   A Model of the Directional Selectivity Circuit in Retina: Transformations by Neurons Singly and in Concert

LYLE J. BORG-GRAHAM[1] and NORBERTO M. GRZYWACZ[1,2]

[1]*Center for Biological Information Processing*
Department of Brain and Cognitive Sciences
Massachusetts Institute of Technology
Cambridge, Massachusetts
[2]*Smith-Kettlewell Eye Research Institute*
San Francisco, California

I. Introduction

Retina is a good candidate for exploring the relationship between neural computation and circuit, in particular given its physically *peripheral* location and its physiologically *central* status. One example of a spatial–temporal computation in the retina is directional selectivity. This computation may rely on interactions within the dendritic tree that are incrementally more complex than the basic *point integration* and fire neuronal response.

In this chapter, we discuss directional selectivity of neurons in the vertebrate retina, including an overview of key experimental findings and an analysis of a model for the underlying circuitry. This analysis will particularly focus on properties of the model that may distinguish it from other model types. Simulations of morphometrically and biophysically detailed cell models will demonstrate model performance, and recent electrophysiological data will be presented that addresses some model predictions. We discuss how this model may work in a developmental context and, finally, we discuss implications for more general multi-dimensional filtering within dendritic trees.

Single Neuron Computation
Neural Nets: Foundations to Applications

II.   Overview of Directional Selectivity and the Retina

A.   *Directional Selectivity versus Directional Difference*

Directional selectivity (DS) is classically defined as the property of a cell that consistently fires more spikes for movement in a specific (preferred) direction as compared to (null) movement that differs only in sign. For our dissection of the DS circuitry, we also consider the broader *directional difference* (DD) response distinction, defined as *any* preferred/null difference in system/cell output. For example, preferred/null (P/N) waveforms with equivalent averages but different shapes would constitute DD, but, by implication from the classical definition, not DS.

The model described in this chapter includes predictions for (at least) DD signals within the circuit that underlie the DS output. These DD signals will identify where in the retina the specific computation of DS first appears. Such a DD finding may be used, in principle, to rule out experimentally models that predict that a strictly DD signal will *not* be found elsewhere in the circuit, under any conditions.

B.   *Structure of the Retina*

The vertebrate retina is organized in several layers of cell bodies and their interacting processes. Signal flow is both direct (perpendicular to the image) and lateral (parallel to the image) at all levels. Light is transduced at the photoreceptor layer, which outputs to bipolar and horizontal cells within the outer plexiform layer. Bipolar cell output impinges on the mesh of amacrine and ganglion cell dendrites within the inner plexiform layer (IPL). Finally, ganglion cell axons form the optic nerve. Each major cell type has several subtypes, classified either anatomically (e.g., according to dendritic tree shape), neurochemically (e.g., cholinergic, GABAergic), or physiologically (e.g.,ON/OFF, directionally selective, red/green opponent).

C.   *Theoretical Requirements for DS and DD*

Motion detection is a computation on *spatially separated* inputs *over time*. Detection of motion *direction* (DD) requires a spatial *asymmetry* in the circuit. Finally, as described by Poggio and Reichardt (1973), DS responses also require a *nonlinearity* in the circuit.

It is useful at this point to define two broad classes of DS models: *ganglionic* models, where the crucial nonlinear interaction occurs in the ganglion cell, and *pre-ganglionic* models, where the interaction occurs prior to the ganglion cell. (See review in Koch *et al.*, 1986. In this chapter, *postsynaptic* means ganglionic, and *presynaptic* means pre-ganglionic.)

Thus, the specific questions we are trying to answer here are:

- What is the anatomy and connectivity of the DS pathway?
- What is the crucial nonlinearity of the DS pathway?
- Where on the DS pathway is the nonlinearity, e.g., is it pre-ganglionic or ganglionic?

D. Experimental Work on Retinal DS

Over the past 30 years, a large body of work has accumulated investigating DS with a variety of preparations. DS retinal output was first described in amphibian (frog, e.g., Maturana *et al.*, 1960) and eventually characterized in insect (fly, e.g., Hausen, 1981), reptile (turtle, e.g., Lipetz and Hill, 1970), bird (pigeon, e.g., Maturana, 1962), and mammal (rabbit, e.g., Barlow and Hill, 1963). We shall now outline some of the key findings that are pertinent to this chapter.

1. Physiology. The early extracellular rabbit experiments of Barlow and Levick (1965) described several phenomena related to the DS response. Using both moving slits and apparent motion protocols, they showed:

- DS Subunits: Small regions within the slit–mapped receptive field that were DS consistent with the response to full field stimuli.
- Inhibitory Mechanisms: A stimulus at a given point in the receptive field inhibited the response to a stimulus at a second point in the receptive field, when the sequence of the two stimuli simulated null-direction motion.

The first finding suggested that the DS circuit elements for a given ganglion cell were replicated many times for that cell. Historically, the most explored interpretation of the second finding was that the DS computation relied on an asymmetric inhibitory pathway. However, these data cannot rule out a model in which excitation is asymmetric and inhibition is symmetric.

Apparent asymmetric inhibition was not the only phenomenon observed in DS; Barlow and Levick (1965) (and later, Grzywacz and Amthor, 1989) also showed:

- Facilitatory Mechanisms: A stimulus at a given point in the receptive field facilitated the response to a stimulus at a second point in the receptive field when the stimuli sequence simulated preferred-direction motion.

Additional spatial and temporal parameters for the DS network may be inferred by the velocity tuning and size of DS receptive fields (e.g., Wyatt and Daw, 1975; Grzywacz and Amthor, 1989a,b; Granda and Fulbrook, 1989). For instance, the maximum length of the facilitatory lateral path on the retinal surface near the visual streak is typically 100 to 200 micrometers, as derived from apparent motion protocols. The minimum path length for a detectable DS is very short, fewer than 10 μm (Amthor and Grzywacz, unpublished data). The velocity of effective DS stimuli ranges from approximately 0.01 to 10 μm/msec (0.1 to 100 degrees/sec in the visual field.

2. Neurochemistry. The neurotransmitters involved in the DS circuit have been investigated by pharmacological protocols, for example Caldwell *et al.* (1978), rabbit, and Ariel and Adolph (1985), turtle. From this work, we can conclude the following:

- Inhibitory Mechanisms: Blockage of GABAergic pathways reduces or eliminates DS.

This result would seem to support the class of models in which inhibition is asymmetric. As we shall see, however, this data is also consistent with models in which the inhibitory pathway is symmetric. In particular, recent results from Smith *et al.* (1991) with GABAergic antagonists in turtle show:

- Inhibitory Mechanisms: For about 50% of all DS cells, DS is maintained or *reversed* when GABAergic pathways are blocked.

This result is similar to that reported previously in fly (Bülthoff and Bülthoff, 1987).

3. Anatomy. Physiologically identified DS ganglion cells have been stained in both rabbit (Amthor *et al.*, 1984) and turtle (Jensen and DeVoe, 1983). A clear result of this work is that:

- DS Morphology: The dendritic trees of DS ganglion cells are not aligned with nor asymmetric along their P/N axes.

Thus, the morphometric substrate for DS is not immediately obvious from the histology.

E.   Retinal DS Models

Inspired in part by the correlation models of Hassenstein and Reichardt (1956) for motion detection in fly, several models have been proposed for the spatial asymmetry, and the sites and biophysical mechanisms for the time–dependent nonlinear interaction (Fig. 1).

Barlow and Levick (1965) considered both asymmetric lateral inhibitory and excitatory pathways in the outer plexiform layer, with a nonlinear interaction at bipolar cells provided (according to them) by the threshold mechanism of the spike (they assumed that bipolar cells were capable of generating spikes). Others, including Torre and Poggio (1978) and Koch *et al.* (1986), suggested that the lateral pathway might be mediated by amacrine cells, among other possibilities. They showed that the interaction between an asymmetric lateral synaptic inhibition of the silent shunting type and symmetric synaptic excitation, possibly on the ganglion cell membrane itself, could provide the necessary nonlinearity for DS. These models do not explicitly define the mechanism of the delay, other than to point out that a mechanism that has a low–pass filter characteristic, or slower inhibitory synaptic kinetics, might suffice.

We note that in these circuit architectures the only directional signal available is strictly DS, under any circumstances. This is because the interaction between the asymmetric and symmetric pathways is immediately nonlinear. If the non-

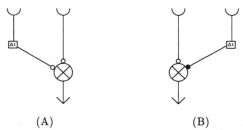

(A)   (B)

FIGURE 1.   Correlation-type models for the computation of DS are typified by that proposed for the fly and vertebrate retina. Versions of this model have considered both an all-excitatory interaction (a) and an excitatory/inhibitory interaction (b) (e.g., asymmetric-veto or AND-NOT models). In the diagram, the nonlinear interaction between the direct and asymmetric delayed inputs is expressed as a multiplication, but this particular nonlinearity is not crucial to the model. In all figures, open and closed circles are excitatory and inhibitory inputs, respectively. Note that a position-dependent delay is applied before the nonlinearity. For both a and b, the preferred response is for motion to the right. For rightward motion in a, the delayed left-hand input is correlated with the undelayed right-hand input; for rightward motion, the delay amplifies the temporal separation at the nonlinearity. For rightward motion in b, the delayed right-hand input is correlated with the undelayed left-hand input, and the inversion of the right-hand input cancels out the left-hand input; for leftward motion, again the delay amplifies the temporal separation at the nonlinearity, in this case allowing passage of the left-hand input (assuming that the nonlinearity is not strictly multiplicative).

linearity is blocked (e.g., blocking inhibition for the inhibitory model), then no interaction can take place: There will be no DD signal at all.

Koch *et al.* (1982) (also O'Donnell *et al.*, 1985) examined the electrotonic structures of ganglion cell morphometrics in detail. They showed that the dendritic tree of the ganglion cell was well-suited for local interactions within the tree between an excitatory input and an inhibitory input that has a strong shunting component. The conclusion was that the computational substrate for subunit response was possible within the tree, supporting a ganglionic model.

Recently, Vaney and co-workers (Vaney *et al.*, 1989; Vaney, 1990) have expanded on ideas from Masland *et al.* (1984) and suggested a specific cell type in rabbit; the starburst cholinergic/GABAergic amacrine cell, as providing an asymmetric excitatory pathway for the DS circuit. They further suggested that starburst dendrites also mediate a lateral inhibitory pathway in the DS circuit, since these cells contain GABA and the tips of adjacent cells are in close proximity. In their *co-transmission* model, inhibitory connectivity from starburst cells is symmetrical, in contrast to the asymmetric excitation. On the other hand, the locus and biophysical mechanisms for the computation of DS are not specified in this model, although they mention both ganglionic and pre-ganglionic alternatives.

III.   A Model of DS Output of Amacrine Cell Dendrite Tips

We now present a pre-ganglionic model for DS that is morphometrically similar to that of Vaney *et al.*, in that the lateral pathway is via individual branches of amacrine cells with tip outputs (Grzywacz and Borg-Graham, 1991). We find as well that given a plausible set of constraints on (a) the distribution and biophysics of synaptic input and output on the branches and (b) the intrinsic cable properties, the outputs of this pathway are at least DD and normally are DS. Thus, the necessary and sufficient conditions for the DS computation occur on the same substrate. The crucial DS element in this circuit is shown in Fig. 2, and the anatomy of the asymmetric pathway and the DS ganglion cell is shown in Fig. 3. We suggest, as Vaney, that multiple-oriented amacrine cell dendritic tips that converge on the DS ganglion cell form the basis for the observed subunit response and, as Torre and Poggio (1978), that the necessary nonlinearity is provided by the interaction of excitatory and inhibitory synaptic inputs. The directional properties of a cable with distributed synaptic conductance input has also been described by Grzywacz and Amthor (1989b). Likewise, Koch *et al.* (1982, Fig. 9a) suggested a similar (ganglionic) arrangement of locally symmetric

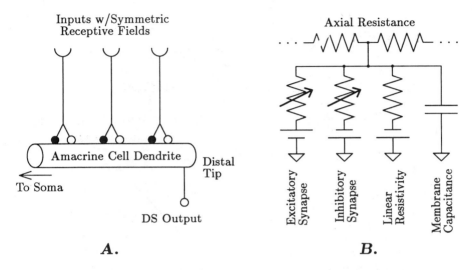

FIGURE 2.   (a) Structure of cable model discussed here. In this model the time element
is provided by the inherent kinetics of the input synapses, (Torre and Poggio, 1978; Koch
et al., 1983) and the nonlinearity by the distributed inhibitory synaptic conductance.
Under certain conditions the location-dependent delay (phase shift) provided by the cable
and an output nonlinearity may also be relevant (Sections IV.D and V.B). The preferred
response of the tip output is for motion to the right in the model version analyzed here.
(b) Equivalent circuit for a section of the dendrite cable, including the axial resistance,
the linear resting membrane resistance and capacitance, and the synaptic inputs. Directionality of the distal tip output arises because in the null direction the delayed inhibition
shunts subsequent excitatory input as the stimulus moves away from the tip. In the
preferred direction, the inhibition is much less effective at attenuating the excitation.

excitatory and delayed inhibitory inputs along the branch of a putative asymmetric
DS ganglion cell.

A.   Start by Finding the Asymmetry

Histological data from retina does not immediately suggest the spatial asymmetry
required for directional selectivity, as mentioned earlier. However, if we consider
putative outputs on the dendritic tips of amacrine cells, the IPL becomes full of
asymmetries, since for a large number of amacrine cells the tips are displaced
with respect to the entire tree. This interpretation deviates from the assumption
of *normal* neuronal polarity: Input to a neuron occurs in the dendritic tree, is
conducted proximally, integrated by the soma membrane, and the output of the

cell is conducted out the somatic axon. Some retinal neurons challenge this dogma, and the distinction of dendrite versus axon becomes blurred. Many retinal cells do not have an axon at all *per se,* and single cell processes may have both input and output.

As Vaney pointed out, detailed studies of starburst amacrine cell dendrites support this potential asymmetry (Famiglietti, 1983a, b; Vaney, 1984; Tauchi and Masland, 1984). Small swellings in the outer third of the dendritic tree have

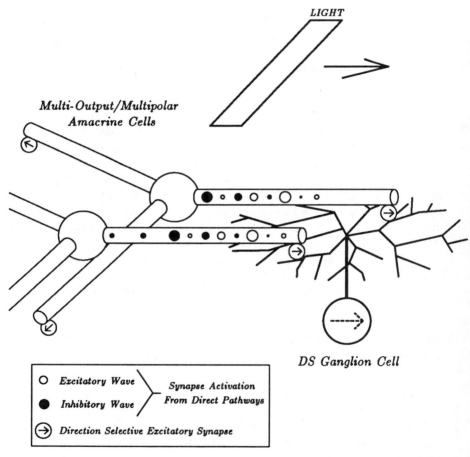

FIGURE 3.   Relationship between the asymmetric pathways in the cable model and the target DS ganglion cell with a preferred direction stimulus. Input onto the asymmetric amacrine cell dendritic paths are from symmetric (direct) pathways (either via bipolar cells or amacrine cells). The DS ganglion cell also receives direct symmetric input from bipolar and amacrine cells (not shown). Note that the inhibition kinetics lag behind the excitation.

been described, and electron micrographs show that synapses with ganglion cells occur in the swellings' zone (Brandon, 1987). In contrast, the general distribution of synaptic inputs along the dendrites is apparently uniform.[1]

B. Add Location Dependence

The morphometric asymmetry is necessary for directional selectivity, but not sufficient. If the dendrite branch is isopotential, then there is no location dependence of branch input, precluding any sort of motion selectivity, much less directional selectivity. When we add the intracellular resistance of the cable, then location dependence results (Koch et al., 1983).

C. Directional Difference for the Linear Case

Let us now examine the response of the cable tip to moving inputs. We begin with the simplest case: distributed excitatory synaptic input onto the (linear) cable. If the synapses are modeled as current sources, e.g., the excitatory synapse injects a depolarizing current, then the tip response can be obtained by the appropriate superposition of responses to individual inputs. The resulting waveforms for motion in the two directions will have different shapes but, since the system is linear, equal areas.

We now have in the dendrite cable the prerequisites—location dependency and asymmetry—for DD (but not DS)—see also Rall (1964). Accentuating the difference is the location–dependent delay of the cable (due to the cable time constant). This tends to increase the output amplitude for motion towards the tip; the EPSPs arriving at the tip are more correlated, since the delay (phase shift) cable counters the motion delay. Otherwise, the two waveforms would simply be reversed in time. We shall return to this phenomenon later in the simulations section.

Now we assume that at each point on the cable there is a paired excitatory/inhibitory input, with the inhibitory kinetics slower than that of the excitation (Torre and Poggio, 1978). Again, if the inhibitory synapse is modeled as a

---

[1] It should be pointed out that there is conflicting support for the starburst amacrine cell specifically in the DS circuit. For example, Linn and Massey (1990) suggest that cholinergic release from starburst cells is not inhibited directly by GABA, which is contrary to the model prediction. However, cable inhibition might be mediated by another system on the starburst cells, and other amacrine cells may have the necessary input–output relationship. Later, we shall show simulations of a cell whose morphometry is based on the starburst cell; despite the evidence against the starburst cell specifically, we feel that the conclusions from this tree geometry are applicable to other amacrine cells.

hyperpolarizing current source, the (linear) responses for the two directions will have equal area. The DD response holds, still without DS.

D.   Synaptic Nonlinearities and Cable Directionality

Synaptic input, however, is more accurately modeled by a conductance change in series with a battery. The circuit is now a nonlinear one; the variable conductance input means that the motion responses cannot be derived from the superposition of a sequence of point responses, and the area for the two directions is generally *not* conserved.

The effect of the inhibitory input is particularly important on the directionality of the cable tip response. Assuming that the reversal potential for the inhibition is close to the cell's resting potential, then the main effect of the inhibition will be to *shunt* locally any excitatory current to the extracellular space. The effect of the shunt on the tip response strongly depends on the relative location of the excitatory and inhibitory inputs with respect to the output. As Rall (1964) and Koch *et al*. (1982, 1983) showed, inhibition is most effective when it is *on the path*, that is, interposed between excitation and the output. Conversely, if the excitation is closer to the output, then the inhibitory shunting is much less effective (Fig. 4).

The inhibitory shunting now supplies the necessary nonlinearity for the DS response. This prediction is in concert with the experimental evidence as to the importance of inhibitory mechanisms for DS, and with evidence that shunting inhibition might mediate DS in rabbit (Amthor and Grzywacz, 1991).

We also note that, while the efficacy of the inhibitory synapse with respect to DS is mainly due to its shunting component, it is also true that the precise value for the inhibitory reversal potential is not crucial (as long as it is in the neighborhood of the resting potential). Locality of interaction is strongest when the inhibitory reversal potential equals the resting potential, but this is not relevant for the geometry of the model presented here.

E.   Considering Tip Output Nonlinearity, Facilitation, and the Sign of the Output

Inhibition onto the cable suffices to make the towards-tip response larger than the opposite direction, so a linear output function would preserve the distinction. It is more likely, though, that the output synapse has a threshold. The resulting supralinear region only amplifies the DS distinction.

The experimental evidence of preferred–direction facilitation data mentioned earlier supports placing a facilitatory mechanism on the DS pathway, as opposed

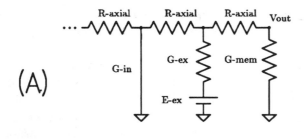

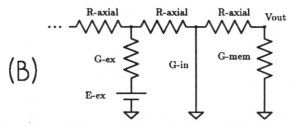

FIGURE 4.   The effect of inhibition on tip output is dependent on the relative position of the excitation and the inhibition with respect to the output. To illustrate, we use an extreme case in which the inhibitory battery is equal to the resting potential, and the activated inhibitory conductance is infinite. In a, the excitatory input is closer to the tip than the inhibitory input. Although the membrane potential at the site of inhibition is clamped to the resting potential (short-circuited), the axial resistance of the cable partially isolates the excitatory location, and the tip (V_{out}) is depolarized. In b, the interposed inhibitory input clamps the cable between the excitatory input and the output to the resting potential, so that the output stays at rest. Adding membrane capacitance, a finite conductance to the activated inhibitory synapse, or making the inhibitory reversal potential not exactly equal to the resting potential does not change the basic interaction. However, a nonzero axial resistance *is* crucial.

to only a symmetric location (e.g., on the ganglion cell soma). We postulate that this could be accounted for by various mechanisms located at the amacrine cell tip. For example, facilitatory mechanisms intrinsic to the synaptic output might suffice. Likewise, we may consider a time- and voltage-dependent K^+ channel at the tip that is normally open at rest but inactivates with depolarization, similar to the I_D channel identified in hippocampal pyramidal cells. (See review by Storm, 1990.) The characteristics of this channel would cause the tip output to be *primed* by distant cable input, such that subsequent excitation near the tip would be unopposed by the now–inactivated K^+ shunt.

Two versions of this model include either that the amacrine DS output forms an excitatory connection with a DS ganglion cell, or the connection is inhibitory. It is also possible to include plausible voltage- and time-dependent nonlinearities in the cable so that the preferred direction for the branch output is *away* from the tip. Also, the tip output may pass through bipolar cells or amacrine cells before the ganglion cell.

While any combination of the preceding polarities will yield a ganglion cell DS response, the fact that directional responses have been recorded in the absence of GABAergic inhibition (Section II.D.1) suggests that the DS pathway has an excitatory component at every junction. Further, data described later in this chapter from turtle ganglion cells with local block of GABAergic inhibitory input supports the excitatory connectivity version. Also, simulations of model cells with likely membrane parameters favor the preferred-towards-tip orientation for the cable output (Section V and unpublished data).

IV.   Predictions of the Model

We now consider testable predictions, namely, those for somatic recordings of amacrine and target ganglion cells.

A.   DS Somatic Recordings?

With respect to amacrine cells, the model predicts that moving stimuli centered on the somatic receptive field would not elicit a directionality, assuming a symmetrical dendritic tree. If either the stimuli or the dendritic tree was asymmetric with respect to the soma, then a directional response would result, perhaps similar to the pre-ganglionic DS/DD recordings of DeVoe *et al.* (1989). While this result supports the proposed mechanism for DS, it does not link the mechanism to an identified DS ganglion cell.

From the point of view of the DS ganglion cell, the model predicts that the input waveforms to the cell are themselves DS; any ganglionic computational mechanisms will be inherently symmetric and serve only to refine the P/N response properties. This result is in contrast with the ganglionic model class, in which the inputs are not DS by themselves.

B.   DS Dependence on Ganglion Cell Membrane Potential

Differences in P/N EPSPs are predicted by the ganglionic model, but since the differences arise from ganglionic interaction of conductances with different reversal potentials, this model predicts that the relationship between the preferred and null EPSPs will depend on the membrane potential.

For example, let us consider the ganglionic AND-NOT circuit: Null direction response reflects a temporal overlap of the excitatory input with inhibitory (mainly shunting) input at the ganglion cell. Since the inhibitory reversal potential might be near the resting potential, there would be few negative portions in either the preferred or null response. A negative portion would result if the membrane potential is artificially raised by injecting current. The *unmasked* IPSP would be expected to be more correlated with the control EPSP in the null direction as compared to the preferred direction (e.g., Marchiafava, 1979). If the ganglion cell was hyperpolarized by the electrode, the *unmasked* inhibitory input would now contribute a component to the EPSP. With sufficient hyper/depolarization (at least such that the entire response stays below the inhibitory reversal potential), the *amplitude* of the null response could become greater than that of the preferred response.

This result is in clear contrast to a pre-ganglionic model. The single reversal potential of the directionally-selective circuit inputs imply that the ratio of the preferred and null EPSPs is *independent* of the membrane potential: Manipulating the membrane potential will not reverse the P/N axis.

C.   Comparing Total Synaptic Input for the P/N Responses

The preceding result is related to measuring the somatic input conductance, $G_{In}(S(x,t))$, of the ganglion cell during a motion stimulus S. For a lumped cell approximation and no voltage-dependent membrane, predictions about $G_{In}(S(x,t))$ are simple. In the ganglionic model, the only difference in the inputs for the preferred versus null responses is in their timing. Thus, the model predicts:

$$\int G_{In}(S(x,t)) = \int G_{In}(S(-x,t)).$$

On the other hand, for the pre-ganglionic model, there is more synaptic input for one direction versus the other; therefore (assuming the lumped cell without voltage dependencies):

$$\int G_{In}(S(x,t)) > \int G_{In}(S(-x,t))$$

for a pre-ganglionic excitatory DS input model, and

$$\int G_{In}(S(x,t)) < \int G_{In}(S(-x,t))$$

for a pre-ganglionic inhibitory DS input model, where $(S(x,t))$ is a stimulus moving in the preferred direction. It can be shown (Borg-Graham, in preparation[a]) that under some constraints on the ganglion cell and the experimental protocol, similar relationships are testable even with distributed inputs on dendritic trees and voltage-dependent membrane.

D.   Dynamic Range of Cable Mechanisms: Saturation and DS Reversal

Real biophysical mechanisms saturate; e.g., a supralinear transfer function will not stay supralinear for unbounded inputs. We now consider possible implications of saturation on the performance of the model.

Saturation of input synaptic conductance onto the cable will not change the basic distinctions between the preferred and null waveforms. Saturation (strictly speaking, a sublinear region) of the output nonlinearity can have quite different effects: For strong enough cable excitation, the null waveform, with its greater temporal support, will eventually yield a final *integrated* output that is equal to the preferred output, despite the larger amplitude of the preferred waveform. Increasing excitation further could cause a *reversal* of the P/N orientation, after the output nonlinearity. This might be observed, for example, with a stimulus contrast that is normally DS, but with inhibition reduced or blocked (e.g., with pharmacological manipulations). Whether DS would be entirely eliminated or reversed would be dependent on circuit and stimulus parameters.

This result is unique to models in which the DS asymmetry includes distributed excitatory input with relative delays along the P/N axis. Such models have been explored theoretically for the interpretation of DS reversal in fly (Ögmen, 1991), and in Section V.B. we show simulations that demonstrate this effect.

V.   Simulations of Morphometrically and Biophysically Detailed Amacrine Cell Models

To investigate neuronal properties that are pertinent to the cable model, we have run simulations of amacrine cells. The model parameters are as constrained as possible; morphometry is obtained from histological data in the literature, and membrane properties are either inferred from experimental data and/or supported by theoretical studies of other cells (Borg-Graham, 1987).

An important aspect of these simulations is that dynamic retinotopic stimuli may be used as input to the model circuit. Since the simulator maintains the three-dimensional structure of the cells, it is straightforward to interpret model response to realistic arbitrary stimuli (Borg-Graham, in preparation[b]).

A.   Simulations of Asymmetric Responses from Symmetric Cells

In Figs. 5 through 7, we show a simulation of an amacrine cell whose morphometry is taken from a rabbit starburst amacrine cell. All membrane and cable properties are uniform, including specific capacitance (C_m), membrane resistivity (R_m), intracellular resistivity (R_i), excitatory synaptic density (G_{ex}), and inhibitory synaptic density (G_{in}). R_m and R_i are fixed at 100 KΩcm^2 and 200 Ωcm (R_i from Shelton, 1985), respectively. C_m is set to 1.0 μFcm^{-2} (τ_m = 100 mS), G_{in} is set to 100 pS μm^{-2}, and G_{ex} to 1 pS μm^{-2}. The resting potential for the cell is -70 mV. The reversal potentials for the excitatory and inhibitory synapses are 50 and -70 mV, respectively. No voltage-dependent membrane is included.

Synaptic transfer functions are fixed. Excitatory response is given by the half-wave-rectified convolution of the light stimulus by the difference of two alpha functions, $\alpha(t)$,

$$\alpha(t) = \frac{t}{\tau^2}e^{-t/\tau},$$

the first with a τ of 10 mS and the second with a τ of 60 mS. This transient response is typical of retinal ON response to a flashing stimulus. The inhibitory response is the half-wave-rectified convolution of the light stimulus with a single alpha function of unit area and τ = 100 mS. This sustained ON response approximates the time course for inhibition to various stimuli (Amthor and Grzy-wacz, 1988). Adding an OFF component to the transfer functions does not change

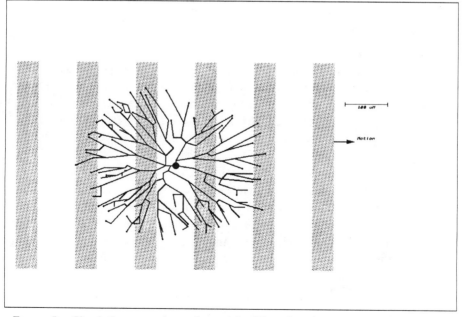

FIGURE 5.   Simulations were done of a model cell based on the morphometry of a rabbit starburst amacrine cell, taken from Tauchi and Masland (1984). Shown here is a flat mount view of cell, with snapshots of the trajectory of the stimulus slit. Node 32211 referred to in Fig. 6 is the distal tip farthest to the right.

the motion-dependent behavior. For simplicity, the spatial impulse response for both synapse types is $\delta(x,y)$. Although the receptive field of candidate bipolar or amacrine cell inputs to the DS dendrite is probably on the order of tens to hundreds of micrometers wide, we were interested in an upper bound on the intrinsic spatial discrimination of the DS dendrite structure.

The large value for R_m is consistent with values measured in our lab and others in various neurons using the whole-cell patch technique (Coleman and Miller, 1989). Synaptic conductance densities and kinetics are less constrained as far as the literature is concerned; the range of values we have chosen produce synaptic potentials that are consistent with available data.

In Fig. 6, we show the response of the soma and a distal node to a bar moving across the entire field of the cell. The soma response is almost identical for opposite motions, while the tip response is highly DS. In Fig. 7, we have plotted the integrated response of each dendritic tip of the cell, scaled by stimulus speed, as a function of angle off the soma, for a slit traveling across the entire breadth of the cell. Despite the overall symmetry of the cell, this functional is highly directional.

B.   Parametric Simulations of Cable Mechanisms

In this section, we present a series of simulations on a simple cell model to illustrate how the different mechanisms in the dendrite cable interact. The basic structure is a one-dimensional symmetrical cell with two opposing unbranched processes (Fig. 8). Cell membrane and synaptic parameters are identical to the simulated cell in Fig. 5, except as follows: We shall vary C_m (1.0 and 1.0^{-5} μFcm^{-2}, or $\tau_m = 100$ mS and 0.001 mS),* and G_{in} (100, 10, 1, 0.1, and 0 pS μm^{-2}). The goal here is to see how directionality depends on the inhibitory input versus the intrinsic cable properties.

As before, the stimulus is a moving slit, 50 μm wide, now with velocities of 0.5, 1, 2, 4, 8, and 16 μm/msec. The response waveforms from the right-hand distal tip for rightward and leftward motion will be compared. In particular, we shall compute a directional index (*DI*) (after Grzywacz and Koch, 1987) for both the linear integral of the waveforms and the integral of the waveforms after being passed through a sigmoidal nonlinearity (representing synaptic transmission):

---

*A cm of 0 μF_{cm}^{-2} was not possible due to the integration technique of the simulator.

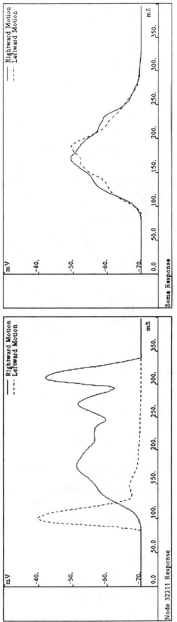

FIGURE 6. Responses of cell in Fig. 5 to 50 μm slit moving from left to right and back again at 2 μm/msec. The voltage at the rightmost distal tip in Fig. 5 (Node 32211) is highly DS, whereas the soma response is barely DD. Model parameters are given in the text. Simulations also show that inclusion of a K⁺ conductance that inactivates with depolarization can both facilitate the preferred-direction response and attenuate the null-direction response. (See Section III.E.)

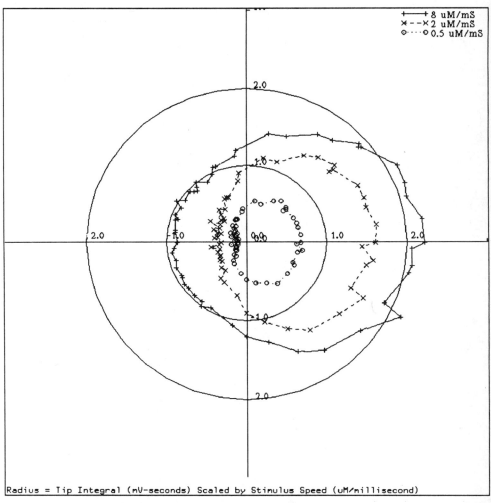

FIGURE 7.   Polar plot of integral of distal tip voltages scaled by the stimulus speed versus angle of tip with respect to the soma, relative to the slit trajectory shown in Fig. 5. Slit speeds include 0.5, 2, and 8 μm/msec. Despite the overall symmetry of the cell, the tip outputs respond asymmetrically to motion.

$$DI = \frac{\int f(V_P) - \int f(V_N)}{\int f(V_P) + \int f(V_N)}$$

$DI = -1$	$DI = 0$	$DI = 1$
Strongly Selective	No	Strongly Selective for
for Motions Away	Directionality	Motions Towards the Tip
from the Tip		

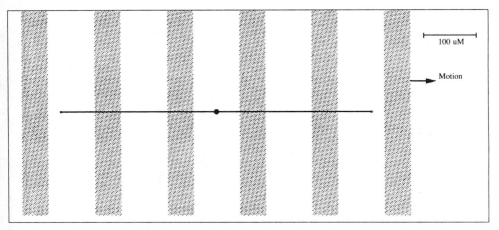

FIGURE 8.   Model cell with two opposing dendrites whose diameters taper linearly from 1.0 to 0.2 μm (proximal to distal). The output used in the later figures is from the right-hand distal tip.

where $f()$ is either the identity ($\Rightarrow DI$(Average)), or the sigmoidal nonlinearity shown in Fig. 11 ($\Rightarrow DI$(Sigmoid Average)). V_P is the voltage waveform for the distal tip for light moving toward the tip, and V_N is the waveform for the opposite direction.

As shown in Fig. 10 and 11, the *normal* directionality of the tip outputs ($DI > 0$) is strongly dependent on the presence of inhibition. However, if a nonlinear functional is applied to the tip waveform, the distributed nature of the cable *without* inhibition generates a directional response, although it is much weaker than the control case. In addition, the particular nonlinearity in Fig. 11 causes a DI reversal. As illustrated in Figs. 9 and 11, the ability of a nonlinear integrator to make this distinction depends on the delay provided by the cable capacitance. This prediction is consistent with the experimental results cited earlier.

VI.   Intracellular DS Recordings with Local Block of Inhibition

We now present recent data from turtle retina that address some of the model predictions.

Interpretation of the ganglion cell membrane potentials may be complicated by inhibitory input onto the cell that might not have a direct link to the directional properties. For this reason, we have recorded from ganglion cells in the intact isolated turtle retina (Borg-Graham and Grzywacz, 1990), using whole-cell patch

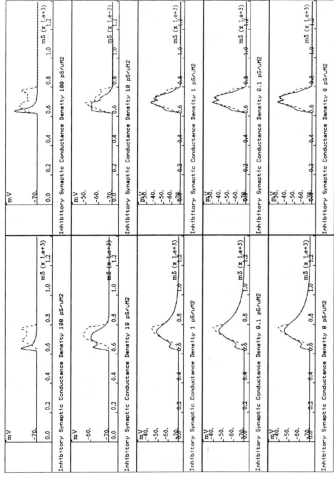

FIGURE 9. Example set of simulations for cell in Fig. 8 for a stimulus moving at 4 μm/msec at various values for G_{in} and C_m. On the left, $C_m = 1.0$ μFcm^{-2}, and on the right, $C_m = 1.0^{-5}$ μFcm^{-2}. Tip output is from the right-hand distal tip in Fig. 8, and the solid and dashed waveforms are for leftward and rightward motion, respectively. When inhibition is absent (bottom simulations), the cable capacitance distorts the equal area waveforms so that a nonlinear integrator may distinguish them (bottom left). If the nonlinearity saturates, then the *fatter* leftward response yields a larger total output than the higher amplitude rightward response (DS reversal). With no inhibition and very low capacitance (bottom right), the waveforms become mirror-symmetric.

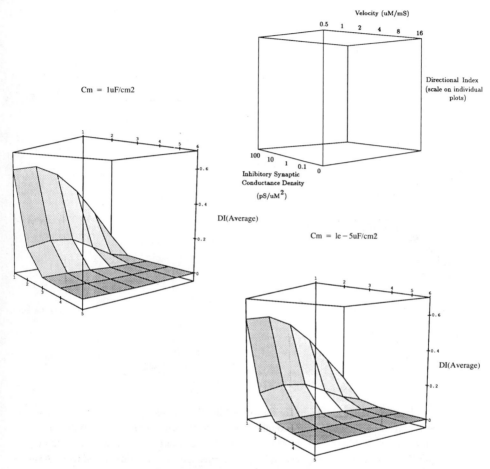

FIGURE 10.   *DI* for the linear average of distal tip waveforms obtained from simulations of cell in Fig. 8, over a range of inhibitory synaptic conductance densities and stimulus velocities, for $C_m = 1.0$ μFcm^{-2} (middle left) and 10^{-5} μFcm^{-2} (lower right). Axes are shown in the upper right. For this linear functional, DS is eliminated as inhibition is lowered, independent of C_m. Furthermore, cable capacitance does not seem to have a significant effect on DS when inhibition is present.

electrodes in which the electrode solution is free of ATP and Mg^{2+}. Given the large bore of the electrode (1–2 μm), it is likely that the cell contents are dialyzed by the electrode solution within several minutes after the start of recording. It has been reported in hippocampus that such conditions block the response of GABA$_A$ receptors (Stelzer *et al.*, 1988); thus, this technique offers a method for selectively blocking an inhibitory component of the synaptic input to the cell being recorded from, without disturbing the rest of the network.

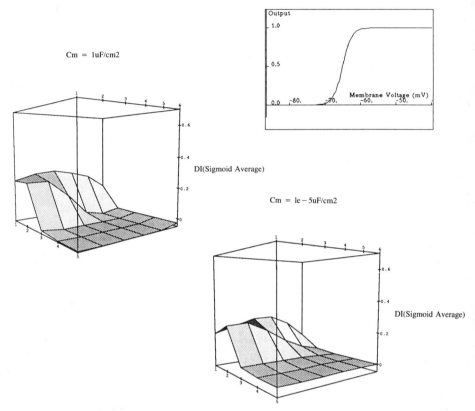

FIGURE 11.　*DI* for the same simulations as in Fig. 10, but taken of the average of distal tip waveforms after being passed through the sigmoid shown in upper right. With normal C_m (middle left), the directionality *reverses* ($DI < 0$) as inhibition is lowered. When C_m and inhibition are both small (lower right), this reversal is dramatically reduced. Axes are as in Fig. 10.

With this technique, ganglion cells show clear light-evoked IPSPs and EPSPs at normal resting potential at the onset of the recording. Within, typically, 10 minutes of recording, light-evoked IPSPs disappear while EPSPs are maintained, suggesting the block described previously. Removal of direct inhibitory input to these cells was verified by depolarizing the cells: Negative synaptic potentials were not observed despite large depolarizations (30 mV above the resting potential). Hyperpolarizing potentials were observed, however, in conjunction with action potentials, suggesting preservation of voltage-dependent hyperpolarizing mechanisms.

In some of these cells, we observe clear directional responses (Fig. 12 and 13), despite the lack of significant inhibitory input onto the ganglion cell. P/N distinctions were stable with respect to both hyperpolarizing and depolarizing current. These results suggest that some turtle ganglion cells receive excitatory input that is already DS (Borg-Graham and Grzywacz, 1991).

VII.   Development of DS: The Problem of Coordination of Asymmetries

A problem for any model of DS is how to break symmetry on a scale significantly larger than that predicted by random distributions of local asymmetries. In the retina, the DS response is correlated over multiple subunits, and several directions

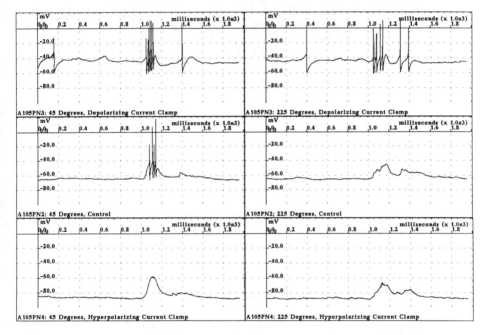

FIGURE 12.   DS intracellular (whole-cell patch) data from an on/off ganglion cell (A105) in isolated intact turtle retina. The response to motion at a 45° orientation (in terms of either the number of spikes or the size of the EPSP) is larger than that for the opposite (225°) orientation, independent of the holding current. IPSPs are not observed, even with depolarizing current, suggesting that the DS response does not require inhibitory input to the recorded cell.

Stimulus is 200 μm square spot, moving at 4 μm/ms, with a path length of 1200 μm. Spot motion is timed so that the spot passes over the receptive field center at 1000 ms for both orientations. The receptive field, mapped with a stationary flashing spot, is approximately 500 μm in diameter. Whole-cell electrode solution lacks ATP and Mg^{++}, which contributes to the attenuation of IPSPs.

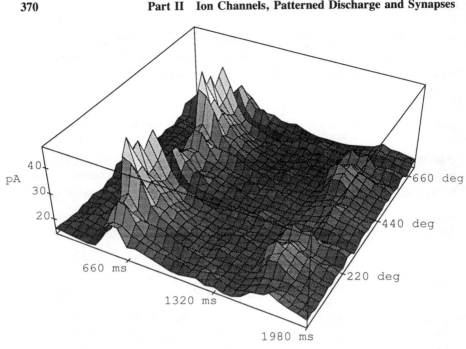

FIGURE 13.   DS voltage clamp data from an ON ganglion cell (A56) in turtle, under the same experimental conditions as Fig. 12. Stimulus is 1 hz grating with a spatial period of 800 μm and a 100 μm square aperture, centered on the spot-mapped receptive field. Grating is presented at 16 orientations (22.5° steps), 1980 ms per orientation. There are 2 averaged trials per orientation, and in this plot the 16 clamp current waveforms are repeated in order to show more clearly the directional selectivity. Holding potential is −80mV, and the clamp current is low pass filtered (f_c = 20 hz) and inverted, so that a positive output represents an inward (or excitatory) current. For all orientations, the stimulus phase was adjusted so that an ON edge appeared at the side of the dark aperture at 0 ms. Extracellular (ON cell patch) recordings of this cell with similar stimuli prior to whole-cell access showed a DS response with a preferred direction of 135°. The strong directionality of the voltage clamp inward current (best at 157.5°) and the low holding potential suggest that the excitatory input to this cell is already DS.

are represented. For the model presented here, this translates into determining how a ganglion cell could selectively connect to amacrine dendrite tips of similar orientation. One solution that seems natural for this model is to postulate a Hebbian correlational process (Hebb, 1949) similar to long-term potentiation in the hippocampus. (See review by Brown *et al.*, 1990). A Hebbian type process strengthens active synapses when the pre- and postsynaptic cells fire simultaneously and may weaken synapses when activity is not correlated.

The initial symmetry in the retina may then be broken as follows: Assume that at an early stage of synaptogenesis, initial DS amacrine connections to a proto-DS ganglion cell have a range of orientations, but overall there is a slight orientation bias. If this bias is strong enough, then a Hebbian mechanism could reinforce it. The connectivity of subsequent synapses would then become stronger or weaker depending on their dendrite's alignment relative to the initial (weak) orientation.

A Hebbian mechanism, at least in this simple form, is suited for a pre-ganglionic DS model because the mechanism requires signals that can be correlated; the postsynaptic cell tends to collect correlatable inputs. Ganglionic computation of DS loses this advantage, since there is no motion asymmetry intrinsic in the ganglion cell afferents.

VIII.   Retinal Directional Selectivity: Exemplar of a Canonical Computational Mechanism?

Thus far, we have discussed the directional properties that may arise intrinsically whenever dendrites have distributed excitatory and inhibitory input, at least in terms of the waveforms at the dendrite end. The key to this property is the asymmetric distribution of the inputs with respect to the output, and we have demonstrated how this sort of asymmetry may be especially effective when the output synapse is on the dendrite tip. Since in retina the salient stimuli features are retinotopic, it is very important to consider the exact geometry of the amacrine cells' dendritic trees.

However, in the more general case, we may consider the cascade of inputs along a dendritic branch of a generic central neuron. The performance of the retinal DS model suggests that a similar directionality will exist under some conditions for non-retinal neurons, with respect to the afferents along each branch. We suggest that with plausible biophysical parameters, the soma of a more general cell, which is asymmetric with respect to each dendritic branch, will see a *directional* response from its branch inputs. Specifically (at least with the version of the model discussed here), for a given set of inputs over time along a branch, the soma EPSP will be greatest when the temporal order of the inputs is from distal to proximal, i.e., stepping in time toward the soma. Conversely, the same set of inputs, only reversed in time, may be much less effective in depolarizing the soma.

As with retinal DS, the net result would be that the dendritic branch functions as more than just a time–independent integrator of inputs. Rather, the branch functions as a nonlinear *spatio*–temporal filter.

IX.   Conclusions

In summary, we have presented a model for directional selectivity in retina whose anatomical structure and simulated performance is consistent with the data. The crucial elements for directionality are as follows:

- Asymmetric input-to-output distribution on the DS dendrite.
- Intracellular resistivity of the dendrite cable allowing on-the-path interactions that depend on stimulus direction.
- Inhibitory nonlinear shunting of sufficient duration to mask subsequent proximal excitation of the distal tip output.

Further, we have showed that other cable mechanisms may generate testable predictions unique to this model.

Acknowledgments

We would like to acknowledge Prof. Tomaso Poggio for his initial suggestion to investigate this problem, and his continuing support of this work. We also thank Randy Smith, John Wyatt, John Lisman and Frank Amthor for their comments.

This chapter describes research done within the Center for Biological Information Processing in the Department of Brain and Cognitive Sciences and at the Smith-Kettlewell Eye Research Institute. This research is sponsored by the grant, EY08921-01A1, from the National Eye Institute; by the grant, BNS-8809528, from the National Science Foundation; a grant from the Office of Naval Research, Cognitive and Neural Sciences Division; the Alfred P. Sloan Foundation; and a grant from the National Science Foundation under contract IRI-8719394. Tomaso Poggio is supported by the Uncas and Helen Whitaker Chair at the Massachusetts Institute of Technology, Whitaker College.

References

AMTHOR, F. R., and GRZYWACZ, N. M. (1988). "The Time Course of Inhibition and Velocity Independence of Direction Selectivity in the Rabbit Retina," *Investigative Opthalmology & Visual Science* **29**(4), 225.

AMTHOR, F. R., and GRZYWACZ, N. M. (1991). "Nonlinearity of the Inhibition Underlying Retinal Directional Selectivity," *Visual Neuroscience* **6,** 95–104.

AMTHOR, F. R., OYSTER, C. W. and TAKAHASHI, E. S. (1984). "Morphology of On–Off Direction-Selective Ganglion Cells in the Rabbit Retina," *Brain Research* **298,** 187–190.

ARIEL, M., and ADOLPH, A. R. (1985). "Neurotransmitter Inputs to Directionally Sensitive Turtle Retinal Ganglion Cells," *Journal of Neurophysiology* **54**(5), 1123–43.

BARLOW, H. B., and LEVICK, W. R. (1965). "The Mechanism of Directionally Selective Units in Rabbit's Retina," *Journal of Physiology* **178**, 477–504.

BARLOW, H. B., and HILL, R. M. (1963). "Selective Sensitivity to Direction of Movement in Ganglion Cells of the Rabbit Retina," *Science* **139**, 412–414.

BORG-GRAHAM, L. (1987). "Simulations Suggest Information Processing Roles for the Diverse Currents in Hippocampal Neurons," in D. Z. Anderson (ed.), *Neural Information Processing Systems*. American Institute of Physics.

BORG-GRAHAM, L. (1992a). "Somatic Determination of Inhibitory Action in the Dendritic Tree," Center for Biological Information Processing Memo, MIT, Cambridge, Massachusetts.

BORG-GRAHAM, L. (1992b). "The SURF–HIPPO Neuron Simulator," Center for Biological Information Processing Memo, MIT, Cambridge, Massachusetts.

BORG-GRAHAM, L., and GRZYWACZ, N. M. (1990). "An Isolated Turtle Retina Preparation Allowing Direct Approach to Ganglion Cells and Photoreceptors, and Transmitted-Light Microscopy," *Investigative Opthalmology & Visual Science* **31**(4), 1039.

BORG-GRAHAM, L., and GRZYWACZ, N. M. (1991). "Whole-Cell Patch Recordings Analysis of the Input onto Turtle Directionally Selective (ds) Ganglion Cells," *Investigative Opthalmology & Visual Science* **32**(4), 2067.

BRANDON, C. (1987). "Cholinergic Neurons in the Rabbit Retina: Dendritic Branching and Ultrastructural Connectivity," *Brain Research* **426**, 119–130.

BROWN, T. H., KAIRISS, E. W., and KEENAN, C. L. (1990). "Hebbian Synapses: Biophysical Mechanisms and Algorithms," in W. M. Cowan, E. M. Shooter, C. F. Stevens, and R. F. Thompson (eds.), *Annual Review of Neuroscience, Vol. 13* (pp. 475–511). Annual Reviews Inc., Palo Alto, California.

BÜLTHOFF, H., and BÜLTHOFF, I. (1987). "Gaba-Antagonist Inverts Movement and Object Detection in Flies," *Brain Research* **407**, 152–158.

CALDWELL, N. W., DAW, J. H., and WYATT, H. J. (1978). "Effects of Picrotoxin and Strychnine on Rabbit Retinal Ganglion Cells: Lateral Interactions for Cells with More Complex Receptive Fields," *Journal of Physiology* **276**, 277–298.

COLEMAN, P. A., and MILLER, R. F. (1989). "Measurement of Passive Membrane Parameters with Whole-Cell Recording from Neurons in the Intact Amphibian Retina," *Journal of Neurophysiology* **61**(1), 218–230.

DEVOE, R. D., CARRAS, P. L., CRISWELL, M. H., and GUY, R. G. (1989). "Not by Ganglion Cells Alone: Directional Selectivity Is Widespread in Identified Cells of the Turtle Retina," in R. Weiler and N. N. Osborne, (eds.) *Neurobiology of the Inner Retina* (pp. 235–246). Springer–Verlag, Berlin.

FAMIGLIETTI, E. V. (1983a)."ON and OFF Pathways through Amacrine Cells in Mammalian Retina: the Synaptic Connection of Starburst Amacrine Cells," *Vision Research* **23**, 1265–1279.

FAMIGLIETTI, E. V. (1983b). "'Starburst' Amacrine Cells and Cholinergic Neurons: Mirror-Symmetric ON and OFF Amacrine Cells of Rabbit Retina," *Brain Research* **261**, 138–144.

GRANDA, A. M., and FULBROOK, J. E. (1989). "Classification of Turtle Retinal Ganglion Cells," *Journal of Neurophysiology* **62**(3), 723–737.

GRZYWACZ, N. M., and AMTHOR, F. R. (1989a). "Computationally Robust Anatomical Model for Retinal Directional Selectivity," in D. S. Touretzky, ed. *Advances in Neural Information Processing Systems, Vol. 1* (pp. 477–484). Morgan Kaufman, Palo Alto, California.

GRZYWACZ, N. M., and AMTHOR, F. R. (1989b). "Facilitation in On–Off Directionally Selective Ganglion Cells of the Rabbit Retina," *Neuroscience Abstracts* **15**, 969.

GRZYWACZ, N. M., and BORG-GRAHAM, L. (1991). "Model of Retinal Directional Selectivity Based on Amacrine Input/Output Asymmetry," *Investigative Opthalmology & Visual Science* **32**(4), 2263.

GRZYWACZ, N. M., and KOCH, C. (1987). "Functional Properties of Models for Direction Selectivity in the Retina," *Synapse* **1**, 417–434.

HASSENSTEIN, B., and REICHARDT, W. E. (1956). "Functional Structure of a Mechanism of Perception of Optical Movement," *Proceedings 1st International Congress Cybernetics Namar,* 797–801.

HAUSEN, K. (1981). "Monocular and Binocular Computation of Motion in the Lobula Plate of the Fly," *Verh. Dtsch. Zool. Ges.* **74**, 49–70.

HEBB, D. O. (1949). *The Organization of Behavior.* John Wiley, New York.

JENSEN, R. J., and DEVOE, R. D. (1983). "Comparisons of Directionally Selective with other Ganglion Cells of the Turtle Retina: Intracellular Recording Staining," *Journal of Comparative Neurology* **217**(3), 271–87.

KOCH, C., POGGIO, T., and TORRE, V. (1982). "Retinal Ganglion Cells: A Functional Interpretation of Dendritic Morphology," *Proceedings of the Royal Society, London B* **298**, 227–264.

KOCH, C., POGGIO, T., and TORRE, V. (1983). "Nonlinear Interactions in a Dendritic Tree: Localization, Timing, and Role in Information Processing," *Proc. Natl. Acad. Sci.* **80**, 2799–2802.

KOCH, C., POGGIO, T., and TORRE, V. (1986). "Computations in the Vertebrate Retina: Gain Enhancement, Differentiation, and Motion Discrimination, *Trends in Neuroscience* **9**(5), 204–211.

LINN, D. M., and MASSEY, S. C. (1990). "GABA Inhibits ACH Release from the Rabbit Retina: A Direct Effect or Bipolar Cell Feedback?" *Society of Neuroscience Abstracts* **16**(297.10), 713.

LIPETZ, L. E., and HILL, R. M. (1970). "Discrimination Characteristics of Turtle's Retinal Ganglion Cells," *Experientia* **26**, 373–374.

MARCHIAFAVA, P. L. (1979). "The Responses on Retinal Ganglion Cells to Stationary and Moving Visual Stimuli," *Vision Research* **19**, 1203–1211.

MASLAND, R. H., MILLS, J. W., and CASSIDY, C. (1984). "The Function of Acetylcholine in the Rabbit Retina," *Proceedings of the Royal Society, London* **223**, 121–139.

MATURANA, H. R., LETTVIN, J. Y., McCULLOCH, W. S., and PITTS, W. H. (1960). "Anatomy and Physiology of Vision in the Frog (*rana pipiens*)," *Journal General Physiology* **43**(Suppl. 2), 129–171.

MATURANA, H. R. (1962). "Functional Organization of the Pigeon Retina, *International Congress Physiology Science, 22nd*, 170–178.

O'DONNELL, P., KOCH, C., and POGGIO, T. (1985). "Demonstrating the Nonlinear Interaction between Excitation and Inhibition in Dendritic Trees Using Computer-Generated Color Graphics: A Film," *Society of Neuroscience Abstracts* **11**, 142.

ÖGMEN, H. (1991). "On the Mechanisms Underlying Directional Selectivity," *Neural Computation*.

POGGIO, T., and REICHARDT, W. E. (1973). "Considerations on Models of Movement Detection," *Kybernetics* **13**, 223–227.

RALL, W. (1964). "Theoretical Significance of Dendritic Tree for Input–Output Relation," in R. F. Reiss (ed.), *Neural Theory and Modeling* (pp. 73–97). Stanford University Press, Stanford, California.

SHELTON, D. P. (1985). "Membrane Resistivity Estimated for the Purkinje Neuron by Means of a Passive Computer Model," *Neuroscience* **14**(1), 111–131.

SMITH, R. D., GRZYWACZ, N. M., and BORG-GRAHAM, L. (1991). "Picrotoxin's Effect on Contrast Dependence of Turtle Retinal Directional Selectivity," *Investigative Opthalmology & Visual Science* **32**(4), 2913.

STELZER, A., KAY, A. R., and WONG, R. K. S. (1988). "GABA-A-Receptor Function in Hippocampal Cells Is Maintained by Phosphorylation Factors," *Science* **241**, 339–341.

STORM, J. F. (1990). "Potassium Currents in Hippocampal Pyramidal Cells," in J. Storm-Mathisen, J. Zimmer, and O. P. Ottersen (eds.), *Progress in Brain Research, Vol. 83* (pp. 161–187). Elsevier Science Publishers B. V. (Biomedical Division).

TAUCHI, M., and MASLAND, R. H. (1984). "The Shape and Arrangement of the Cholinergic Neurons in the Rabbit Retina," *Proceedings of the Royal Society, London B* **223**, 101–119.

TORRE, V., and POGGIO, T. (1978). "A Synaptic Mechanism Possibly Underlying Directional Selectivity to Motion," *Proceedings of the Royal Society, London* **202**, 409–416.

VANEY, D. I. (1990). "The Mosaic of Amacrine Cells in the Mammalian Retina," in N. N. Osborne and G. Chader (eds.), *Progress in Retinal Research, Vol. 9.* (pp. 49–100). Pergamon Press, Elmsford, New York.

VANEY, D. I. (1984). "'Coronate' Amacrine Cells in the Rabbit Retina Have the 'Starburst' Dendritic Morphology," *Proceedings of the Royal Society, London B* **220**, 501–508.

VANEY, D. I., COLLIN, S. P., and YOUNG, H. M. (1989). "Dendritic Relationships between Cholinergic Amacrine Cells and Direction–Selective Retinal Ganglion Cells," in R. Weiler and N. N. Osborne (eds.), *Neurobiology of the Inner Retina* (pp. 157–168). Springer–Verlag, Berlin.

WYATT, H. J., and DAW, N. W. (1975). "Directionally Sensitive Ganglion Cells in the Rabbit Retina: Specificity for Stimulus Direction, Size, and Speed," *Journal of Neurophysiology* **38**, 613–626.

PART III

NEURONS IN THEIR NETWORKS

This section examines the interaction of neuron and network properties. The issues here include (a) selection of the basic, essential neuron types and combining them into prototypical circuits as building blocks for large-scale biological network simulations, and (b) the match or interplay of neuron and network properties. What neuronal biophysical mechanisms are essential for network function? Does modulation of these neuron properties alter network function in a useful or pathological manner?

Douglas and Martin (Chapter 14) reduce the diversity of neuron types in the visual cortex to three essential types: layer 3 small pyramidal neurons, layer 5 large pyramidal neurons, and nonspiny inhibitory neurons. These neurons are combined in a canonical microcircuit. In their view, the cortex is a quasi-crystalline structure with this basic subunit repeated over and over. They present evidence derived from experimental data and computer simulations showing that the predominant communication within the cortex is among the excitatory pyramidal neurons, which form a recurrent cortical amplifier. Inhibition functions both to control access to this amplifier and control the gain of the neurons to suprathreshold excitatory inputs. They consider how feature selectivity operates in such a system and contrast it with the Barlow–Levick model of directional selectivity.

Mahowald (Chapter 15) describes the evolution of analog very-large-scale integrated (VLSI) neurons in silicon. She describes the neuron biophysics that have been implemented, the communications among these neurons, and a real-time on-chip learning synapse that uses a Hebbian learning rule. The communication among neurons uses an address-event representation such that both the identity (address) of a neuron and the action potential event are communicated among neurons. She demonstrates the integration of these subunits into a simple network that learns to preferentially respond to a temporal sequence.

Bower (Chapter 16) considers the relations between the dynamical properties of single neurons and their networks in piriform cortex. He argues that to understand the functional organization of single cells, it will be necessary to consider their structure in the context of the network in which they are embedded. His large-scale simulation results suggest that cerebral cortical networks may be arranged to assure that the synaptic influences on pyramidal cells are timed to occur in regular and repeatable spatial/temporal patterns. In his view, system-level oscillations and associative memory capability emerge as combinations of network and cellular properties, including the structure and synaptic locations on pyramidal neurons.

Traub and Miles (Chapter 17) examine neuronal population oscillations observed in the hippocampus that depend on particular neuronal biophysical properties. These oscillations may be pathological and relevant to seizure-related mnemonic phenomena such as *déja vu*. Alternatively, the oscillations may be an exaggeration of a normal population activity. Based on experimental manipulations of synchronized multiple bursts in disinhibited hippocampal slice preparations, the authors added particular synaptic currents and membrane properties to realistic network simulations. Based on this combined experimental and simulation approach, they argue that this population oscillation depends on the long-duration synaptic interaction of NMDA-activated calcium currents and calcium intracellular kinetics in the dendrites of pyramidal neurons.

Chapter 14

Exploring Cortical Microcircuits: A Combined Anatomical, Physiological, and Computational Approach

RODNEY J. DOUGLAS[1] and KEVAN A. C. MARTIN[1,2]

[1]*MRC Anatomical Neuropharmacology Unit*
Oxford, England
[2]*Neurobiology Research Center*
University of Alabama at Birmingham
Birmingham, Alabama

I. Introduction

A cardinal feature of the receptive fields of neurons in the primary visual cortex is that they are tuned to specific characteristics of the visual stimuli. The primary source of input to the cortex is from the thalamus, where the relay neurons are far more broadly tuned to the characteristics of the visual stimulus. Thalamic neurons respond to all orientations of a bar or edge, cortical neurons respond to a very restricted range of orientations; thalamic neurons respond to motion in all directions, cortical neurons may only respond to motion in one general direction. Similar differences are found for parameters like contrast, size, velocity, and location in visual space. Thus, cortical neurons seem to be involved in a process of data reduction, whereby a subset of the information arising from the thalamic relay neurons is 'extracted' by individual cortical neurons. However, this is not their only role, for cortical neurons also integrate information from different sources. Each cortical neuron receives thousands of synapses from a great variety of sources, including brain stem, basal forebrain, claustrum, thalamus, and cortex. (See reviews by Martin, 1988; Douglas and Martin, 1990a.) Not surprisingly, physiologists have made great efforts to understand how cortical receptive fields are constructed from the raw material provided by the thalamus. While many intriguing possibilities have been suggested, it is true to say that little definitive has arrived by way of a description of the basic structure of the cortical microcircuitry (Martin, 1988) or the mechanisms underlying the function of the circuits.

A renewed attempt has been made in recent years to understand the function of cortical circuits by using *neural network* models to replicate the phenomenology of the physiological and psychophysical investigations (e.g., Zipser and Anderson, 1988; Lehky and Sejnowski, 1990). Self-evidently, the structures of the networks are not based on the circuits of the neocortex; but even relatively successful simulations have left open the question of how these processes might be implemented by biological neurons and the circuits they form. One clear fact that has emerged for the neural network simulations is that neural networks are not a cure-all. The artificial networks perform less well than their biological counterparts, and they are unlikely to compete with their biological counterparts in the foreseeable future unless some major technological problems are solved (Hopfield, 1990; Faggin and Mead, 1990). In particular, the artificial networks do not scale well with the complexity of the problems they are required to solve. Simply enlarging the number of nodes in a network may not be adequate to make the solution of a particular problem more rapid, more accurate, or even soluble at all. Large networks carry with them potential complications if they are to be implemented in hardware, because there are a restricted number of connections that can physically be made between the nodes of the network. Thus, we are left in the uncomfortable position of attempting to model brain networks using artificial neural networks that bear only a passing resemblance to the brain networks and that, in any event, may be inherently incapable of solving the problems they are addressing.

We are currently mapping out a route that might take us around this impasse. Evolution, it seems to us, has made more progress than we have in solving problems of architectures of neural networks and their implementation. Thus, by studying the biology, we should be able to learn a great deal about strategies that could provide new design principles for the architectures of artificial neural networks. This has been our course of action for over a decade. We have been studying biological and synthetic neurons and nerve nets in the context of the mammalian visual system. The results obtained from our biological investigations have been extensively reviewed elsewhere (Martin, 1988; Douglas and Martin,

1990a; Somogyi, 1989). From this experimental work, we have been able to identify the different structural and functional elements of the cortical circuits and their interconnections. These data form the impetus and database for our theoretical work. In this chapter, we concentrate on showing how we have used this experimental work to extend and shape our efforts in the theoretical arena. We show how the close linkage of experiment and theory has given us novel insights into the biology, as well as providing an accurate biological basis for simplified artificial neurons and neural networks.

II.   Abstraction of Single Cortical Neurons

In developing our models of cortical circuits, the first task was to specify a generic *cortical neuron*. There are obvious advantages in simplifying the structure of these neurons. In particular, we wished to avoid unnecessary detail that added significantly to the computing load and the subsequent time spent on analyzing these details. Thus, we eliminated detail that did not make a significant contribution to the solution of the circuit problem in question. On the other hand, the simplification had to permit translation between levels of complexity. This meant that, in some cases, we had to arrange that certain features of the problem could be simulated in increased detail if necessary. For example, we could study the properties of the synaptic input from a single neuron in some detail before generalizations were made about such input (Koch et al. 1990). This ensured that anatomical and physiological characteristics that were of great significance at the level of single neurons, or even parts of neurons, were not lost or obscured when we moved to the level of microcircuits. In some cases, the detailed model was a necessary step in obtaining and setting the values of parameters for the simplified model. Thus, we could move from detailed experimental data to detailed model neuron to simplified neuron, and the assumptions made at every stage of the theoretical work could be validated by comparing the behavior of the models with the actual experimental data. (See next section.)

A.   Simplification of Neuronal Morphology

The morphological components of our model neurons were derived from cortical neurons that had been labeled with a marker, horseradish peroxidase, which was injected intracellularly after physiological recordings were made from the neuron. The marker fills the dendrites and axon of the single identified neuron in its entirety (see color insert, Figs. 1, 2). The three-dimensional coordinates and dendritic diameters of the dendritic trees were measured from 80 μm-thick

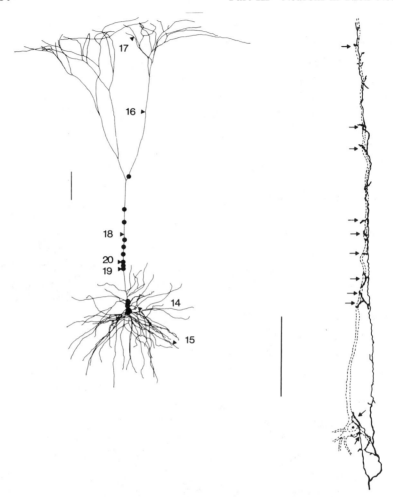

FIGURE 3.   Use of anatomical reconstructions to investigate the inhibitory control of pyramidal cells. We simulated the effect of layer 5/6 basket cell inhibition of a layer 5 pyramidal neuron, and compared these results against our intracellular data, to explain the unexpected absence of strong inhibition in these cells under certain conditions of visual stimulation (Koch *et al.*, 1990; Douglas *et al.*, 1988). (a) Reconstructed layer 5 pyramidal neuron in cat visual cortex (area V1) shown here in wire mode. Filled circles indicate the locations of 16 smooth (inhibitory) cell synapses obtained from the reconstruction shown in b. Four of the synapses are located on the soma. Numbered arrowheads indicate the locations of spines that received excitatory inputs. Scale bar = 100 μm. (b) Afferent of layer 5/6 basket cell making synaptic contact with a layer 5 pyramidal cell (redrawn from Kisvárday *et al.*, 1987). Arrows point to the synaptic contacts that were transposed to the cell shown in a. Scale bar = 100 μm.

histological sections, and the neurons reconstructed by a computer-assisted method. A reconstructed layer 5 pyramidal neuron is shown in Fig. 3a. Models that incorporate this degree of detailed morphology are useful for examining questions at the neuronal level, such as synaptic interactions (Koch *et al.*, 1990). However, efficient circuit models require a simplified morphology. Simplification of neuronal structure has conventionally been achieved by collapsing the dendritic tree into a single equivalent dendritic cylinder (Rall and Rinzel, 1973; Rinzel and Rall, 1974). This method of simplification requires, among other things, that all terminal dendritic branches should end at the same electrotonic length from the soma and that the sum of the 3/2 power of the diameters of daughter branches should equal the 3/2 power of the diameter of the parent branch. Unfortunately, we found experimentally that these two conditions do not apply for cortical pyramidal neurons. Thus, to simplify the pyramidal neurons, we had to develop a different, empirical, strategy.

We transform the detailed 3D structure of the soma and dendrites into a simplified neuron consisting of an ellipsoidal somatic compartment and a small number of cylindrical compartments that represent the dendritic arbor. The diameters of all the dendrites are summed together at intervals of 10 μm displacement from the soma to yield the diameter profile of a single dendrite whose surface area is approximately equivalent to the total of the component dendrites (Fig. 4). This profile is simplified further into separate cylinders by averaging over regions in which the diameter remains fairly constant. The resulting set of cylinders, and the ellipsoidal somata, yield the simplified compartmental representation of the different types of cortical neurons used in the model. The exact number of compartments depends on the complexity of the source neuron's morphology and the anatomical detail of the problem that we wish to model. In the case of pyramidal neurons, the summation is performed separately for the basal dendrites and the apical dendrites.

In performing this simplification, we have tried to conserve the following properties: Firstly, we have preserved the dendritic surface area as a function of distance from the soma, because this permits translation of anatomical measurements of synaptic locations and densities directly to the model. Secondly, the dendritic impedance load seen by the active conductances in the soma has been preserved, because this permits agreement between experimental and model parameters for those conductances. Finally, the transfer impedance between synapses located at various sites on the dendrite and the soma has been preserved, because this affects the contribution of the various synapses to neuronal output.

The impedance characteristics of each cylinder are set by scaling its axial resistance so that its electrotonic length matched that of the detailed neuron.

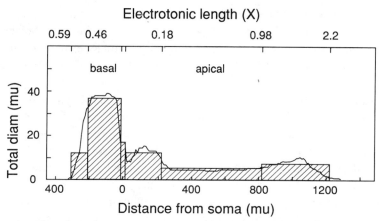

FIGURE 4.   Procedure used to reduce the detailed three-dimensional structure of the layer 5 pyramidal neuron shown in Fig. 3a, to the simplified model neuron shown in Fig. 7. The total diameter of the basal and apical dendritic components was measured at 10 μm increments of distance from the soma (solid line). The dendritic profile was decomposed into a number of cylinders that preserved approximately the dendritic surface area of the original neuron. The average electrotonic displacements at the boundaries of the cylinders were measured in the dendrites of the actual layer 5 pyramid.

This matching is achieved as follows: The value of the membrane resistance of the model neuron is obtained by comparing the experimentally measured somatic input resistance (R_{in}) to that computed from the detailed morphology. Using the algorithm of Koch and Poggio (1985), the R_{in} is adjusted until the computed R_{in} converges on the observed R_{in}. The membrane resistance (R_m) is assumed to have the same constant value for the entire neuron. The value obtained for a particular neuron is used to compute the average electrotonic length to locations in the detailed neuron that corresponded to the boundaries of the cylinders of the simple neuron (Fig. 4). In this way, the equivalent electrotonic length of each simple cylinder is obtained, and used to calculate the effective axial resistance (R_a) for that cylinder.

B.   Validation of a Simplified Model Neuron

The validity of the simplification procedure outlined in the preceding section was assessed by comparing the somatic input impedance of the detailed and simple representation as a function of the frequency of the injected current (Fig. 5). There is remarkable agreement between the behavior of the detailed and the simplified cells. The corner frequency is approximately 10 Hz for both cases. The initial roll-off is about 15 dB/decade. This is less than the 20 dB/decade anticipated for the low-pass RC filter offered by the local membrane alone, because of the contribution of the dendritic cylinder. As the frequency increases, so the length constant shortens and the dendrites become electrotonically longer. At high frequencies, the roll-off reaches nearly 10 dB/decade, which is the slope expected for an infinite cable. In this region, the simplified model behaves slightly less like an infinite cable than does the detailed neuron. This small difference is probably due to the truncation of the dendrites inherent in the simplification procedure (Fig. 4).

These cases consider the impedance characteristics of the cell viewed from the soma, which is the load seen by the action potential generation and adaption mechanisms located in the vicinity of the soma. However, another important aspect of the simple model's performance is the effect on the soma of synaptic input located at various sites on the dendritic tree. Assessing the fidelity of the simple model in this case is more complicated because the detailed dendritic morphology has been lost to simplification. Nevertheless, we find good agreement between the behavior of the detailed neuron and that of the simple model for dendritic input (Fig. 6). We computed the transfer impedances between an excitatory synapse on the apical dendrite and the soma for both models. The excitatory synapse was located at site 18 of Fig. 3a. The transfer impedances were remarkably similar (not shown). The minor differences that were present were probably due to the fact that the parameters of the simple model had been rounded off for convenience in other simulations and so were not exactly the best fit to the detailed cell. In Fig. 6, the transfer functions from two synaptic

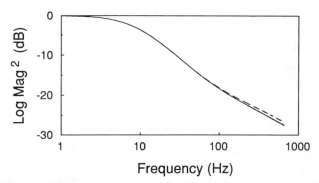

FIGURE 5.   The input impedance at the soma in the detailed layer 5 neuron (broken line; see Fig. 3) and its simplified form (solid line; see Fig. 7). The initial roll-off is approximately 15 dB per decade, expressing the combined low-pass characteristics of the local membrane and the electronically short dendritic cable. At higher frequencies, the slope approaches 10 dB per decade as the response becomes dominated by the increasing electrotonic length of the dendritic cable.

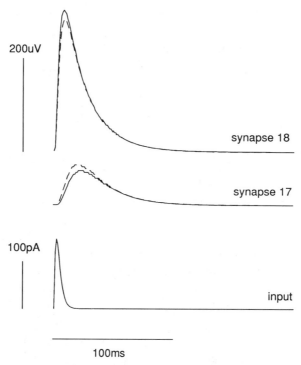

FIGURE 6.   Comparison of voltage responses in the somata of the detailed layer 5 neuron
(broken line) and simple (solid line) models, evoked by current injections at synapses on
the apical dendrite. The responses were obtained by computing the transfer impedances
to the soma from synapses 17 and 18 (Fig. 3), and then convolving these with the alpha
function shown.

sites on the apical dendrite have been convolved with a current input, which had
the form of an alpha function. The resulting voltage signals in the soma of the
detailed and simple neurons are closely comparable.

C.   Excitability and Adaptation

Neurons with passive membrane properties and relatively linear behavior are
useful for resolving specific questions about the interactions between neurons.
However, nonlinear properties, such as excitability and adaptation of the action
potential discharge, are likely to be important factors governing the performance
of the cortical microcircuits. Simulation methods that are able to take into account
the behavior of passive, active, and ligand-gated conductances thus become
necessary.

Most of the excitability and adaptation characteristics found in intracellular

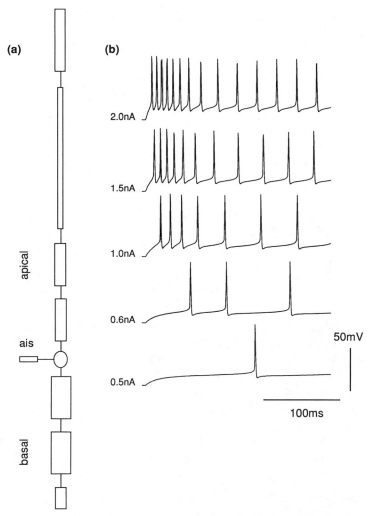

FIGURE 7. Incorporation of anatomical and biophysical data into a simplified compartmental model of a layer 5 pyramidal cell (Fig. 3a). The model simulates the excitability
of these neurons with remarkable accuracy. (a) Structure of the compartmental model
used to represent cortical pyramidal neurons. The lower three cylinders represent the
basal dendritic arborization, and the upper four cylinders represent the apical dendrite.
The dimensions of the cylinders were obtained by the procedure outlined in Fig. 4, but
the initial 200 μm of both the apical and basal dendrites have been further subdivided.
The small cylinder to the left of the ellipsoidal soma represents the axon initial segment
(ais). Each cylinder is allocated a particular profile of passive, active and ligand-gated
conductances. (b) Simulation in this model of a regular firing in response to intra-somatic
current injection. The latency to first spike, adaptation, and the membrane trajectory
during the successive interspike intervals agrees with intracellular records derived from
real cortical neurons.

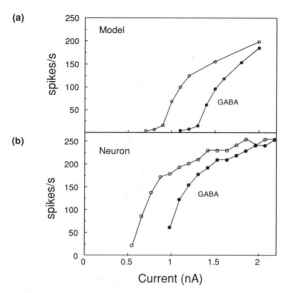

FIGURE 8.   Effect of GABA on the current-discharge relation of cortical neurons. (a) Response of model neuron (similar to Fig. 7). The effect of GABA was modeled as a sustained 2 mS/cm^2 increase in somatic chloride conductance, with $E_{Cl} = -70$ mV. (b) Response of cortical neuron recorded *in vitro*. GABA was applied iontophoretically in the vicinity of the recorded neuron (Berman *et al.*, 1991a). Open circles, control response; filled circles, response to GABA.

recordings of cortical cells can be achieved with a limited set of conductances. We routinely use a sodium spike conductance (G_{na}) and potassium spike conductance (delayed rectifier, G_{kd}) to generate the action potential. These conductances are supplemented by additional conductances that have been observed experimentally and that underlie phenomena like adaptation and bursting in pyramidal neurons (McCormick, 1990). A potassium *A* conductance (G_{ka}) controls the low-frequency discharge in response to peri-threshold current injections (Figs. 7, 8a). A persistent sodium conductance (G_{nap}) also acts in the peri-threshold region. It provides a current amplifier for synaptic currents reaching the soma, and so offsets the current sink offered by the passive dendritic load. The major conductance-controlling adaptation is a medium time-constant calcium-dependent potassium conductance (G_{kca}). In some cases, a medium time-constant ligand-modulated potassium current (G_{km}) is also incorporated. Calcium enters the cell by a voltage-sensitive conductance. Its accumulation and removal from the intracellular compartment by various mechanisms is modeled with a leaky integrator.

The active conductances have the following Hodgkin–Huxley-like dynamics:

$$G_i = G_{max,i} \cdot f_i \cdot (V - E_i),$$

where $G_{max,i}$ is the maximum conductance attributable to the channels of the ith ion, V is the trans-membrane potential, and E_i is the reversal potential of the ionic species. The fraction of channels that are open is given by $f_i = m^a h^b$. The m and h obey

$$\frac{dm}{dt} = \frac{m_{ss} - m}{\tau},$$

and steady-state value m_{ss} (or h_{ss}) is a sigmoid of the form,

$$m_{ss} = \frac{1}{1 + e^{(V-V1/2)/\rho}},$$

where $V_{1/2}$ is the trans-membrane potential when $m_{ss} = 0.5$, and ρ is a shape parameter.

Notice that the m and h particle time-constants do not depend on voltage. This simplification improves simulation speed and reduces the number of free parameters that govern the conductances, with no significant loss in fidelity. The abstract neuron provides a remarkably accurate simulation of the excitability characteristics of actual cortical neurons recorded both *in vitro* and *in vivo* (Fig. 7).

D.   Simulation Tools

We have developed a general network simulation program that provides a useful compromise between detail and efficiency at the level of AT-type computers. This program, CANON, permits the morphologies of neurons to be specified as sets of interconnected compartments. The passive, active, and ligand-gated conductances of each compartment can be specified, as well as the synaptic linkages and transmission delays between the output of a neuron and the compartments of its target neurons. The program is suitable for simulating single neuron behavior in some detail and also small networks of simplified neurons. It was the simulation tool used in the following examples. Similar programs have been described in Koch and Segev (1989).

III.   Exploring Neuronal Interactions

One of the most controversial aspects of cortical processing is the role of inhibition in generating the tuning of cortical neurons. (See Martin, 1988.) The simplified neurons described previously provided an ideal means of investigating aspects of inhibition that we were unable to study directly using experimental techniques. There is strong circumstantial evidence that inhibition in the neocortex is mediated by the neurotransmitter gamma-amino butyric acid (GABA). The cortical neurons that contain GABA form a distinct morphological subclass, called *smooth neurons*. (See Somogyi, 1989; Douglas and Martin, 1990a.) The smooth neurons can be further differentiated into at least seven different types, based on their connectivity. Of these seven, two are of particular concern: the chandelier cell and the basket cell.

The *chandelier* (Szentágothai and Arbib, 1974) or *axoaxonic cell* (Somogyi *et al*. 1982; Freund *et al*. 1983) makes synaptic contact exclusively with the initial segment of the axon of pyramidal neurons. On average, each pyramidal cell in the superficial cortical layers receives about 40 synapses from the convergent input of about five chandelier cells (Somogyi *et al.,* 1982; Freund *et al.,* 1983). Since action potentials are initiated at the axon initial segment (AIS), the synapses of the chandelier cell seem well-placed to quench action potential output (Peters, 1984). Unfortunately, very little is known about the physiology and synaptic actions of these cells, since they are rarely recorded during physiological experiments (Martin and Whitteridge, unpublished). The second type, the *basket cell,* appears to be the most common smooth neuron in the neocortex. About 85% of its targets are pyramidal cells, and it connects to the soma, proximal dendritic shafts, and dendritic spines, but not the AIS. About 10–30 basket cells provide a convergent input of about 30–300 synapses to each target neuron (Fariñas and De Felipe, 1991). The clear difference in the location of the basket cell and chandelier cell synapses suggests that they have distinctly different functional roles. However, the rarity with which these types are encountered suggests that, with present technology, any insights into what these differences in connectivity mean are unlikely to emerge from the present generation of experimental neuroscientists. In the interim, we have used simulations of cortical neurons to explore the inhibitory effects of the chandelier cells and basket cells on pyramidal cells.

Previous theoretical analyses of inhibition (Koch *et al.,* 1982, 1983, 1990) considered only the subthreshold condition, and assumed that the output of the cell is proportional to the somatic voltage. However, the output from neocortical neurons is in the form of action potentials. The frequency of discharge is determined by the current arriving at the axon initial segment. Injecting more current produces more action potentials. We expected that the effect of chandelier or basket cell inhibition would be to increase the threshold before action potential threshold were reached. Also, if the inhibition was not strong enough to completely suppress the excitation, then the peak discharge would be reduced relative to the non-inhibited case.

We used the model to evaluate the effects of basket cell and chandelier cell inhibition on the action potential output of the pyramidal cell. Using the connection statistics outlined previously and a range of conductances for the GABA synapses, our simulations showed that the action of initial segment inhibition was not as we intuitively expected (Douglas and Martin, 1990b). Postsynaptic inhibition was effective in inhibiting action potential output only if the amount of excitatory current was small. With strong excitation, the inhibition, whether by chandelier or basket cells or both combined, was relatively ineffective against the excitatory input (Fig. 8a).

An important aspect of this finding is that it applies generally to all cases of postsynaptic inhibition; it is not a peculiarity of the siting of the chandelier or basket cell synapses. In examining the model in detail, it became obvious why the result was obtained: The increase in inhibitory conductance shunts the excitatory current, and so the threshold for discharge is higher than in the non-inhibited case. However, the increase in conductance also has the effect of reducing the membrane time constant. So, if the excitatory current is sufficient to exceed threshold, the membrane will recharge more quickly following each action potential. This means that the inhibited neuron will discharge faster for a given suprathreshold current, and this offsets some of the effects of inhibition. Consequently, the current-discharge curve of the inhibited neuron is steeper and eventually achieves approximately the same discharge rate as the control case for high-input currents. The performance of the neuron at these high rates is dominated by the relaxation of the spike conductances, which are usually much larger than adaptation and synaptic conductances.

Because we could now use the theoretical analysis to generalize the case to encompass all postsynaptic GABA-mediated inhibition, not just the chandelier and basket cell inhibition, we could devise a means of testing the theory experimentally (Berman *et al.*, 1991a). We recorded from slices of visual cortex *in vitro*. Single neurons were recorded intracellularly and the current discharge curves were determined by injecting current into the soma through the intracellular micropipette. GABA was then iontophoresed onto the neuron through a second pipette and the current-discharge curve was replotted. We found that direct application of GABA affects the current discharge relation of neurons in the way predicted by the model neuron (Fig. 8). The effects of synaptic inhibition were also consistent with the predictions of the model. Electrical stimulation of the optic radiation evokes a 100–300 ms GABA-mediated IPSP in cortical neurons (Kelly and Krnjevic, 1969; Dreifuss *et al.* 1969; Berman *et al.*, 1991b; Douglas *et al.*, 1989; Douglas and Martin, 1991). We measured the effect of this IPSP on the current-discharge relation of single neurons, and found that it could block the action potential discharge elicited by small (0.4–0.8 nA) current injections, but had little effect on the initial spikes in the response to larger excitatory currents.

The combined experimental and simulation results suggest that cortical inhibition acts to increase the small signal (noise) threshold of the neuron, but does not affect its response to large transient signals. Since large transient signals are always transmitted, significant novel stimuli are always able to switch the cortex to a new response state—the cortex is not bound by inhibition. This is a further caution from neuronal simulation data that we should not depend too much on inhibition to suppress the strong excitatory currents that are in evidence during the response of cortical neurons to natural stimuli.

A. The Canonical Microcircuit

In the foregoing sections, we have shown how simplified neurons can be a formidable tool in understanding the intrinsic properties of neurons and their responses to synaptic input. Neurons, of course, are the basic elements of the cortical circuits. The natural next step is to extend these techniques to an analysis of the cortical microcircuits; but what are the cortical microcircuits? Over 100 years of study of the cortical machinery has not produced a consensus on what the cortical microcircuits are, or even how many of them there are (Douglas and Martin, 1990a). On some accounts, there appear to be a large number of special-purpose circuits, each designed to perform a different function, like orientation tuning or direction selectivity. On other views, the cortex is seen to be quasi-crystal-like in its structure with the same basic circuits being repeated again and again. This is much more our view: Indeed, the relative uniformity of the structure of neocortex led to it being referred to as an *isocortex*. Detailed examination of the different cortical areas has revealed that the same basic neuronal types are found in all the cortical areas and that they form the same specific connections.

In the visual cortex, we have devoted considerable effort to mapping out the connections between the elements. (See Somogyi, 1989; Martin, 1988.) On the basis of this, we have suggested that there is, indeed, a simple basic circuit, which we refer to as a *canonical microcircuit* (Douglas *et al.*, 1989). This same basic circuit we suggest may be used with relatively minor modifications to perform many of the different functions that have been observed in visual cortex, including orientation and direction selectivity, etc. In the following section, we show how these basic cortical elements can be assembled into a simple micro-circuit that provides novel explanations of the mechanisms underlying the operation of the cortical microcircuits.

The organization of these elements is derived directly from our structural and immunochemical studies. From intracellular injections of markers into single neurons, we were able to determine the projection patterns of the axons of the different neuronal types in the visual cortex (Somogyi *et al.*, 1983; Freund *et al.*, 1983; Martin and Whitteridge, 1984; Martin and Somogyi, 1985; Kisvárday *et al.*, 1987; Gabbott *et al.*, 1987; Martin, 1988; Somogyi, 1989). One of the major findings of this work was that the pyramidal cells, which form the majority of cortical neurons, are the major recipients of synaptic input from other cortical neurons. Over 85% of the intracortical output from pyramidal cells is to other pyramidal cells. A similar proportion of the output of GABAergic neurons is to pyramidal cells. Consequently, intracortical connections are likely to play a dominant role in shaping the cortical response. The circuit (Fig. 9) consists of three basic populations of neurons: a population of superficial layer pyramidal neurons; a population of deep layer pyramidal cells; and a population of smooth cells. These neurons have the simplified morphology and excitability characteristics described in Section II.A. The pyramidal neurons excite their postsynaptic targets. In the model, members of a pyramidal population excite one another, and also members of the other two populations. Smooth neurons inhibit their postsynaptic targets. Members of this population inhibit one another, and also the activity of both pyramidal populations. Thalamic afferents are excitatory and exert their influence particularly on the superficial pyramidal and smooth cell populations. The extent of the synaptic coupling within and between populations was estimated from the available anatomical and physiological data. At present, all the GABAergic inhibitory neurons in our canonical microcircuit are lumped into a single basket cell type. Other types exist in cortex (Somogyi and Martin, 1985), but too little is known about their functional differences to justify discrimination of inhibitory neurons within the model. In its present form, the intracortical excitatory connections are simple and do not employ NMDA mechanisms.

In developing the canonical microcircuit, it was convenient to use a simple pulse stimulus to examine the response properties of the microcircuit. A similar stimulus can be used experimentally: an electrical pulse stimulation applied to the thalamocortical afferents evokes in most cortical neurons a marked hyperpolarization, lasting a few hundred milliseconds (Phillips, 1959; Kelly and Kŕñjevic, 1969; Dreifuss *et al.*, 1969; Berman *et al.*, 1990; Douglas *et al.*, 1989; Douglas and Martin, 1991). Moreover, the apparent magnitude of the inhibitory potentials is in marked contrast to our observations in neurons responding to normal visual stimuli. Thus, one important test of the microcircuit was whether it could account both for the characteristic response to electrical pulse stimuli and the response to normal visual stimuli.

The intracellular responses to pulse stimulation of pyramidal neurons indicated that there were distinct laminar differences within the cortical circuit (Fig. 10) (Douglas *et al.*, 1989; Douglas and Martin, 1991). Pyramidal neurons in layers 2 and 3 exhibited an early phase of excitation that lasted a few tens of milliseconds, and this was followed by a hyperpolarization that reached its maximum

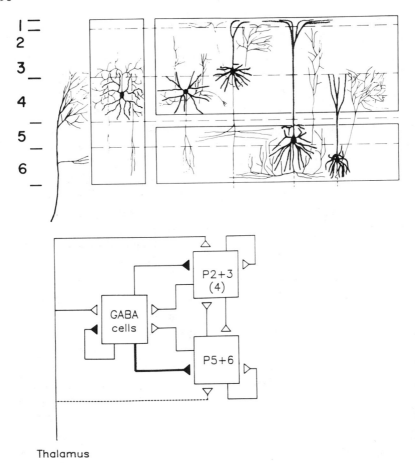

FIGURE 9.   The canonical microcircuit that successfully predicts the intracellular responses of cortical neurons to stimulation of thalamic afferents (Douglas *et al.,* 1989). (a) Anatomical characteristics of neurons that comprise microcircuit. The three rectangles surrounding the various cell types map onto the three squares in the circuit schematic below. (b) Three populations of neurons interact with one another: One population is inhibitory (GABA cells, solid synapses), and two are excitatory (open synapses), representing superficial (P2 + 3) and deep (P5 + 6) layer pyramidal neurons. Long-range pyramidal projections (dashed axons in a) are not incorporated in the microcircuit because their functions are not modeled. The layer 4 spiny stellate cells are incorporated with the superficial group of pyramids. Each population receives excitatory input from the thalamus. The inhibitory inputs activate both $GABA_A$ and $GABA_B$ receptors on pyramidal cells. The thick line connecting GABA to P5 + 6 indicates that the inhibitory input to the deep pyramidal population is relatively greater than that to the superficial population. However, the increased inhibition is due to enhanced $GABA_A$ drive only. The $GABA_B$ input to P5 + 6 is similar to that applied to P2 + 3.

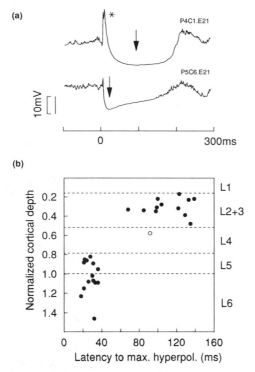

FIGURE 10.   (a) Averaged intracellular responses to electrical stimulation of the optic radiation revealed that hyperpolarization evolved more slowly in the pyramidal cells of layers 2 and 3 (upper trace) than those located in layers 5 and 6 (lower trace) (Douglas and Martin, 1991). The latencies to maximum hyperpolarization are indicated by arrows. Superficial pyramids always exhibited marked early excitation (asterisk), which was less prevalent in deep layers. (b) Relationship between latency to maximum hyperpolarization and cortical depth for 26 identified pyramidal neurons and one spiny stellate neuron (open circle).

more than 100 ms after the electrical stimulus. The pyramidal neurons in layers 5 and 6 exhibited very little early excitation, and the hyperpolarization evolved more quickly, reaching a maximum about 30–40 ms after the stimulus. We examined these differences using a combination of intracellular recording and iontophoresis of agents that affected the inhibition of the neurons (Figs. 11, 12). The early phase of the response was sensitive to the $GABA_A$ blocker, bicuculline, while the later phase was not. Baclofen hyperpolarized the neurons, but electrical stimulation could still evoke strong excitation during the early phase of the response; but this response was prevented by application of GABA. These results

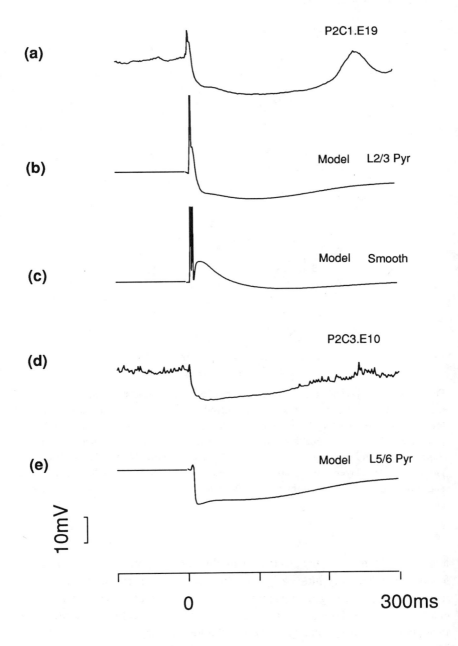

suggested that the early phase of the response was mediated by $GABA_A$ receptors and the later part by $GABA_B$. From these structural and anatomical results, we could deduce that the fundamental principles of operation in the canonical microcircuit are that thalamic input evokes strong intracortical re-excitation, and that this evolution is restricted by both $GABA_A$ and $GABA_B$ inhibition. The inhibition is more severe in the deep pyramids than in the superficial pyramids; $GABA_A$ inhibition is two times stronger in the deep cells. The responses of superficial and deep cells in the canonical microcircuit to pulse stimuli were very similar to those observed experimentally (Douglas and Martin, 1991), and provide an explanation of the experimental results. Activation of the superficial pyramids by the geniculocortical afferents results in a suprathreshold depolarization and the generation of one or two action potentials (Figs. 10, 11a). The depolarization is followed by a sustained hyperpolarization that reaches a maximum about 110 ms after the stimulus. This hyperpolarization comprises two inhibitory components: an early transient $GABA_A$-mediated component and a more sustained $GABA_B$ component. The inhibition is initiated by the smooth cells, which receive feedforward excitation from the geniculocortical afferents, and feedback excitation from the spiny neurons. The deep layer pyramids receive only 10–20% of the thalamic input and are more strongly inhibited by their $GABA_A$ input. The depolarization in response to the pulse input is smaller and the latency to maximum hyperpolarization is shorter than is seen in the pyramidal

---

FIGURE 11.   Intracellular responses of real and model cortical neurons to brief pulse stimulation of the geniculocortical afferents. The neurons were identified by intracellular injection of horseradish peroxidase. In this and the following Figures, intracellular traces of real neurons are averaged, and so individual action potentials are attenuated unless they have a constant latency; the *model* traces were obtained from neurons in the canonical microcircuit (Fig. 9). They are single sweeps, but action potentials have been cropped at -30mV. (a) Response of real layer 3 pyramidal neuron. Early depolarization evoked an action potential. The depolarization was followed by sustained hyperpolarization that reached maximum amplitude at a latency of about 100 ms. In this, and some other neurons a rebound depolarization accompanied by a burst of action potentials occurred during relaxation of the hyperpolarization. (b) Response of model superficial pyramidal cell was similar to real cell. No attempt was made to model the rebound excitation. (c) Response of model smooth cell consists of multiple spikes because these cells lack adaptation currents. No smooth cells were positively identified in these experiments. (d) Response of actual layer 6 pyramidal cell. Hyperpolarization occurred rapidly, and reached a maximum at a latency of about 25ms. (e) Response of model deep pyramid was similar to real cell.

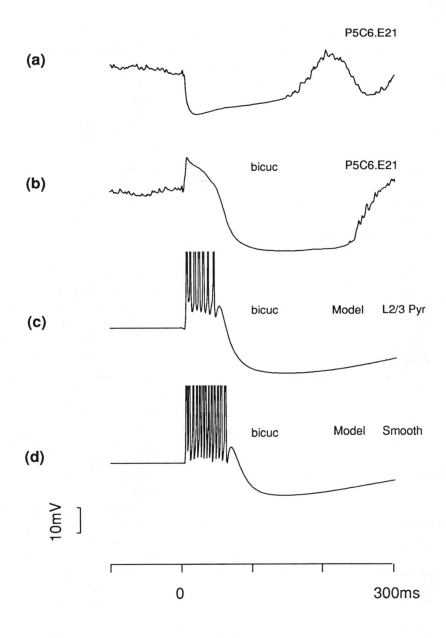

cells of the superficial layers (Figs. 10, 11d). As expected, partial blockade of the $GABA_A$ mechanism in deep pyramids modifies their response so that they resemble those found in superficial cells (Douglas and Martin, 1991). As the blockade becomes more severe, the early excitatory response becomes more pronounced (Fig. 12) because the smooth cells can no longer restrict the early phase of intracortical re-excitation. Eventually, however, the smooth cells are able to quench the spiny cell discharge via the $GABA_B$-mediated inhibition. The time constants for $GABA_B$ activation and inactivation are relatively slow, and so the $GABA_B$ inhibition becomes stronger with time. This integration in the inhibition of the spiny neurons is reflected in the increased amplitude and duration of the hyperpolarization that eventually terminates the intracortical re-excitation (Fig. 12b).

There is good agreement between the performance of the model and the experimental data (Figs. 11, 12). This suggests that the organization of the canonical microcircuit does reflect a basic level of organization of the actual cortical circuitry. Clearly, it is not a comprehensive description of cortical microcircuits. Rather, it offers the minimum set of connections from which more complex patterns of interconnection might be elaborated. In deriving the model, we have made the fewest possible assumptions about the cortical connections and the weighting of the synaptic connections. The simplicity of this circuit is a great virtue; it is closely matched to the quality of real anatomical and physiological data, and so does not depend on brave speculation. However, despite its seeming simplicity, exploration of the model has given us new insights into general principles of cortical processing that are important to computational theories of neocortex. For example, one notable aspect of the canonical microcircuit is that the thalamic afferents provide only a small fraction (10–20%) of the excitatory synapses of the neurons they contact. This is consistent with the

---

FIGURE 12.   Effect of the $GABA_A$ receptor blocker bicuculline on the intracellular responses of actual and model cortical neurons to brief pulse stimulation of the geniculocortical afferents. (a) Control response of real neuron. (b) Application of bicuculline to the actual neuron increased the magnitude and duration of the early depolarization. Multiple action potentials rode on the crest of the depolarization, but were attenuated by the averaging procedure. The depolarization was terminated by a hyperpolarization of enhanced amplitude and duration. (c) Simulation of bicuculline application to a model superficial pyramidal neuron gave a similar response to that seen in actual neurons. (d) Response of a member of the smooth cell population that interacts with the bicuculline-affected pyramidal cells. Train of action potentials is due to excitation by the hyperexcitable pyramidal cells.

anatomy (White, 1981; Freund *et al.*, 1985; LeVay, 1986; White, 1989), but is at variance with the vast majority of cortical feedforward network models. In the canonical microcircuit, the major drive to the cortical neurons is from the excitatory cortico-cortical circuits.

B.   Direction Selectivity

The possibility that thalamic input serves only to *ignite* the cortical circuits, which are designed for strong positive feedback, may explain the paradoxical absence of strong inhibition in direction- and orientation-selective neurons in the visual cortex when they are stimulated with non-optimal stimuli (Douglas *et al.*, 1988, 1991). The classical explanation of direction selectivity is that of Barlow and Levick (1965). In their scheme, two detectors are connected to an AND-NOT logic gate, one via a delay (Fig. 13a). The same excitatory and inhibitory input signals arrive at the gate in both the preferred and the non-preferred direction. The selective output of the gate arises because of the difference in the phase relationships of inhibition and excitation induced by the delay line. Translated into biology, we can consider the gate to be a piece of circuitry where a neuron receives input from two non-directional sources, one direct and the other via an inhibitory interneuron (Fig. 13b). The loop through the inhibitory neuron contributes the delay and completes the AND-NOT gate. In the preferred direction of motion, the inhibition arrives after the excitation and the target neuron produces an output. In the non-preferred direction, the excitation and inhibition arrive coincidently at the target neuron and cancel each other, resulting in no output.

We have used intracellular recording to study directionality because it provides a direct view of the subthreshold synaptic activity, as well as the spike discharge seen in conventional extracellular recordings. Our intracellular recordings showed that direction-selective neurons respond with strong depolarization in the preferred direction, and little or no response in the non-preferred direction (Fig. 14). This was the basic pattern predicted by the Barlow–Levick model. The one small fly in the ointment was the magnitude of the observed inhibition. For the model to work, inhibition must cancel excitation for stimuli moving in the non-preferred direction. In the optimal direction of motion, the inhibition that followed excitation was characterized by a membrane hyperpolarization of only a few millivolts, and this inhibition was not associated with large shunting conductances. Similarly, in the non-preferred direction, there was little change in either the membrane potential, or the input conductance. How, then, is the excitation restricted during non-optimal stimulation? The simple explanation that emerges from the canonical microcircuit is that the control of cortical excitation hinges around the control of an initially weak excitatory input, in this case the

(a)

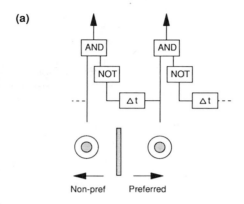

(b)

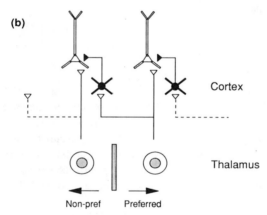

FIGURE 13.   Barlow and Levick (1965) model of direction selectivity. (a) Two non-directional receptors (e.g., thalamic relay cells) are connected to a logical AND-NOT gate, one via a delay, Δt. In the non-preferred direction, excitation AND inhibition (i.e., NOT-excitation) combine temporally, resulting in no output. (b) Barlow and Levick (1965) model used to explain how the direction selectivity of cortical cells is derived from the non-directional, circular symmetric receptive fields of thalamic neurons (Barlow, 1981). Pyramidal neurons receive direct excitation from thalamic neurons, and a synaptically delayed input from inhibitory cells (filled).

thalamic input. If this initial signal can be inhibited, then the pathways responsible for cortical re-excitation will not be engaged. Thus, rapid inhibition, mediated by the GABAergic neurons that receive direct thalamic excitation, can shape the cortical response by controlling access to the cortical amplifier. If no such inhibition is present or if the inhibition arrives too late—as, for instance, for motion of a bar in the preferred direction—inhibition will fail to block the weak

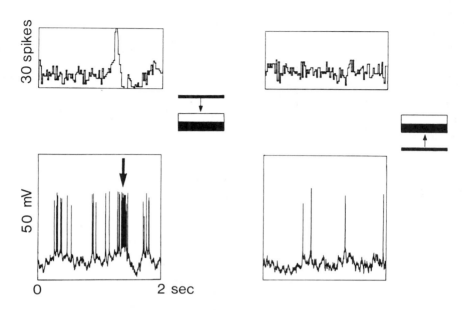

FIGURE 14.   Direction-selective simple cell, recorded in layer 4 of the primary visual cortex of the cat. The neuron was excited monosynaptically by non-directional thalamic afferents. Top panels show extracellular peri-stimulus time histograms for a stimulus moving in the preferred (left panel) and non-preferred directions (right panel). Lower panels show equivalent recordings made intracellularly in the same neuron. The strong excitatory response (arrowed) is followed by hyperpolarization and inhibition of action potential discharge. In the non-preferred direction, there is no inhibition of spontaneous discharge, and no change in membrane potential.

geniculate excitation and the neuron will begin to fire. This firing will then be amplified by the local excitatory circuitry among pyramidal cells, eventually producing the strong excitation associated with optimal stimulation.

This can be modeled formally by connecting together two canonical micro-circuits in the manner illustrated in Fig. 15. For simplicity of simulation, only the superficial layers were considered, but the principle of operation applies to the entire circuit. Each subcircuit consisted of a population of L2/3/4 spiny cells, and a population of smooth cells that interacted according to the canonical principles. The populations were modeled using the simplified neurons discussed in Section II.A. The thalamic input provided a slightly asymmetric drive to the

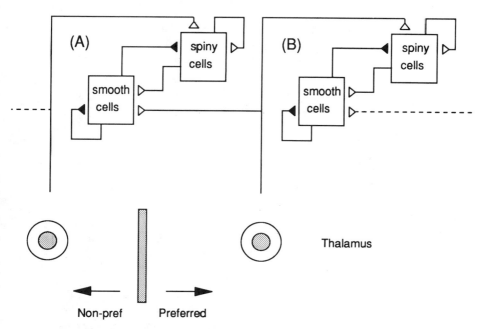

FIGURE 15.   Schematic of circuit that models the observed intracellular response of cortical neurons during direction-selective behavior (Douglas and Martin, 1991). Thalamic afferents stimulate two subcircuits (A,B) derived from the canonical model shown in Fig. 9. In each subcircuit, a population of inhibitory cells (smooth cells, solid synapses) interacts with a population of excitatory cells (spiny cells, open synapses). The thalamic afferents excite the smooth cell populations of subcircuits to the left, and the spiny cell populations of subcircuits to the right. When activity in the left thalamic afferent precedes that in the right (preferred direction), the spiny population (A) expresses recurrent excitation that is restricted by its associated smooth cell population. When activity in thalamic afferent B precedes that in A (non-preferred direction), the smooth population (B) suppresses its spiny population during the arrival of thalamic excitation (A), and no recurrent excitation occurs. In this model, the major component of excitation that is observed in spiny cells during the preferred response is due to cortical re-excitation. The thalamic input provides a relatively small initial component that is easily controlled by inhibition in the non-preferred direction.

subcircuits. When the visual stimulus moved in the preferred direction, the thalamic input excited the spiny population ahead of the associated inhibitory population. When the bar moved in the non-preferred direction, the thalamic input activated the inhibitory population ahead of the associated excitatory population.

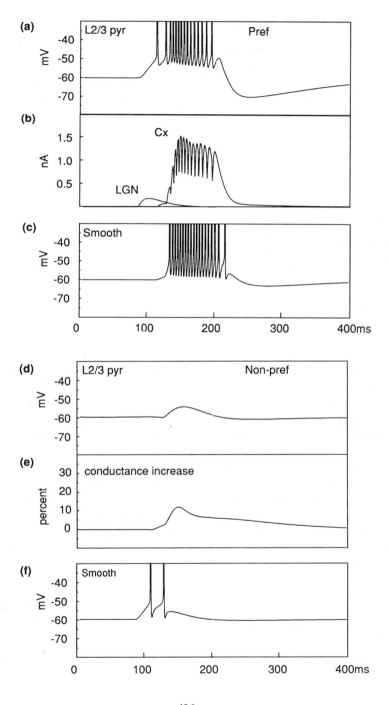

The response of a spiny neuron to preferred direction of movement (Fig. 16a) showed that the membrane depolarized to a level where spike discharge was initiated. The rate increased over the first few spikes, despite the fact that the neuron had the intrinsic mechanisms for adaptation. Figure 16b shows the reason for the increase in rate: The initial excitatory current is delivered by the geniculate afferents, but when the spiny population reached threshold, cortical re-excitation generated a dramatic increase in net excitatory current, which reached about 1.5 nA. In the preferred direction, the smooth neurons were excited only by the recurrent collateral of the spiny cells. The spiny discharge generated a growing $GABA_B$-mediated inhibition, as described in Section III.A, which contributed to the eventual termination of the action potential discharge (Fig. 16c). The spiny discharge was followed by a hyperpolarizing phase that was partly due to the cumulative $GABA_B$ effect and partly due to the activation of the calcium-dependent potassium current.

The response of a model spiny neuron to stimulation in the non-preferred direction (Fig. 16d) consisted of a small subthreshold depolarization. The reason that it remained subthreshold is seen in Figs. 16e and f. In the null direction, activation of the smooth cell population by the geniculate afferents preceded spiny activation. The smooth cell population provided only a small amount of inhibition (compare discharge rate in the non-preferred direction with that in the preferred direction), but the inhibition was enough to keep the thalamic excitation from reaching threshold.

Although the phase relation between excitation and inhibition necessarily remains important, the process described here is quite different from the Barlow and Levick model. Here, the *veto* depends on the control of excitatory gain rather than the inhibition of large excitatory currents. Because the excitatory

---

FIGURE 16.   Intracellular responses of a model spiny neuron (population A, Fig. 15) during preferred (pref) and non-preferred (non-pref) stimulation. (a,b) During stimulation in the preferred direction, the neuron generates action potentials in response to excitatory current that is largely a result of cortical re-excitation. (c) Response of smooth cell from the same subcircuit. The strong discharge reflects the control that the smooth cells impose on the maximum discharge of the pyramidal cells. (d) During non-preferred stimulation, there is only a small depolarization of the membrane potential, and the net excitatory current is only a small fraction of that in the preferred direction. (e) The increase in neuronal input conductance during non-preferred stimulation is small (12%), showing that large inhibitory conductance changes are not required to achieve direction preference. (f) Response of a smooth cell in the same subcircuit, showing that only a small inhibitory signal is required to control the pyramidal cells in the non-preferred direction.

currents in the null direction are small, the inhibitory conductance change can be small, too. Figure 16e shows that the maximum increase in somatic input conductance during null stimulation was only 12%. Thus, the performance of the model circuit is consistent with our experimental finding that shunting inhibition is not a prominent feature of direction selectivity (Douglas *et al.*, 1988; Berman *et al.*, 1991b).

IV. Conclusion

In this chapter, we have shown that rather simple model neurons can abstract the most important anatomical and physiological properties required for exploring cortical circuits. We have also shown that analysis of the behavior of model neurons and circuits can reveal principles of cortical operation that were not evident in the original data alone.

The detailed morphology, passive electrical properties, and the conductances governing excitability and adaptation of cortical neurons were abstracted to generate simple and efficient model neurons. These model neurons preserve many of the anatomical and functional properties of cortical neurons. Because the computational overhead is relatively small, these neurons are particularly suitable for exploring the characteristics of cortical microcircuits. We have used these models together with experimental anatomy and physiology to develop a new understanding of the role of inhibition in generating the selective responses of cortical neurons. In theoretical studies, detailed neuronal morphology was used to examine the visibility from the soma of inhibitory conductance change applied to cortical pyramidal neurons. The simulations confirmed that a major fraction of the inhibitory conductance change should be visible to our recording pipette located in the soma. In the experimental work, however, we did not detect any large inhibitory conductances during normal visual stimulation. This suggested that shunting inhibition was not a feature of the selective mechanism. Simulations with abstract neurons were used to examine the possibility that inhibition of the axon initial segment might provide an alternative source of strong inhibition. The results were surprising. The GABAergic mechanism was rather ineffective against strong transient excitation and this casts further doubt on the role of inhibition in selective behavior. Concern about the role of an obvious cortical inhibitory population led us to establish what the minimum organizational principles of cortical neurons might be. A combination of intracellular labeling, intracellular recording, iontophoresis, and circuit modeling using abstract neurons enabled us to specify a canonical microcircuit for neocortex.

The central doctrine of the canonical model is that cortical amplification is controlled by inhibitory neurons. Despite its seeming simplicity, exploration of the model has given us new insights into general principles of cortical processing that are important to computational theories of neocortex. Firstly, the thalamic input does not provide the major source of excitation in cortical neurons. Instead, the intracortical excitatory connections provide most of the excitation. This suggests that intracortical connections may have a dominant role in shaping cortical responses. Secondly, inhibition and excitation are not separable events. Activation of the cortex inevitably sets in motion a sequence of excitation and inhibition in every neuron. In particular, cortical re-excitation should be seen as a process affecting a population of cells, rather than a sequential activation of separate neurons as in traditional models of visual cortex. Finally, the time evolution of excitation and inhibition is far longer than the synaptic delays of the circuits involved. This means that cortical processing cannot rely on precise timing between individual synaptic inputs, as has been suggested. Instead, the precise timing required for many perceptual processes must arise as a property of the microcircuits.

We have found that microcircuit modeling is a fine form of explanation and prediction. Subsequent experiments have confirmed and elaborated the findings predicted from the simulations. Successful exploitation of the interaction between simulation and experiment has depended on two factors: firstly, biologically rich and accurate specification of the model; secondly, selection of a level of abstraction in the model that matches the biological question being addressed. Application of both these principles ensures that the predictions of the model will be cast at a level where experimental confirmation and progression are feasible.

Acknowledgments

We acknowledge John Anderson's technical assistance. We thank Christof Koch and Misha Mahowald for creative discussions. RJD acknowledges the support of the Human Frontier Science Program (T. Tsumoto, P. I.) and the Mellon Foundation. KACM is the Henry Head Fellow of the Royal Society.

References

BARLOW, H. B. (1981). "Critical Factors Limiting the Design of the Eye and Visual Cortex (The Ferrier Lecture)," *Proceedings of the Royal Society of London B* **212,** 1–34.

BARLOW, H. B., and LEVICK, W. R. (1965). "The Mechanism of Directionally Selective Units in the Rabbit's Retina," *Journal of Physiology* **178,** 477–504.

BERMAN, N. J., DOUGLAS, R. J., and MARTIN, K. A. C. (1990). "The Conductances Associated with Inhibitory Postsynaptic Potentials Are Larger in Visual Cortical Neurones than in Similar Neurones in Intact, Anaesthetised Rats," *Journal of Physiology* **41,** 107P.

BERMAN, N. J., DOUGLAS, R. J., and MARTIN, K. A. C. (1991a). In preparation.

BERMAN, N. J., DOUGLAS, R. J., MARTIN, K. A. C., and WHITTERIDGE, D. (1991b). "Mechanisms of Inhibition in Cat Visual Cortex," *Journal of Physiology,* **440,** 697–722.

DOUGLAS, R. J., and MARTIN, K. A. C. (1990a). "Neocortex," in G. Shepherd (ed.), *The Synaptic Organization of the Brain (3rd ed.)* (pp. 220–243). Oxford University Press, New York.

DOUGLAS, R. J., and MARTIN, K. A. C. (1990b). "Control of Neuronal Output by Inhibition at the Axon Initial Segment," *Neural Computation* **2,** 283–292.

DOUGLAS, R. J., and MARTIN, K. A. C. (1991). "A Functional Microcircuit for Cat Visual Cortex," *Journal of Physiology,* in press.

DOUGLAS, R. J., MARTIN, K. A. C., and WHITTERIDGE, D. (1988). "Selective Responses of Cortical Cells Do Not Depend on Shunting Inhibition," *Nature* **332,** 642–644.

DOUGLAS, R., MARTIN, K., and WHITTERIDGE, D. (1989). "A Canonical Microcircuit for Neocortex," *Neural Computation* **1,** 480–488.

DOUGLAS, R. J., MARTIN, K. A. C., and WHITTERIDGE, D. (1991). "An Intracellular Analysis of the Visual Responses of Neurones in Cat Visual Cortex," *Journal of Physiology,* **440,** 735–769.

DREIFUSS, J. J., KELLY J. S., and KRŃJEVIC K. (1969). "Cortical Inhibition and Gamma-Aminobutyric Acid," *Experimental Brain Research* **9,** 137–154.

FAGGIN, F., and MEAD, C. (1990). "VLSI Implementation of Neural Networks," in S. F. Zornetzer, J. L. Davis, and C. Lau (eds.), *An Introduction to Neural and Electronic Networks* (pp. 275–292).

FARIÑAS, I., and DE FELIPE, J. (1991). "Patterns of Synaptic Input on Corticocortical and Corticothalamic Cells in the Visual Cortex. I. The Cell Body," *Journal of Comparative Neurology* **304,** 53–69.

FREUND, T. F., MARTIN, K. A. C., SMITH, A. D., and SOMOGYI, P. (1983). "Glutamatedecarboxylase-Immunoreactive Terminals of Golgi-Impregnated Axo-Axonic Cells and of presumed Basket Cells in Synaptic Contact with Pyramidal Neurons of the Cat's Visual Cortex," *Journal of Comparative Neurology* **221,** 263–278.

FREUND, T. F., MARTIN, K. A. C., SOMOGYI, P., and WHITTERIDGE, D. (1985). "Innervation of Cat Visual Areas 17 and 18 by Physiologically Identified *X* and *Y*-Type Thalamic Afferents: II. Identification of Postsynaptic Targets by GABAimmunocytochemistry and Golgi Impregnation," *Journal of Comparative Neurology,* **242,** 275–291.

GABBOTT, P. L. A., MARTIN, K. A. C., and WHITTERIDGE, D. (1987). "The Connections between Pyramidal Neurons in Layer V of Cat Visual Cortex (Area 17)." *Journal of Comparative Neurology* **259,** 364–381.

HOPFIELD, J. J. (1990). "The Effectiveness of Analogue 'Neural Network' Hardware," *Network*, 27–40.

KELLY, J. S., and KRŇJEVIC, K. (1969). "The Action of Glycine on Cortical Neurons," *Experimental Brain Research* **9**, 155–163.

KISVÁRDAY, Z. F., MARTIN, K. A. C., FRIEDLANDER, M. J., and SOMOGYI, P. (1987). "Evidence for Interlaminar Inhibitory Circuits in Striate Cortex of the Cat," *Journal of Comparative Neurology* **260**, 1–19.

KOCH, C., and POGGIO, T. (1985). "A Simple Algorithm for Solving the Cable Equation in Dendritic Trees of Arbitrary Geometry," *Journal of Neuroscience Methods* **12**, 303–315.

KOCH, C., and SEGEV, I. (1989). *Methods in Neuronal Modelling*. Bradford.

KOCH, C., DOUGLAS, R., and WEHMEIER, U. (1990). "Visibility of Synaptically Induced Conductance Changes: Theory and Simulations of Anatomically Characterized Cortical Pyramidal Cells," *Journal of Neuroscience* **10**, 1723–1744.

KOCH, C., POGGIO, T., and TORRE, V. (1982). "Retinal Ganglion Cells: A Functional Interpretation of Dendritic Morphology," *Philosophical Transactions of the Royal Society of London B* **298**, 227–264.

KOCH, C., POGGIO, T., and TORRE, V. (1983). "Nonlinear Interaction in a Dendritic Tree: Localization Timing and Role in Information Processing," *Proceedings of the National Academy Science USA* **80**, 2799–2802.

LEHKY, S. R., and SEJNOWSKI, T. J. (1990). "Neural Network Model of Visual Cortex for Determining Surface Curvature from Images of Shaded Surfaces," *Proceedings of the Royal Society of London B* **240**, 251–278.

LEVAY, S. (1986). "Synaptic Organization of Claustral and Geniculate Afferents to the Visual Cortex of the Cat," *Journal of Neuroscience* **6**, 3564–3575.

MARTIN, K. A. C. (1988). "The Wellcome Prize Lecture: From Single Cells to Simple Circuits in the Cerebral Cortex," *Quarterly Journal of Experimental Physiology* **73**, 637–702.

MARTIN, K. A. C., and SOMOGYI, P. (1985). "Local Excitatory Circuits in Area 17," in D. Rose and V. G. Dobson (eds.), *Models of the Visual Cortex* (pp. 504–513). John Wiley, New York.

MARTIN, K. A. C., and WHITTERIDGE, D. (1984). "Form, Function, and Intracortical Projections of Spiny Neurons in the Striate Visual Cortex," *Journal of Physiology* **353**, 463–504.

McCORMICK, D. A. (1990). "Membrane Properties and Neurotransmitter Actions," in G. Shepherd (ed.), *The Synaptic Organization of the Brain (3rd ed.)* (pp. 220–243). Oxford University Press, New York.

PETERS, A. (1984). "Chandelier Cells," in E. G. Jones and A. Peters (eds.), *Cerebral Cortex, Vol. 1: Cellular Components of the Cerebral Cortex* (pp. 361–380). Plenum, New York.

PHILLIPS, C. G. (1959). "Intracellular Records from Betz Cells in the Cat," *Quarterly Journal of Experimental Physiology* **44**, 1–25.

RALL, W., and RINZEL, J. (1973). "Branch Input Resistance and Steady Attenuation for

Input to One Branch of a Dendritic Neurone Model," *Biophysics Journal* **13,** 648–688.

RINZEL, J., and RALL, W. (1974). "Transient Response in a Dendritic Neurone Model for Current Injected at One Branch," *Biophysics Journal* **14,** 759–790.

SOMOGYI, P. (1989). "Synaptic Organization of GABAergic Neurons and GABA$_A$ Receptors in the Lateral Geniculate Nucleus and Visual Cortex," in D. K.-T. Lam and C. D. Gilbert (eds.), *Neural Mechanisms of Visual Perception* (pp. 35–62). Portfolio Publishing Co., Houston, Texas.

SOMOGYI, P., FREUND, T. F., and COWEY, A. (1982). "The Axo-Axonic Interneuron in the Cerebral Cortex of the Rat, Cat and Monkey," *Neuroscience* **7,** 2577–2608.

SOMOGYI, P., and MARTIN, K. A. C. (1985). "The Role of Inhibitory Interneurones in the Function of Area 17," in D. Rose and V. G. Dobson (eds.), *Models of the Visual Cortex* (pp. 514–523). John Wiley, New York.

SOMOGYI, P., KISVÁRDAY, Z. F., MARTIN, K. A. C., and WHITTERIDGE, D. (1983). "Synaptic Connections of Morphologically Identified and Physiologically Characterised Large Basket Cells in the Striate Cortex of Cat," *Neuroscience* **10,** 261–294.

SZENTÁGOTHAI, J., and ARBIB, M. B. (1974). "Conceptual Models of Neural Organization," *Neurosciences Research Program Bulletin* **12,** 306–510.

WHITE, E. L. (1981). "Thalamocortical Synaptic Relations," in G. Adelman, S. G. Dennis, F. O. Schmitt, and F. G. Worden (eds.), *The Organization of the Cerebral Cortex* (pp. 153–161). MIT Press, Cambridge, Massachusetts.

WHITE, E. L. (1989). *Cortical Circuits*. Birkhauser, Boston.

ZIPSER, D., and ANDERSON, R. A. (1988). "A Back-Propagation Programmed Network that Simulates Response Properties of a Subset of Posterior Parietal Neurons," *Nature (London)* **331,** 679–684.

Chapter 15   Evolving Analog VLSI Neurons

M. A. MAHOWALD

*Department of Computation
and Neural Systems
California Institute of Technology
Pasadena, California*

I. Introduction

We are using very large scale integrated circuit (VLSI) technology to develop artificial nervous systems. Although the nervous system can perform many specialized tasks, at a gross level its primary function is to gather sensory data and to translate them into effective action. Animals learn from experience so that their responses become more appropriate. We hope to capture the essential nature of biological nervous systems by evolving our artificial system in a real-time sensorimotor context.

Real-time sensorimotor processing as complex as that performed by the common housefly is unattainable even with today's fastest digital computers. The computational ability of the fly is incomparable to that of the digital computer because the principles of digital computation are fundamentally unlike those used in the nervous system. The computer reduces the information on its wires to 1 bit and combines the information in a sparsely connected array of logic gates. In contrast, neurons communicate in analog values and are richly interconnected.

CMOS VLSI is an analog electronic computational medium that has many properties in common with nervous tissue, and has the potential to achieve real-time sensorimotor processing. Although we are a long way from realizing an autonomous artificial neural system in this medium, we have made progress in key areas. Fast, high-density sensory processing (Lazzaro and Mead, 1989; Mead and Mahowald, 1988; Mead, 1989b) and simple sensorimotor feedback systems (DeWeerth and Mead, 1988; Delbrück, 1989) now exist in CMOS. Analog computation allows these chips to perform complicated functions in real time. Some of these chips are able to modify themselves based on their past history. For example, the adaptive retina (Mead, 1989a) adapts to a long time-average intensity to center itself in the correct operating range. Ideally, a system that incorporates real-time sensory and motor processing and on-chip learning will be able to learn directly from experience and optimize its own performance in a changing environment.

In this chapter, we describe the components of an analog neuron: the *axon*, which is a communications network, the body of the neuron, and a real-time learning synapse. As a demonstration of the integration of these subunits, we present a simple network that learns to respond preferentially to a temporal sequence.

II. Interface

Application of neural networks to real-world problems requires an interface between the world and the network. Vision is an important sensory modality that provides a link between the nervous system and the external world. The silicon retina (Mead and Mahowald, 1988; Mead, 1989b), which has been integrated into sensory (Mahowald and Delbrück, 1989) and sensorimotor processing chips (Delbrück, 1989), can provide real-time sensory input to our artificial nervous system.

The silicon retina shares several features with its biological counterpart. Transducers and computational elements (silicon neurons) are arrayed in a thin layer over a two-dimensional surface. The lens focuses the image directly on the transducers. The system is small and mobile. Like the biological retina, the silicon retina acts to reduce the bandwidth needed to communicate reliable information (Laughlin, 1987). The need for data compression arises because ambient light intensity varies over many orders of magnitude, yet the local intensity variation in a single image is usually small. In the presence of noise, communication of absolute image intensity would require a large dynamic range to encode reliably small differences in image intensity over the full possible illumination range. The retina reduces the bandwidth by subtracting average intensity levels from the image and reporting only spatial and temporal changes. This process has the further advantage of providing sensory invariance. The contrast of white square on a black background is invariant under changes in illumination, even though the photon flux from the black background in bright light may be larger than the photon flux from the white square in dim illumination. The need to communicate information has generated an abstraction that provides the representation for further information processing.

Animals depend on motor control to acquire sensory information actively. DeWeerth *et al.* (1990) have used a retina-like array of photodetectors (DeWeerth and Mead, 1988) as input to a controller that generates eye movements. In this tracker system, a simple transformation between retinotopic space and motor space is performed directly on the chip. Normalized intensity is aggregated independently in the horizontal and vertical dimensions. In each dimension, neurons arrayed in retinotopic space synapse onto two antagonistic motor neurons. The coupling strengths between the retinotopic neurons and the motor neurons are a function of position. These motor neurons generate action potentials that drive an antagonistic pair of motors that move the eye in one dimension. Action potentials are a useful output representation because pulses can overcome static friction in the motors, which can then move the eye at very slow speeds.

We have had some success building sensory arrays and motor outputs. However, the complexity of computation in these subsystems has been severely limited because the computational machinery is confined to the two-dimensional surface of a single chip. The high bandwidth of information in two-dimensional arrays of dynamical signals makes interchip communication difficult. Because we cannot easily integrate individual chips into multichip systems, each project tends to develop its own unique computational paradigm; the computational element of one project thus bears little relation to the computational element of another. These problems of communication and compatibility have inhibited the development of more complex representations for image analysis and oculomotor control. Interfacing and extending silicon subsystems requires a computational framework that can accommodate multiple chips.

The choice of representation for interchip communication is established by the requirement for direct integration of a sensory input and motor output into the network. In our communications framework, we have chosen to use an action potential representation like that used in the nervous system. The communications framework preserves the time of individual action potentials and so implicitly preserves the average firing frequency of the neuron. We believe that this representation will best facilitate the interface of existing sensory and sensorimotor subsystems. Because it emphasizes timing information, it is particularly suited for signaling dynamic events. These dynamic events are an integral part of real-time sensorimotor processing.

III. Communication

The action potential is a digital amplitude pulse. The advantages of pulse-based techniques for analog–digital hybrid neural network designs have been demonstrated (Brownlow *et al.*, 1990; Murray, 1989). One important property of a fully restored digital signal is noise immunity. Digital signals are also immune to interchip parameter variation. Since silicon technology has been highly optimized for implementation of fast digital computers, digital representations are particularly well-suited for high-bandwidth communication in silicon. A digital output representation allows many neuron outputs to be multiplexed rapidly onto the same wire. Multiplexing signals is essential if several thousand neurons are to be placed on the same chip, as is done in the silicon retina.

We have fabricated and tested prototype chips for self-timed interchip communication (Sivilotti, 1990). The sender chip multiplexes the outputs of 16 action-potential generators onto a single bus. Our system has two unique features: It preserves as much as possible the timing of individual action potentials and it uses what we call an *address-event data representation*. Using this representation, the sending chip transmits the identities (addresses) of neurons issuing action potentials, rather than literally transmitting the action potential as a pulse. The receiving chip decodes the address into an action potential. The diagrams in Fig. 1 illustrate the communication protocol. Each cell in the network has a digital word as its address. At the time that a cell fires an action potential, its address is broadcast on a bus. The broadcast of the address is an event that corresponds to the generation of an action potential from the sending neuron. If several neurons try to generate action potentials simultaneously, an arbitration circuit decides which neuron controls the bus. After receiving an *acknowledge* signal, the arbiter removes the selected address from the bus and resets the state of the selected neuron. The neurons whose addresses were not broadcast continue to issue requests until they are acknowledged by the arbiter. Although the arbitration is not *fair,* each neuron is acknowledged eventually if the refractory period of a neuron is longer than the number of neurons sharing the bus multiplied by the data transfer period. Information from this test chip can be transmitted at rates of 2 million addresses per second. If a typical neuron had a peak firing rate of 200 spikes per second, 10,000 neurons simultaneously firing at peak rates could share the same bus.

This communications framework takes advantage of the abstraction of the visual image computed by the retina. The silicon retina transmits information only from areas in the image where there is spatial or temporal change in the image. For this reason, areas of uniform illumination do not contribute to the communication load. In contrast, raster-scanning mechanisms sample these areas

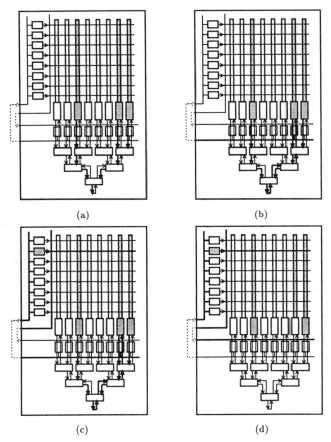

(a)   (b)

(c)   (d)

FIGURE 1.   Neural-network architecture using address-event scheme. Both the sending
and receiving mechanisms have been illustrated on the same chip, although this feedback
arrangement is not required. (Prototype chips that we have fabricated and tested separate
these two functions.) In the diagrams, a heavy line or shaded box represents an active
signal. Dashed lines represent signals that go off chip. (a) Several neurons activated
simultaneously are issuing requests. Requests propagate along the arbitration tree, being
steered by each arbitration element. Because arbitration takes place in parallel, addition
of more neurons affects the data rate only logarithmically by addition of more arbitration
elements in the binary tree. (b) The arbiter *acknowledge* signal is propagated back down
along the selected path to the neuron. It resets that neuron's state, but does not remove
the request. It also switches the selected neuron's address on the bus. (c) The input address
decoder receives the address. Upon receiving the address, the decoder activates an input
to the neural array. It also generates a reset signal for the transmitter indicating that it
has received the address. (d) The reset signal from the receiver resets the request lines.
The reset removes the address from the bus. After the address lines are reset, the receiver
terminates its signal, and the remaining active neurons are free to reinstate their requests.

regardless of whether there is information present or not. In addition, the self-timed nature of the communications framework minimizes temporal aliasing. Events are transmitted as they occur, not when the raster scan selects them. At low data rates, the bandwidth of the bus is completely devoted to accurate transmission of event timing. The preservation of event timing is crucial in auditory localization, and is significant in visual motion processing.

We believe that the address-event representation will have many advantages when used in future artificial neural systems. By preserving the integrity of individual action potentials, we can capture information in the detailed dynamics of the network. Neural network models in which the presynaptic cells transmit only their average firing rates do not differentiate between synchronized and unsynchronized neuronal firing. In our network, the *postsynaptic* cell computes the average of pulse events. The synchronization of these events, which occur with constant frequency at each presynaptic terminal, provides a strong stimulus to the postsynaptic cell. The same set of presynaptic inputs that is not synchronized at the postsynaptic cell may provide only a weak stimulus.

In the nervous system, the time of arrival of an action potential at postsynaptic target depends on the conduction delay. Although conduction delays are realized most obviously in a system with dedicated wires for each signal, this solution is not always practical. A digital event representation can facilitate the modeling of conduction delays. Compact asynchronous digital delay lines (Lazzaro and Mead, 1989) can be used to model conduction delays with the bandwidth necessary to preserve individual events. Cascading these delays through chips placed in a meaningful spatial array would allow the conduction delay to be a natural function of the distance between the presynaptic cell and its target. The address-event communication framework can be extended to include conduction delays provided that appropriate local addressing schemes are developed. Statistics of the distribution of axon collaterals in the area of the nervous system to be modeled is required to formulate optimal addressing schemes.

We believe that the address-event representation will facilitate the development of realistic neuronal systems due to the flexibility introduced by the substitution of an address for a *labeled-line* (or labeled-time-slot) approach. The areas in which this addressing communications method may be well-applied include modeling the topography of axonal projections and modeling dendritic arborizations.

The pattern of axonal projection between two areas of the nervous system performs transformations between the input space and the output space (Schwartz, 1977). The address-event representation allows tremendous flexibility of interconnection. It frees the designer from routing wires in complex patterns or arranging shift registers with complicated phase relationships. In the address-event data representation, there is no frame rate; postsynaptic chips monitor the bus continuously and pick out signals as they choose. Mappings from the presynaptic neurons to the postsynaptic neurons are specified digitally as a set of addresses stored on the postsynaptic chip. We can fix the set of addresses at the time of fabrication by using a digital address decoder, or we can choose the set of addresses using programmable static digital latches.

At the cost of some silicon area, mapping from presynaptic cell to postsynaptic target can be accomplished on a synapse-by-synapse basis. In a fully reconfigurable network, each synapse would store the address of its presynaptic cell with a programmable digital latch. In this scenario, all the synapses monitor the bus, and those synapses storing the address matching the one that appears on the bus receive a signal of the action-potential event. A neuron has a fixed number of synapses, but the presynaptic element can be selected from a larger number of possible inputs. The ability to configure the input on a synapse-by-synapse basis is important if synapses are to be placed along the dendritic tree of a model neuron.

IV.   Neurons

A neuron model is an essential part of a neural network. The development of our silicon (CMOS) neuron model is still in its preliminary stages. Our modeling effort has focused on system-compatibility requirements that constrain the input and output representations of the neuron. We plan to elaborate intermediate processing within the neuron model once the system constraints have been met. A design style well-suited to modeling the internal workings of a neuron has been described by Mead (1989a). Primary circuit elements include: the transistor, which, when operated in subthreshold, has an exponential current-voltage relation; the transconductance amplifier, whose conductance is controlled by a bias voltage; a resistor with variable resistance; and a capacitor. Using these elements, we will be able to model both the passive and active electronic properties of a neuron.

We have adopted a simple and compact model that is easy to implement. About 100 such model neurons can be integrated onto a single chip. The model is depicted schematically in Fig. 2. The model preserves two key computational elements: input (synaptic integration) and output (action-potential generation). We have included simplified passive electrical properties in our neuron model. We have separated the neuron into two compartments, each with its own state that is stored on a capacitor. The first compartment models the postsynaptic membrane; the second compartment is primarily a mechanism for generating action potentials. The two compartments are separated by a transconductance

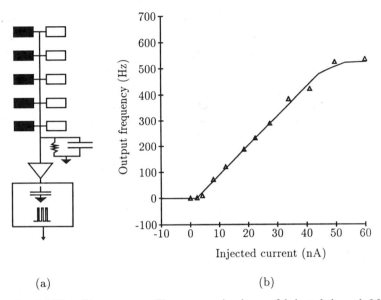

(a)   (b)

FIGURE 2.   (a) The silicon neuron. The neuron circuit was fabricated through MOSIS in a double-poly 2-micron CMOS process. The neuron sums input current from its 10 synapses on a capacitor that represents the postsynaptic membrane capacitance. There is an adjustable leakage resistor to the cell's resting potential. A spike generation mechanism is activated through a transconductance amplifier that compares the voltage on the postsynaptic-membrane capacitance to a stationary threshold. The spike generator integrates charge supplied by the transconductance amplifier and converts it into pulses. (b) Data taken from the chip show the response of the silicon neuron to tonic current injection. Within a limited range, the response curve can be shifted along the x axis by adjustment of $V_{\text{threshold}}$, and the peak firing rate (as well as the gain and the delay) can be adjusted via I_b.

amplifier that acts as a nonlinear resistive element. The transconductance amplifier conducts in only one direction, so synaptic inputs affect the spike-generating mechanism but spike generation does not affect the synapses. The separation of the postsynaptic membrane potential and the spike-generation mechanism allows the action potential to be terminated without the state of the postsynaptic membrane being eradicated. The transconductance amplifier and the capacitance of the spike generator introduces a delay between when the postsynaptic membrane crosses threshold and when the first spike occurs.

We have chosen to model the postsynaptic membrane as a leaky integrator comprising a capacitor and a leakage conductance to the resting potential. The capacitance is fixed at the time of fabrication. The leakage conductance is set by a variable bias. The conductance can be varied over several orders of magnitude, giving the membrane time constants that range from tens of milliseconds to tens of nanoseconds. The leakage conductance is linear for small differences between the resting potential and the potential on the integrator. For larger voltage differences, the dynamics of the membrane are slower than linear because the leakage element is in its slew-rate limit. The postsynaptic-membrane capacitor collects charge from the synapses as they are activated by presynaptic input.

In keeping with the protocol for interchip communication, the output of the neuron is in the form of action potentials. The mechanism for generating action potentials is activated only when the voltage on the postsynaptic membrane capacitor exceeds a threshold voltage set-off chip. In this particular model, the threshold is a stationary nonlinearity; it is independent of the past history of the neuron. Action potentials transmit a nonlinear function of the analog voltage of the integrator via a pulse-frequency encoding. The steady-state pulse frequency generated in response to a tonic current input is given by $I_b * \max(0, \tanh((I_{in}/G_{leak}) - V_{threshold}))$, where voltages are in units of $kT/q\kappa$ and are referenced to the resting potential of the cell.

Another way of thinking of the action potential is as an event that is digital in amplitude but analog in time. This property is useful for processing dynamic sensory information. An example is shown in Fig. 3. When the rate and number of presynaptic action potentials are sufficient to drive the integrator voltage above threshold, the neuron generates an action-potential output. However, if the same

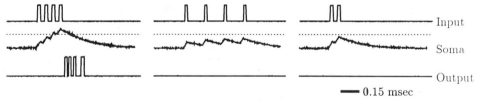

FIGURE 3.   Data taken from the silicon neuron. In each panel, the presynaptic input is shown in the top trace, the response of the postsynaptic membrane capacitor is shown in the middle trace (the threshold level indicated by a dotted line), and the neuron's action potential output is shown in the lower trace. The first panel shows the summation of the presynaptic action potentials on the postsynaptic-membrane capacitor, driving it above threshold for the generation of action potentials. Notice that the action-potential output of the neuron is delayed relative to the postsynaptic-membrane potential. This delay is the result of the transconductance amplifier and the capacitance of the spike generator. The second panel demonstrates that the same number of presynaptic action potentials presented at lower frequency is insufficient to drive the neuron above threshold. The third panel demonstrates that fewer action potentials, presented at the original rate, are also insufficient to drive the neuron above threshold.

number of action-potential inputs are spread out in time, the leaky integrator fails to reach threshold and the neuron produces no output. Although, in this example, we provided the input through a single synapse, we can achieve the same effect by presenting temporally correlated events through separate synapses. The simple neuron model is thus capable of discriminating temporally correlated events from events that are uncorrelated in time.

V.   Synapses

Synaptic input is supplied through a conductance modulated by presynaptic action potentials. We are using the simplest possible synaptic element: two transistors in series. The action potential drives the gate of the first transistor, and the synaptic weight is the control voltage on the gate of the second transistor. Each action potential increases the synaptic conductance and places an increment of charge on the postsynaptic-membrane capacitor. The size of the increment is determined by the conductance of the synaptic weight transistor, the action-potential duration, and the postsynaptic-membrane potential. We assume that all action potentials have the same duration; therefore, any variation in duration in the actual circuit is *noise*. The current–voltage relation of the synaptic weight transistor is similar to that of the leakage conductance. It is linear with a relatively high conductance when the postsynaptic-membrane potential is close to the reversal potential. When the postsynaptic-membrane potential is far from the reversal potential, the conductance decreases and is dominated by channel-length modulation (the Early effect). In this regime, the synapse acts more like a current source. The nonlinearity of the conductance puts the silicon model somewhere between the *realistic neuron model,* in which synapses are modeled as linear conductances, and the *artificial neural network model,* in which the synaptic weight multiplied by the presynaptic state is independent of the state of the postsynaptic neuron.

This synapse model has the advantage that it is small. It produces qualitatively the correct response; the postsynaptic potential can be driven only to the reversal potential for the ion, and the extent to which it is driven there is a monotonic function of the presynaptic weight. This simple synapse is used only for ions whose reversal potentials are on one extreme or the other of the neuron's operating range. Ions like chloride, with an intermediate reversal potential, require a slightly more complex synapse similar to the leakage-conductance circuit.

The synaptic weight is the repository of memory in artificial neural networks. Several techniques for synaptic-weight storage in silicon have been developed; many of them require special fabrication processes, such as the EEPROM process

(Holler *et al.*, 1989). (See Card and Moore (1989) for a review.) Neural networks incorporating a variety of weight-storage techniques have been implemented in standard CMOS VLSI. These weight-storage techniques include digital (Moopenn *et al.*, 1990; Mueller *et al.*, 1989; Boahen *et al.*, 1989) and dynamic analog weight storage (Satyanarayana *et al.*, 1990); Brownlow *et al.*, 1990). Dynamic analog weight storage requires periodic update of the stored value but requires less silicon area that does digital weight storage.

The silicon neural-network systems mentioned have been designed to implement general learning algorithms, and synaptic weights are programmed by off-chip controllers. This requirement reduces the autonomy of the resulting neural network. On-chip weight modification has been used to implement specific learning algorithms. For example, Alspector *et al.* (1989) have built a stochastic learning chip that can perform a supervised Boltzmann machine learning algorithm that takes advantage of noise. Their design uses an external clock that tells the synapses when to measure correlations and when to update the digitally stored weight. Due to temporal aliasing of the weight-update procedure, this design is not optimal for learning temporal patterns in which the correlation is a continuously changing function of time.

The technology we have been exploring for on-chip learning is the ultraviolet programmable floating gate (Glasser, 1985), available in a standard double polycrystalline-silicon CMOS process. The floating gate stores analog values statically and is small enough to be used in high-density arrays. The ultraviolet programmable floating gate has been used to create a silicon retina (Mead, 1989a) that adapts continuously in time based on on-chip signals. This technology is well-suited for building autonomous neural networks that must learn temporally changing patterns of correlation in high-density, real-time sensory input.

We have designed and characterized an ultraviolet programmable silicon synapse. The silicon synapse autonomously implements a simple Hebbian learning rule. The Hebbian learning rule is based on local variables that are readily accessible to the circuitry at the synapse itself, eliminating the need for off-chip control. Each synapse has its own floating gate and includes a circuit that performs synaptic update independently. Synaptic modification is enabled globally on each chip, and proceeds continuously in time as the system operates. Because the design is fully analog, it is small; in the conservative design used in this implementation, each synapse and associated update circuitry occupies an area of 86-by-100 square microns.

A block diagram of the silicon Hebbian synapse is shown in Fig. 4. The heart of the synapse is the floating mode, depicted as a capacitor, which stores the synaptic weight. The floating-node element is an electrically isolated piece of polycrystalline silicon. The target weight capacitor, which is also made of

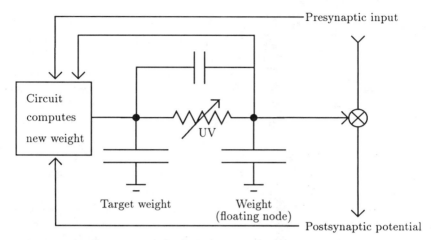

FIGURE 4.   Conceptual diagram of a modifiable synapse using an ultraviolet programmable floating gate. The floating node stores a voltage that represents the synaptic weight:
the coupling coefficient between the presynaptic and postsynaptic neurons. The floating
node is the gate on the conductance-controlling transistor in the synapse circuit. In the
presence of ultraviolet light, the voltage on the floating node slowly approaches the voltage
on the adaptation capacitor, which stores a target weight. The value of the target weight
changes continuously over time. It is computed by circuitry local to the synapse. The
target weight is a function of the spike output of the presynaptic neuron, the postsynaptic
potential, and the present value of the synaptic weight.

polycrystalline silicon, is not isolated electrically; rather, it is connected to
circuitry that computes the target weight. The target weight capacitor stores an
amplified version of what the synaptic weight would be if integration time for
experience were limited to a single event. The target weight is a function of the
presynaptic input, the postsynaptic potential, and the value of the synaptic weight.

Long-term modification of the synaptic weight is enabled by application of
ultraviolet light. The ultraviolet light excites electrons in the polysilicon to ener-
gies that allow electrons to pass through the silicon oxide. We can think of the
excited electrons as flowing through a somewhat nonlinear resistor (M. Maher,
personal communication) whose strength is determined by the intensity of the
ultraviolet light. This resistor allows the synaptic weight to slowly approach the
target weight. When the ultraviolet light is turned off, the resistance is infinite
and charge is trapped on the floating gate. It remains there even if power to the
system is interrupted. Modulation of the ultraviolet light controls the learning
rate and thus provides a kind of attention signal for learning. When the light is
turned on, learning is enabled at all illuminated synapses.

Our silicon synapse shares several properties with real modifiable synapses. The Hebbian learning rule enforced by the silicon synapse is consistent with conditions for long-term potentiation (LTP) observed in real synapses (Brown *et al.*, 1990). In addition to providing a long-term synaptic modification, the silicon synapse executes a short-term change in synaptic conductance using a capacitive-coupling mechanism. The floating node is separated from the target weight capacitor by a thin layer of silicon-oxide insulator. Because the oxide is so thin, the target weight capacitor and the floating node are capacitively coupled. The capacitive coupling causes the synaptic weight to increase briefly when the target weight increases. This increased conductance is similar to that observed in biological synapses with NMDA channels. When the depolarization of the postsynaptic membrane at such a synapse is correlated with presynaptic activity, NMDA channels are unblocked and there is a short-term conductance increase at the synapse (Forsythe and Westbrook, 1988). The magnitude of the effect in the silicon synapse is controlled by the strength of capacitive coupling and by the gain in the computation of the target weight. Unless the gain in the computation of the target weight is set to zero, this short-term increase in synaptic conductance occurs even in the absence of ultraviolet light, when LTP is disabled.

The learning rule that is actually used in biological systems is the subject of much investigation. Since the correct choice of rule is unclear, we have chosen a rule that was natural to implement in silicon. The circuit enforces a learning rule of the form: $\delta w = \alpha f(V_{\mathrm{pre}}) \tanh(V_{\mathrm{post}} - V_{\mathrm{ref}}) - \varepsilon w$. The function $f(V_{\mathrm{pre}})$ is zero when there are no presynaptic action potentials and is a monotonically increasing function of the presynaptic spike rate. This is a Hebbian learning rule, because the change in the weight is proportional to the product of the output of the presynaptic neuron and the state of the postsynaptic neuron. In this case, the postsynaptic cell can be either depolarized, leading to increased synaptic coupling, or hyperpolarized, leading to decreased synaptic coupling. Variations on this rule could be implemented easily by modifying the circuit that computes the target weight. Although the change in the synaptic weight can be negative, the weight represents only a magnitude of coupling; the weight itself cannot change sign. Whether a synapse is excitatory or inhibitory is determined when the synapse is designed. The learning-rule circuit can be used with either excitatory or inhibitory synapses.

The learning rule includes a general decay term that is proportional to the synaptic weight. This decay can stabilize the magnitude of synaptic weight. One way to examine the stability of synaptic coupling is to assume that the synaptic weights on the neuron have reached equilibrium. In this case, the response of a neuron to a given input pattern is fixed. We have mimicked this condition by repeatedly activating a single presynaptic cell while clamping the postsynaptic potential to a depolarized state using other inputs. We have maximally activated the postsynaptic cell so that the weight equilibrates at its maximum value for this presynaptic input pattern. The presynaptic input is correlated with the state of the postsynaptic neuron. The input is never activated when the postsynaptic cell is hyperpolarized, so the magnitude of the weight is never selectively decreased. (In other words, there is no LTD.) Because the hyperbolic tangent is a saturating function and the output of the presynaptic cell is a saturating function, the maximum value of weight increase is finite and can be balanced by the decay term if α is small enough.

We controlled the asymptotic value of the weight of coupling our single input to the neuron by varying the parameter α. The results of this experiment are shown in Fig. 5. The rate at which the weight stabilized in this experiment was not limited by the ultraviolet resistor as much as by the particular circuit implementation and the particular experiment. If the input pattern is repeated indefinitely, the decay of the weight must just balance the upward learning in response to the input pattern for the weight to be stable. Since the circuit was designed to have a slow rate of decay (forgetting), learning proceeds slowly.

Before the system has reached equilibrium, the change in synaptic coupling affects the postsynaptic neuron's response to a given input pattern. The effect on learning of the changing postsynaptic response is shown in Fig. 6. Changing the postsynaptic response causes an inherent instability in the coupling strength of excitatory synapses. As the learning progresses, the response of the postsynaptic cell becomes larger, so the synaptic strength increases more quickly. The magnitude of the synaptic weight increases until it reaches its maximum value. In contrast, when the strength of inhibitory coupling is learned using the same learning rule, the magnitude of the weight is self-stabilizing. The inhibitory input couples more strongly to the postsynaptic cell until the inhibition just cancels the excitatory input that was driving the learning process.

A method of weight stabilization based on Hebb-modifiable inhibitory connections in a system of linear neurons is discussed by Easton and Gordon (1984). They show that feedforward inhibitory input is able to cancel an excitatory input of fixed value, but does not stabilize synaptic weights when the excitatory input is free to change magnitude. Feedback inhibition, however, is able to stabilize synaptic weights if the neurons are linear. Further investigation is needed to determine whether these results can be applied usefully to our system, which includes a threshold nonlinearity and a delay. The separation of the postsynaptic potential on which learning is based and the output of the neuron (which is a presynaptic input to an inhibitory feedback loop) leads to interesting dynamic behaviors.

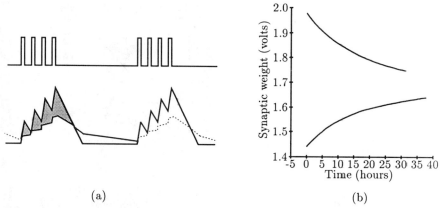

(a)   (b)

FIGURE 5.   (a) The condition for stability of a synaptic weight is illustrated schematically. The presynaptic input is repeated periodically. The response of the postsynaptic cell is held constant. The voltage on the target weight capacitor is shown in the solid line. The voltage on the floating gate is shown on the same axis as a dotted line. The change in the voltage on the floating gate has been magnified greatly. A current proportional to the difference between the target weight and the floating gate flows through the ultraviolet resistor, continuously changing the voltage on the floating gate. Since the resistance is large, the actual change in the voltage on the floating gate is small over one stimulus presentation. In the period during which the presynaptic input is inactive, the difference between the target weight and the floating gate is proportional to the magnitude of the weight. The change in voltage on the floating gate over the repetition period is given by the integral of the difference between the target weight and the floating gate weighted by the value of the ultraviolet resistor (which is a nonlinear function of voltage). When the upward integral balances the downward integral, the synaptic weight is at equilibrium because there is zero net charge transfer onto the floating gate. In the illustration, the dark shaded region represents the positive integral when the weight is increasing and the lightly shaded integral represents the decay of the weight. The equilibrium point depends on the amplitude of the target weight (which is set by the gain parameter, α) and the frequency with which the input is presented. (b) Data from the chip. The synaptic weight was given an initial starting value and allowed to equilibrate while the input (a burst of four action potentials) was repeated. If the starting weight is lower than the equilibrium value, the weight increases. If the starting weight is higher than the equilibrium value, the weight decreases. The rate at which the weight equilibrates is determined by ε. Since the ultraviolet resistor is nonlinear, the decay is a function of voltage. The time constant of equilibration in this experiment increases as the weight approaches its asymptotic value.

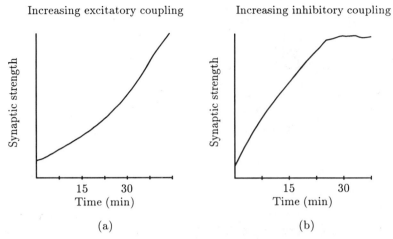

FIGURE 6.   Change in synaptic weight when the postsynaptic response is allowed to change. Two paired inputs were presented to a single cell. The first input was a control input of fixed magnitude, which excited the postsynaptic cell slightly. The other input was presented presynaptically to a learning synapse that was (a) excitatory or (b) inhibitory. The weight of the learning synapse was initialized such that the presynaptic input had no effect on the postsynaptic membrane. The synaptic weight increased initially because the presynaptic input was paired with the control excitatory input. When the presynaptic input was excitatory (a), the weight increase proceeded more and more quickly as the presynaptic input contributed to the excitation of the postsynaptic cell. When the presynaptic input was inhibitory (b), the weight increased until it reached the point where the presynaptic input inhibition just canceled the control excitation.

VI.   Neurons that Learn Sequence

As a demonstration of the potential of these system components, we combined our silicon neurons with Hebbian synapses in a small neural network. The goal of the network was to learn to prefer a repeatedly presented temporal sequence. The computational ability of our simple analog neuron model when it is placed in the context of real-time inputs and continuous-time learning is explored in this experiment.

Our sequence-selective network comprised four chips. Each chip contained a single neuron, and each neuron was connected to its two nearest neighbors in a symmetric pattern. The neurons used an action-potential output representation as described in the neuron section. Because only one neuron was integrated on each chip, the address-event arbitration scheme was not used. The connection pattern is shown in Fig. 7. The connections between neurons were made via learning synapses. The synaptic weights of all the learning synapses were initialized to zero. Two control inputs to each neuron—one excitatory and the other inhibitory—were presented in the form of computer-generated pulse bursts. The pattern of stimulation to each neuron was excitation followed by inhibition, giving the neuron a biphasic response. Inputs were presented sequentially to each of the four neurons. Inputs to adjacent neurons were presented in quadrature phase, shifted by a quarter of the period of the biphasic response.

Under these conditions, an asymmetric pattern of weights developed. Neurons formed strong connections in the preferred direction and remained decoupled in the null direction. How can an asymmetric connectivity pattern develop from a Hebbian update rule that appears symmetric at first inspection? The essential feature is that the spike-train output of the neuron is separated from the post-synaptic-membrane potential. Consider the interactions between two adjacent cells in the array, illustrated in Fig. 8. Examine the neurons in the same order as the preferred input sequence, neuron 2 presynaptic to neuron 3: The presynaptic

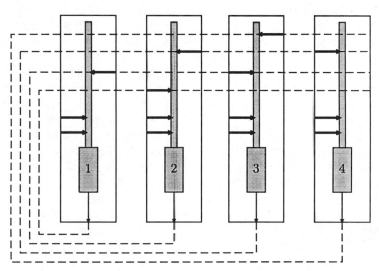

FIGURE 7.   Four-neuron network. Each neuron is on a separate chip. Connections between neurons are indicated by arrows. The two arrows at the base of each dendrite indicate two control inputs generated by computer: one excitatory, the other inhibitory. During learning, the control inputs were always presented in the same sequence: neuron 1, neuron 2, neuron 3, neuron 4. Connections between neurons in which the control stimulus to the presynaptic neuron precedes the control stimulus to the postsynaptic neuron are indicated by rightward arrows. Connections between neurons in which the control stimulus to the presynaptic neuron follows the control stimulus to the postsynaptic neuron are indicated by leftward arrows.

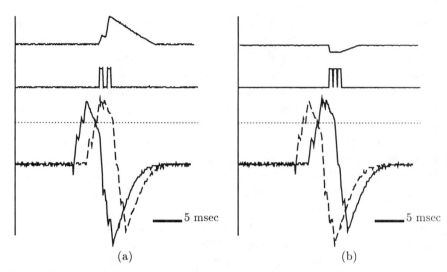

(a)   (b)

FIGURE 8.   Synaptic interactions leading to asymmetric coupling. During learning, only one sequence (neuron 1 stimulated first, neuron 4 stimulated last) was presented repeatedly. The sequence speed was adjusted so that the previous neuron was firing when the sequence was onset at the new location. Postsynaptic potentials of neuron 2 and neuron 3 in response to presentation of the preferred sequence are shown in the bottom traces. The postsynaptic neuron of the pair is shown in a dashed line. (Threshold is indicated by a dotted line.) (a) The spike output of neuron 2 is shown in the middle trace. Neuron 2 is presynaptic to neuron 3 at a synapse. The target weight computed by this synapse is shown in the top trace. When the preferred sequence is presented, neuron 2 is firing when the postsynaptic potential of neuron 3 is elevated. The target weight forces the synaptic coupling between these neurons to increase. (b) The spike output of neuron 3 is shown in the middle trace. Neuron 3 is presynaptic to neuron 2 at a synapse. The target weight computed by this synapse is shown in the top trace. When the preferred sequence is presented, neuron 3 is firing when the postsynaptic potential of neuron 2 is depressed. The target weight forces the synaptic coupling between these neurons to decrease. The target weight, in this case, saturates at the minimum possible synaptic weight.

cell is firing while the postsynaptic cell is being excited but has not yet begun spiking. Therefore, the coupling between the two cells is increased. Switching the roles of the neurons so that neuron 3 is presynaptic to neuron 2, we see that the presynaptic cell is firing when the postsynaptic potential is in the hyperpolarizing phase of its response. The synaptic coupling is driven towards zero. The separation of postsynaptic potential and spike output is likely to be significant in a real neuron in which the synaptic inputs can be electrotonically distant.

    The effects of learning are visible in the response of the system, shown in

Fig. 9. Prior to learning, the response of a neuron to the null sequence and the preferred sequence were the same, because the neurons were not coupled. To probe the system response after learning, we decreased the magnitude of the control input such that it produced a subthreshold response except in the first neuron of the sequence. Threshold of each neuron is above input magnitude except in the first neuron of the sequence. In this case, we saw no response in the null direction, because there was no coupling from the first neuron to the other neurons. In contrast, when the sequence that the system was trained on was presented, the neurons responded vigorously. We decreased the activation of the neurons in the middle of the chain to below threshold levels because we inadvertently set the α parameter too high during learning. As a consequence, the synaptic coupling in the preferred direction became so strong that presentation of any element of the learned sequence was sufficient to reconstruct the remainder of the sequence. The experiment could not be replicated with a smaller maximum weight due to accidental damage to the chips.

These experiments illustrate the importance of the method of weight stabilization employed by a learning rule. In our learning rule, the parameter α, which controls the learning rate, must be small because the decay term ε is very small. The synaptic weight is stabilized by counterbalancing the stimulus-dependent learning rate, α, against the general forgetting rate, ε, that proceeds continuously in time. If ε is not very small, the system forgets what it has learned more quickly. In this system, learning and forgetting are inextricably linked. One alternative for controlling maximum synaptic coupling is to set a hard limit on

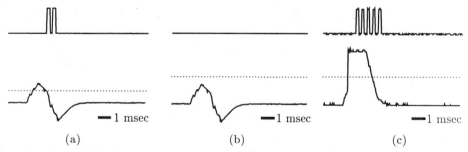

(a)   (b)   (c)

FIGURE 9.   Response of the system to the preferred and null sequence. (a) Before learning, a neuron shows equal response in preferred sequence (neuron 1 stimulated first and neuron 4 stimulated last) and null direction (neuron 4 stimulated first and neuron 1 stimulated last). The threshold level is indicated by a dotted line. The top trace shows spike output. The bottom trace shows postsynaptic potential. (b) After learning, a neuron shows no response in the null direction. (c) Also after learning, the same neuron shows a brisk response in the preferred direction. Threshold of each neuron is raised above input magnitude except in the first neuron of the sequence.

the maximum weight. This limit would permit a fast learning rate. However, the use of a hard limit does not ensure that the synaptic weights will stabilize at intermediate analog values. We believe that a preferable method of weight stabilization is to make the learning rate, α, be a function of the absolute value of the weight. The rate of synaptic increase should vary inversely with the synaptic weight. This variation would allow the system to learn quickly when it was not storing any information already. The rate of synaptic decrease might vary proportionally with the magnitude of the weight. With this type of learning rule, a synaptic weight stabilizes based on the statistics of the training set, independently of the temporal presentation frequency of the stimulus.

VII. Summary

We are evolving the components of a silicon neuron for making neural networks that can perform real-time sensorimotor tasks. These components include a communications framework, a neuron body, and an on-chip modifiable synapse. All the components were fabricated through MOSIS in a standard, 2 micron, double-poly CMOS process.

The desire to perform real-time sensorimotor tasks influenced the development of the neuron components. We chose to use an address-event representation in our communication framework to provide an interface between our network and existing VLSI sensory and motor subsystems. This representation takes advantage of the abstraction computed by the silicon retina and is consistent with the output representation of the silicon oculomotor system. The neuron body has two stages: a postsynaptic integrator and a mechanism for generating action potentials. These two stages are separated by a nonlinear element. The analog circuitry naturally provides dynamics. The neuron-body circuitry is able to detect temporally coincident events. The synapse, which computes a weight-update signal according to a Hebbian rule, provides continuous on-chip learning in response to real-time signals. The synapse uses an action-potential representation and, therefore, can use correlations between temporal events to update the synaptic weight.

Test results from these circuits are promising. In a preliminary experiment to test the behavior of the neuron and Hebbian synapse in a real-time context, four neurons were integrated into a system that learned to prefer a repeated temporal sequence. Using an action-potential representation, the system was able to learn to prefer the repeated temporal sequence over the inverted sequence in spite of mismatches between neurons on different chips. The system was able to learn an asymmetric weight pattern using a symmetric learning rule because the structure of the neuron separated the postsynaptic potential and the spike output.

The performance of these components on simple tasks drives their evolution. For example, the difficulties encountered in learning a simple temporal pattern prompted the design of a new learning rule. The next generation learning synapse incorporates a learning rule in which the learning rate is a function of the synaptic weight. Hopefully, this improved synapse will learn more quickly than its predecessor. Our future goal is to interface our learning systems to real sensory input and test them on real motor output tasks. This interface will supply the criterion for evolutionary selection.

We chose to model the nervous system in a medium that shares many of the properties of nervous tissue and we ask the model to perform real-time sensory motor tasks. In this way, we hope to discover some of the essential computational strategies employed by the nervous system and embody them in silicon neural networks.

Acknowledgments

I wish to thank Massimo Sivilotti for collaboration on the communications interface. I also thank David Gillespie for fruitful discussions and technical support. MaryAnn Maher, Doug Kerns, and Carver Mead provided valuable advice on ultraviolet programmable floating-gate structures. Fabrication was provided by DARPA through the MOSIS service. This work was supported by the ONR, the NIH, the Systems Development Foundation, and the State of California Department of Commerce Competitive Technologies Program.

References

ALSPECTOR, J., GUPTA, B., and ALLEN, R. (1989). "Performance of a Stochastic Learning Microchip," in D. Touretzky (ed.), *Advances in Neural Information Processing Systems 1* (pp. 748–760). Moran Kaufmann, San Mateo, California.

BOAHEN, K., POULIQUEN, P., ANDREOU, A., and JENKINS, R. (1989). "A Heteroassociative Memory Using Current-Mode MOS Analog VLSI Circuits," *IEEE Transactions on Circuits and Systems* **36**, 747–755.

BROWN, T. H., KAIRISS E. W., and DEEMAN, C. W. (1990). "Hebbian Synapses: Biophysical Mechanisms and Algorithms," *Annual Review of Neuroscience* **44**, 447–511.

BROWNLOW, M., TARASSENKO, L., MURRAY, A., HAMILTON, A., HAN, I. L., and REEKIE, H. M. (1990). "Pulse-Firing Neural Chips for Hundreds of Neurons," in D. Touretzky (ed.), *Advances in Neural Information Processing Systems 2* (pp. 785–792). Morgan Kaufmann, San Mateo, California.

CARD, H., and MOORE, W. (1989)."VLSI Devices and Circuits for Neural Networks," *International Journal of Neural Systems* **1**, 149–165.

DELBRÜCK, T. (1989). "A Chip that Focuses an Image on Itself," in C. Mead and M. Ismail (eds.), *Analog VLSI Implementation of Neural Systems* (pp. 171–188). Kluwer, Boston.

DEWEERTH, S., and MEAD, C. (1988). "A Two-Dimensional Visual Tracking Array," in *Advanced Research in VLSI: Proceedings of the Fifth MIT Conference* (pp. 259–275). MIT Press, Cambridge, Massachusetts.

DEWEERTH, S., NIELSEN, L., MEAD, C., and ASTRÖM, K. (1990). "A Neuron-Based Pulse Servo for Motion Control," *Proceedings of the IEEE Conference on Robotics and Automation* (pp. 1698–1703). IEEE Computer Society Press, Los Alamitos, California.

EASTON, P., and GORDON, P. (1984). "Stabilization of Hebbian Neural Nets by Inhibitory Learning," *Biological Cybernetics* **51**, 1–9.

FORSYTHE, I., and WESTBROOK, G. (1988). "Slow Excitatory Postsynaptic Currents Mediated by N-methyl-D-aspartate Receptors in Cultured Mouse Central Neurons," *Journal of Physiology (London)* **396**, 515–533.

GLASSER, L. (1985). "A UV Write-Enabled PROM," in H. Fuchs (ed.), *Chapel Hill Conference on VLSI* (pp. 61–65). Computer Science Press, Rockville, Maryland.

HOLLER, M., TAM, S., CASTRO, H., and BENSON, R. (1989). "An electrically Trainable Artificial Neural Network Etann with 10240 'Floating Gate' Synapses," in *IJCNN International Joint Conference on Neural Networks, Vol. 2* (pp. 191–196). Publishing Services IEEE, New York.

LAUGHLIN, S. (1987). "Form and Function in Retinal Processing," *Trends in Neuroscience* **10**, 478–483.

LAZZARO, J., and MEAD, C. (1989). "Silicon Models of Auditory Localization," *Neural Computation* **1**, 41–70.

MAHER, M. (1990). Personal communication.

MAHOWALD, M., and DELBRÜCK, T. (1989). "Cooperative Stereo Matching Using Static and Dynamic Image Features," in C. Mead and M. Ismail (eds.), *Analog VLSI Implementation of Neural Systems* (pp. 213–246). Kluwer, Boston.

MEAD, C., and MAHOWALD, M. (1988). "A Silicon Model of Early Visual Processing," *Neural Networks* **1**, 91–97.

MEAD, C. A. (1989a). *Analog VLSI and Neural Systems*. Addison-Wesley, Reading, Massachusetts.

MEAD, C. (1989b). "Adaptive Retina," in C. Mead and M. Ismail (eds.), *Analog VLSI Implementation of Neural Systems* (pp. 239–246). Kluwer, Boston.

MOOPENN, A., DUONG, T., and THAKOOR, A. (1990) "Digital–Analog Hybrid Synapse Chips for Electronic Neural Networks," in D. Touretzky (ed.), *Advances in Neural Information Processing Systems 2* (pp. 769–776). Morgan Kaufmann, San Mateo, California.

MUELLER, P., VAN DER SPIEGEL, J., BLACKMAN, D., CHIU, T., CLARE, T., DONHAM, C., HSIEH, T., and LOINAZ, M. (1989). "Design and Fabrication of VLSI Components for a General Purpose Analog Neural Computer," in C. Mead and M. Ismail (eds.), *Analog VSLI Implementation of Neural Systems* (pp. 135–169). Kluwer, Boston.

MURRAY, A. (1989). "Pulse Arithmetic in VLSI Neural Networks," *IEEE Micro* **9:6,** 64–74.

SATYANARAYANA, S., TSIVIDIS, Y., GRAF, H. P. (1990). "A Reconfigurable Analog VLSI Neural Network Chip," in D. Touretzky (ed.), *Advances in Neural Information Processing Systems 2* (pp. 758–768). Morgan Kaufmann, San Mateo, California.

SCHWARTZ, E. (1977). "Spatial Mapping in the Primate Sensory Projection: Analytic Structure and Relevance to Perception," *Biological Cybernetics* **25,** 181–194.

SIVILOTTI, M. (1990). *Wiring Consideration in Analog VLSI Systems, with Application to Field-Programmable Networks*. Doctoral Dissertation. Department of Computer Science, California Institute of Technology, Pasadena, California.

Chapter 16   Relations between the Dynamical Properties of Single Cells and Their Networks in Piriform (Olfactory) Cortex

JAMES M. BOWER

Computation and Neural Systems Program
Division of Biology
California Institute of Technology
Pasadena, California

I. Introduction

Understanding the relationship between the biophysical properties of single neurons and how they process information is clearly central to any effort to understand how brains work. Information within the nervous system is, after all, encoded by the outputs of single neurons. Thus, the question of how the activity converging on a particular neuron affects its output is crucial to understanding overall brain function. The problem, of course, is that the complex biophysical structure of single neurons makes determining such input–output relationships extremely difficult.

Using modern experimental techniques, neurobiologists are rapidly amassing large amounts of data concerning the anatomical and physiological properties of single cells. In fact, most experimental methods used by neurophysiologists and neuroanatomists focus on the properties of single neurons. Over the last 20 years, for example, there has been an explosive growth in biophysical descriptions of the ionic conductances governing neuronal electrical activity. A great deal is also now known about the physiology of the synapses that influence these conductances. In addition, improved anatomical procedures have provided more and more detailed descriptions of neuronal structure, while the largest amount of electrophysiological data available about the nervous system comes from single neuron recordings made under a wide variety of experimental conditions. Yet despite the massive amount of data now available, a complete understanding of any single cell's function has not yet been obtained. While considerable progress has been made in some invertebrate (Nelson and Bower, 1991) and peripheral mammalian neurons (Yamada *et al.*, 1989), within the central nervous system of mammals, our understanding of neuronal structure/function relationships can probably at best be described as primitive.

Given the complexity of neuronal anatomy and physiology, it seems likely that our understanding of neuronal function will increasingly rely on the construction and exploration of computational models (Koch and Segev, 1989). In fact, single cell modeling has always constituted a central focus in computational neurobiology. While traditionally most models of this type have attempted to simplify neuronal structure into a mathematically more manageable form (Rall, 1989), with the increasing availability of substantial computer resources, there has been a marked increase in the use of so-called *compartmental models* that can reflect more detailed features of single cells (Segev *et al.*, 1989). As a result, it is increasingly common to find neurophysiologists constructing and using single cell simulations in an attempt to understand the organization of the cells they study.

Both abstract and compartmental models of single cells can provide useful insights into neuronal function. However, we believe that ultimately, to understand the functional organization of single cells, it will be necessary to consider their structure in the context of the network in which they are embedded. This derives from the fact that the activity of any neuron is not only dependent on its anatomical and physiological properties, but also on the spatial and temporal pattern of inputs it receives. In other words, while it is possible to generate extremely detailed dendritic models, in the absence of information about how that dendrite is normally activated, the significance of dendritic structure may not be obvious. Given the complexity and numbers of inputs received by a typical mammalian neuron, for example, simply generating random combinations of different inputs and looking for interesting effects is unlikely to be a particularly efficient approach. Accordingly, we believe that single cell neuronal modeling must eventually be performed in the context of larger-scale network models. This chapter describes our own initial efforts to take this approach to study pyramidal cells within the mammalian olfactory cortex. These results suggest that cerebral cortical networks may be organized to assure that the synaptic influences on pyramidal cells are timed to occur in regular and repeatable spatial/temporal patterns.

II.   The Olfactory System as a Model Cerebral Cortical Sensory Network

We believe that the *olfactory* or *piriform cortex,* which is the primary region of the cerebral cortex devoted exclusively to processing olfactory information (Haberly and Price, 1978; Haberly, 1985), provides an outstanding opportunity to explore the general relationships between cerebral cortical neuronal and network organization. First, the overall structure of piriform cortex is somewhat less complex than other cerebral cortical sensory regions (Haberly, 1985; Haberly and Bower, 1989). For example, while the basic cell type in this cortex, like all other cortical regions, is the pyramidal cell (Haberly and Bower, 1984), there are fewer other anatomically distinct cell types (Haberly, 1983). Piriform cortex also contains only three layers rather than the six found in neocortical areas (Haberly, 1985). Further, and of particular importance to this chapter, the organization of fiber systems within the piriform cortex is less complex and more regular than in other cerebral cortical sensory areas. In particular, there is a sharp anatomical separation between the terminations of afferent inputs on the distalmost apical dendrites of pyramidal cells and the terminations of the intrinsic association fiber system on more proximal dendritic regions (Schwob and Price, 1978).

Beyond its relatively simple structure, the study of piriform cortex also benefits from its unusual position with respect to the periphery and deeply buried cortical structures (Haberly, 1985; Haberly and Bower, 1989). In this system, peripheral sensory information from olfactory receptor neurons in the nasal epithelium is processed by only one intervening structure, the olfactory bulb, before it is sent to the piriform cortex (Fig. 1). In all other sensory systems, peripheral sensory information reaches primary cerebral cortical areas only through the thalamus (Price, 1985). With respect to its outputs, a primary projection of piriform cortex activity is through the multisensory entorhinal cortex directly to the hippocampus (Luskin and Price, 1983; Price, 1985; Insausti *et al.,* 1987). The hippocampus, which is believed to be crucial for general properties of memory storage in the nervous system (Squire, 1986a, b; McNaughton and Morris, 1987), is influenced by other sensory modalities through entorhinal cortex, but only after many more processing stages (Van Essen *et al.,* 1991). Thus, the olfactory system effectively represents a shortcut from the sensory periphery to deeply buried forebrain structures. Because ultimately any understanding of the relationship between

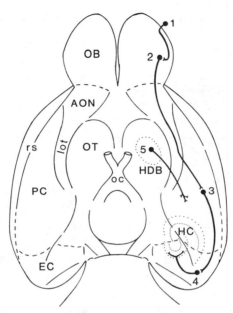

FIGURE 1.   Diagram demonstrates the principal pathways providing input to and output from the piriform cortex of the rat. Neural signals originate in the olfactory epithelium (1), whose sensory neurons project to the olfactory bulb (2). Outputs from the olfactory bulb project to the olfactory cortex (3), which, in turn, influences the enterhinal cortex (4) and, finally, the hippocampus (HC). This figure also shows a cholinergic neuromodulatory pathway that projects from the horizontal band of Brocca (HDB) to the piriform cortex. Reprinted with permission from Hasselmo *et al.*, 1991a).

neuronal activity and network organization must be obtained in the context of natural sensory activation (Bhalla and Bower, 1988, 1991), the relative lack of processing stages in the olfactory information stream may make such a connection easier to obtain in this system.

III.   Modeling Olfactory Cortex

For the last several years, we have been using computer modeling techniques to explore the functional organization of the piriform cortex (Wilson *et al.*, 1986; Wilson and Bower, 1988, 1989; Nelson *et al.*, 1989; Bower, 1990, 1991a, b; Hasselmo *et al.*, 1991a, b; Wilson, 1991). A major part of this effort has involved the construction of realistic network simulations based on what is currently known of the actual anatomical and physiological features of this network. Being realistic, these models also generate physiologically measurable signals like neuronal spike trains, EEGs, and evoked potentials. For this reason, they provide a means to explore the functional consequences of known network components while also generating physiological predictions (Bower, 1990, 1991a, b; Hasselmo *et al.*, 1991b). These predictions serve to motivate and identify those additional experiments necessary to advance our understanding of the system (Hasselmo and Bower, 1990, 1991a, b, Hasselmo *et al.*, 1991b).

A. Model Structure

While the models we build are realistically based, any computer model by definition is an abstraction. In the current case, for example, the initial focus of our modeling effort on network properties resulted in the use of relatively simple cellular components (Wilson and Bower, 1988). However, over the last several years, as the questions addressed with the model have become more sophisticated, so have our simulated cells (Wilson, 1991). From the point of view of the network itself, the model simulates the lateral olfactory tract (LOT) input to the piriform cortex from the olfactory bulb as a sparse, non-topographic distributed set of excitatory axonal connections (Figs. 2, 3; also Devor, 1976). This input

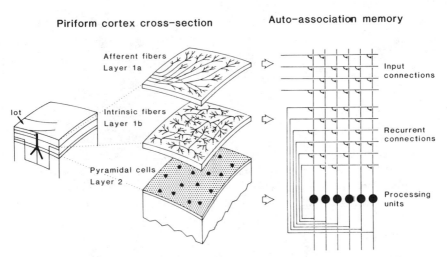

FIGURE 2. Comparison of the anatomical structure of piriform cortex with that of an auto-association matrix memory. In layer 1a, afferent input from the lateral olfactory tract (LOT) forms synapses on the distal dendrites of piriform cortex pyramidal cells in a diffuse, non-topographic manner. This input may be analogous to the distributed input connections of an auto-association memory. In layer 1b, intrinsic fibers arising from piriform cortex pyramidal cells in layers 2 and 3 synapse on proximal portions of other pyramidal cell dendrites in a diffuse, non-topographic manner. These synaptic connections may be analogous to the recurrent connections of an auto-association memory. Reprinted with permission from Hasselmo *et al.*, 1991a.

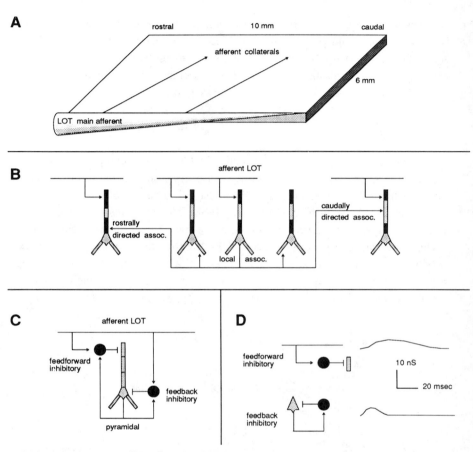

FIGURE 3.   Schematic representation of the piriform cortex model. (a) Spatial pattern of afferent input from the lateral olfactory tract (LOT) (b) Basic circuitry of intrinsic connections arising from piriform cortex pyramidal cells, forming synapses in layers 1b and 3. (c) Pattern of interconnection between pyramidal cells and two classes of inhibitory interneurons modeled. (d) Temporal properties of inhibitory synaptic conductances induced by the two classes of inhibitory interneurons. (Reprinted with permission from Bower, 1990).

spreads across the surface of the cortex rather than entering the cortex vertically, as is the case in neocortical sensory regions, e.g., geniculo-cortical input (Shepherd, 1979). The model also replicates the extensive set of so-called association fiber connections between pyramidal cells within the cortex itself (Figs. 2, 3). As in the real cortex, this fiber system makes sparse, distributed excitatory connections with other pyramidal cells throughout the cortex (Haberly and Bower, 1984, 1989; Haberly and Presto, 1986). Figure 4 shows the pattern of pyramidal cell connectivity viewed from a single neuron. Physiological data has been used to establish parameters for the conduction velocities, dendritic termination zones, and the relative amounts of influence of each fiber system in different cortical areas (Haberly and Shepherd, 1973; Haberly, 1973; Schwob and Price, 1978).

Typically these models consists of several thousand neurons of each of three types: the pyramidal cells and two types of inhibitory interneurons (Figs. 3, 4). All neurons are constructed using standard compartmental modeling techniques (Segev *et al.*, 1989). The two types of inhibitory neurons are modeled as single

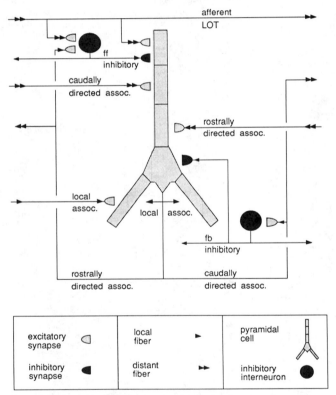

FIGURE 4.   Simplified diagram of the local circuitry contained in the model of piriform cortex. The cell in the center represents a pyramidal cell, which is the primary excitatory cell type in the piriform cortex. Two other cells types are shown as darkened circles adjacent to the pyramidal cell. The small semi-ellipses that appear next to the pyramidal and inhibitory cells indicate the locations of synaptic connections. Darkened connections are inhibitory, while lightened connections are excitatory. Arrows indicate the direction of propagation of signals originating from various cell types as well as the distance traveled. Reprinted with permission from Wilson and Bower, 1989.

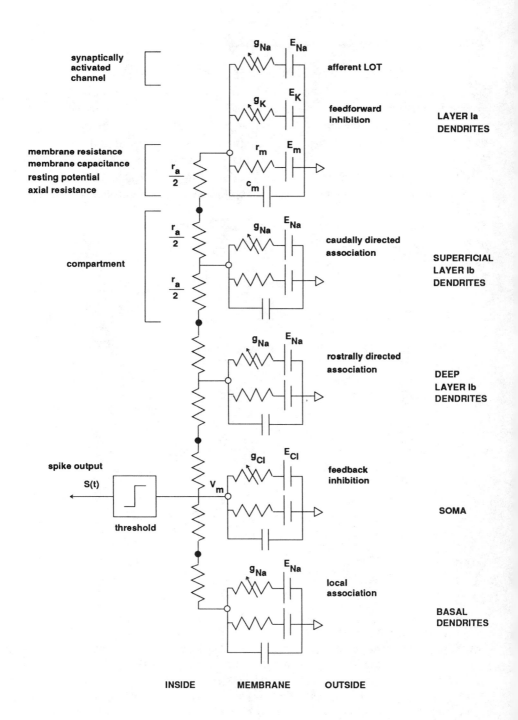

synaptically
activated
channel

g_{Na} E_{Na}

afferent LOT

g_K E_K

feedforward
inhibition

LAYER Ia
DENDRITES

membrane resistance
membrane capacitance
resting potential
axial resistance

$\dfrac{r_a}{2}$

r_m E_m

c_m

compartment

$\dfrac{r_a}{2}$

$\dfrac{r_a}{2}$

g_{Na} E_{Na}

caudally directed
association

SUPERFICIAL
LAYER Ib
DENDRITES

g_{Na} E_{Na}

rostrally directed
association

DEEP
LAYER Ib
DENDRITES

spike output

S(t)

V_m

threshold

g_{Cl} E_{Cl}

feedback
inhibition

SOMA

g_{Na} E_{Na}

local
association

BASAL
DENDRITES

INSIDE   MEMBRANE   OUTSIDE

444

compartments. One type, the feedforward inhibitory neuron, has a long-latency, long-duration effect on pyramidal cells and is primarily influenced itself by direct input from the LOT (Figs. 3c, d). The other type of inhibitory neuron has a short-latency, short-duration inhibitory effect on pyramidal cells and primarily receives input from, and projects to, nearby pyramidal cells (Figs. 3c, d). Both types of inhibitory influences have been shown to exist in this cortex (Tseng and Haberly, 1987; Sato et al., 1982).

The pyramidal cells are the principal neurons both in our model and in the cortex itself, where they provide the sole cortical output. Figure 5 summarizes the more complex compartmental structure of the pyramidal cells. As can be seen, they consist of several compartments linked by intercompartmental resistances. Our basic network simulations model these cells as five compartments (Figs. 4, 5), four associated with the four distinct regions of the cell receiving synaptic terminations from the different fiber systems, and one compartment for the soma. For some of the results described in this chapter, however, a more complete model of the pyramidal cell was used. This model consisted of several hundred compartments and was used principally to simulate experimentally obtained current source density measurements (Rodriguez and Haberly, 1989), allowing a more detailed description of the pattern of inputs to these cells. Regardless of the complexity of a particular modeled neuron, all synaptic inputs were modeled as conductance changes that were then integrated to produce transmembrane voltage. When these voltages exceed specific threshold values, spike output is generated and propagated along simulated axons. All models were constructed within the GENESIS simulation environment (Wilson et al., 1989; Bower and Hale, 1991). Additional mathematical details of this model can be found in Wilson and Bower (1989).

B.   Simulating Network Dynamics

The earliest stage in our modeling effort involved an attempt to replicate cortical responses to different types of afferent stimuli. It has been known for some time that piriform cortex generates distinct periodic patterns of neuronal activity under

FIGURE 5.   Circuit representation of the multicompartmental pyramidal cell used in the basic piriform cortex model described in the text. There are five compartments, each containing a membrane resistance, membrane capacitance, and resting potential, as well as one or more synaptically activated conductances. In some model runs, this basic cell model was expanded to include more than a hundred compartments. Reprinted with permission from Wilson and Bower, 1989.

a variety of experimental conditions (Freeman 1970; Bower, 1990). In the awake behaving animal, for example, the EEGs recorded from the surface of the cortex demonstrate distinct patterns of oscillatory behavior centered around 7–10 Hz and 40 Hz (Fig. 6; also Freeman, 1970, 1975; Freeman and Schneider, 1982). Under light anesthetic conditions, weak direct electrical shocks of the primary input pathway to the olfactory cortex (the LOT) evoke a multi-phase prolonged oscillatory surface field potential with a prominent 40 Hz frequency component (Fig. 6; also Haberly, 1973; Freeman, 1968a, b). A strong shock delivered to the same input pathway, however, evokes only a short-duration biphasic evoked potential (Fig. 6).

Using a model based on relatively simple neurons, we have been able to replicate each of these typical response patterns by modifying only the type of input presented to the cortex (Fig. 6; also Wilson and Bower, 1988, 1991b; Bower, 1990). Figure 7 shows graphically the sequence of simulated events that generates evoked potential profiles in response to direct weak electrical activation. During the first few msec after shocking the LOT, afferent activity sweeps into the cortex depolarizing the rostral pyramidal cells it first encounters (7 msec, Fig. 7). The resulting pyramidal cell action potentials spread excitation locally to the feedback inhibitory interneurons, which, in turn, set up a short-latency, relatively short-duration current shunting-type inhibition in these same pyramidal

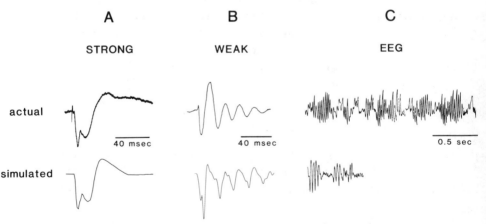

FIGURE 6. Comparison of actual physiological records with those generated by our simulation. In a and b, field potentials evoked by large-amplitude shocks of the LOT (a) are compared to potentials evoked by weak shocks (b). In c, EEG recordings from awake behaving animals are compared to simulated results with continuously presented random input to the model. Real field potentials in a are from Haberly, 1973; those in b from Freeman, 1968a. EEG recordings shown in c are from Bressler, 1984. Reprinted with permission from Bower, 1990.

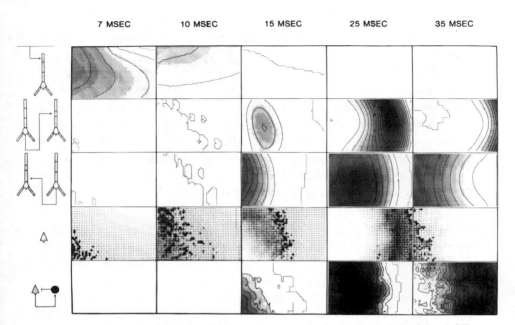

FIGURE 7.   Chart details the model response to weak shock simulation of the LOT. Network activity is shown at five different time steps. At each time step, five different features of the simulation are shown. Each plot represents the activity at each time step across the full two-dimensional extent of the cortex. Rostral cortex is to the left and caudal to the right. The icons at the left of the figure indicate the feature being displayed in each row of plots. The first row indicates the conductance changes due to the afferent input to the cortex. The next two rows represent conductance changes due to the influence of the rostral-to-caudal and caudal-to-rostral association fiber systems, respectively. The third row displays the level of depolarization of the pyramidal cell somas where the size of each box indicates the level of depolarization. Full black boxes indicate spiking activity. Finally, the last row shows conductance changes in local pyramidal cells due to the feedback inhibitory interneurons. As discussed in the text, the principal feature of interest in this figure is the reactivation found in rostral cortex at $T = 35$ msec in response to the weak shock strength. Figure modified with permission from Bower, 1990.

cells (15 msec, Fig. 7). This inhibitory influence serves to selectively suppress cell firing in response to the activation of local association fiber system synapses. As time proceeds in the simulation, cells in the caudal regions of the cortex begin to fire (25 msec, Fig. 7). However, unlike the rostral cells, the primary excitatory influence on these neurons arises from intrinsic association inputs, not the original afferent input. This is because the influence of the LOT is, in general, much less in caudal regions of cortex (Haberly, 1973; Schwob and Price, 1978). Thus, caudal pyramidal cells are activated as a secondary consequence of the afferent activation of rostral pyramidal cells. Because the rostral-to-caudal association fiber system has approximately half the conduction velocity of the LOT fibers (Haberly, 1973), this dependence on association fiber activity results in a delay in the activation of caudal cortical regions. The model suggests that one of the more significant consequences of this delay is to prolong the reciprocal influence of caudal cortex on the rostral regions of cortex first activated by the weak shock (35 msec, Fig. 7). This delay allows rostral regions of the cortex to come out from under the influence of local inhibitory neurons while there is still enough activity in the network to produce a reactivation of rostral cortex. In the simulation, this reactivation of rostral cortex produces a second rostral-to-caudal sweep of neuronal activity even in the absence of another LOT input. With the proper stimulus parameters, this rostral-to-caudal-to-rostral activation pattern can repeat multiple times (Fig. 6). In both the model and the real cortex, as the amplitude of the LOT shock is decreased, the number of these reverberating cortical cycles increases (Ketchum and Haberly, 1988; Freeman, 1962, 1968b). In each case, the frequency of the oscillations remains at around 40 Hz.

Using precisely the same model parameters but varying the nature of the input to the model, we have also been able to replicate the naturally occurring patterns of cortical oscillation seen in the EEG of behaving animals (Fig. 6; also Freeman, 1970). In this case, when the model is presented with low levels of phasic afferent input of the sort that is naturally generated by the olfactory bulb (Freeman and Schneider, 1982), the cortical simulation replicates both the low-frequency, 7–10 Hz, and higher-frequency, 40 Hz, components of the EEG. This is perhaps not surprising given that the bulbar activity patterns being simulated have similar frequency components (Bressler, 1984; Freeman and Schneider, 1982; Ketchum and Haberly 1988). More interestingly, however, we find exactly the same model behavior in response to low levels of continuous afferent input with no phasic components, suggesting that these oscillatory patterns may be intrinsic properties of piriform cortical circuitry and not simply driven by the bulbar input (Bower, 1990). A similar conclusion has been drawn from experimental data (Freeman, 1968b). It turns out in the model that the tendency for the network to oscillate at 40 Hz relies on the same mechanism that generates the similar frequency responses to weak shock stimulation just discussed. Specifically, the 40 Hz oscillation pattern in the EEG also emerges from a combination of network and cellular properties including the conduction velocities of axonal systems and the time constants of local inhibition. The pattern of rostral reactivation by more caudal neurons as seen in response to weak shock stimulation (Fig. 7) is also present at the beginning of each phase of EEG 40 Hz bursting (Fig. 8). In both responses, the location of the activating neurons reflects quite closely the regions of the cortex that have been removed from inhibitory influence. Figure 8 also

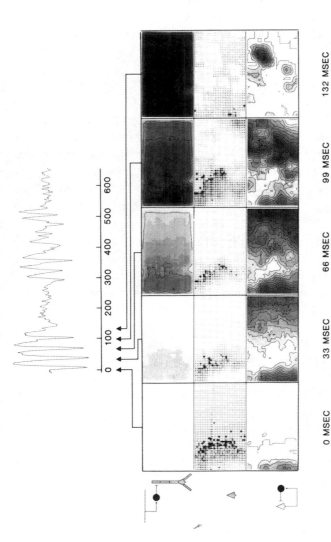

FIGURE 8. Chart demonstrates the patterns of pyramidal cell activity found in simulation at the beginning of each cycle of oscillation in the simulated EEG. The arrows and lines indicate which column of plots is associated with which cycle of the EEG oscillation. The icons again indicate the feature being displayed in each row of plots. The top and bottom rows indicate the conductance changes in the local pyramidal cells due to the feedforward and feedback neurons, respectively. The middle row shows the depolarization of the pyramidal cells across the cortex at each time indicated. Note the changing pattern of single neuron activity associated with each cycle of the EEG. Reprinted with permission from Bower, 1990.

449

demonstrates that the tendency of the model to oscillate at 7–10 Hz regardless of the structure of the input is also related to the interaction of inhibition and network activity. In this case, it is the longer time constant of the feedforward inhibitory neurons that principally regulates the effect, although we have also shown that the faster-acting feedback neurons appear to contribute as well (Wilson, 1991; Wilson and Bower, 1991b).

C.   Tuning in Network and Neuronal Properties

Our models have suggested that the dynamical behavior of piriform cortex is dependent on the interaction of a number of different network components. In particular, there appears to be a kind of *tuning* between different cortical components that serves to assure that the network oscillates at frequencies relevant to the natural pattern of afferent input this cortex receives (Bower, 1990). Thus, for example, it has been shown that the 7–10 Hz oscillation in cortical EEGs is correlated with the sniffing rate of the animal (Freeman and Schneider, 1982; Bressler, 1984). Within the model, the tendency to oscillate at this frequency and at 40 Hz is principally dependent on relationships between the conduction velocities of intrinsic axonal pathways, the total horizontal extent of this cortical region, the relative strengths of afferent versus association fiber connections, and the time constants of local inhibitory neurons (Bower, 1990; Wilson, 1991b; Wilson and Bower, 1991b). As we will now describe, the model also suggests that these patterns of network activity may, in addition, be fundamentally related to the basic structure of the pyramidal cell itself.

Our analysis of the mechanisms underlying the periodic behavior seen in piriform cortex has relied on our ability to use the model to separately examine the contributions made by different network components. In particular, simulated network activity immediately draws attention to the relative contributions of the different fiber systems. In this regard, the model suggests that there is a highly repeatable temporal sequence of activation of these different components. For example, early in each oscillation, cortical behavior is dominated by the afferent input (Fig. 7). Subsequently, the dominant influence shifts to the rostrally-to-caudally directed axons, followed by a shift to the caudal-to-rostral system before the afferent input again becomes dominant (Fig. 7). We have recently suggested that this pattern may be related to an iterative computational process involving the repeated comparison of the current state of afferent input to the response of the network to previous afferent patterns (Bower, 1990). However, we believe this repeated cycle of influence is also relevant to the functional organization of the pyramidal cells themselves. Specifically, in piriform cortex, there is a distinct laminar organization in the synaptic terminations of each of these fiber systems

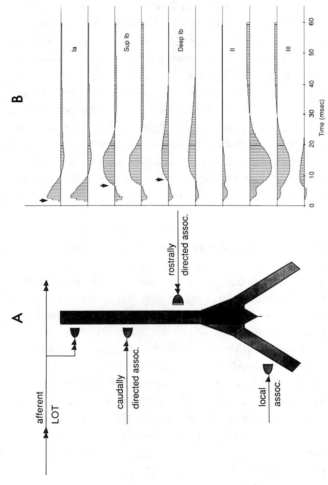

FIGURE 9.   Diagram shows simulated current source density profiles occurring at different depths within piriform cortex in response to afferent activation of the LOT with a weak electric shock. The pyramidal cell on the left (a) is aligned in depth with the CSD profiles on the right (b). Inward currents (sinks) are indicated as positive deflections on all traces. The three arrows indicate the onset of current sinks associated with the three different fiber systems in the simulations. These simulated CSD profiles closely resemble those described by Rodriguez and Haberly (1989).

(Fig. 4). Afferent synapses are located on the distalmost regions of the apical dendrite, rostral-to-caudal fibers synapse on intermediate dendritic regions, while caudal-to-rostral fibers synapse proximal on the dendrite (Haberly and Behan, 1984). When this anatomical distribution is coupled with the dynamical properties of the network, the result is a regular and repeated synaptic activation of the pyramidal cell apical dendrite. Thus, during each 40 Hz oscillation cycle, activation progresses from the distalmost to the proximalmost dendritic regions (Fig. 7). This result has also been experimentally demonstrated in response to weak shock stimulation using current source density analysis (Fig. 9; also Ketchum and Haberly, 1988; Rodriguez and Haberly, 1989). We have recently simulated these current source density results (Fig. 9), and have further suggested that, while more complex than with shock stimulation, natural activation patterns may also induce similar effects (Wilson, 1991). Further, with the model, we have been able to examine this behavior for neurons throughout the cortex. The results suggest that this cerebral cortical network is organized to assure that this general pattern of dendritic activation is reproduced in pyramidal cells regardless of their location in the cortex.

IV.   Functional Significance of Patterns of Dendritic Activation

While it seems reasonable to assume that this distinct spatial and temporal pattern of activation of the apical dendrite is functionally significant, determining this significance inevitably raises the question of the function of the network as a whole. As with the physiological results just described, we believe that this question can also best be addressed using modeling techniques. Accordingly, for the last several years we have used the piriform cortex model to explore the possible role of this cortex in olfactory object recognition (Wilson and Bower, 1988; Bower, 1990; Hasselmo et al., 1991a, b). Our results do, in fact, suggest that the spatial–temporal pattern of afferent activation of the apical dendrites of pyramidal cells may be critically important to this function. To understand the significance of this pattern, however, it is first necessary to briefly discuss the computational context for this work.

We have previously suggested that the process of olfactory object recognition fundamentally involves an associative memory process (Haberly and Bower 1989; Bower, 1990, 1991a, b; Hasselmo et al., 1991a, b). This view is based on both a general consideration of the computational task faced by this system as well as on the overall structure of its neural circuits. At the computational level, olfactory object recognition involves identifying the chemically diverse blends of airborne molecules emitted by different objects (Lancet, 1986; Laing *et al.*, 1989). While very little is yet known about the natural structure of the *olfactory stimulus space*, olfactory discrimination certainly requires that the olfactory system be capable of recognizing diverse subsets of a large number of airborne molecules under difficult environmental conditions (Bower, 1991b). Abstract associative memory models have been shown to be capable of similar functions (Palm, 1980; Hopfield, 1982; Kohonen, 1984; Grossberg, 1988). In addition, these abstract associative memory models bear a striking anatomical resemblance to the olfactory system (Haberly and Bower, 1989; Bower, 1990, 1991a, b; Hasselmo *et al.*, 1991a, b; Kauer, 1991). In particular, the extensive divergence and convergence of neuronal processes in both the afferent and association fiber systems may very well provide a neural substrate for associating activity evoked by many different combinations of chemically diverse molecules. In addition, the broad tuning of olfactory neurons in response to olfactory stimuli (Kauer and Moulton, 1974; Holley and Doving, 1977) suggest the kind of distributed representation characteristic of many abstract models (Bower, 1990; Kauer, 1991).

While the associative functioning of the olfactory system, just like the natural olfactory stimuli it interprets, is likely to be highly complex and varied (Bower, 1991b), our initial investigations of the associative memory capacity of the piriform cortex model have concentrated on two relatively simple aspects of associative learning (Wilson and Bower, 1988; Hasselmo *et al.*, 1991a, b). First, we have explored the model's ability to generate consistent patterns of neuronal activity in response to specific input patterns. Presumably, if the olfactory cortex is responsible for odor recognition, it should be able to generate consistent neuronal output in the presence of consistent sensory input (Bower, 1991b). Second, we have investigated the ability of the model to generate a stable pattern of neuronal activity in the presence of slight changes in the input stimulus. Because the mix of molecules being emitted by any object can vary with, for example, its age or environmental circumstances, it is presumed that olfactory recognition must, to some extent, be insensitive to such variations (Bower, 1991b).

To explore these questions, the olfactory cortex model was provided with input intended to represent the activity of single neurons in the olfactory bulb. Synaptic connections from these putative bulbar neurons to pyramidal cells in the cortical model were assigned completely randomly, as were the initial weights of each connection (Fig. 10). To explore learning in this network, a Hebb-type correlation learning rule was also introduced to govern activity-dependent changes in the synaptic strengths of modeled connections (Hebb, 1949). More details on the actual structure of these simulations can be found in Wilson and Bower (1988) and Hasselmo *et al.*, (1991b).

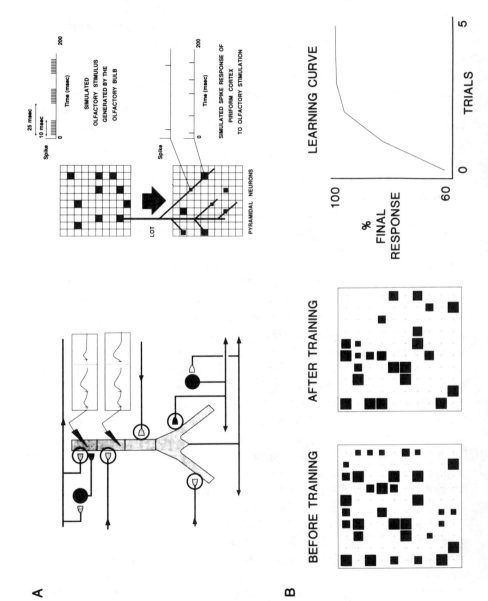

A

25 msec

10 msec

Spike

Time (msec)
0   200

SIMULATED
OLFACTORY STIMULUS
GENERATED BY THE
OLFACTORY BULB

LOT

PYRAMIDAL NEURONS

Spike

Time (msec)
0   200

SIMULATED SPIKE RESPONSE OF
PIRIFORM CORTEX
TO OLFACTORY STIMULATION

B

BEFORE TRAINING

AFTER TRAINING

LEARNING CURVE

100

%
FINAL
RESPONSE

60

0   5

TRIALS

Results of this modeling effort demonstrate clearly that the piriform cortex model is capable of learning to generate a stable output when presented with a consistent input pattern, and is also capable of reconstructing an original output even if the current input pattern is degraded (Wilson and Bower, 1988; Hasselmo *et al.*, 1991a, b). With respect to the subject of this chapter, however, the most interesting aspect of this capacity has to do with the relative roles and the relative timing of the afferent and association fiber systems in this model behavior. First, we have found that associative memory performance is only possible if the primary site of synaptic modification is in the association fiber system (Wilson and Bower, 1988; Bower, 1990; Bower, 1991a, b). While this was first a model prediction, subsequent experimental work has suggested that this may, in fact, be the case (Kanter and Haberly, 1989; Hasselmo and Bower, 1990). If correct, both the model and physiology suggest that the afferent and association fiber systems may play different roles during the acquisition of the memory itself. Second, we have found that the capacity of the network to store memories is enhanced if the association fiber system is suppressed during the initial storage process (Hasselmo *et al.*, 1991a, b). This model-based result follows

---

FIGURE 10.   Results from simulations in which the memory properties of the piriform cortex model were explicitly studied. (a) Circles on the left indicate those synapses undergoing activity-dependent changes in connection strengths in these simulations. (See text and Wilson and Bower, 1988.) As shown by the simulated intracellular records attached to the diagramed electrodes, changes in the afferent synapses occurred only as a consequence of paired stimuli and were not permanent. All other circled synapses were capable of long-term Hebbian modifications as discussed in the text. The diagram at the top right illustrates several features of these simulations. First, the random projection pattern of one of 10 active bulbar neurons (100 total in the simulation) to cortical pyramidal cells is shown. The stereotyped pattern of activity in the active bulbar neurons used in these simulations is also shown. The cortical section of this diagram demonstrates the display convention for cortical activity. As shown, the size of the black box overlying the position of each pyramidal cell in the two-dimensional simulated cortical array corresponds to the total number of action potentials the cell produced during 200 msec of bulbar activity. (b) Response of the simulated cortex to activity in a random set of 10 bulbar neurons. On the far left is shown the response of the cortex to the first presentation of the stimulus, i.e., effectively before training. The middle diagram indicates the final stable pattern of activity induced in the cortex by this particular active set of bulbar neurons in the presence of synapse modification. The graph on the far right indicates the changes in the pattern of activity over the five trials necessary for the simulated cortex to converge on a fixed final response pattern. One trial in this case corresponds to one 200 msec period of stimulation as shown in a. Figure modified with permission from Wilson and Bower, 1988.

experimental evidence that the strengths of synapses in these two fiber systems are differently regulated neuropharmacologically by acetylcholine (Hasselmo and Bower, 1990, 1991a, b). Third, the model has also allowed us to distinguish between two distinctly different phases during memory recall itself that are differentially influenced by afferent and association fiber connections. Specifically, during the early stages of accessing a previously stored memory, it is important that the initial patterns of pyramidal cell activity primarily reflect the structure of the raw afferent input. Once this initial activity pattern is established, the association fiber system appears to be responsible for further refining the responses of neurons to generate the stable output patterns we take to indicate object recognition. This secondary effect of the association fiber system is particularly important in the presence of noise or a degraded input signal (Hasselmo et al., 1991a, b). Finally, the learning model has also suggested that the relationship between inhibitory feedback and excitatory synaptic input is important for both memory storage and recall. This is particularly true with respect to the association fiber system. (See following text; also Hasselmo et al., 1991a, b.)

In each of the examples just discussed, our associative memory models of piriform cortex suggest that the afferent and association fiber synapses may be performing distinctly different functions within the network. At the same time, the dynamical network models reveal distinct temporal and spatial differences in the activation of these two synaptic populations. It turns out that the models also suggest several relationships between these different results. First, the distal-to-proximal activation of the apical dendrite on each oscillatory cycle mirrors quite closely the switch from afferent to association fiber-driven activity that we find is optimal for memory recall. Thus, the pattern of activation of these two synaptic populations may reflect and be ideal for the recall process. Second, the model-based suggestion that the balance between association fiber system excitation and local inhibition is important for learning reflects quite closely the temporal juxtaposition of these two synaptic effects during oscillatory behavior (Wilson, 1991; Wilson and Bower, 1991b). Again, the dynamics of the network seem tuned to temporally coincide with association fiber excitation and feedback inhibition. Third, as mentioned previously, the multi-cycle mixing of raw afferent information with association fiber activity in response to earlier afferent input may also be crucial to an iterative process of memory recall. In this regard, it should be noted that several cycles of the 40 Hz oscillation occur during each sniff cycle (Fig. 8). Finally, we suspect that the 25 msec period of the 40 Hz oscillation may also be related to the integration time constant of cortical pyramidal cells. While recent results in vitro suggest that the baseline time constants for pyramidal cells may be in the hundreds of msec, it is quite likely that the actual time constants are highly influenced by the patterns of activity being received by the dendrite of the cell (Holmes and Woody, 1989). In this sense, this critical cell parameter may be under dynamical regulation. Current modeling efforts in the laboratory are devoted to understanding how such effects could be influenced by the regular spatial–temporal patterns of activation we have described here.

V.   Conclusion

In summary then, we have used our network models to begin to explore the functional significance of dendritic processing in cortical pyramidal cells. The results suggest that this dendrite is normally activated in a particular temporal–spatial pattern over multiple cortical oscillations. Our efforts to model the possible associative memory function of this network suggest that these same temporal–spatial patterns may be important to the ability of the network to initially store memories and to later perform memory recall. An important extension of these ideas involves the possible interaction of these spatial–temporal patterns with the process of activity-dependent synaptic modification itself. Considering these interactions in detail has required that our pyramidal cell models include more of the detailed anatomical branching patterns of real cells. Based on this preliminary modeling and physiological results, we suspect that the dynamical properties of the overall network may very well have a significant influence on the way local synaptic properties are modified with activity.

Based on our results, we also suspect that the spatial–temporal coordination of projections onto the apical dendrite of cortical pyramidal cells may be a generally important feature of cerebral cortical networks (Wilson and Bower, 1990, 1991a; Wilson, 1991). Pyramidal cell dendrites span all cortical layers in all regions of cerebral cortex. In fact, the dependence of cortical processing on the properties of this dendrite may provide an explanation for the relative lack of interspecies variation in cortical thickness while the horizontal extent of cortex varies by five orders of magnitude. Further, as in our models of piriform cortex, the importance of temporally coordinating input to pyramidal cell dendrites may very well be directly related to the dynamical properties of cerebral cortical networks in general. In this regard, we have recently suggested that the 40 Hz oscillations that are now being reported in other regions of cerebral cortex (Eckhorn et al., 1988; Gray et al., 1989) primarily reflect such a coordinating mechanism (Wilson and Bower, 1990, 1991a). In any event, it seems likely that the relationship between the dynamical properties of networks and the neurons they are built of will be an increasingly active area of research. While this makes the process of studying mammalian networks much more difficult, this chapter has attempted to demonstrate that models based on realistic network and cellular features can provide a means to approach these complex relationships.

Acknowledgments

The author is indebted to the members of his laboratory for performing these experiments as well as contributing substantially to the ideas presented here. In particular, the contributions of Matt Wilson, Mike Hasselmo, and Upinder Bhalla must be acknowledged. The author is doubly indebted to Mike Hasselmo for his comments on this manuscript. In addition, the continuing contribution of Lew Haberly to this work is substantial. This research was supported by Office of Naval Research contract N00014-88-K-0513.

References

BHALLA, U. S., WILSON, M. A., and BOWER, J. M. (1988). "Integration of Computer Simulations and Multi-Unit Recording in the Rat Olfactory System," *Soc. Neurosci. Abstr.* **14,** 1188.

BHALLA, U. S., and BOWER, J. M. (1991). "Multiple-Single-Unit Recording from the Olfactory Bulb of Awake Behaving Rats," *Soc. Neurosci. Abst.* **17,** 636.

BOWER, J. M. (1990). "Reverse Engineering the Nervous System: An Anatomical, Physiological, and Computer Based Approach," in S. Zornetzer, J. Davis, and C. Lau (eds.), *An Introduction to Neural and Electronic Networks* (pp. 3–24). Academic Press, San Diego, California.

BOWER, J. M. (1991a). "Associative Memory in a Biological Network: Structural Stimulations of the Olfactory Cerebral Cortex," in V. Multinovic and P. Antognetti (eds.), *Neural Networks, Vol. II* Prentice Hall, New Jersey (in press).

BOWER, J. M. (1991b). "Piriform Cortex and Olfactory Object Recognition," in J. Davis and H. Eichenbaum (eds.), *Olfaction as a Model System for Computational Neuroscience*. MIT Press, Cambridge, Massachusetts, 265–285.

BOWER, J. M., and HABERLY, L. B. (1986). "Facilitating and Nonfacilitating Synapses on Pyramidal Cells: A Correlation between Physiology and Morphology," *Proc. Natl. Acad. Sci. USA.* **83,** 1115–1119.

BOWER, J. M., and HALE, J. (1991). "Exploring Neural Circuits on graphics Workstations," *Scientific Computing and Automation* **7,** 35–46.

BRESSLER, S. L. (1984). "Spatial Organization of EEGs from Olfactory Bulb and Cortex," *Electroenceph. Clinical Neurophysiol.* **57,** 270–276.

DEVOR, M. (1976). "Fiber Trajectories of Olfactory Bulb Efferents in the Hamster," *J. Comp. Neurol.* **166,** 31–48.

ECKHORN, R., BAUER, R., JORDON, W., BROSCH, M., KRUSE, W., MUNK, M., and REITBOECK, H. J. (1988). "Coherent Oscillations: A Mechanism of Feature Linking in Visual Cortex," *Biol. Cybern.* **60,** 121–130.

FREEMAN, W. J. (1962). "Alterations in Prepyriform Evoked Potential in Relation to Stimulus Intensity," *Exp. Neurol.* **6,** 70–84.

FREEMAN, W. J. (1968a). "Relation between Unit Activity and Evoked Potentials in Prepiriform Cortex of Cats," *J. Neurophysiol.* **31,** 337–348.

FREEMAN, W. J. (1968b). "Effects of Surgical Isolation and Tetanization of Prepyriform Cortex in Cats," *J. Neurophysiol.* **31,** 349–357.

FREEMAN, W. J. (1970). "Amplitude and Excitability Changes of Prepyriform Cortex Related to Work Performance in Cats," *J. Biomed. Systems.* **1,** 3–29.

FREEMAN, W. J. (1975). *Mass Action in the Nervous System.* Academic Press, New York.

FREEMAN, W. J., and SCHNEIDER, W. (1982). "Changes in Spatial Patterns of Rabbit Olfactory EEG with Conditioning to Odors," *Psychophysiol.* **19,** 44–56.

GRAY, C. M., KONIG, P., ENGEL, A. K., and SINGER, W. (1989). "Oscillatory Responses in Cat Visual Cortex Exhibit Inter-Columnar Synchronization which Reflects Global Stimulus Properties," *Nature* **338,** 334–337.

GROSSBERG, S. (ed). (1988). *Neural Networks and Natural Intelligence.* MIT Press, Cambridge, Massachusetts.

HABERLY, L. B. (1973). "Unitary Analysis of Opossum Prepyriform Cortex," *J. Neurophysiol.* **36,** 762–774.

HABERLY, L. B. (1983). "Structure of the Pirifrom Cortex of the Opossum: I. Description of Neuron Types with Golgi Methods," *J. Comp. Neurol.* **213,** 163–187.

HABERLY, L. B. (1985). "Neuronal Circuitry in Olfactory Cortex: Anatomy and Functional Implications," *Chemical Senses* **10,** 219–238.

HABERLY, L. B., and BEHAN, M. (1984). "Structure of the Piriform Cortex of the Opossum: III. Ultrastructural Characterization of Synaptic Terminals of Association and Olfactory Bulb Afferent Fibers," *J. Comp. Neurol.* **219,** 448–460.

HABERLY, L. B., and BOWER, J. M (1984). "Analysis of Association Fiber System in Piriform Cortex with Intracellular Recording and Staining Techniques," *J. Neurophysiol.* **51,** 90–112.

HABERLY, L. B., and BOWER, J. M. (1989). "Olfactory Cortex: Model Circuit for Study of Associative Memory?" *Trends in Neurosci.* **12,** 258–264.

HABERLY, L. B., and PRESTO, S. (1986). "Ultrastructural Analysis of Synaptic Relationships of Intracellular Stained Pyramidal Cell Axons in Piriform Cortex," *J. Comp. Neurol.* **248,** 464–474.

HABERLY, L. B., and PRICE, J. L. (1978). "Association and Commissural Fiber Systems of the Olfactory Cortex of the Rat: I. Systems Originating in the Piriform Cortex and Adjacent Areas," *J. Comp. Neurol.* **178,** 711–740.

HABERLY, L. B., and SHEPHERD, G. M. (1973). "Current Density Analysis of Opossum Prepyriform Cortex," *J. Neurophysiol.* **36,** 789–802.

HASSELMO, M. E., and BOWER, J. M. (1990). "Afferent and Association Fiber Differences in Short-Term Potentiation in Piriform (Olfactory) Cortex," *J. Neurophysiol.* **64,** 179–190.

HASSELMO, M. E., and BOWER, J. M. (1991a). "Cholinergic Suppression of Synaptic

Transmission Is Specific to Intrinsic but Not Afferent Excitatory Fiber Pathways in the Piriform (Olfactory) Cortex of the Rat," *J. Neurophysiol.* (in press).

HASSELMO, M. E., and BOWER, J. M. (1991b). "Selective Suppression of Afferent but Not Intrinsic Fiber Synaptic Transmission by 2-amino-4-phosphonobutyric Acid (AP4) in Rat Piriform Cortex," *Brain Res.* **548**, 247–255.

HASSELMO, M. E., WILSON, M. A., ANDERSON, B., and BOWER, J. M. (1991a). "Associative Function in Piriform (Olfactory) Cortex: Computational Modeling and Neuropharmacology," in *The Brain, Cold Spring Harbor Symposium*. Cold Spring Harbor Press, Cold Spring Harbor, New York, 599–610.

HASSELMO, M. E., ANDERSON, B. P., and BOWER, J. M. (1991b). "Cholinergic Modulation Selective for Intrinsic Fiber Synapses May Enhance Associative Memory Properties of Piriform Cortex," in D. Touretzky (ed.), *Advances in Neural Information Processing Systems, Vol. 3*. Morgan Kaufmann, San Mateo, California, 46–52.

HEBB, D. O. (1949). *The Organization of Behavior*. John Wiley, New York.

HOLLEY, A., and DOVING, K. B. (1977). "Receptor Sensitivity, Acceptor Distribution, Convergence and Neural Coding in the Olfactory System," in J. Le Magnen and P. MacLeod (eds.), *Olfaction and Taste* (pp. 113–123). IRL Press, Oxford, U.K.

HOLMES, W. R., and WOODY, C. D. (1989). "Effects of Uniform and Non-Uniform Synaptic 'Activation Distributions' on the Cable Properties of Modeled Cortical Pyramidal Neurons," *Brain Res.* **505**, 12–22.

HOPFIELD, J. J. (1982). "Neural Networks and Physical Systems with Emergent Collective Computational Abilities," *Proc. Natl. Acad. Sci. USA*. **79**, 2554–2558.

INSAUSTI, R., AMARAL, D. G., and COWAN, W. M. (1987). "The Entorhinal Cortex of the Monkey: II. Cortical Afferents," *J. Comp. Neurol.* **264**, 356–395.

KANTER, E. D., and HABERLY, L. B. (1989). "APV Dependent Induction of Long Term Potentiation in Piriform (Olfactory) Cortex Slices," *Soc. Neurosci. Abst.* **15**, 929.

KAUER, J. S. (1991). "Contributions of Topography and Parallel Processing to Odor Coding in the Vertebrate Olfactory Pathway," *Trends in Neurosci.* **14**, 79–85.

KAUER, J. S., and MOULTON, D. G. (1974). "Responses of Olfactory Bulb Neurons to Odor Stimulation of Small Nasal Areas in the Salamander," *J. Physiol. (London)* **243**, 717–737.

KETCHUM, K. L., and HABERLY, L. B. (1988). "CSD Analysis of Oscillatory Responses in Rat Piriform Cortex Reveals Stereotyped Cyclical Components Mediated by Afferent and Intrinsic Association Fibers," *Soc. Neurosci. Abst.* **14**, 1188.

KOCH, C., and SEGEV, I. (eds.). (1989). *Methods in Neuronal Modeling: From Synapses to Networks*. MIT Press, Cambridge, Massachusetts.

KOHONEN, T. (1984). *Self-Organization and Associative Memory*. Springer-Verlag, Berlin.

LAING, D. G., CAIN, W. S., MCBRIDE, R. L., and ACHE, B. W. (1989). *Perception of Complex Smells and Tastes*. Academic Press, New York.

LANCET, D. (1986). "Vertebrate Olfactory Reception," *Ann. Rev. Neurosci.* **9**, 329–355.

LUSKIN, M. B., and PRICE, J. L. (1983). "The Topographic Organization of Associational Fibers of the Olfactory System in the Rat, including Centrifugal Fibers to the Olfactory Bulb," *J. Comp. Neurol.* **216**, 264–291.

McNAUGHTON, B. L., and MORRIS, R. G. (1987). "Hippocampal Synaptic Enhancement and Information Storage within a Distributed Memory System," *Trends Neurosci.* **10**, 408–415.

NELSON, M., FURMANSKI, W., and BOWER, J. M. (1989). "Simulating Neurons and Neuronal Networks on Parallel Computers," in C. Koch and I. Segev (eds.), *Methods in Neuronal Modeling: From Synapses to Networks* (pp. 397–438). MIT Press, Cambridge, Massachusetts.

NELSON, M. E., and BOWER, J. M. (1991). "Dynamics of Neural Excitability," in L. Nadel and D. Stein (eds.), *Lectures in Complex Systems, SFI Studies in the Sciences of Complexity, Lect. Vol. III.* Addison-Wesley, San Diego (in press).

PALM, G. (1980). "On Associative Memory," *Biol. Cybernetics* **36**, 19–31.

PRICE, J. L. (1985). "Beyond the Primary Olfactory Cortex: Olfactory Related Areas in the Neocortex, Thalamus and Hypothalamus," *Chem. Senses* **10**, 239–258.

RALL, W. (1989). "Cable Theory for Dendritic Neurons," in C. Koch and I. Segev (eds.), *Methods in Neuronal Modeling: From Synapses to Networks* (pp. 9–62). MIT Press, Cambridge, Massachusetts.

RODRIGUEZ, R., and HABERLY, L. B. (1989). "Analysis of Synaptic Events in the Opossum Piriform Cortex with Improved Current Source Density Techniques," *J. Neurophysiol.* **61**, 702–718.

SATO, M., MORI, K., TAZAWA, Y., and TAKAGI, S. F. (1982). "Two Types of Postsynaptic Inhibition in Pyriform Cortex of the Rabbit: Fast and Slow Inhibitory Postsynaptic Potentials," *J. Neurophysiol.* **48**, 1142–1156.

SCHWOB, J. E., and PRICE, J. L. (1978). "The Cortical Projection of the Olfactory Bulb: Development in Fetal and Neonatal Rats Correlated with Quantitative Variation in Adult Rats," *Brain Res.* **151**, 169–174.

SEGEV, I., FLESHMAN, J. W., and BURKE, R. E. (1989). "Compartmental Models of Complex Neurons," in C. Koch and I. Segev (eds.), *Methods in Neuronal Modeling: From Synapses to Networks* (pp. 63–96). MIT Press, Cambridge, Massachusetts.

SHEPHERD, G. (1979). *The Synaptic Organization of the Brain.* Oxford University Press, New York.

SQUIRE, L. R. (1986a). "Mechanisms of Memory," *Science* **232**, 1612–1619.

SQUIRE, L. R. (1986b). "Memory: Brain Systems and Behavior," *Trends Neurosci.* **11**, 170–175.

TSENG, G.-F., and HABERLY, L. B. (1987). "Characterization of Synaptically Mediated Fast and Slow Inhibitory Processes in Piriform Cortex in an *in vitro* Slice Preparation," *J. Neurophysiol.* **59**, 1352–1376.

VAN ESSEN, D. C., FELLEMEN, D. J., DEYOE, E. A., OLAVARRIA, J., and KNIERIM, J. (1991). "Modular and Hierarchical Organization of Extrastriate Visual Cortex in the Macaque Monkey," in *The Brain, Cold Spring Harbor Symposium.* Cold Spring Harbor Press, Cold Spring Harbor, New York, 679–696.

WILSON, M. A. (1991). "An Analysis of Olfactory Cortical Behavior and Function Using Computer Simulation Techniques." Doctoral Dissertation, California Institute of Technology, Pasadena, California.

WILSON, M., and BOWER, J. M. (1988). "A Computer Simulation of Olfactory Cortex with Functional Implications for Storage and Retrieval of Olfactory Information," in D. Anderson (ed.), *Neural Information Processing Systems* (pp. 114–126). American Institute of Physics, New York.

WILSON, M., and BOWER, J. M. (1989). "The Simulation of Large-Scale Neuronal Networks," in C. Koch and I. Segev (eds.), *Methods in Neuronal Modeling: From Synapses to Networks* (pp. 291–334). MIT Press, Cambridge, Massachusetts.

WILSON, M. A., and BOWER, J. M. (1990). "Computer Simulation of Oscillatory Behavior in Cerebral Cortical Networks," in D. Touretzky (ed.), *Advances in Neural Information Processing Systems, Vol. 2* (pp. 84–91). Morgan Kaufmann, San Mateo, California.

WILSON, M. A., and BOWER, J. M. (1991a). "A Computer Simulation of Oscillatory Behavior in Primary Visual Cerebral Cortex," *Neural Computation* **3** (in press).

WILSON, M. A., and BOWER, J. M. (1991b). "Simulating Cerebral Cortical Networks: Oscillations and temporal Interactions in a Computer Simulation of Piriform (Olfactory) Cortex," *J. Neurophysiol.* (in press).

WILSON, M., BOWER, J. M., and HABERLY, L. B. (1986). "Computer Simulation of Piriform Cortex," *Soc. Neurosci. Abs.* **12,** 1358.

WILSON, M., BHALLA, U., UHLEY, J., and BOWER, J. M. (1989). "GENESIS: A System for Simulating Neural Networks," in D. Touretzky (ed.), *Advances in Neural Network Information Processing Systems* (pp. 485–492). Morgan Kaufmann, San Mateo, California.

YAMADA, W. M., KOCH, C., and ADAMS, P. (1989). "Multiple Channels and Calcium Dynamics," in C. Koch and I. Segev (eds.), *Methods in Neuronal Modeling: From Synapses to Networks* (pp. 97–134). MIT Press, Cambridge, Massachusetts.

Chapter 17
Synchronized Multiple Bursts in the Hippocampus: A Neuronal Population Oscillation Uninterpretable without Accurate Cellular Membrane Kinetics

ROGER D. TRAUB

IBM Research Division
IBM T.J. Watson Research Center
Yorktown Heights, New York

RICHARD MILES

Institut Pasteur
Laboratoire de Neurobiologie Cellulaire
INSERM
Paris, France

I. Introduction

Oscillatory behaviors in populations of neurons are widespread in the vertebrate central nervous system (CNS). In some cases, the oscillation has a clear behavioral concomitant, somewhat analogous to the case of an invertebrate central pattern generator, although perhaps the underlying oscillating network is large and contains no identifiable neurons. An example of this would be the systems of brain stem neurons that regulate respiration (Feldman and Ellenberger, 1988). In other cases, population rhythms may be involved in sensory (Gray and Singer, 1989) or motor (Komisaruk, 1970) information processing. In still other cases, the behavioral significance of a striking population oscillation may be unclear.

An example of this is thalamic and reticular thalamic nuclear spindle waves, extensively reviewed by Steriade and Llinás (1988).

Cellular intrinsic properties (Llinás, 1988), together with intercellular interaction properties, will determine what types of population oscillations are expressed under what conditions. Sometimes neurons possess an intrinsic oscillatory capability at precisely the frequency at which a neuronal ensemble oscillates. This is true for certain entorhinal cortical neurons (Alonso and Llinás, 1989), inferior olivary neurons (Llinás and Yarom, 1981a,b, 1986), and thalamocortical relay cells (Jahnsen and Llinás, 1984a,b). Several, but not all (Alonso and Llinás, 1989), such cases involve cells with low-threshold calcium currents (Coulter *et al.*, 1989), a matter reviewed by Gutnick and Yarom (1989). In contrast, there are other examples where participation of cellular intrinsic properties in the population oscillation is more subtle. Two examples of this have been analyzed with computer models.

In the first example, the slightly disinhibited hippocampal slice exhibits a population oscillation at about 1 to 4 Hz, manifest as low-amplitude apical dendritic field potentials (Schneiderman, 1986) or as synchronized synaptic potentials (Schwartzkroin and Haglund, 1986; Miles and Wong, 1987a). This oscillation is interesting because few cells actually fire during any one population wave, and the firing of individual cells can be quite irregular. A very detailed circuitry model of this system indicates that long-duration afterhyperpolarizations (AHPs), slow $GABA_B$ IPSPs, and recurrent pyramidal cell excitation are all critical for the phenomenon (Traub *et al.*, 1989). The single-cell model used in this study contained four active currents and a number of dendritic compartments, but the current kinetics were not based on voltage-clamp data.

A second example concerns EEG oscillations at about 40 Hz, modulated in spindle-like fashion at lower frequencies, in piriform cortex. A model of this structure suggests that conduction delays of excitatory activity through long loops in the system, as well as IPSPs, contribute to the oscillation (Wilson and Bower, 1989). This circuitry model was able to exhibit several types of realistic population behaviors even though each model neuron contained a small number of compartments and no representation of the details of membrane current kinetics.

In this chapter, we shall describe still another type of observed neuronal population oscillation that appears to depend on the properties of long-duration synaptic interactions (specifically, the NMDA type of glutamate-mediated excitation), as well as on some rather intricate intrinsic membrane properties. The detailed kinetics of intracellular calcium regulation also appear to play a role. The relevant membrane conductance in this case does not appear to be a low-threshold calcium conductance.

II.   Synchronized Multiple Bursts (Afterdischarges) in Disinhibited Hippocampal Slices

An interesting type of behavior, *synchronized multiple bursts* (SMB), occurs in the CA3 region of hippocampal slices bathed in $GABA_A$-blockers such as pic-trotoxin (Miles *et al.*, 1984; Hablitz, 1984). While SMB may be pathological and probably may represent a seizure fragment, it is important to observe that they can be produced by repetitive stimulation of CA3 afferent fibers (Miles and Wong, 1987b); by analogy with long-term potentiation (LTP), therefore, SMB may be relevant to learning and memory: if not in their complete form, then in the partially expressed types of SMB that can be recorded after few tetanic stimulations. We shall, for the sake of simplicity, consider primarily the fully expressed SMB generated in picrotoxin.

SMB in picrotoxin occur spontaneously, with interevent intervals up to 10 or 12 seconds, or they can be evoked by local stimulation to a small group of cells or even to one cell (Miles *et al.*, 1984; Miles and Wong, 1983). At the cellular level, each SMB appears as a relatively long (on the order of 150 ms) initial or 1° burst, followed by one or more briefer 2° bursts (Fig. 1, bottom). The 2°

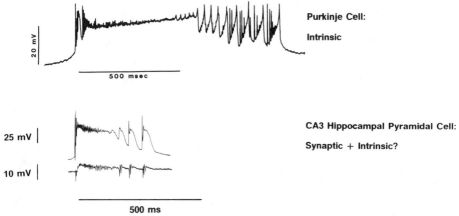

FIGURE 1.   A long initial burst followed by a series of shorter secondary bursts, in two cell types. (Top) Spontaneous event recorded intracellulary in a guinea pig Purkinje cell, in an *in vitro* slice bathed in standard medium. (Data kindly supplied by Dr. William N. Ross.) In view of the absence of recurrent excitation in this system, the event is almost certainly generated intrinsically. (Bottom) Synchronized multiple burst, guinea pig CA3 region in the *in vitro* slice bathed in picrotoxin (0.1 mM). Synchrony is indicated by concordance between the pyramidal cell intracellular record (above) and the local stratum pyramidale field potential (below). (R. Miles and R.K.S. Wong, unpublished data.)

bursts occur at 15 to 20 Hz (50 to 65 ms intervals). The whole event is usually followed by a prolonged afterhyperpolarization. Dual recordings indicate that all of the cells participate synchronously in the SMB, so that any one cell is tightly locked to the local field potential (Fig. 1).

When cells are held at potentials near 0 mV, by injecting current through a Cs^+-containing electrode, one can uncover the synaptic excitation that drives the event; this excitation appears to have both slow and fast components, with the fast component varying at the same frequency as do the 2° bursts (Miles *et al.*, 1984). Participation of excitatory synapses in the event is also indicated by the effects of the nonspecific glutamate-blocker, γ-D-glutamylglycine: The 2° bursts drop out one by one, in dose-dependent fashion, eventually leaving only a shortened initial burst.

A *strictly local* population of neurons is all that is required to generate SMB. This was established by microdissecting the slice with ophthalmic scissors. It was estimated that approximately 1,000 pyramidal cells were sufficient (Miles *et al.*, 1984).

III.   Considerations on the Mechanisms of SMB

There are several issues here that can be enumerated:

(1) Where does the primary burst come from? We believe that the primary burst can be explained in the same fashion as synchronized bursting that occurs in the presence of weaker $GABA_A$-blockers such as penicillin (Traub and Wong, 1982). The critical factors are these: CA3 pyramidal cells can generate intrinsic bursts (Kandel and Spencer, 1961; Wong and Prince, 1981). They are furthermore interconnected by recurrent excitatory synapses, enough so that each cell contacts about 20 or more other cells within 200 μ to each side (Miles and Wong, 1986, 1987a; Traub and Miles, 1991). Finally, the recurrent connections are powerful enough that bursting can be induced in one cell by bursting in a monosynaptically connected precursor, with probability on the order of 0.5 (Traub and Miles, 1991). Thus, even allowing for the stochastic nature of burst propagation, each cell should be able to evoke bursting in approximately 10 others. Normally, synaptic inhibition (the rapid or $GABA_A$ type) limits the spread of excitation, but in the presence of a $GABA_A$-blocker, a chain reaction of burst propagation can take place, leading to the synchronized firing of all of the cells. This event appears to be terminated by outward currents, both intrinsic and synaptic (i.e., $GABA_B$-elicited K currents).

(2) Could electric-field-induced synchronization explain SMB (Jefferys and Haas, 1982; Taylor and Dudek, 1984; Traub *et al.*, 1985)? The large population spikes associated with each 2° burst suggest that field effects are operative under these conditions, as they are in synchronizing action potentials during single picrotoxin-induced bursts (Snow and Dudek, 1984). Nevertheless, SMB require synaptic transmission to occur, whereas so-called field bursts occurring in the absence of synaptic transmission have a very different appearance than SMB.

(3) Could picrotoxin, in addition to blocking $GABA_A$ synapses, also alter the intrinsic properties of the cells? If one assesses the intrinsic properties by the response to currents injected at the soma, then no alterations of intrinsic properties have been defined (Hablitz, 1984).

(4) Under what conditions does the 1982 network model (Traub and Wong, 1982) generate SMB? The single-cell model used in that study was capable of generating realistic bursts, but not somatic or dendritic bursts at 15 Hz. The excitatory synaptic interactions were of short duration, corresponding approximately to the so-called quisqualate type of glutamate effect. In this case, extensive variation of the system parameters never led to SMB unless either of two very restrictive assumptions were made. In the first case, one can construct a special topology for the connections so that all of the cells lie on cycles of the same length, say, 5. Such a network can generate, by re-excitation, a series of network oscillations, but five different oscillation phases will be observed in the different cells. This does not agree with experiment. Furthermore, the localized bidirectional nature of the synaptic connections (Miles *et al.*, 1988), together with the fact that afterdischarges can propagate smoothly along CA3 in either direction (Knowles *et al.*, 1987), does not fit with this special topology. The other restrictive assumption is to postulate an activity-dependent block of axonal conduction with recovery time course of about 30 ms (Traub *et al.*, 1984). SMB can then occur provided the excitatory synapses are extremely powerful. While conduction block most likely occurs in CA3 axons—for example, when extracellular K^+ concentration is sufficiently high (Poolos *et al.*, 1987)—we are not familiar with a system that exhibits the correct recovery time course.

(5) Since the NMDA conductance is both long-lasting (Lester *et al.*, 1990; Forsythe and Westbrook, 1988) and voltage-dependent (Mayer *et al.*, 1984; Jahr and Stevens, 1990a), the 1° burst could activate NMDA receptors that then, in effect, add a new voltage-dependent current to CA3 pyramidal dendrites. This current could be the *intrinsic* oscillator underlying SMB. It is difficult to make hard-and-fast comments about this idea,

but our experience has been that the NMDA current by itself, without appropriate intrinsic currents, is not sufficient to account for SMB.

(6) Could GABA$_B$ IPSPs, which are not blocked by picrotoxin, account for SMB? The time course of the underlying GABA$_B$ conductance relaxes with a time constant of more than 100 ms (Hablitz and Thalmann, 1987), so that these IPSPs are probably too slow to explain the 50 to 65 ms period between 2° bursts.

(7) Is there a precedent for cellular intrinsic properties leading to potentials resembling SMB? Purkinje cells can produce strikingly similar potentials (Fig. 1, top). These cells are embedded in a network that, *in vitro,* contains no recurrent excitation onto the Purkinje cell. Therefore, the behavior of the cell is almost certainly a result of its own membrane conductances, including Na plateau and dendritic calcium conductances (Llinás and Sugimori, 1980a,b).

The problem with using cellular intrinsic properties to account for SMB in the hippocampus has been the difficulty in recognizing potentials in CA3 cells that resemble multiple bursts in Purkinje cells, unless inhibition has been blocked and the CA3 oscillation becomes collective. Nevertheless, recordings such as the top of Fig. 1 suggested that an accurate cell model might provide clues concerning the collective behavior. We proceeded, using voltage-clamp data from isolated hippocampal neurons (Kay and Wong, 1986) for five of the currents, and from bullfrog sympathetic neurons for the C-current, a voltage- and calcium-dependent current (Adams *et al.,* 1982). This model is described in Traub *et al.* (1990, 1991). Some of the critical elements of this model include the following:

(1) There is a high-voltage-activated calcium conductance whose activation kinetics are known (Kay and Wong, 1987), which is probably located mainly in the dendrites. This latter conclusion is based on the usual absence of calcium spikes in isolated hippocampal neurons (Wong *et al.,* 1986), on the recording of large calcium spikes in dendrites (Wong *et al.,* 1979), and on calcium imaging studies (Regehr *et al.,* 1989).

(2) The conductance of C-current channels parallels that of the calcium channels. This is a hypothesis of the model.

(3) The removal of submembrane calcium from the submembrane shell (from which it interacts with K-conductance channels) occurs in the dendrites with rapid kinetics, specifically with a time constant of less than 20 ms. This is also a hypothesis of the model, but one that seems possible in view of data from Purkinje cells (Miyakawa *et al.,* 1988).

This single-cell model gave realistic responses to somatic stimulation: the spike depolarizing afterpotential, intrinsic bursting, and the transition from repetitive bursting to higher frequency rhythmic action potentials as a steady depolarizing current was increased (Wong and Prince, 1981). This gives a certain confidence in the model. We could not, however, generate potentials similar to Fig. 1 with any paradigm of somatic stimulation that we tried. Nevertheless, we must note that the soma of hippocampal pyramidal cells is not believed to contain excitatory synapses, the latter occurring for the most part onto dendritic spines. The relevant form of stimulation must be into the dendrites. As Fig. 2 demonstrates, phasic excitation superimposed on a tonic background, when applied to both basilar and apical dendrites (shaded compartments), does lead to a train of dendritic calcium spikes and somatic bursts at 20 Hz. We emphasize that this behavior is a *prediction* of the model that remains to be verified experimentally.

The next issue concerns how the intrinsic properties of hippocampal neurons

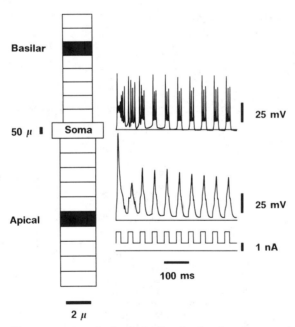

FIGURE 2.   Oscillatory response in dendritically stimulated compartmental CA3 pyramidal cell model. Electrotonic structure is shown on the left. The shaded compartments were excited with current pulses added to a steady background (lowest trace on right). In response, the middle dendrites produce a series of calcium spikes (middle trace) and the soma produces a series of bursts. Details of this model are given in Traub *et al.* (1990, 1991).

can be expressed as collective oscillations. Here we have postulated that recurrent synapses contain both quisqualate (QUIS) and NMDA components, a reasonable notion but one that needs to be established definitively. Because synchronized bursting can occur in the presence of APV (Dingledine *et al.*, 1986), we make the QUIS synapses powerful enough alone so that stimulation of one cell can lead to synchronized firing, i.e., to the 1° burst. We have also used a Boltzmann form for the voltage dependence of the NMDA-activated cation channel (C.E. Jahr and C.F. Stevens, personal communication, 1990b). Simulations have been run in networks of 100 or 1,000 pyramidal cells, sometimes (but not always) including a separate population of inhibitory neurons that mediates GABA$_B$ IPSPs. Figure 3 shows that a randomly connected network (in this case of 100 pyramidal cells) can generate SMB. The NMDA conductance, which we assume to saturate, produces a tonic dendritic excitation leading to expression of a series of dendritic calcium spikes. The latter, in turn, are correlated with somatic bursting whose output is delivered to other cells. The QUIS synaptic input in each cell samples the mean firing of the population. Note that this simulation accounts not only for oscillating cellular potential but also for the fast and slow components of the synaptic drive (Miles *et al.*, 1984). (Nevertheless, the 1° burst is too short and the underlying burst depolarizations may be too small.) It

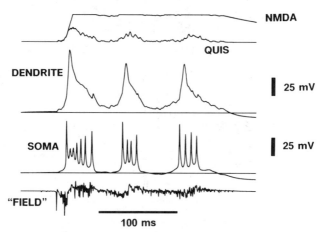

FIGURE 3.   Collective oscillation in a randomly connected network of 100 model pyramidal neurons. Each neuron has 20 synaptic inputs, and each input has a QUIS (fast, voltage-independent) and an NMDA (slow, voltage-dependent) component. Stimulus was to one cell. Synchrony develops through the rapid propagation of bursting. Mutual excitation then unmasks cellular oscillatory properties. The simulated field is a sum of compartmental transmembrane currents weighted by inverse distance. The simulation did not include field effects. (R.D. Traub and R. Miles, unpublished data, 1990. See also Traub *et al.*, 1991.)

will be interesting to see how this proposed mechanism for SMB accords with future experiments.

IV.   Hypotheses as to the Biological Significance of SMB

Because SMB require the participation of a population of neurons, and because they have not (to our knowledge) been observed except after inhibitory blockade or intense afferent stimulation, the question remains as to the functional importance of SMB. We would like to offer two possibilities. First, the calcium conductance in hippocampal neurons is endowed with special properties to enable these cells to burst. Bursting, in turn, is important as a special form of intercellular communication that allows a single cell to dominate the firing of a single other cell, and as a form of cellular interaction that may be particularly plastic; but one could hypothesize that it is only the *single* dendritic calcium spike that matters, and that a *series* of calcium spikes only occurs because of the rapid clearing of submembrane dendritic calcium ion. SMB would then be an epiphenomenon that appears in certain cell populations under special conditions. Second, one could speculate that (localized) SMB actually do occur *in vivo* in the normal brain, but are not recognized. SMB could be the biological substrate of déja vu and other experiences that can either appear as the aura of a complex partial seizure or that can occur in individuals without epilepsy. These intense subjective experiences are undoubtedly present for some reason connected with memory.

V.   Conclusion

We have described an interesting, albeit probably pathological, collective oscillation that has been observed in the disinhibited hippocampal slice: synchronized multiple bursts, or SMB. We have been able to produce a reasonable and self-consistent hypothesis for the generation of SMB only by first developing a detailed model of single CA3 pyramidal cells that incorporates, to the best of our ability, voltage-clamp data. Even so, the aspects of the model that involve calcium regulation and dendritic properties require considerable further experimental investigation. Because our SMB hypothesis involves the interaction between intrinsic cellular oscillatory properties and the different types of recurrent excitation (fast and voltage-independent versus slow and voltage-dependent), it was also essential to incorporate into the network model recent data on synaptic excitation. In particular, the detailed kinetics of quisqualate and NMDA types of synapses appear to be important.

References

ADAMS, P. R., CONSTANTI, A., BROWN, D. A., and CLARK, R. B. (1982). "Intracellular Ca²⁺ Activates a Fast Voltage-Sensitive K⁺ Current in Vertebrate Sympathetic Neurones," *Nature* **296,** 746–749.

ALONSO, A., and LLINÁS, R. R. (1989). "Subthreshold Na⁺-Dependent Theta-Like Rhythmicity in Stellate Cells of Entorhinal Cortex Layer II," *Nature* **342,** 175–177.

COULTER, D. A., HUGUENARD, J. R., and PRINCE, D. A. (1989). "Calcium Currents in Rat Thalamocortical Relay Neurones: Kinetic Properties of the Transient, Low-Threshold Current," *J. Physiol.* **414,** 587–604.

DINGLEDINE, R., HYNES, M. A., and KING, G. L. (1986). "Involvement of N-methyl-D-aspartate Receptors in Epileptiform Bursting in the Rat Hippocampal Slice," *J. Physiol.* **380,** 175–189.

FELDMAN, J. L., and ELLENBERGER, H. H. (1988). "Central Coordination of Respiratory and Cardiovascular Control In Mammals," *Ann. Rev. Physiol.* **50,** 593–606.

FORSYTHE, I. D., and WESTBROOK, G. L. (1988). "Slow Excitatory Postsynaptic Currents Mediated by N-methyl-D Asparate Receptors on Cultured Mouse Central Neurones," *J. Physiol.* **396,** 515–533.

GRAY, C. M., and SINGER, W. (1989). "Stimulus-Specific Neuronal Oscillations in Orientation Columns of Cat Visual Cortex," *Proc. Natl. Acad. Sci.* **86,** 1698–1702.

GUTNICK, M J., and YAROM, Y. (1989). "Low Threshold Calcium Spikes, Intrinsic Neuronal Oscillation and Rhythm Generation in the CNS," *J. Neurosci. Meth.* **28,** 93–99.

HABLITZ, J. J. (1984). "Picrotoxin-Induced Epileptiform Activity in the Hippocampus: Role of Endogenous versus Synaptic Factors," *J. Neurophysiol.* **51,** 1011–1027.

HABLITZ, J. J., and THALMANN, R. H. (1987). "Conductance Changes Underlying a Late Synaptic Hyperpolarization in Hippocampal CA3 Neurons," *J. Neurophysiol.* **58,** 160–179.

JAHNSEN, H., and LLINÁS, R. (1984a). "Electrophysiological Properties of Guinea-Pig Thalamic Neurones: An *in vitro* Study," *J. Physiol.* **349,** 205–226.

JAHNSEN, H., and LLINÁS, R. (1984b). "Ionic Basis for the Electroresponsiveness and Oscillatory Properties of Guinea-Pig Thalamic Neurones *in vitro*," *J. Physiol.* **349,** 227–247.

JAHR, C. E., and STEVENS, C. F. (1990a). "A Quantitive Description of NMDA Receptor-Channel Kinetic Behavior," *J. Neurosci.* **10,** 1830–1837.

JAHR, C. E., and STEVENS, C. F. (1990b). Personal communication.

JEFFERYS, J. G. R., and HAAS, H. L. (1982). "Synchronized Bursting of CA1 Hippocampal Pyramidal Cells in the Absence of Synaptic Transmission," *Nature* **300,** 448–450.

KANDEL, E. R., and SPENCER, W. A. (1961). "Electrophysiology of Hippocampal Neurons II: Afterpotentials and Repetitive Firing," *J. Neurophysiol.* **24,** 243–259.

KAY, A. R., and WONG, R. K. S. (1986). "Isolation of Neurons Suitable for Patch-

Clamping from Adult Mammalian Central Nervous Systems," *J. Neurosci. Meth.* **16,** 227–238.

KAY, A. R., and WONG, R. K. S. (1987). "Calcium Current Activation Kinetics in Isolated Pyramidal Neurones of the CA1 Region of the Mature Guinea-Pig Hippocampus," *J. Physiol.* **392,** 603–616.

KNOWLES, W. D., TRAUB, R. D., and STROWBRIDGE, B. W. (1987). "The Initiation and Spread of Epileptiform Bursts in the *in vitro* Hippocampal Slice," *Neuroscience* **21,** 441–455.

KOMISARUK, B. R. (1970). "Synchrony between Limbic System Theta Activity and Rhythmical Behavior in Rats," *J. Compar. Physiol. Psychol.* **70,** 482–492.

LESTER, R. A. J., CLEMENTS, J. D., WESTBROOK, G. L., and JAHR, C. E. (1990). "Channel Kinetics Determine the Time Course of NMDA Receptor-Mediated Synaptic Currents," *Nature* **346,** 565–567.

LLINÁS, R. (1988). "The Intrinsic Electrophysiological Properties of Mammalian Neurons: Insights into Central Nervous System Function," *Science* **242,** 1654–1664.

LLINÁS, R., and SUGIMORI, M. (1980a). "Electrophysiological Properties of *in vitro* Purkinje Cell Somata in Mammalian Cerebellar Slices," *J. Physiol.* **305,** 171–195.

LLINÁS, R., and SUGIMORI, M. (1980b). "Electrophysiological Properties of *in vitro* Purkinje Cell Dendrites in Mammalian Cerebellar Slices," *J. Physiol.* **305,** 197–213.

LLINÁS, R., and YAROM, Y. (1981a). "Electrophysiology of Mammalian Inferior Olivary Neurones *in vitro:* Different Types of Voltage-Dependent Ionic Conductances," *J. Physiol.* **315,** 569–584.

LLINÁS, R., and YAROM, Y. (1981b). "Properties and Distribution of Ionic Conductances Generating Electroresponsiveness of Mammalian Inferior Olivary Neurones *in vitro,*" *J. Physiol.* **315,** 569–584.

LLINÁS, R., and YAROM, Y. (1986). "Oscillatory Properties of Guinea-Pig Inferior Olivary Neurones and Their Pharmacological Modulation: An *in vitro* Study," *J. Physiol.* **376,** 163–182.

MAYER, M. L., WESTBROOK, G. L., and GUTHRIE, P. B. (1984). "Voltage-Dependent Block by Mg^{2+} of NMDA Responses in Spinal Cord Neurones," *Nature* **309,** 261–263.

MILES, R., TRAUB, R. D., and WONG, R. K. S. (1988). "Spread of Synchronous Firing in Longitudinal Slices from the CA3 Region of the Hippocampus," *J. Neurophysiol.* **60,** 1481–1496.

MILES, R., and WONG, R. K. S. (1983). "Single Neurones Can Initiate Synchronized Population Discharge in the Hippocampus," *Nature* **306,** 371–373.

MILES, R., and WONG, R. K. S. (1985). Unpublished data.

MILES, R., and WONG, R. K. S. (1986). "Excitatory Synaptic Interactions between CA3 Neurones in the Guinea-Pig Hippocampus," *J. Physiol.* **373,** 397–418.

MILES, R., and WONG, R. K. S. (1987a). "Inhibitory Control of Local Excitatory Circuits in the Guinea-Pig Hippocampus," *J. Physiol.* **338,** 611–629.

MILES, R., and WONG, R. K. S. (1987b). "Latent Synaptic Pathways Revealed after Tetanic Stimulation in the Hippocampus," *Nature* **329,** 724–726.

MILES, R., WONG, R. K. S., and TRAUB, R. D. (1984). "Synchronized Afterdischarges in the Hippocampus: Contribution of Local Synaptic Interactions," *Neuroscience* **12,** 1179–1189.

MIYAKAWA, H., LEV-RAM, V., and ROSS, W. N. (1988). "Synaptically Evoked Calcium Transients in the Dendrites of Cerebellar Purkinje Cells *in vitro*," *Soc. Neurosci. Abstr.* **14,** 759.

POOLOS, N. P., MAUK, M. D., and KOCSIS, J. D. (1987). "Activity-Evoked Increases in Extracellular Potassium Modulate Presynaptic Excitability in the CA1 Region of the Hippocampus," *J. Neurophysiol.* **58,** 404–416.

REGEHR, W. G., CONNOR, J. A., and TANK, D. W. (1989). "Optical Imaging of Calcium Accumulation in Hippocampal Pyramidal Cells during Synaptic Activation," *Nature* **341,** 533–536.

SCHNEIDERMAN, J. H. (1986). "Low Concentrations of Penicillin Reveal Rhythmic, Synchronous Synaptic Potentials in Hippocampal Slice," *Brain Res.* **398,** 231–241.

SCHWARTZKROIN, P. A., and HAGLUND, M. M. (1986). "Spontaneous Rhythmic Synchronous Activity in Epileptic Human and Normal Monkey Temporal Lobe," *Epilepsia* **27,** 523–533.

SNOW, R. W., and DUDEK, F. E. (1984). "Electrical Fields Directly Contribute to Action Potential Synchronization during Convulsant-Induced Epileptiform Bursts," *Brain Res.* **323,** 114–118.

STERIADE, M., and LLINÁS, R. R. (1988). "The Functional States of the Thalamus and the Associated Neuronal Interplay," *Physiol. Rev.***68,** 649–742.

TAYLOR, C. P., and DUDEK, F. E. (1984). "Excitation of Hippocampal Pyramidal Cells by an Electrical Field Effect," *J. Neurophysiol.* **52,** 126–142.

TRAUB, R. D., DUDEK, F. E., TAYLOR, C. P., and KNOWLES, W. D. (1985). "Simulation of Hippocampal Afterdischarges Synchronized by Electrical Interactions," *Neuroscience* **14,** 1033–1038.

TRAUB, R. D., KNOWLES, W. D., MILES, R., and WONG, R. K. S. (1984). "Synchronized Afterdischarges in the Hippocampus: Simulation Studies of the cellular Mechanism," *Neuroscience* **12,** 1191–1200.

TRAUB, R. D., and MILES R. (1990). Unpublished data.

TRAUB, R. D., and MILES, R. (1991). "*Neuronal Networks of the Hippocampus*. Cambridge University Press, New York.

TRAUB, R. D., MILES, R., and WONG, R. K. S. (1989). "Model of the Origin of Rhythmic Population Oscillations in the Hippocampal Slice," *Science* **243,** 1319–1325.

TRAUB, R. D., and WONG, R. K. S. (1982). "Cellular Mechanism of Neuronal Synchronization in Epilepsy," *Science* **216** 745–747.

TRAUB, R. D., WONG, R. K. S., and MILES, R. (1990). "A Model of the CA3 Hippocampal Pyramidal Cell Based on Voltage-Clamp Data," *Abstr. Soc. Neurosci.* **16,** 1297.

TRAUB, R. D., WONG, R. K. S., MILES, R. and MICHELSON, H. (1991). "A Model of a CA3 Hippocampal Pyramidal Neuron Incorporating Voltage-Clamp Data on Intrinsic Conductances," *J. Neurophysiol.* **66,** 635–650.

WILSON, M. A., and BOWER, J. M. (1989). "The Simulation of Large-Scale Neural

Networks," in C. Koch and I. Segev (eds.), *Methods in Neuronal Modeling* (pp. 291–333). MIT Press, Cambridge, Massachusetts.

WONG, R. K. S., and PRINCE, D. A. (1981). "Afterpotential Generation in Hippocampal Pyramidal Cells," *J. Neurophysiol.* **45,** 86–97.

WONG, P. K. S., PRINCE, D. A., and BASBAUM, A. I. (1979). "Intradendritic Recordings from Hippocampal Neurons," *Proc. Nat. Acad. Sci.* **76,** 986–990.

WONG, R. K. S., TRAUB, R. D. and MILES, R. (1986). "Cellular Basis of Neuronal Synchrony in Epilepsy," in A. V. Delgado-Escueta, A. A., Ward, Jr., D. M. Woodbury, and R. J. Porter (eds.), *Basic Mechanisms of the Epilepsies. Molecular and Cellular Approaches. Advances in Neurology, Vol. 44* (pp. 583–592). Raven Press, New York.

PART IV

MULTISTATE NEURONS AND
STOCHASTIC MODELS OF
NEURON DYNAMICS

This section consists of formal models of neurons based on stochastic approximations, physical system analogies, and nonlinear dynamical systems.

Tam (Chapter 18) applies discrete function (point process) approximations to neural spike trains for neurons with multiple thresholds. He reviews the evidence for neurons with two thresholds for action potential generation and points out that the low-threshold spikes can be considered as reflections of delayed excitability following a prolonged hyperpolarization. Discrete models of multi-threshold delayed activation are developed and discussed in relation to delayed Hebbian synaptic modification, classical conditioning, neuron computation of cross-correlation functions, signal multiplexing and demultiplexing, and the generation of topographic maps coding temporal signal properties.

Bulsara, Schieve, and Moss (Chapter 19) develop a model neuron based on the Hopfield model, exploring analogies to physical systems. They show that multiplicative noise in such neurons can introduce new forms of critical behavior into the dynamics of the neuron, and produce multiple stable states. This phenomenon should reveal itself in the distribution of multimodal peaks in the interspike interval histograms of neurons. They consider the case of a noisy neuron with a sinusoidal input that can extract signals from noise by use of the phenomenon of stochastic resonance.

Selz and Mandell (Chapter 20) emphasize the temporal complexity evident in neural systems and present formal analyses of multiple-time-scale phenomena observed in the discharge patterns of single neurons and in neuronal population dynamics. They provide a brief history of physical models of action potential generation, which leads to the view that the strong nonlinearities on neurons can produce strange attractor phenomena. Selz and Mandell go on to examine interspike interval data from brain stem neurons for evidence of bursting patterns, and present non-statistical measures for describing the behavior of neurons over multiple time scales. They hypothesize that a range of temporal complexity is required for normal functioning in biological systems, as measured in entropy per unit time.

Smith (Chapter 21) reviews a range of stochastic approximation models of neurons. These models emphasize the time and membrane potential variables but not the neuron geometry. The subthreshold membrane potential is considered as a continuous-time random process with either a continuous-state space or first-order discontinuities. The time for the membrane potential to reach action potential threshold is then considered as the first passage time (FPT). The value of such models comes from realization that the FPT distribution is equivalent in some sense (with caveats) to the interspike interval distribution of neuron firings. For many neurons in the CNS, we have this type of data available under a range of conditions. Since stochastic models can generate a variety of FTP distributions which match the observed ISI's of real neurons, this provides some basis for inferring neural mechanisms underlying different discharge patterns in the absence of direct intracellular recording. Smith considers four models: the Ornstein–Uhlenbeck process (Wiener process with restoring force), the Stein model (Poisson train inputs), Stein model with synaptic reversal potentials, and a stochastic after-hyperpolarization model. He presents the ranges of behavior of the FTP for these models, and discusses how plotting the relations among the moments of interspike intervals can be diagnostic for determining the nature of the stochastic activity of a neuronal spike train.

Teich (Chapter 22) presents evidence that the discharge pattern of auditory nerve axons exhibits long-term correlations that are fractal in nature; that is, the spike rate fluctuations are self-similar over a large range of integration times. He considers the biophysical origins and functional significance of this phenomenon. (Since many naturally occurring sounds are fractal patterns, this may be an effective means of sampling these sounds.) Teich goes on to identify the mathematical point processes underlying the observed spike train fluctuations.

Chapter 18

Signal Processing in Multi-Threshold Neurons

DAVID C. TAM

Division of Neuroscience
Baylor College of Medicine
Houston, Texas

I. Introduction

The signal processing capabilities of biological neurons can be investigated by considering the input/output (I/O) relationship between the electrical signals received by the dendrites and/or soma of a neuron and the electrical signals generated by the axon of the neuron. The electrical signals are represented by the voltage difference between the inside and outside of the membrane. The direction of signal flow can generally be considered to be from the point of generation of the postsynaptic membrane potential at the dendrites and/or soma to the generation of the action potential at the axon initial segment. Once the electrical signal (transmitted as action potentials) arrives at the axon terminal, the signal is converted into a chemical signal by the release of neurotransmitters (in a chemical synapse), where this signal will be converted back into an electrical signal in the postsynaptic neuron in the form of a postsynaptic potential. The conversion of a chemical signal to an electrical signal is done by the ligand-gated receptor channels in the postsynaptic membrane. Thus, signals are relayed from neuron to neuron, where intraneuron signals are propagated electrically and interneuron signals are propagated chemically. This is a greatly simplified view of signal processing in neurons, since chemical signal processing does occur within a neuron, such as ionic signaling and second-messenger signaling.

Single Neuron Computation
Neural Nets: Foundations to Applications

481

The primary signal can be considered as electrical, however, since the major role of a neuron is to process the postsynaptic potential to produce, or not to produce, a threshold event represented by an action potential.

II.   Representation of Neuronal Signals

The input and output signals of a neuron can be represented by the voltage signal across the membrane. The signal processing capabilities of a neuron can, therefore, be studied by considering the processing of the electrical signals alone. In other words, we may consider the signal to be processed by a neuron to be entirely represented by the electrical signal. This also implies that signals are encoded as well as decoded by neurons. Signals are encoded through the generation of threshold events consisting of individual action potentials. Signals are decoded through the processing of the postsynaptic potentials. Thus, the signal processing performed by a neuron consists of decoding the signal from the input (presynaptic) neurons and encoding the output signal to be transmitted to the next (postsynaptic) neurons down the line. The signal content encoded by a presynaptic neuron (action potentials) will then be decoded by the succeeding presynaptical neurons through the processing of the resulting postsynaptic potentials (PSPs).

Although the voltage signal can be represented by a continuous function, it can also be represented by a discrete function as an approximation to elucidate the signal processing capabilities of a neuron. Since the thresholding mechanism of action potential generation produces an all-or-none event, represented by a spike, the PSP can, therefore, be represented by a discrete finite event pulse, with an amplitude given by some multiple of the minimum discrete voltage step. The output of a neuron is a sequence of action potentials, which, in turn, can be considered as a time series of event pulses. Such a time series is also called a *spike train*. A spike train is essentially a point process, where discrete events occur in continuous time (Perkel *et al.* 1967a,b). Such a spike train can be represented by a function $s(t)$ consisting of a series of Dirac delta functions (δ functions) representing the occurrence of spikes at time τ_j:

$$s(t) = \sum_{j=1}^{N} \delta(t - \tau_j), \tag{1}$$

where N is the total number of spikes in this spike train. An explicit representation of the Dirac delta function is given by:

$$\delta(t) = \lim_{\Delta t \to 0} \begin{cases} 1/\Delta t, & -\Delta t/2 \leqslant t \leqslant \Delta t/2 \\ 0, & \text{otherwise}, \end{cases} \tag{2}$$

which satisfies the condition:

$$\int_{-\infty}^{\infty} \delta(t) \, dt = 1. \tag{3}$$

We may further discretize the time variable t, which is represented by $t_i = i\Delta t$, where i is an integer. Representation of the signal by a discrete function instead of a continuous function also allows for a more convenient mathematical analysis of the signal, since integration can be replaced by summation and differential equations replaced by difference equations. Furthermore, discrete functions are more efficient computationally when implemented on digital computer than are continuous functions. The discretization of time is equivalent to synchronizing the processing over a small time increment Δt. The continuous representation is then recovered by letting $\Delta t \to 0$. Thus, a discrete spike train can be represented by:

$$s(t_i) = \sum_{j=1}^{N} \delta'(t_i - \tau_j), \tag{4}$$

where

$$\delta'(t) = \begin{cases} 1, & 0 \leqslant t < \Delta t \\ 0, & \text{otherwise} \end{cases}. \tag{5}$$

The effective pulse duration of the spike is Δt in this case, although the pulse duration can last longer than the time quantization variable Δt, in general. Using this formulation, the signal encoded by a neuron can be represented by the temporal sequence of spike events. Although the notation of t_i may be needed to represent the discretization of the time variable, we will use the plain notation t without subscript for the discrete time variable, where $t = t_i = i\Delta t$ is assumed, to avoid confusion with other indexing subscript notations in the following sections.

III. Spike Codes in Neurons

Since spikes are discrete events of constant amplitude and duration, the important variable is the time between occurrences of spikes in the spike train representing the temporal codes. The pulse height and pulse width of an action potential may vary by small amounts in biological neurons, which, in turn, may affect the signal content carried by an action potential. For instance, spike-width broadening may increase the amount of neurotransmitter release, resulting in an increase to the quantal signal received by the postsynaptic neurons (Tam and Perkel, 1989a). However, we will assume that all spikes are indistinguishable from one another as a simplification of the model. One of the temporal variables representing the signal in the spike train is the interspike interval, defined as the time interval between successive spikes. The first-order interspike interval is defined as the time interval between consecutive spikes τ_j and τ_{j-1}:

$$I'_j = \tau_j - \tau_{j-1}. \tag{6}$$

The second-order interspike interval is defined as the time interval between any two spikes with one intervening spike in between the two:

$$I''_j = \tau_j - \tau_{j-2}. \tag{7}$$

The sum of two consecutive first-order interspike intervals is, therefore, a second-order interval:

$$I''_j = I'_j + I'_{j-1}. \tag{8}$$

Similarly, the nth-order interspike interval is given by:

$$I^n_j = \tau_j - \tau_{j-n}, \tag{9}$$

$$= \sum_{k=j-n+1}^{j} I'_k. \tag{10}$$

Thus, the signal encoded in the spike train can be represented by these interspike interval measures, since the time intervals are relative time measures independent of the absolute time of occurrence of the spikes. The higher-order interspike intervals capture the firing history of the neuron, such as bursting patterns. Thus, the codes generated by a neuron can be represented by the interspike interval parameters.

IV.   Multiple Thresholds in Neurons

A non-pacemaking neuron usually stays at a membrane potential called *resting potential* when it is not stimulated. Most neurons fire action potentials when the membrane potential is depolarized to a threshold that is above the resting potential. The firing threshold of a neuron is usually thought of as a constant, but, in fact, changes depending on the firing history of the neuron. For some neurons, there are two separate thresholds that can trigger the generation of action potentials. The thresholds occur not only at a depolarized membrane potential (above the resting potential), but also at a hyperpolarized potential (below the resting potential). The second threshold appears when the membrane potential is held at a hyperpolarized potential for a prolonged period of time. The bi-threshold phenomenon in biological neurons has been reported in a number of experimental preparations, including, for example, the giant squid axons (Hodgkin and Huxley, 1952), thalamic neurons (Jahnsen and Llinás, 1984a,b; Mc-Cormick et al., Chapter 10 in this volume); McCormick and Feeser, 1990; Deschênes et al., 1984; Steriade and Deschênes, 1984), inferior olivary neurons (Yarom and Llinás, 1987), and hippocampal neurons (Stasheff and Wilson, 1990). The phenomenon of triggering the firing of action potentials at a membrane potential below the resting potential level following prolonged hyperpolarization has been observed under different conditions in different neurons, such as during the anodal break after a prolonged voltage-clamped hyperpolarization (Hodgkin and Huxley, 1952). They are called *low-threshold spikes* by Yarom and Llinás (1987) and *baseline spikes* by Stasheff and Wilson (1990), who reported that these spikes can be elicited naturally during the after-hyperpolarization period (a.h.p.). This phenomenon is also exhibited as the post-inhibitory rebound observed in various central neurons (e.g., Steriade and Deschênes, 1984) and spinal neurons. Thus, the existence of multiple thresholds in neurons is well-documented.

The generation of low-threshold spikes is a voltage- and time-dependent process occurring after a prolonged hyperpolarization that de-inactivates the ionic conductances. Such a bi-threshold phenomenon has been studied and modeled by a number of investigators using various mathematical models. They include, for example, Hodgkin and Huxley's model (1952), FitzHugh–Nagumo's model (FitzHugh, 1960; Nagumo et al., 1962), Rose and Hindmarsh's model (1985), and Goldbeter and Moran's model (1988). These are continuous variable models using sets of differential equations to describe the bi-threshold phenomena in terms of the dynamics of the underlying ionic conductances.

Alternatively, such a bi-threshold characteristic can be described by a discrete model rather than a continuous model. If the output of a neuron at time t is denoted by $y(t)$, where $y(t) = 1$ denotes the presence of a spike and $y(t) = 0$ the absence of a spike, and the membrane potential of the neuron is represented by $V(t)$, then the firing condition of a single threshold neuron is given by:

$$y(t) = \begin{cases} 1, & \text{if } V(t) > \theta_{\text{high}} \\ 0, & \text{otherwise,} \end{cases} \tag{11}$$

where

$$\theta_{high} > V_{resting}. \tag{12}$$

θ_{high} is the high threshold and $V_{resting}$ denotes the resting potential. The firing condition of a bi-threshold neuron is given by:

$$y(t) = \begin{cases} 1, & \text{if } V(t) > \theta_{high} \qquad\qquad\qquad \text{or} \\ & \text{if } V(t - j\Delta t) \le \theta_{low} \text{ and } V(t) > \theta_{low}, \forall\, 0 < j \le h, \\ 0, & \text{otherwise} \end{cases} \tag{13}$$

where

$$\theta_{low} \le V_{resting}. \tag{14}$$

θ_{low} denotes the low threshold, and $h\Delta t$ represents the minimum duration of hyperpolarization such that the neuron will fire when depolarized from this hyperpolarization potential, and j is an integer. Alternatively, Eq. (13) can be rewritten in the form similar to Eq. (11), for which the threshold is a function of voltage and time, $\theta = \theta(V,t)$:

$$y(t) = \begin{cases} 1, & \text{if } V(t) > \theta(V,t) \\ 0, & \text{otherwise} \end{cases}, \tag{15}$$

where

$$\theta(V,t) = \begin{cases} \theta_{low} & \text{if } V(t - j\Delta t) \le \theta_{low}, \forall j; 0 < j \le h \\ \theta_{high}, & \text{otherwise} \end{cases}, \tag{16}$$

and j is an integer. The voltage- and time dependence of the threshold is captured by Eq. (16) such that the threshold is at θ_{high} unless the membrane potential is held below θ_{low} for a duration of $h\Delta t$ or longer.

The preceding equations do not take into account the refractoriness of firing in neurons. To ensure that the neuron will fire a spike with a constant pulse width of Δt, an *absolute refractory period* may be imposed by the following condition:

$$y(t) = 1$$

$$\Rightarrow y(t + j\Delta t) = 0, \qquad \forall j; 0 < j \le a,$$

where $a\Delta t$ is the absolute refractory period, and j is an integer. The width of the pulse may be multiples of Δt, such as $d\Delta t$ (for $d > 0$), but for simplicity in the derivation, pulse width of Δt will be used. The absolute refractory period is the time period during which a neuron will not generate a spike regardless of the voltage potential. The *relative refractory period* is the time period during which a neuron will have a higher threshold than normal. In this case, the high threshold is no longer a constant, but a function of time, $\theta_{high}(t)$. The relative refractory period may be approximated by the following condition:

$$y(t) = 1$$

$$\Rightarrow \theta_{high}(t + j\Delta t) = \theta_{high} + c\, e^{-b\Delta t}, \qquad \forall j;\ 0 < j \leqslant b,$$

where c is a positive constant, b is the time constant of the relative refractory period, and j is an integer.

To summarize the preceding, the spike-firing threshold in neurons is not a constant but a dynamical variable that can be expressed as a function of both time and membrane voltage. The threshold during the absolute and relative refractory period is a function of time; the threshold for a bi-threshold neuron is a function of voltage and time.

V.   Functional Significance of Multi-Threshold Neurons

Given this bi-threshold for triggering the firing of spikes, a neuron can function in two modes of operations: one at depolarization potential and the other at hyperpolarization potential. Thus, when the neuron is depolarized from the resting potential, the neuron will process signals based on the high threshold, and when the neuron is hyperpolarized for a prolonged duration, the neuron will process signals based on the low-threshold.

A.   Delayed Excitability

When a single threshold neuron is depolarized, it becomes more excitable, since the membrane potential is closer to the high threshold (θ_{high}). When the same neuron is hyperpolarized, it becomes less excitable, since the membrane potential is further from the high threshold (θ_{high}); but when a bi-threshold neuron is hyperpolarized for a sufficiently prolonged period of time, it then becomes more excitable again, since the low threshold becomes activated. More importantly, the membrane potential does not need to be depolarized to the resting level before triggering an action potential.

Thus, a bi-threshold neuron becomes more excitable after a delayed period after the hyperpolarization. The period of delayed excitability is given by $h\Delta t$, the minimum duration of hyperpolarization before a spike can be activated by crossing the low threshold, θ_{low}. This delayed excitability may be used for suppressing firing activity for a period of time and allowing spikes to go through only after this delayed period is over. Some of the signal processing consequences due to the delayed excitability will be addressed next.

B. Delayed Hebbian Synaptic Modification

Hebbian synapses are usually thought to be changed when there is simultaneous activation of presynaptic and postsynaptic neurons (Hebb, 1949); but non-simultaneous (or delayed) activation of presynaptic and postsynaptic neurons can be achieved when the high and low thresholds for firing in neurons are considered.

The rule for conventional Hebbian synaptic weight change can be described as follows. Let $w(t)$ be the synaptic weight, $\Delta w(t)$ the synaptic weight change, $x(t)$ the input to a neuron, and $y(t)$ the output from the neuron; the simultaneous activation Hebbian synaptic weight is modified by the following relationship:

$$w(t + \Delta t) = w(t) + l \, \Delta w(t), \tag{17}$$

where

$$\Delta w(t) = x(t) \, y(t), \tag{18}$$

and l is the learning rate constant. For a delayed Hebbian synaptic modification rule (Tam and Perkel, 1989b; Tam and McMullen, 1989), Eq. (18) becomes:

$$\Delta w(t,\tau) = x(t - \tau) \, y(t) \tag{19}$$

when the input signal is delayed by time τ. The Hebbian synapse can be generalized to include another input, $z(t)$, called the reinforcement signal (Barto and Sutton, 1981; Barto et al., 1981; Sutton and Barto, 1981; Tam, 1989). In such a case, the delayed and reinforced Hebbian synaptic weight change rule becomes:

$$\Delta w(t,\tau) = x(t - \tau) \, y(t) \, z(t). \tag{20}$$

The delay in the delayed input $x(t - \tau)$ can be accomplished by various means, such as conduction delay, synaptic transmission delay in multiple synapses, delay loops, and via a delay-line architecture in artificial neural networks (Tam and Perkel, 1989b). Conduction delays in biological neurons can only account for short delays, so other mechanisms may be needed to account for longer delays.

The delayed excitation in bi-threshold neurons described before may account for delaying mechanisms of longer duration. In such cases, when the neuron receives a hyperpolarization input $x(t - h\Delta t)$ at time $t - h\Delta t$, there is a delay period of $h\Delta t$ before the neuron is still hyperpolarized. If after this delay period, at time t, another input from the reinforcement $z(t)$ activates the neuron to cross the low threshold (θ_{low}), it causes the neuron to fire a spike at the output $y(t)$. This satisfies the preceding Hebbian condition in Eq. (20), which allows modification of the synaptic weight under this delayed activation condition. The delayed excitable reinforcement Hebbian learning rule becomes:

$$\Delta w(t, h\Delta t) = x(t - h\Delta t) \, y(t) \, z(t) \tag{21}$$

for a bi-threshold neuron without the need of an external delay-line circuit for the input producing the delayed Hebbian learning. The time delay of the input signal is produced internally by the delayed excitation of the neuron. This implicit delay is elegantly implemented without requiring any delay circuitry, and it is self-contained within a single neuron. This also addresses some of the computational capability of a single neuron when signal processing can be done implicitly via the multi-threshold delayed activation phenomena.

C. Classical Conditioning

When the classical conditioning paradigm is viewed at the level of neuronal input and output analogous to an organism's input and output, the neuronal mechanisms for producing the classical conditioning effects may be explained—providing there is a delayed Hebbian synaptic mechanism. In classical conditioning, an unconditioned stimulus (US) always produces an unconditioned response (UR). A conditioned stimulus (CS) may not produce a response initially, but after paired training with the US, the CS will produce a conditioned response (CR).

Consider the case in which a reinforcement input signal $z(t)$ is coupled with the output $y(t)$ such that a spike in the input $z(t)$ always produces a spike in the output $y(t)$; the probability of producing an action potential in the postsynaptic neuron at that synaptic connection is one. When such a causal relationship exists between the input $z(t)$ and the output $y(t)$, $z(t)$ can be considered as the US, $y(t)$ the UR, and $x(t)$ the CS to be paired with the US signal $z(t)$ in a classical conditioning paradigm. Since the US input $z(t)$ always produces an UR output $y(t)$, the association between the presynaptic CS input $x(t)$ and the postsynaptic CR output $y(t)$ is necessary for Hebbian weight change. This can alternatively be considered as the association between the CS and US ($x(t)$ and $z(t)$) instead.

The ordered temporal pairing of CS–US is necessary for classical conditioning to occur. The CS signal needs to precede the US signal for effective learning to occur. In the neuronal analog, the input $x(t)$ needs to precede the reinforcement input $z(t)$ by a time τ. In other words, the signals $x(t - \tau)$ and $z(t)$ would arrive simultaneously at the neuron. When all the signals, $x(t - \tau)$, $z(t)$, and $y(t)$, contain a spike, the delayed reinforced Hebbian condition of Eq. (20) is satisfied. Thus, classical conditioning can be explained by the neuronal equivalent of the Hebbian synapses with a delayed and a reinforcement signal.

Under this classical conditioning paradigm, synaptic weight changes occur *only* in the presence of the US spike $z(t)$ that produces the firing of $y(t)$, but not in the presence of the CS spike $x(t)$ *alone,* which may coincide with the firing of $y(t)$ in the postsynaptic neuron. Thus, the weight change rule of Eq. (20) can be reduced to:

$$\Delta w(t,\tau) = x(t - \tau) \, z(t). \tag{22}$$

Again, since the delay between the occurrence of the US and CS signals is much longer than a millisecond, the delaying mechanism may rely on the delayed excitability in bi-threshold neurons as discussed in Section B:

$$\Delta w(t,h\Delta t) = x(t - h\Delta t) \, z(t). \tag{23}$$

D.   Computing Cross-Correlation Function

It will be shown that the cross-correlation function can be computed by a neuron by storing the values of the function as the synaptic weights of the delayed Hebbian synapses (Tam and McMullen, 1989). Let us consider the accumulated Hebbian synaptic weight $w(t,\tau)$, which has gone through n discrete training steps (of Δt) with a learning rate constant $l = 1$:

$$w(n\Delta t,\tau) = w(0,\tau) + \sum_{j=1}^{n} \Delta w(j\Delta t,\tau), \tag{24}$$

where $w(0,\tau)$ is the initial synaptic weight before training and $w(n\Delta t,\tau)$ the final weight after time $n\Delta t$. For a delayed Hebbian synapse, the final synaptic weight at time $n\Delta t$ in Eq. (24) becomes

$$w(n\Delta t,\tau) = w(0,\tau) + \sum_{j=1}^{n} x(j\Delta t - \tau) \, z(j\Delta t), \tag{25}$$

using Eq. (22). Now consider the cross-correlation function $r_{xy}(\tau)$ between inputs $x(t)$ and $z(t)$, and assuming $x(t)$ and $z(t)$ are stationary and ergodic:

$$r_{xy}(\tau) = \lim_{T \to \infty} \frac{1}{T} \int_{0}^{T} x(t - \tau) \, z(t) \, dt. \tag{26}$$

Discretizing the time step, Δt, it becomes

$$r_{xy}(\tau) = \lim_{\Delta t \to 0} \lim_{n \to \infty} \frac{1}{n\Delta t} \sum_{j=1}^{n} x(j\Delta t - \tau)\, z(j\Delta t)\, \Delta t. \tag{27}$$

The cross-correlation function then can be expressed in terms of the initial weight and the final weight after n learning steps by combining Eqs. (25) and (27):

$$r_{xy}(\tau) = \lim_{\Delta t \to 0} \lim_{n \to \infty} \frac{1}{n} [w(n\Delta t, \tau) - w(0, \tau)]. \tag{28}$$

The preceding derivation results suggest that the cross-correlation function between the inputs $x(t)$ and $z(t)$ can be deduced from the synaptic weight accumulated during the training period. It also suggests that classical conditioning in neurons can be modeled by computing the cross-correlation function between the US and CS input signals. In other words, the synaptic weight established by the delayed Hebbian synapse under a classical conditioning paradigm can be considered as establishing correlation between the US and CS signals mathematically.

E.   Regulating Firing Rate

Since hyperpolarization is followed by delayed excitation in bi-threshold neurons, there is a period of suppressed firing. The period during which the neuron is inhibited, until the release of such inhibition by the delayed excitation mechanism, can be used to limit the frequency of firing. When such an inhibition period is over, a small depolarization can trigger firing of the neuron, thus producing the so-called post-inhibitory rebound phenomenon. This suppression period thus sets the limit of minimal firing interspike intervals to $h\Delta t$. It effectively limits the firing rate by the delayed excitation following the suppression period.

F.   Synchronizing Firing

When a bi-threshold neuron is inhibited (hyperpolarized), most of the input from various neurons having excitatory synaptic contact with this neuron will be suppressed. Thus, depolarizing (excitatory) signals can go through only when the inhibition period is over, even though the membrane potential is still below the resting level. This provides a mechanism for inputs that are synchronized at the end of the inhibition period to go through without waiting for the membrane potential to return to the resting level. Desynchronized inputs that arrive at the neuron during the inhibition period are essentially filtered out, which, in effect, allows the more synchronized inputs to go through. The effects of inhibition in the phasing of spontaneous firing in thalamo-cortical neurons has been observed (Andersen and Sears, 1964), and synchronized firing in hippocampal neurons (Michelson and Wong, 1991).

G.   Epileptogenesis

The hyper-excitability at the end of after-hyperpolarization, which results in post-inhibitory rebound, may account for the reported susceptibility to epileptic kindling at hyperpolarized potentials in hippocampal neurons. Epileptogenesis often occurs even when the hippocampus is presumably *inhibited* by hyperpolarization (e.g., Stasheff and Wilson, 1990). This contradicts the normal intuitive interpretation for single threshold neurons that hyperpolarization is usually interpreted as inhibition, not excitation; but when the bi-threshold firing phenomenon is considered, such hyperpolarization input may, in effect, become a prelude for subsequent hyper-excitation. Most importantly, hyperpolarization can be considered as excitation to a neuron rather than inhibition, since under these hyperpolarized conditions, the neuron actually becomes more excitable (i.e., closer to the low threshold) after the delayed period. Under this circumstance, the neuron is more susceptible to synchronized activation.

H.   Rhythms Generation and Periodic Bi-Stable Oscillation

It has been suggested that bi-threshold phenomena may be involved in generating two different rhythms produced by a periodic bi-stable oscillator, such as the generation of α and θ rhythms of the EEG and the oscillation in Parkinson's tremor (Jahnsen and Llinás, 1984a,b) and multiple modes of oscillations (Rose and Hindmarsh, 1985; Goldbeter and Moran, 1988; McCormick, 1991; McCormick and Feeser, 1990; Deschênes et al., 1984; Steriade and Deschênes, 1984). The repetitive firing of neurons, burst activation, and the switching of firing frequencies are often emphasized as the consequences of such multi-threshold neurons. Such interpretations are revealed by considering the neuron as a pacemaking regenerative bi-stable periodic generator at a single neuron level without synaptic interactions from other neurons.

I.   Signal Switching

Most neurons are non-pacemakers. They require input from other neurons for activation. When a neuron is considered as a signal processor, the input/output relationship of the neuron is important for understanding how signals are processed by decoding the input codes, processing the information, and encoding the processed result as the output signal. The multi-threshold phenomena may be effectively used for signal processing by neurons, which require the interaction of signals generated by various neurons having synaptic contacts. Such signal processing requires interaction of multiple neurons to produce the desired results, whereby the signal is decoded, processed, re-encoded, and transmitted to the next processing stage.

Multiplexing of signals is also suggested in bi-threshold neurons with high and low thresholds for switching modes of operation in signal processing (Tam, 1990c, 1991). Thus, two types of signal (one depolarizing, the other hyperpolarizing) can be processed by a neuron, which would result in switching the mode of firing. The hyperpolarization signal is not only an inhibitory signal but also a switching signal that changes the mode of operation in the neuron to utilize the low threshold for generation of spikes, which has different dynamics from the potential at or near the high threshold.

Neurons are often thought to encode a single type of code or signal, but more than one type of signal may be encoded by a neuron. One of the mechanisms for embedding multiple types of signals to be processed is *multiplexing*. Temporally coded signals in neurons can be multiplexed to increase the transmission capacity. If the signals are multiplexed, they also need to be demultiplexed to extract the useful information transmitted by neurons. Suggestive evidence for such multiplexing and demultiplexing schemes for signal processing by neurons will be given shortly. It has also been suggested that the multiple modes of discharge in thalamic neurons switch the neurons between the oscillatory mode (rhythmic bursting discharge), where signals from peripheral stimuli are suppressed while discharging rhythmically, and the transfer mode (phasic discharge), where signals from peripheral stimuli can be relayed through the neurons (McCormick and Feeser, 1990).

J.   Multiplexing/Demultiplexing

Encoding and decoding of signals requires establishing coherent input and output relationships among different neurons. The understanding of the signal processing function requires not only detailed knowledge of the single neuron properties locally, but also intimate appreciation of global interactions among neurons in a population (or network).

As an example, we describe how a network of neurons can extract the firing intervals of temporally modulated codes embedded in a spike train. The firing intervals of the encoded signal can be extracted by a network of neurons such that the firing of these output neurons will decode the interspike intervals of the input signal. The input signal is demultiplexed by distributing the output according to the interspike interval of the input spike train. In this network, the temporal codes of the input spike train will be converted into a spatially distributed topographical code where each output neuron represents a particular firing interval with a specific bandwidth. Thus, the input codes are demultiplexed by mapping the input firing intervals into the firing of specific neurons based on the spatial location of the neurons in the output layer.

The architectural circuitry of this network of neurons utilizes time delays for signal processing (Reiss, 1964; Tam, 1990a,b). Examples of delay-line architecture used for signal processing can be found in the cerebellar cortex (Eccles *et al.*, 1967), inferior colliculus (Yin *et al.*, 1987, 1986, 1985; Chan *et al.*, 1987), and cochlear nucleus (Carr and Konishi, 1990).

The time-delayed network can be described as follows. Let the single input to the network be a spike train $x(t)$ given by Eq. (4). There are k neurons in the first input layer of the network. The input is split into multiple branches, each of which is connected to all k neurons in the first layer. In addition to the direct connection between the input and the first layer neurons, each input branch to the first layer neuron is also split into multiple branches each with successive incremental time delays. In other words, the kth neuron in the first layer has $k + 1$ input lines; each input is successively delayed by a time delay Δt relative to the previous one; that is, the ith input to this kth neuron in the first layer at time t is given by $x(t - i\Delta t)$. Thus, the weighted sum of the input to this kth neuron is given by:

$$X_k(t) = \sum_{i=0}^{k} w_k(t)\, x(t - i\Delta t), \qquad (29)$$

where $w_k(t)$ is the synaptic weight at the kth synapse. If the synaptic weights are all equal to one, then the weighted sum becomes a simple arithmetic sum of the inputs:

$$X_k(t) = \sum_{i=0}^{k} x(t - i\Delta t). \qquad (30)$$

The sum of inputs can be interpreted analogously as the membrane potential resulting from the bombardment of all the synaptic inputs. Although the neuron in this case is considered to be a *point neuron,* where all inputs arrive at a single point in space, the neuron can be generalized to take into account the spatial processing contributed by the spatial properties of a neuron. In addition to spatial properties, the spatio-temporal processing phenomena can be accounted for by including a spatio-temporal parameter $s_k(t - \sigma\Delta t)$ associated with each synapse k. Thus, the spatio-temporal weighted sum of inputs becomes:

$$X_k(t) = \sum_{i=0}^{k} s_k(t - \sigma\Delta t) \, w_k(t) \, x(t - i\Delta t), \tag{31}$$

where $\sigma\Delta t$ is the time delay for the synaptic signal to reach the triggering site of the spike (or the initial segment of a myelinated axon). The spatio-temporal parameter s_k is analogous to the function describing attenuation of signal due to cable properties and space constants in dendrites.

1. Bandpass Filtering.   Neurons can act as bandpass filters for the incoming spike train input. Using our delay-line network as an illustrative example, band-pass filtering can be accomplished by the processing performed at the first layer of neurons. It will be shown that such signal processing capability can be accomplished with a simple point neuron even without adaptive synaptic weights. In fact, the network can perform interesting signal processing functions with fixed synaptic weights using a hard-wired architecture. This illustrates some of the built-in properties of networks that are a consequence of the systematic design of the connections and judicial use of appropriate thresholds for processing.

Now consider the sum of inputs of a point neuron described by Eq. (30). If the threshold for the generation of an output spike for the kth neuron is set at one, then this neuron will fire only when the interspike interval I_j of the input spike train is within the time-delay window, $k\Delta t$; that is, the output of this kth neuron is given by:

$$y_k(t) = \begin{cases} 1, & \text{if } X_k > 1 \\ 0, & \text{otherwise} \end{cases}. \tag{32}$$

Therefore, the kth neuron can be considered as encoding a bandpass-filtered-input interspike interval, $0 < I_j \le k\Delta t$. Thus, the kth neuron in the first layer essentially captures the input interspike interval firing of less than $k\Delta t$, the bandpassed interspike interval. To ensure that the neuron will fire a spike of Δt in duration, we introduce a refractory period of $(k - 1)\Delta t$ after the firing of a spike for the kth neuron to suppress continual activation of the neuron due to the phase differences of the incoming delayed signal. Note that the output of the neuron not only extracts first-order interspike intervals but also any higher-order intervals.

2. Higher-Order Interspike Interval Processing.   Extraction of higher-order interspike intervals can be performed by further processing of the signal downstream to the neurons described previously in the network model. Higher-order interspike intervals can be shown to be eliminated by the second-layer neurons. If the second-layer neurons receive excitatory input from the corresponding neuron with a threshold, $\theta_{high} > 1$, and inhibitory input from the corresponding neuron with a threshold, $\theta_{high} > 2$, then the higher-order intervals are eliminated, with the output of the second layer (double-primed) neuron given by:

$$y_k''(t) = y_k(t) - y_k'(t) = \begin{cases} 1, & \text{if } 2 \geqslant X_k > 1 \\ 0, & \text{otherwise} \end{cases}. \tag{33}$$

where

$$y_k'(t) = \begin{cases} 1, & \text{if } X_k > 2 \\ 0, & \text{otherwise} \end{cases}. \tag{34}$$

This requires that an additional input layer of neurons—which we call the *first-parallel layer*, whose input/output relationship is given by Eq. (34)—be added to the network. In other words, there are k first-layer neurons and k first-parallel-layer neurons serving as the input layers of the network. The kth neuron in the first layer and the kth neuron in the first-parallel layer are similar in their inputs, but the thresholds for producing an output spike are different. The difference between the outputs of the first set of neurons (first layer) in the first layer and the primed set of neurons (first-parallel layer) is computed by the *second* layer by making excitatory connections from the first-layer neuron and inhibitory connections from the first-parallel-layer neuron for each corresponding kth neuron, respectively, as described by Eq. (33). This will ensure the estimation of only a first-order interspike interval, $0 < I_j \leqslant k\Delta t$, within the time-delay window, $k\Delta t$.

3. Bandwidth Processing.   Specific windows of bandwidth of interspike intervals can be created through processing by yet another layer of the network. The third-layer neurons will filter the input signal by distributing the frequency (or interval) of firing of neurons within a specific bandwidth. Since the kth neuron in the second layer detects the bandpassed first-order interspike interval $(0 < I_j \leqslant k\Delta t)$, and the hth neuron detects another bandpassed interspike interval $(0 < I_j \leqslant h\Delta t)$, then the difference between these two neurons will detect first-order interspike intervals with a bandwidth of $(k - h)\Delta t$. In other words, it will detect the first-order interspike interval between $k\Delta t$ and $h\Delta t$, i.e., $h\Delta t < I_j \leqslant k\Delta t$. Effectively, it uses two low-pass filters to create a bandpass filter by combining the difference between the outputs of the two low-pass filters.

This requires that the *third*-layer neurons derive their inputs from two sources:

one excitatory and the other inhibitory from the second layer. The output of the kth neuron in the third layer, $y'''k(t)$, is obtained from the difference between the outputs of kth and hth neurons in the second layer:

$$y'''_{kh}(t) = y''_k(t) - y''_h(t) = \begin{cases} 1, & \text{if } 2 \geqslant \sum_{i=h}^{k} x(t - i\Delta t) > 1 \\ 0, & \text{otherwise} \end{cases}. \tag{35}$$

4.   Topographical Mapping.   A two-dimensional topographical map of the bandpassed interspike intervals of the input spike train can be represented by arranging the third-layer neurons in a two-dimensional array, with one axis (the horizontal axis) representing the k index (the bandpassed interspike interval) of Eq. (35), and the other axis (the vertical axis) representing the $(k - h)$ index (the bandwidth interspike interval). Thus, the firing of the third-layer neurons represents the bandpassed filtered version of the original input spike train, extracting the firing interspike interval of the input signal. The *coordinate* of the neuron in the third layer represents the bandpassed interspike interval ($0 < I_j \leqslant k\Delta t$) and the bandwidth interspike interval ($h\Delta t < I_j \leqslant k\Delta t$) of the original input spike train signal. The bandwidth can be used to detect the variations (or jittering) in the timing for firing of spikes in the input spike train, since the timing of firing of spikes in biological neurons can be very variable. Thus, the network maps the firing intervals of the input spike train into a topographically mapped firing at the output of the network representing the bandpassed versions of the original input spike train.

5.   Extracting Embedded Signals.   When a neuron receives a synaptic input resulting in a hyperpolarized response, that signal is most often considered as lost, since that synaptic input will not produce an output spike and the signal will not be relayed to the next neuron downstream; but hyperpolarization may not necessarily imply suppression of signal, since that signal can be retrieved given the delayed excitation. The history of being hyperpolarized can be readout by another input that can depolarize the potential to cross the low threshold, thus enabling firing of an output spike that can be detected by the next neuron downstream.

For example, if the neurons in the second and third layers in the time-delayed network described before are bi-threshold neurons where the high threshold is at a positive value and the low threshold is at a negative value, then additional information may be extracted based on the level of firing threshold. Since the neurons in the second and third layers receive inhibitory inputs from the preceding layer, the connections can be arranged such that there are instances where the sum of the inputs to the neuron are negative, or *hyperpolarized*. Such a condition occurs when the neurons detecting the first-order interspike intervals are subtracted from the neurons detecting the order of interspike intervals higher than one. In other words, the higher-order interspike interval signal is embedded in the hyperpolarization, which is normally suppressed from generating a spike when there is only one threshold for firing at the *depolarized* level (θ_{high}); but for bi-threshold neurons where there is another threshold at the hyperpolarized level (θ_{low}), such an embedded signal encoded as hyperpolarization can be extracted by sending an external depolarizing signal to this neuron, causing the neuron to fire at the low threshold. Thus, the hyperpolarization signal can be *readout* by an external input to the bi-threshold neuron. In other words, the hyperpolarization signal can still get through the neuron provided that there is a read-out signal sent to the neuron to extract the *hidden* (hyperpolarized) signal, which normally will not produce spike firing.

K.   Long-Term Potentiation

The induction of long-term potentiation (a physiological mechanism for increasing synaptic strength in synaptic plasticity) may be accounted for not only by the glutamatic excitatory synaptic inputs to a hippocampal neuron but also by the inhibitory GABAergic synaptic inputs. The hyperpolarization effects of GABAergic inputs onto the hippocampal neuron may not necessarily be inhibitory but excitatory (or delayed excitatory), whereby long-term potentiation is induced not only by simultaneous activation of the pre- and postsynaptic neurons but also by delayed activation of the postsynaptic neuron by an "inhibitory" presynaptic neuron. The mechanisms for activation of such multi-threshold neuron are discussed in the preceding sections. Experimental evidence for excitatory synaptic responses mediated by GABAergic inhibitory neurons in hippocampus has been observed (Michelson and Wong, 1991).

VI.   Summary

A discretized model for neuronal signal processing of spike trains is described. Signal processing is performed by utilizing the properties of multi-threshold phenomena found in neurons and the pulse-coded nature of the signal. The possible functional significance of such multi-threshold neurons for signal processing has been discussed, which includes the role in delayed excitation, post-inhibitory rebound, delayed Hebbian synaptic modification, classical conditioning, computing the cross-correlation function, regulating firing rate, synchronizing firing, epileptogenesis, rhythms generation, long-term potentiation, signal switching, multiplexing/demultiplexing, and extracting embedded signals.

The discretized model provides simplicity in mathematical description and computational implementation. In fact, the description of the model provides an algorithmic representation of the functional properties of the neuron in addition to the mathematical formulation. It also provides an intuitive representation of the processing capabilities of a neuron algorithmically as well as its computational capabilities. Also, since such representation is algorithmic and discrete, it can be implemented on a digital computer much more efficiently than a continuous model.

Acknowledgment

This work was supported by ONR contract N00014-90-J-1353.

References

ANDERSEN, P., and SEARS, T. A. (1964). "The Role of Inhibition in the Phasing of Spontaneous Thalamo-Cortical Discharge," *Journal of Physiology (London)* **173,** 459–480.

BARTO, A. G., and SUTTON, R. S. (1981). "Landmark Learning: An Illustration of Associative Search," *Biological Cybernetics* **42,** 1–8.

BARTO, A. G., SUTTON, R. S., and BROUWER, P. S. (1981). "Associative Search Network: A Reinforcement Learning Associative Memory," *Biological Cybernetics* **40,** 201–211.

CARR, C. E., and KONISHI, M. (1990). "A Circuit for Detection of Interaural Time Differences in the Brain Stem of the Barn Owl," *Journal of Neuroscience* **10,** 3227–3246.

CHAN, J. C., YIN, T. C., and MUSICANT, A. D. (1987). "Effects of Interaural Time Delays of Noise Stimuli on Low-Frequency Cells in the Cat's Inferior Colliculus: II. Responses to Band-Pass Filtered Noises," *Journal of Neurophysiology* **58,** 543–561.

DESCHÊNES, M., PARADIS, M., ROY, T. P., and STERIADE, M. (1984). "Electrophysiology of Neurons of Lateral Thalamic Nuclei in Cat: Resting Membrane Properties and Burst Discharges," *Journal of Neurophysiology* **51,** 1196–1219.

ECCLES, J. C., ITO, M., and SZENTÁGOTHAI, J. (1967). *The Cerebellum as a Neuronal Machine.* Springer-Verlag, New York.

FITZHUGH, R. (1960). "Impulses and Physiological States in Theoretical Models of Nerve Membrane," *Biophysics Journal* **1,** 445–466.

GOLDBETER, A., and MORAN, F. (1988). "Dynamics of a Biochemical System with Multiple Oscillatory Domains as a Clue for Multiple Modes of Neuronal Oscillations," *European Biophysical Journal* **15,** 277–287.

HEBB, D. O. (1949). *The Organization of Behavior.* John Wiley, New York.

HODGKIN, A. L., and HUXLEY, A. F. (1952). "A Quantitative Description of Membrane Current and Its Application to Conduction and Excitation in Nerve," *Journal of Physiology (London)* **117**, 500–544.

JAHNSEN, H., and LLINÁS, R. (1984a). "Electrophysical Properties of Guinea-Pig Thalamic Neurones: An *in vitro* Study," *Journal of Physiology (London)* **349**, 205–226.

JAHNSEN, H., and LLINÁS, R. (1984b). "Ionic Basis for the Electroresponsiveness and Oscillatory Properties of Guinea-Pig Thalamic Neurones *in vitro*," *Journal of Physiology (London)* **349**, 227–247.

MCCORMICK, D. A. (1991). "Determination of State Dependent Processing in Thalamus by Single Neuron Properties and Neural Modulation," in T. McKenna, J. Davis, and S. F. Zornetzer (eds.), *Single Neuron Computation*. Academic Press, Boston.

MCCORMICK, D. A., and FEESER, H. R. (1990). "Functional Implications of Burst Firing and Single Spike Activity in Lateral Geniculate Relay Neurons," *Neuroscience* **39**, 103–113.

MICHELSON, H. B., and WONG, R. K. S. (1991). "Excitatory Synaptic Responses Mediated by GABA$_A$ Receptors in the Hippocampus," *Science* **253**, 1420–1423.

NAGUMO, J. S., ARIMOTO, S., and YOSHIZAWA, S. (1962). "An Active Pulse Transmission Line Simulating Nerve Axon," *Proceedings of Inst. Radio Eng.* **50**, 2061–2070.

PERKEL, D. H., GERSTEIN, G. L. and MOORE, G. P. (1967a). "Neuronal Spike Trains and Stochastic Point Processes: I. The Single Spike Train," *Biophysics Journal* **7**, 391–418.

PERKEL, D. H., GERSTEIN, G. L., and MOORE, G. P. (1967b). "Neuronal Spike Trains and Stochastic Point Processes: II. Simultaneous Spike Trains," *Biophysics Journal* **7**, 419–440.

REISS, R. F. (1964). "A Theory of Resonant Networks," in R. F. Reiss (ed.), *Neural Theory and Modeling: Proceedings of the 1962 Ojai Symposium*. Stanford University Press, Stanford, California.

ROSE, R. M., and HINDMARSH, J. L. (1985). "A Model of a Thalamic Neuron," *Proceedings of Royal Society of London* **225**, 161–193.

STASHEFF, S. F., and WILSON, W. A. (1990). "Increased Ectopic Action Potential Generation Accompanies Epileptogenesis *in vitro*," *Neuroscience Letter* **111**, 144–150.

STERIADE M., and DESCHÊNES, M. (1984). "The Thalamus as a Neuronal Oscillator," *Brain Research Review* **8**, 1–63.

SUTTON, R. S., and BARTO A. G. (1981). "Toward a Modern Theory of Adaptive Networks: Expectation and Prediction," *Psychological Review* **88**, 135–170.

TAM, D. C. (1989). "A Positive/Negative Reinforcement Learning Model for Associative Search Network," *Proceedings of the First Annual IEEE Symposium on Parallel and Distributed Processing*, 300–307.

TAM, D. C. (1990a). "Temporal–Spatial Coding Transformation: Conversion of Frequency-Code to Place-Code via a Time-Delayed Neural Network," *Proceedings of the International Joint Conference on Neural Networks, Jan. 1990* **1**, 130–133.

TAM, D. C. (1990b). Decoding of Firing Intervals in a Temporal-Coded Spike Train

Using a Topographically Mapped Neural Network," *Proceedings of International Joint Conference on Neural Networks, June 1990* **3**, 627–632.

TAM, D. C. (1990c). "Functional Significance of Bi-Threshold Firing of Neurons," *Society for Neuroscience Abstract* **16**, 1091.

TAM, D. C. (1991). "Signal Processing by Multiplexing and Demultiplexing in Neurons," in D. S. Touretzky and R. Lippman (eds.), *Advances in Neural Information Processing Systems 3*. Morgan Kaufmann, San Mateo, California. (in press).

TAM, D. C., and McMULLEN, T. A. (1989). "Hebbian Synapses as Cross-Correlation Functions in Delay Line Circuitry," *Society for Neuroscience Abstract* **15**, 777.

TAM, D. C., and PERKEL, D. H. (1989a). "Quantitative Modeling of Synaptic Plasticity," in R. D. Hawkins and G. H. Bower (eds.), *The Psychology of Learning and Motivation: Computational Models of Learning in Simple Neural Systems, Vol. 23* (pp. 1–30). Academic Press, San Diego.

TAM, D. C., and PERKEL, D. H. (1989b). "A Model for Temporal Correlation of Biological Neuronal Spike Trains," *Proceedings of the IEEE International Joint Conference on Neural Networks* **1**, 781–786.

YAROM, Y., and LLINÁS, R. (1987). "Long-Term Modifiability of Anomalous and Delayed Rectification in Guinea Pig Inferior Olivary Neurons," *Journal of Neuroscience* **7**, 1166–1177.

YIN, T. C., CHAN, J. C., and CARNEY, L. H. (1987). "Effects of Interaural Time Delays of Noise Stimuli on Low-Frequency Cells in the Cat's Inferior Colliculus: III. Evidence for Cross-Correlation," *Journal of Neurophysiology* **58**, 562–583.

YIN, T. C., CHAN, J. C., and IRVINE, D. R. (1986). "Effects of Interaural Time Delays of Noise Stimuli on Low-Frequency Cells in the Cat's Inferior Colliculus: I. Responses to Wideband Noise," *Journal of Neurophysiology* **55**, 280–300.

YIN, T. C., HIRSCH, J. A., and CHAN, J. C. (1985). "Responses of Neurons in the Cat's Superior Colliculus to Acoustic Stimuli: II. A Model of Interaural Intensity Sensitivity," *Journal of Neurophysiology* **53**, 746–758.

Chapter 19   Cooperative Stochastic Effects in a Model of a Single Neuron

ADI R. BULSARA

Naval Ocean Systems Center
Materials Research Branch
San Diego, California

WILLIAM C. SCHIEVE

Physics Department and Center
for Studies in Statistical Mechanics
University of Texas
Austin, Texas

FRANK E. MOSS

Physics Department
University of Missouri
St. Louis, Missouri

I. Introduction

Past investigations of artificial neural networks have typically been carried out using noise-free neurons. While such networks possess many interesting properties, they are usually taught using noisy data sets. These noisy sets are used to speed up learning and to help avoid local minima on the memory surface. However, despite the benefits of noise in these otherwise noise-free systems, and the fact that most biological systems are based on noisy elements, there has, to date, been a distinct lack of interest in the property of noisy neurons.

Recently, Buhmann and Schulten (1986, 1987) investigated an artificial neural network with added noise. Their network is composed of interconnected neurons whose behavior is described by deterministic equations in which the neurons fire whenever their input membrane potential exceeds a threshold value, then go through a refractory period. This network has been shown to be capable of, among other tasks, associative storage and adaptive filtering. They have found that, in the presence of added noise, the network learns faster and yields a greatly improved performance. Indeed, they concluded that:

> *(T)he noise . . . is an essential feature of the information processing abilities of the neural network, and not a mere source of disturbance better suppressed.*

Hopfield (1982, 1984) showed that continuous response neurons (in which the firing rate is the state variable), when connected properly, can possess associative memory, pattern recognition, and other collective computational properties; the Hopfield model, while not a model of a real (i.e., biological) neuron, nevertheless reproduces several properties of real neurons and often provides a convenient starting point in the mathematical modeling of real neuron behavior. A simple model of a single Hopfield neuron (as well as a simple two-neuron network) has been studied by Babcock and Westervelt (1986, 1987). They started with an optimized neuron modeled as an RC circuit in which a nonlinear input–output transfer function (represented by a hyperbolic tangent) was inserted. Above a threshold value of the gain parameter, this system may be described from the standpoint of a particle in a double-well potential. Below the threshold, the potential is approximately parabolic. To simulate the departure (due, often, to hardware implementations) from the ideal neuron, Babcock and Westervelt introduced an additional, inductance-dependent inertial term. This has the effect of transforming the original RC model into a two-dimensional LCR model with a nonlinear transfer function. In the presence of an external periodic forcing term, this system displays extremely complex behavior that includes spontaneous oscillations, intertwined basin boundaries, and chaos. Since noise is used (as pointed out before) to teach sets rapidly find the global minimum, and since Buhmann and Schulten found that intrinsic noise produced results equivalent to those obtained through noisy teaching sets, it seems reasonable to conjecture that the removal of the inertial term from the Babcock–Westervelt model might be compensated for by the inclusion of noise. Clearly, then, it is of interest to study the response of the Babcock–Westervelt single neuron model (in the non-inertial limit) to additive (i.e., external or Langevin-type) noise as well as multiplicative noise, which can arise due to the fluctuations of the system parameters themselves.

The work of Babcock and Westervelt reflects the recent upsurge of interest in *single* or few-neuron nonlinear dynamics (Tuckwell, 1988; Li and Hopfield,

1989; Aihara *et al.*, 1990; Paulus *et al.*, 1989). However, the precise relationship between the many-neuron connected model and a single effective neuron dynamics has not been examined in detail. Recently, Schieve *et al.* (1990) have considered a network of N symmetrically interconnected neurons. Through an adiabatic elimination procedure, they have obtained, in closed form, the dynamics of a single neuron (termed the single *effective* neuron) from the system of coupled differential equations describing the N-neuron problem. The problem has been treated both deterministically and stochastically (through the inclusion of additive and multiplicative noise terms). It is important to point out that, in contrast to the suggestion by Hopfield (1984) and the work of Babcock and Westervelt, the work of Schieve *et al.* does not include *a priori* a self-coupling term (although the inclusion of such a term can be readily implemented in their theory; this has been done by Bulsara and Schieve (1991)). Rather, their theory results in an explicit form of the self-coupling term, in terms of the parameters of the remaining neurons in the network. This term, in effect, renormalizes the self-coupling term in the Babcock–Westervelt and Hopfield models.

In Section II, we outline the adiabatic elimination procedure and obtain the effective single neuron equation for the deterministic as well as the stochastic cases. In the latter case, the calculation is detailed for the case of two neurons and the extension to many neurons outlined. We consider the case when the neuron is influenced by additive (i.e., Langevin) as well as multiplicative noise. Such noise processes have recently received a great deal of attention (Stratonovich, 1963; Bulsara *et al.*, 1979; West *et al.*, 1979; Schenzle and Brand, 1979; Horsthemke and Lefever, 1984; Moss and McClintock, 1989) because they can qualitatively change the physical state of the system. In the presence of multiplicative noise, the deterministic description of the system dynamics proves to be inadequate and one must, of necessity, invoke a stochastic description. The method of choice is to construct a Fokker Planck equation for the conditional probability density; in contrast to the additive noise case, however, this Fokker Planck equation has a non-constant diffusion term. Solving the Fokker Planck equation in the long time limit yields the stationary probability density function from which one recovers a deterministic description of the system through a computation of the mean value $\langle x(t) \rangle$ of the system response. Multiplicative noise is introduced by allowing the self-connection coefficient, i.e., the nonlinearity parameter, to fluctuate. The extrema and turning points of the stationary probability density (or, equivalently, the macroscopic *potential* associated with it) have been examined in some detail (Bulsara *et al.*, 1989). It has been found that the multiplicative noise can, *on average,* actually introduce new forms of critical behavior in the dynamics of the neuron. Specifically, the multiplicative noise introduces multistability, characterized by a multimodal probability density function, in the output of the neuron for parameter values that would not ordinarily support such behavior. In some cases, the noise may also suppress hysteresis effects that are known to occur in its absence, i.e., a potential that was bi- or multi-stable in the deterministic case is rendered monostable by the multiplicative noise. (In the presence of multiplicative noise, the "potential" is, of course, the macroscopic potential associated with the long-time solution of the Fokker Planck equation.) The multiplicative noise can, in other words, introduce bifurcations in the most probable value of the random variable x characterizing the system response. It can also suppress such bifurcations in regimes where they are known to occur in the presence of solely additive noise. The preceding effects have been extensively discussed by Bulsara *et al.* (1989) and will not be repeated here. They are similar to those observed by the authors of the rf SQUID (Bulsara *et al.*, 1987) and in nonlinear optical systems (Bulsara *et al.*, 1978; Englund *et al.*, 1984). (An excellent review of the subject of noise-induced critical behavior, together with a compendium of experimental results, may be found in the book of Horsthemke and Lefever (1984).) It is seen that including a dc component in the external input plays a crucial role in determining the response of the system; this dc term plays the role of a symmetry-breaking *order parameter* in certain regimes of interest.

In Section III, we consider the noisy neuron in the presence of a weak periodic external modulation. The simple Babcock–Westervelt model (without the re-normalization of the self-coupling term derived in Section II) is used. The modulation introduces a correlated switching between the bistable states, driven by the noise with the signal-to-noise ratio (SNR) obtained from the power spectrum, being taken as a measure of the information content in the neuron response. As the additive noise variance increases, the SNR passes through a maximum. This effect has been called *stochastic resonance* (multiplicative fluctuations tend to degrade the effect) and describes a phenomenon in which the noise actually enhances the information content, i.e., the observability of the signal. The effect seems to indicate that a small amount of noise might actually enhance the information processing capabilities of the neuron, in keeping with the conjecture of Buhmann and Schulten (1986, 1987). The phenomenon of stochastic resonance was first investigated in connection with a possible explanation of the observed periodicities in the recurrences of the ice ages (Nicolis and Nicolis, 1981; Nicolis, 1982). Recent interest in the phenomenon has been rekindled by its demonstration in a ring laser cavity (McNamara *et al.*, 1988; Vemuri and Roy, 1989). The experiment was followed by an upsurge of theoretical interest in the problem (Jung, 1989; Jung and Hanggi, 1989, 1990; McNamara and Wiesenfeld, 1989; Presilla *et al.*, 1989; Gammaitoni *et al.*, 1989; Debnath *et al.*, 1989) as well as analog simulation experiments (Zhou and Moss, 1989, 1990). Recently, a new theoretical approach to stochastic resonance, based on the probability density of residence times (in one of the two stable states of the potential) has been developed (Zhou *et al.*, 1990). In this work, we summarize the recent work of Bulsara et al. (1990c) on the response of a single neuron to a weak periodic signal superimposed on a white noise background. An approximate theory (based on the work of McNamara and Wiesenfeld (1989)) has been developed in the so-called *adiabatic regime* corresponding to weak driving amplitudes and frequencies. This theory has been found (Bulsara *et al.*, 1990c) to yield results that compare very favorably with analog simulation studies.

Finally, it is worth pointing out that recent (and still ongoing) work by the authors (Longtin *et al.*, 1991) on the probability density function characterizing residence times in the potential wells of the noisy bistable neuron model considered in this work has yielded theoretical interspike-interval probability density functions that compare very favorably to experimental data, taken some 25 years apart, on two different animals. Indeed, this work, which explores the fundamental symmetries available to elementary two-state systems of the type considered in this review, has, for the first time, been able to provide an explanation of the results of these classical experiments. Further, this work (which we outline briefly in Section IV), gives additional credence to the essential role of noise in information processing by neurons, a property that forms the backbone of much of our current research.

II.   The Single Effective Neuron

The deterministic Hopfield model has the form (Hopfield, 1982, 1984),

$$
C_i \frac{dU_i}{dt} = \sum_{j \neq i = 1}^{N} J_{ij} \tanh U_j - \frac{U_i}{R_i}, \tag{1}
$$

where U_i is the potential of the ith neuron having input capacitance C_i and a leakage current due to the inter-membrane resistance R_i. The first term on the right-hand side represents the input to the soma from the other neurons with the characteristic saturation with potential of their firing rates, taken to be a hyperbolic tangent function. A good discussion of this equation is given by Shamma (1989). We assume the neuron connectivity to be symmetric and neglect self-excitation:

$$
J_{ij} = J_{ji}; \qquad J_{ii} = 0. \tag{2}
$$

For simplicity of notation, we will consider a three-neuron network; the results may be readily generalized to N neurons. We focus on the neuron $i = 1$ and assume that the other two neurons may relax to a steady state on a much shorter time scale than the $i = 1$ neuron. To do this, we assume $R_i \ll R_1 (i = 2,3)$. Thus, we may set

$$C_i \frac{dU_i}{dt} \approx 0, \quad i = 2,3.$$

The resulting equations are solved (Schieve et al., 1990) for tanh U_2 and tanh U_3, which are then substituted into the $\dot{U}_1$ equation. The result (neglecting terms of $O(R_2 R_3)$) is a closed equation for U_1 in which the variables U_2 and U_3 have been adiabatically eliminated:

$$C_1 \dot{U}_1 = -\frac{U_1}{R_1} + \{J_{12}^2 R_2 + J_{13}^2 R_3\}\tanh U_1, \tag{3}$$

which may be readily generalized to yield,

$$C_1 \dot{U}_1 = -\frac{U_1}{R_1} + \sum_{j=2}^{N} J_{1j}^2 R_j \tanh U_1 \equiv -\frac{U_1}{R_1} + J_{11} \tanh U_1, \tag{4}$$

where the self-connection may be defined as:

$$J_{11} = \sum_{j=2}^{N} J_{ij}^2 R_j > 0. \tag{5}$$

This is the single-neuron equation written by Babcock and Westervelt (1986, 1987). It should be noticed that the self-term need not be assumed but depends on the assignment of the parameters of connectivity and resistance to the *other* neurons. If a self-connection J_{11} is assumed initially, however, the term in Eq. (5) renormalizes it; clearly, this results in an effective self-connectivity dependent on the parameters of the remaining neurons in the network.

As mentioned in the preceding section, noise is important in the modeling of neural networks (Holden, 1976; Peretto, 1984; Buhmann and Schulten, 1986, 1987), particularly for biological cases. This has also led to the fruitful use of spin-glass analogies (Sampolinsky, 1988; Amit, 1989). In what follows, we consider a two-neuron system, for simplicity, and carry out the stochastic analog of the deterministic adiabatic elimination (Haken, 1977, 1983). The details of the calculation, as well as the extension to more complicated networks, are given by Schieve et al. (1990). In Eq. (1), we allow the connectivity of the neurons to fluctuate:

$$J_{12}(t) = J + \delta J_2(t),$$

$$J_{21}(t) = J + \delta J_1(t), \tag{6}$$

where $\delta J_1(t)$ and $\delta J_2(t)$ are taken to be white noise terms having zero mean and uncorrelated:

$$\langle \delta J_{1,2}(t) \rangle = 0 = \langle \delta J_1(t)\delta J_2(0) \rangle,$$

$$\langle \delta J_1(t)\delta J_1(0) \rangle = \sigma^2 \delta(t) = \langle \delta J_2(t)\ \delta J_2(0) \rangle, \tag{7}$$

$\delta(t)$ being the Dirac delta function. In addition, we assume a Langevin background noise term, uncorrelated with the multiplicative fluctuations:

$$\langle F(t) \rangle = 0 = \langle F(t)\ \delta J_i(t) \rangle,$$
$$\langle F(t)\ F(0) \rangle = \sigma_a^2\ \delta(t). \tag{8}$$

We then have the two-neuron coupled stochastic differential equations:

$$C_{1,2}\ \dot{U}_{1,2} = -\frac{U_{1,2}}{R_{1,2}} + (J + \delta J_{2,1}) \tanh U_{2,1} + F(t). \tag{9}$$

In the Ito interpretation, we may write down a two-dimensional Stratonovich Fokker Planck equation (Stratonovich, 1963; Bulsara *et al.*, 1979; West *et al.*, 1979; Gardiner, 1985; Risken, 1989) for the probability density function $P(U_1,U_2,t)$:

$$\frac{\partial P}{\partial t} = -\frac{\partial}{\partial U_1}\{F_1(U_1,U_2)P\} - \frac{\partial}{\partial U_2}\{F_2(U_1,U_2)P\} \tag{10}$$
$$+ \frac{1}{2}B_1(U_2)\frac{\partial^2 P}{\partial U_1^2} + \frac{1}{2}B_2(U_1)\frac{\partial^2 P}{\partial U_2^2} + \frac{\sigma_a^2}{C_1 C_2}\frac{\partial^2 P}{\partial U_1\ \partial U_2},$$

with

$$F_1(U_1,U_2) \equiv -\frac{U_1}{R_1 C_1} + \frac{J}{C_1}\tanh U_2,$$

$$F_2(U_1,U_2) \equiv -\frac{U_2}{R_2 C_2} + \frac{J}{C_2}\tanh U_1, \tag{11}$$

and

$$B_1(U_2) \equiv \frac{\sigma^2}{C_1^2}\tanh^2 U_2 + \frac{\sigma_a^2}{C_1^2},$$

$$B_2(U_1) \equiv \frac{\sigma^2}{C_2^2}\tanh^2 U_1 + \frac{\sigma_a^2}{C_2^2}. \tag{12}$$

We assume neuron 2 to be statistically rapidly varying and *enslaved* by neuron 1 (Haken, 1977, 1983). Let

$$P(U_1,U_2,t) = h(2|1,t)g(1,t), \tag{13}$$

with both h and g being normalized to unity and h being interpreted as a conditional probability density function. Substituting Eq. (13) into the original Fokker Planck equation (10), one may separate this equation into Fokker Planck equations for h and g. (The procedure is described by Haken (1977).) The Fokker Planck equation for the single neuron probability density function, $g(U_1,t) \equiv g(1,t)$ is:

$$\frac{\partial}{\partial t} g(1,t) = -\frac{\partial}{\partial U_1} \{\overline{F}_1(U_1) \, g(1,t)\} + \frac{1}{2} \frac{\partial^2}{\partial U_1^2} \{D_1(U_1)g(1,t)\}, \qquad (14)$$

where we introduce the quantities,

$$\overline{F}_1(U_1) \equiv -\frac{U_1}{R_1 C_1} + \frac{J}{C_1} G(U_1), \qquad (15a)$$

$$D_1(U_1) \equiv C_1^{-2} \{\sigma_a^2 + \sigma^2 H(U_1)\} \qquad (15b)$$

$$H(U_1) \equiv \int_{-\infty}^{\infty} h \, (2|1)\tanh^2 U_2 \, dU_2, \qquad (15c)$$

$$G(U_1) \equiv \int_{-\infty}^{\infty} h(2|1)\tanh U_2 \, dU_2. \qquad (15d)$$

The corresponding Fokker Planck equation for the conditional probability density function $h(2|1,t)$ may be written down and, since neuron 2 is the enslaved neuron, solved (Schieve et al., 1990) in the steady state (i.e., the $t \to \infty$ limit), keeping U_1 constant. Then we obtain the following Gaussian distribution for $h(2|1) = h(2|1,r \to \infty)$:

$$h(2|1) = K^{-1} \exp \left[\frac{2C_2^2}{\sigma^2 \tanh^2 U_1 + \sigma_a^2} \left\{ -\frac{U_2^2}{2R_2 C_2} + \frac{JU_2}{C_2} \tanh U_1 \right\} \right], \qquad (16)$$

K being the normalization constant. This solution is used in the evaluation of the integrals in Eq. (15). The hyperbolic functions in Eqs. (15c) and (15d) are expanded about the steady-state value $\overline{U}_2 = JR_2 \tanh U_1$, which also coincides with the maximum of the Gaussian (16). To second-order, we find

$$\overline{F}_1(U_1) = -\frac{U_1}{R_1 C_1} + \frac{J}{C_1} \left\{ \tanh \overline{U}_2 - \frac{1}{2A} \tanh \overline{U}_2 \, \text{sech}^2 \overline{U}_2 \right\} \qquad (17)$$

and

$$D_1(U_1) = \frac{\sigma_a^2}{C_1^2} + \frac{\sigma^2}{C_1^2} \left\{ \tanh^2 \overline{U}_2 + \frac{1}{2A} \text{sech}^2 \overline{U}_2(\text{sech}^2 \overline{U}_2 - 2\tanh^2 \overline{U}_2) \right\}, \qquad (18)$$

where

$$A \equiv \frac{C_2/R_2}{\sigma^2 \tanh^2 U_1 + \sigma_a^2}. \tag{19}$$

The higher terms in the preceding expansion increase as powers of $\sigma^2 R_2/C_2$ and are neglected.

It is evident that the procedure just described yields a Fokker Planck equation (14) (decoupled from the *fast* variable U_2) for the *effective* or *slow* neuron. The drift and diffusion terms in the Fokker Planck equation are non-constant and depend on the parameters of both neurons. The procedure described in the preceding has been extended (Schieve et al., 1990) to three or more neurons. In this case, however, one cannot readily write down a closed-form solution analogous to Eq. (16) for the steady-state probability density function $h(2,3 \ldots N|1)$ corresponding to the fast neurons, because of the detailed balance conditions (Gardiner, 1985). However, using a local equilibrium *ansatz* allows one to write down, formally, a Gaussian solution for the $N - 1$ fast neurons in terms of a set of undetermined multipliers. These parameters are calculated following a procedure of Stratonovich (1989). The details have been given by Schieve et al. (1990). Corresponding to the Fokker Planck equation (14) and its N-body analog, we may write down, by inspection, the single-neuron Stratonovich stochastic differential equation for the N-neuron case (with the $i = 1$ neuron assumed to be the slow neuron):

$$\dot{U}_1 = -\frac{U_1}{R_1 C_1} + C_1^{-1} \sum_{j=2}^{N} \bar{J}_{1j} A_{1j}(U_1) - \frac{1}{4C_1^2} \sum_{j=2}^{N} \sigma_{1j}^2 \frac{\partial}{\partial U_1} B_{1j}(U_1)$$
$$+ \frac{1}{C_1} \left\{ \sigma_a^2 + \sum_{j=2}^{N} \sigma_{1j}^2 B_{1j}(U_1) \right\}^{1/2} \xi(t), \tag{20}$$

where $\xi(t)$ is white noise having zero mean and unit variance, and we have defined:

$$A_{1j}(U_1) \equiv \int h(2, 3, \ldots N|1) \, \tanh U_j \, dU_2 dU_3 \ldots dU_N,$$
$$B_{1j}(U_1) \equiv \int h(2, 3, \ldots N|1) \, \tanh^2 U_j \, dU_2 dU_3 \ldots dU_N. \tag{21}$$

In writing down Eqs. (20) and (21), we have assumed the following statistics for the multiplicative noise:

$$J_{ij} = \bar{J}_{ij} + \delta J_{ij}; \quad \bar{J}_{ij} = \bar{J}_{ji},$$
$$\langle \delta J_{ij}(t) \rangle = 0,$$
$$\langle \delta J_{ij}(t) \, \delta J_{kl}(0) \rangle = \delta_{ik} \, \delta_{jl} \, \sigma_{ij}^2 \, \delta(t); \quad \sigma_{ij}^2 = \sigma_{ji}^2,$$

where δ_{ab} is the Kroenecker delta, and the Langevin noise terms $F(t)$ are defined in a manner analogous to Eq. (8), they are assumed to be the same in each neuron. The preceding restrictions are not essential to the calculation; however, they do simplify it considerably.

Before concluding this section, a few remarks are in order concerning the validity of the approximations used in the evaluation of the integrals $H(U_1)$ and $G(U_1)$ in Eqs. (16c,d). In both cases, the full approximations, given by the terms in chain brackets in Eqs. (17) and (18), agree extremely well (in the limit $\sigma^2 R_i/C_i$ used throughout the calculation) with the results of an evaluation, via numerical integration, of G and H. Further, the drift term $\overline{F}(U_1)$ is well-approximated by the first term in the approximation, i.e., the second term in Eq. (17). In this case, we may write

$$G(U_1) \approx \tanh\overline{U}_2 \approx J R_2 \tanh U_1. \tag{22}$$

For the diffusion term $D_1(U_1)$, however, one may make a similar approximation only in the case of very weak multiplicative noise, i.e., small σ^2 or, for a given σ^2, for small R_2. In this case, we have:

$$H(U_1) \approx \tanh^2\overline{U}_2 \approx J^2 R_2^2 \tanh U_1. \tag{23}$$

The conditions leading to approximations (22) and (23) have been further quantified by Schieve et al. (1990). In the limit where these approximations are justified, however, one may write the general one-neuron stochastic differential equation (20) in the form:

$$
\begin{aligned}
\dot{U}_1 = {}& -\frac{U_1}{R_1 C_1} + \frac{1}{C_1}\left[\sum_{j=2}^{N} \overline{J}_{1j}^2 R_j\right] \tanh U_1 \\
& -\frac{1}{2C_1^2}\left[\sum_{j=2}^{N} \sigma_{1j}^2 \overline{J}_{1j}^2 R_j^2\right] \tanh U_1 \operatorname{sech}^2 U_1 \\
& +\frac{1}{C_1}\left[\sigma_a^2 + \left\{\sum_{j=2}^{N} \sigma_{1j}^2 \overline{J}_{1j}^2 R_j^2\right\} \tanh^2 U_1\right]^{1/2} \xi(t),
\end{aligned}
\tag{24}
$$

where $\xi(t)$ is white noise having zero mean and unit variance. This equation has the *form* of the stochastic differential equation assumed by Bulsara et al. (1989) in their construction of a stochastic generalization to the single-neuron model of Babcock and Westervelt (1986, 1987). As mentioned earlier, however, a self-connection term with fluctuation strength $J_{11}(t)$ was included *a priori* in the dynamics by Bulsara et al. (1989). Such a term has been excluded from the current analysis. It is apparent from Eq. (24) that there are important additional terms in the effective one-neuron *self-diffusion*. Let us note, in passing, that one may examine the effective or macroscopic *potential* U_s corresponding to the steady-state single-neuron probability density function (14). This has been done

(Bulsara and Schieve, 1991) and the effects of Langevin and multiplicative noise treated in detail. By introducing *a priori* a self-coupling term in the neuron dynamics and assuming that the strength of the fluctuations in this term exceeds those in the cross-coupling terms, one recovers, rigorously, the simple model of Babcock–Westervelt and its stochastic generalization (Bulsara *et al.*, 1989). In general, the effective single-neuron self-diffusion terms derived previously would renormalize the self-coupling term. We shall not consider the precise nature of this renormalization in the remainder of this work. Rather, we shall, in the following section, return to the intuitive single-neuron model of Babcock–Westervelt (1986, 1987), ignoring the renormalization of the self-coupling term that has been derived earlier. Our purpose in the next section is to describe important cooperative phenomena in which the noise (both additive and multiplicative) plays a dominating role.

III. Response to Weak Modulation: Stochastic Resonance

The Babcock–Westervelt model consists of a single Hopfield-type computational element, consisting of an L-C-R series circuit with inertial feedback provided by an operational amplifier having a sigmoid transfer function. The circuit is described by Babcock and Westervelt (1986, 1987). The input voltage $U(t)$ obeys the rate equation,

$$LC\ddot{U} + RC\dot{U} + U - \alpha \tanh (\beta U) = F(t) + U_0, \qquad (25)$$

where β is the gain coefficient and the dots denote differentiation with respect to time, U_0 and $F(t)$ being externally applied dc and time-dependent inputs, respectively. We consider the case of large damping, assuming that the first term in Eq. (25) is negligible. In addition, we make the simple change of variables, $x(t) = \beta U(t)$, and take $F(t)$ to be Gaussian delta-correlated Langevin noise with zero mean and variance σ_a^2. Equation (25) then reduces to:

$$\dot{x} + \zeta x - \zeta \eta \tanh x = \zeta x_0 + \zeta \beta F(t), \qquad (26)$$

where we set $\zeta \equiv (RC)^{-1}$ and $\eta \equiv \alpha\beta$. The analogy of this equation to Eq. (4) is evident.

A complete analysis of Eq. (26), based on the Fokker Planck equation for the probability density function $P(x,t)$ has been given by Bulsara *et al.* (1989). This analysis takes into account the effects of multiplicative fluctuations (in the non-linearity parameter η) as well as the Langevin noise $F(t)$, and will not be repeated here. Instead, we turn to the overdamped Eq. (26) with $\zeta = 1$ for simplicity. In addition, we assume that no dc driving is present ($x_0 = 0$) and there are no multiplicative fluctuations. (This condition will be relaxed later.) In this case, the neuron may be treated as a particle in a one-dimensional potential given by:

$$U(x) = \frac{x^2}{2} - \eta \ln (\cosh x), \tag{27}$$

x being the one-dimensional state variable representing the firing rate. This potential is bimodal for $\eta > 1$, with the extrema occurring at:

$$c = 0, \pm \left[1 - \frac{1 - \tanh\eta}{1 - \eta \operatorname{sech}^2\eta} \right] \approx \eta \tanh\eta, \tag{28}$$

the approximation holding for large η. For a stationary potential, and for $D \ll \Delta U$ (we redefine $D \equiv \sigma_d^2/2$), where ΔU is the depth of the deterministic potential, the probability that a switching event will occur in unit time, i.e., the switching rate, is given by the Kramers formula (Kramers, 1940; Gardiner, 1985; Risken, 1989),

$$T_0 = (2\pi)^{-1}[|U^{(2)}(0)|U^{(2)}(c)]^{1/2} \exp\left(-\frac{\Delta U}{D}\right), \tag{29}$$

where $U^{(2)}(x) \equiv d^2U/dx^2$. It is important to realize that the Kramers formula (29) is only valid for large damping and relatively weak noise strength (as quantified before). For other ranges of parameters, various modifications to Eq. (29) have been proposed (Büttiker, 1989).

We now include a periodic modulation term $\varepsilon\sin\omega t$ on the right-hand side of Eq. (26). This leads to a modulation of the potential (27) with time: An additional term, $-x\varepsilon\sin\omega t$, is now present on the right-hand side of Eq. (27). In this case, the Kramers rate (29) becomes time-dependent:

$$r(t) \approx r_0 \exp\left(-\frac{x\varepsilon\sin\omega t}{D}\right), \tag{30}$$

which is accurate only for $\varepsilon \ll \Delta U$ and $\omega \ll \{U^{(2)}(\pm c)\}^{1/2}$. The latter condition is referred to as the *adiabatic approximation*. It ensures that the probability density corresponding to the time-modulated potential is approximately stationary. (The modulation is slow enough that the instantaneous probability density can *adiabatically* relax to a succession of quasi-stationary states.) This approximation was first investigated by Caroli *et al.* (1981) and first applied to stochastic resonance in climate models by Nicolis (1982).

We now follow the work of McNamara and Wiesenfeld (1989), developing a two-state model by introducing a probability of finding the particle in the left or right well of the potential. A rate equation is constructed based on the Kramers rate $r(t)$ given by Eq. (30). Within the framework of the adiabatic approximation, this rate equation may be integrated to yield the time-dependent conditional probability density function for finding the particle in a given well of the potential. This leads directly to the autocorrelation function, $\langle x(t)\, x(t + \tau)\rangle$, and finally, via the Wiener–Khinchine theorem, to the power spectral density $P(\Omega)$. The details are given by Bulsara *et al.* (1990c):

$$
P(\Omega) = \left[1 - \frac{2r_0^2\varepsilon^2 c^2}{D^2(4r_0^2 + \Omega^2)} \right]\left[\frac{8c^2 r_0}{4r_0^2 + \Omega^2} \right]
$$
$$
+ \frac{4\pi c^4 r_0^2 \varepsilon^2}{D^2(4r_0^2 + \Omega^2)}\, \delta(\omega - \Omega),
$$

(31)

where the first term on the right-hand side represents the noise background, the second term being the signal strength. We note that the presence of the delta function in the second term implies an infinite ratio of signal amplitude to noise amplitude. In practice, however, the delta function is reduced to a finite-amplitude, peaked function by the finite bandwidth of the measuring system (Zhou and Moss, 1989). To compare theoretical results with experiment, one must integrate over the lineshape of this function. This is tantamount to replacing the delta function in Eq. (31) by the quantity $(\Delta\omega)^{-1}$, where $\Delta\omega$ is the width of a frequency bin in the (experimental) Fourier transformation. We introduce (for the purpose of comparison with experiment) the signal-to-noise ratio, SNR = $10 \log R$ in decibels, where R is given by:

$$
R \equiv 1 + \frac{4\pi c^4 r_0^2 \varepsilon^2}{D^2(4r_0^2 + \omega^2)}\,(\Delta\omega)^{-1}\left[1 - \frac{2r_0^2\varepsilon^2 c^2}{D^2(4r_0^2 + \omega^2)} \right]^{-1}\left[\frac{4r_0^2 + \omega^2}{8c^2 r_0} \right].
$$

(32)

The location of the maximum of the SNR is found by differentiating the preceding equation; it depends on the amplitude ε and the frequency ω of the modulation, as well as the additive noise variance D.

In Fig. 1, we show the signal-to-noise ratio (SNR) as a function of the additive noise variance D for different modulation frequencies ω. As mentioned in the introduction, the SNR curve passes through a maximum as the noise strength is increased. The effect of lowering the driving frequency is to sharpen the resonance peak and shift it toward lower noise intensities.

The phenomenon quantified in the preceding may be readily explained. The effect of the weak modulating signal is to alternately raise and lower the potential well with respect to the barrier height ΔU. In the absence of noise and for $\varepsilon \ll \Delta U$, the system cannot switch states, i.e., no information is transferred to the output. In the presence of noise, however, the system can switch states

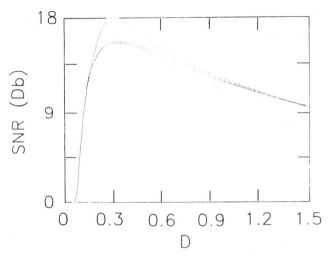

FIGURE 1.   SNR calculated from Eq. (32) with $(\eta, \varepsilon, \Delta\omega) \equiv (2, 0.2, 0.00153)$ and modulation frequency $\omega \equiv 0.031, 0.155, 031$ reading from the topmost curve downward.

through stochastic activation over the barrier. Although the switching process is statistical, the transition probability is periodically modulated by the external signal. Hence, the output will be correlated, to some degree, with the input signal.

The case of multiplicative noise (in addition to the additive noise considered previously) has been considered by Bulsara *et al.* (1990c). The calculation is similar to the preceding except that the Kramers rate r_0 in the absence of the modulation must now be changed. The potential $U(x)$ is now the *thermodynamic potential* obtained by writing down and solving (in the steady state) a Fokker Planck equation, with the nonlinearity parameter η now fluctuating as in the preceding section. One may then derive an expression for the signal-to-noise ratio R_m, which reduces to Eq. (32) in the absence of multiplicative noise. The dependence of R_m on the multiplicative noise variance D_m has been explored by Bulsara *et al.* (1990c). It has been found that increasing D_m leads to a decrease in the SNR as well as a shift in its maximum to lower values of the additive noise variance D. These effects are easily explained using the work of Bulsara *et al.* (1989), which predicts that increasing the multiplicative noise decreases the effective height ΔU of the potential barrier. In addition, the location of the peak in the SNR depends also on the multiplicative noise variance D_m.

The results of this section have been compared with those obtained from an analog simulation of the original equation (26) (Bulsara *et al.*, 1990c). The agreement is found to be very good. Analog simulations have also been carried out in the case of an inertial system (i.e., including an $\ddot{x}$ term in Eq. (26)), for which we do not at present have a theoretical result. It is found that decreasing the damping in an inertial system sharpens the SNR peak and shifts it toward smaller D values; these effects are also mimicked by decreasing the modulation frequency ω.

IV. Discussion

In this chapter, we have attempted to summarize the results of our work in the generic area of nonlinear stochastic processes in the simple one-neuron model. While much of our work has been based on the intuitive model, Eqs. (25) and (26), we have seen, in Section I, how such a model may be obtained, from the N-neuron model of Hopfield, under certain conditions; these conditions comprise the adiabatic elimination technique, also known as the *slaving principle* (Haken, 1977, 1983). One may obtain a deterministic equation of the form (26) through this procedure whether we assume an initial self-connection term in the Hopfield model or not (as has been done in this work). The extension to noisy neurons has also been shown to lead to qualitatively similar results; in this case, we obtain a single-neuron Fokker Planck equation from which an equivalent stochastic differential equation (20) may be deduced. Under certain conditions, one recovers an equation of the form (24) that is reminiscent of the stochastic generalization (Bulsara *et al.*, 1989) of the simple Babcock-Westervelt model introduced in section three. In any case, if we include *a priori* the self-connection term, then the theory of Section II describes how this term is modified or *renormalized* by the interactions with the other neurons in the network.

The implications of fluctuations in the dynamics of a simple neuron (and, possibly, in simple neural networks) are likely to be profound. It is well-known that the optimization of a single neuron, or a network of such neurons, depends on the ease with which the system can find the absolute minimum (corresponding to the state of minimum free energy) on its energy envelope. We have seen (Bulsara *et al.*, 1989) that, for the case of pure additive noise, one obtains a bimodal distribution above a certain critical value of the nonlinearity parameter so that although a high-gain system may well be desirable for certain applications, such a system can (for appropriate choices of the circuit and forcing parameters) admit of a double-well potential with differing probabilities (depending on the relative well depths) of finding it in one or the other stable state. By carefully selecting the dc driving term x_0, one well may be made very deep at the expense of the other, so that the system might find the global minimum more rapidly. At the same time, increasing the variance of the additive noise has the effect of increasing the halfwidth of the probability distribution function (while simultaneously decreasing the peak heights), so that the noise itself is instrumental in driving the system to its long-time equilibrium state. When multiplicative fluctuations are present (in addition to the external additive noise), they may prove to be a mixed blessing, since, on the one hand, they can suppress the deterministic bifurcation in the most probable value of the system response for certain values of the system parameters, thereby making the system appear monomodal even for $\eta > \eta_c$. (Of course, the choice of the dc driving term x_0 is important as well; this point has been elucidated by Bulsara *et al.* (1989).) On the other hand, the multiplicative fluctuations can introduce additional minima in the *thermodynamic* or *effective* potential (defined via the long-time solution of the Fokker Planck equation). By and large, the implications of multiplicative fluctuations in a single neuron and in simple networks, together with a clear understanding of their origin in a real system, are not well-known. Clearly, this is an area that merits future study.

We have concluded with the study of the (nonequilibrium) response of the single neuron to a weak external modulation superimposed on a noisy background. Under normal conditions, the power spectrum from a nonlinear oscillator such as Eq. (26), driven by a combination of noise and periodic modulation, consists of a broadband noise background upon which are superimposed odd harmonics of the modulation frequency ω. (The even harmonics of ω do not appear because of the internal symmetries of the system (Jung and Hanggi, 1989).) In the parameter range described in Section III, one obtains a maximum in the signal-to-noise ratio, as illustrated in Fig. 1. This effect has been (somewhat misleadingly) termed *stochastic resonance* in the literature. In this regime, the system behaves similarly to a nonlinear filter. One finds that the total power of the signal plus noise (the power being obtained by calculating the area under the power spectral density curve) is approximately constant in a wide bandwidth system. (It is absolutely constant in an infinite bandwidth system.) If one plots the power spectral density corresponding to the variable $x(t)$ in Eq. (26) in the absence of modulation, one obtains a broadband noise spectrum. When the modulation is turned on (in the parameter range for which one observes stochastic resonance), the broadband background is lowered and the peaks corresponding to the harmonics of ω appear. *The power in the signal spike, therefore, grows at the exact expense of the power in the background noise.* This property of gathering a fraction of the noise power and concentrating it into the signal power is a unique feature of stochastic resonance. It illustrates the ability of the noise to assist information flow to the output.

We have made no mention, in this chapter, about the theory of residence times, recently developed by Zhou *et al.* (1990). In their work, the probability density of escape times has been taken to be the physical quantity of interest, in contrast to the power spectrum. Within the framework of the same adiabatic approximation used in Section III, an analytic theory has been developed for the escape time probability density. One obtains, for the probability, a sequence of peaks located at times $t = nT/2$, where T is the period of the modulation and n is an odd integer. The amplitudes of the peaks decay exponentially with increasing time. On a semi-logarithmic scale, the slope of this decay line yields (to leading order in our adiabatic approximate theory) the ratio of signal strength to noise strength, i.e., ε/D in the notation of the preceding section. This technique should, in general, prove extremely useful for the detection/quantification/ analysis of a period signal whose frequency is its most important aspect. Recently, however, the concept of residence times probability density functions has been applied by the authors (Longtin *et al.* 1991) in a model description of experimental interspike-interval (isi) histograms. In this work, experimental isi data taken from two different animals (from two different sensory organs) some 25 years apart (Rose *et al.*, 1967, 1969; Siegel (1990)) have been compared with results obtained from a simple bistable model of a neuron. In our model (developed in the spirit of the classical and seminal work of Gerstein and Mandelbrot (1964)), the neuron is assumed to be inherently noisy and is then subject to a deterministic modulating signal at a known frequency. The residence time in each well of the potential has been measured and the data assembled into a histogram. This yields the characteristic sequence of peaks with an exponentially decaying envelope. By assuming that a counting interval (i.e., a given particle trajectory) starts in the left-hand well, say, proceeds to the right-hand well and then terminates in the left-hand well, the peaks in the isi histogram occur at multiples of the modulation period as is the case in the experimental histograms. This case should be contrasted with the case referred to earlier in this paragraph, wherein the peaks are located at odd-integer multiples of one-half the modulation period. These two cases point up two possible symmetries in the model system and, as mentioned in the introduction, an understanding of these symmetries, fundamental to bistable systems, has pointed the way to an explanation of the experimental isi histogram data, as well as pointing out a link to the experimentally well-established concept of a refractory/reset mechanism in real neurons. The preceding sequence has been observed in a *hard potential* model (the Schmitt trigger) of a real neuron as well as in a model with a finite slope in the transfer characteristic, of the type considered throughout this work. What is important is the fact that *this sequence of peaks cannot exist in the absence of noise,* leading us to conclude that these sensory organs have evolved in such a way as to exploit noise in the transmission of information.

We conclude with some comments regarding possible future directions for this research. Although a vast literature exists on the subject of neural networks, relatively little attention has been focused on the details of individual neuron models. On first glance, the simple bistable model treated in Section III might appear to be naive, although the work of Section II provides some basis for assuming such a model. Real neurons are known to exhibit chaos (e.g., Guevara *et al.*, 1983; Holden, 1986; Milton *et al.*, 1989) under certain conditions, either when coupled to other neurons or when periodically modulated. The inertial model of Babcock and Westervelt (1986, 1987) has attempted to explain these effects. The interplay between chaos and high-dimensional noise, as investigated, for example, by Crutchfield *et al.* (1982), Bulsara *et al.*, (1990a,b), and Schieve and Bulsara (1990) in connection with other nonlinear dynamic systems, has yet to be investigated in neural models. Moreover, one is led to speculate on the possibilities for stochastic resonance, or similar phenomena, in slowly modulated chaotic systems with multiple strange attractors. Finally, real neurons are known to exhibit time delay (Longtin and Milton, 1989; MacDonald, 1989; Aihara *et al.*, 1990; Longtin *et al.*, 1990); in such cases, even a one-degree-of-freedom system may be chaotic. Virtually nothing is known about delay-differential systems subject to noise. The questions raised in these topical areas would seem to offer fruitful opportunities for research with more realistic applications to neural modeling.

Acknowledgment

The work described in this chapter was supported in part by the Office of Naval Research under grant N000149AF00001.

References

AIHARA, K., TAKABE, T., and TOYODA, M. (1990). "Chaotic Neural Networks," *Physics Letters* **144,** 333–340.

AMIT, D. (1989). *Modeling Brain Function.* Cambridge University Press, London.

BABCOCK, K., and WESTERVELT, R. (1986). "Stability and Dynamics of Simple Electronic Neural Networks with Added Inertia," *Physica* **23D,** 464–469.

BABCOCK, K., and WESTERVELT, R. (1987). "Dynamics of Simple Electronic Neural Networks," *Physica* **28D,** 305–316.

BUHMANN, J., and SCHULTEN, K. (1986). "Influence of Noise on the Behavior of an Autoassociative Neural Network, in J. Denker (ed.), *Neural Networks for Computing AIP Conference Proceedings, Vol. 151.* AIP, New York.

BUHMANN, J., and SCHULTEN, K. (1987). "Influence of Noise on the Function of a 'Physiological' Neural Network," *Biological Cybernetics* **56,** 313–327.

BULSARA, A., SCHIEVE, W., and GRAGG, R. (1978). "Phase Transitions Induced by White Noise in Bistable Optical Systems," *Physics Letters* **68A,** 294–296.

BULSARA, A., LINDENBERG, K., SESHADRI, V., SHULER, K., and WEST, B. (1979). "Stochastic Processes with Non-Additive Fluctuations: II. Some Applications of the Ito and Stratonovich Calculus," *Physica* **97A**, 234–243.

BULSARA, A., SCHIEVE, W., and JACOBS, E. (1987). "Noise-Induced Critical Behavior in a Multistable System," *Physica* **146A**, 126–150.

BULSARA, A., BOSS, R., and JACOBS, E. (1989). "Noise Effects in an Electronic Model of a Single Neuron," *Biological Cybernetics* **61**, 211–222.

BULSARA, A., SCHIEVE, W., and JACOBS, E. (1990a). "Homoclinic Chaos in Systems Perturbed by Weak Langevin Noise," *Physical Review* **A41**, 668–681.

BULSARA, A., JACOBS, E., and SCHIEVE, W. (1990b). "Noise Effects in a Nonlinear Dynamic System: The rf Superconducting Quantum Interference Device," *Physical Review* **A42**, 4614–4621.

BULSARA, A., JACOBS, E., ZHOU, T., MOSS, F., and KISS, L. (1990c). "Stochastic Resonance in a Single Neuron Model: Theory and Analog Simulation," *Journal of Theoretical Biology* (in press).

BULSARA, A., and SCHIEVE, W. (1991). "The Single Effective Neuron: Macroscopic Potential and the Effects of Noise," *Physical Review,* in press.

BÜTTIKER, M. (1989). "Escape from the Underdamped Potential Well," in F. Moss and R. McClintock (eds.), *Noise in Nonlinear Dynamic Systems, Vol. 2*. Cambridge University Press, London.

CAROLI, B., CAROLI, C., ROULET, B., and SAINT-JAMES, D. (1981). "On Fluctuations and Relaxation in Systems Described by a 1-D Fokker Planck Equation with Time-Dependent Potential," *Physica* **A108**, 233–256.

CRUTCHFIELD, J., FARMER, D., and HUBERMAN, B. (1982). "Fluctuations and Simple Chaotic Dynamics," *Physics Reports* **92**, 45–82.

DEBNATH, G., ZHOU, T., and MOSS, F. (1989). "Remarks on Stochastic Resonance," *Physical Review* **A39**, 4323–4326.

ENGLUND, J., SNAPP, R., and SCHIEVE, W. (1984). "Fluctuations, Instabilities and Chaos in the Laser-Driven Nonlinear Ring Cavity," in E. Wolf (ed.), *Progress in Optics, Vol. 21*. North-Holland, Amsterdam.

GAMMAITONI, L., MENCHILLA-SAETTA, E., SANTUCCI, S., MARCHESONI, F., and PRESILLA, C. (1989). "Periodically Modulated Bistable Systems: Stochastic Resonance," *Physical Review* **A40**, 2114–2119.

GARDINER, C. (1985). *Handbook of Stochastic Processes*. Springer, New York.

GERSTEIN, G., and MANDELBROT, B. (1964). "Random Walk Models for the Spike Activity of a Single Neuron," *Biophys. J.* **4**, 41–64.

GUEVARA, M., GLASS, L., MACKEY, M., and SHRIER, A. (1983). "Chaos in Neurobiology," *IEEE Transactions on Systems, Man and Cybernetics* **SMC-13**, 790–798.

HAKEN, H. (1977). *Synergetics*. Springer, New York.

HAKEN, H. (1983). *Advanced Synergetics*. Springer, New York.

HOLDEN, A. (1976). *Models of the Stochastic Activity of Neurons*. Springer, Berlin.

HOLDEN, A. (1986). *Chaos*. Manchester University Press, Manchester, U.K.

HOPFIELD, J. (1982). "Neural Networks and Physical Systems with Emergent Collective Computational Abilities," *Proceedings of the National Academy of Sciences* **79**, 2554–2558.

HOPFIELD, J. (1984). "Neurons with Graded Response Have Collective Computational Properties like Those of Two-State Neurons," *Proceedings of the National Academy of Sciences* **81**, 3088–3092.

HORSTHEMKE, W., and LEFEVER, R. (1984). *Noise-Induced Transitions*. Springer, New York.

JUNG, P. (1989). "Thermal Activation in Bistable Systems under External Periodic Forces," *Zeitschrift für Physik* **B16**, 521–535.

JUNG, P., and HANGGI, P. (1989). "Stochastic Nonlinear Dynamics Modulated by External Periodic Forces," *Europhysics Letters* **8**, 505–511.

JUNG, P., and HANGGI, P. (1990). "Resonantly Driven Brownian Motion: Basic Concepts and Exact Results," *Physical Review* **A41**, 2977–2988.

KRAMERS, H. (1940). "Brownian Motion in a Field of Force and the Diffusion Model of Chemical Reactions," *Physica* **7**, 284–304.

LI, Z., and HOPFIELD, J. (1989). "Modeling the Olfactory Bulb and Its Neural Oscillatory Processings," *Biological Cybernetics* **61**, 379–392.

LONGTIN A., and MILTON, J. (1989). "Modeling Autonomous Oscillations in the Human Pupil Light Reflex Using Nonlinear Delay-Differential Equations," *Bulletin of Mathematical Biology* **51**, 605–624.

LONGTIN, A., MILTON, J., BOS, J., and MACKEY, M. (1990). "Noise and Critical Behavior in the Pupil Light Reflex at Oscillation Onset," *Physical Review* **A41**, 6992–7005.

LONGTIN, A., BULSARA, A., and MOSS, F. (1991). "Time Interval Sequences in Bistable Systems and the Noise-Induced Transmission of Information by Sensory Neurons," *Phys. Rev. Lett.* **67**, 656–659.

MACDONALD, N. (1989). *Biological Delay Systems*. Cambridge University Press, London.

MCNAMARA, B., and WIESENFELD, K. (1989). "Theory of Stochastic Resonance," *Physical Review* **A39**, 4854–4869.

MCNAMARA, B., WIESENFELD, K., and ROY, R. (1988). "Observation of Stochastic Resonance in a Ring Laser," *Physical Review Letters* **60**, 2626–2629.

MILTON, J., LONGTIN, A., BEUTER, A., MACKEY, M., and GLASS, L. (1989). "Complex Dynamics and Bifurcations in Neurobiology," *Journal of Theoretical Biology* **138**, 129–147.

MOSS, F., and MCCLINTOCK, P. (1989). *Noise in Nonlinear Dynamic Systems*. Cambridge University Press, London.

NICOLIS, C., and NICOLIS, G. (1981). "Stochastic Aspects of Climatic Transitions—Additive Fluctuations," *Tellus* **33**, 225–237.

NICOLIS, C. (1982). "Stochastic Aspects of Climatic Transitions—Response to a Periodic Forcing," *Tellus* **34**, 1–9.

PAULUS, M., GASS, S., and MANDELL, A. (1989). "A Realistic 'Middle Layer' for Neural Networks," *Physica* **D40**, 135–155.

PERETTO, P. (1984). "Collective Properties of Neural Networks: A Statistical Physics Approach," *Biological Cybernetics* **50**, 51–62.

PRESILLA, C., MARCHESONI, F., and GAMMAITONI, L. (1989). "Periodically Time-Modulated Bistable Systems: Nonstationary Statistical Properties," *Physical Review* **A40**, 2105–2113.

RISKEN, R. (1984). *The Fokker Planck Equation.* Springer, New York.

ROSE, J., BRUGGE, J., ANDERSON, D., and HIND, J. (1967). "Phase-Locked Response to Low-Frequency Tones in Single Auditory Nerve Fibers of the Squirrel Monkey," *J. Neurophysiol.* **30**, 769–793.

ROSE, J., BRUGGE, J., ANDERSON, D., and HIND, J. (1969). "Some Possible Neural Correlates of Combination Tones," *J. Neurophysiol.* **32**, 402–423.

SAMPOLINSKY, H. (1988). "Statistical Mechanics of Neural Networks," *Physics Today* **41**, 70–80.

SCHENZLE, A., and BRAND, H. (1979). "Multiplicative Stochastic Processes in Statistical Physics," *Physical Review* **A20**, 1628–1647.

SCHIEVE, W., and BULSARA, A. (1990). "Multiplicative Noise and Homoclinic Crossing (Chaos)," *Physical Review* **A41**, 1172–1174.

SCHIEVE, W., BULSARA, A., and DAVIS, G. (1990). "The Single Effective Neuron," *Physical Review* **43**, 2613–2623.

SHAMMA, S. (1989). "Spatial and Temporal Processing in Central Auditory Networks," in C. Koch and I. Segev (eds.), *Methods in Neuronal Modeling.* MIT Press, Cambridge, Massachusetts.

SIEGAL, R. (1990). "Non-Linear Dynamical System Theory and Primary Visual Cortical Processing," *Physica* **42D**, 385–395.

STRATONOVICH, R. (1963). *Topics in the Theory of Random Noise.* Gordon and Breach, London.

STRATONOVICH R. (1989). "Some Markov Methods in the Theory of Stochastic Processes in Nonlinear Dynamical Systems, in F. Moss and P. McClintock (eds.), *Noise in Nonlinear Dynamic Systems, Vol. 1.* Cambridge University Press, London.

TUCKWELL, H. (1988). *Introduction to Theoretical Neurobiology.* Cambridge University Press, London.

VEMURI, G. and ROY, R. (1989). Stochastic Resonance in a Bistable Ring Laser," *Physical Review* **A39**, 4668–4674.

WEST, B., BULSARA, A. LINDENBERG, K., SESHADRI, V., and SHULER, K. (1979). "Stochastic Processes with Non-Additive Fluctuations: I. Ito and Stratonovich Calculus and the Effects of Correlations," *Physica* **97A**, 211–233.

ZHOU, T., and MOSS, F. (1989). "Remarks on Stochastic Resonance," *Physical Review* **A39**, 4323–4326.

ZHOU, T., and MOSS, F. (1990). "Analog Simulations of Stochastic Resonance," *Physical Review* **A41**, 4255–4264.

ZHOU, T., MOSS, F. and JUNG, P. (1990). "Escape Time Distributions of a Periodically Modulated Bistable System with Noise," *Physical Review* **A42**, 3161–3169.

Chapter 20

Critical Coherence and Characteristic Times in Brain Stem Neuronal Discharge Patterns

KAREN A. SELZ and ARNOLD J. MANDELL
Laboratory of Experimental and Constructive Mathematics
Departments of Mathematics and Psychology
Florida Atlantic University
Boca Raton, Florida

I. Introduction

Pathophysiology can often serve to elucidate normal physiological mechanisms. For a number of years, our group has pursued the hypothesis that a broad class of *dynamical disorders* (i.e., ones that occur in the absence of established neuroanatomical or chemical defects) are present in systems that demonstrate a loss of dynamical complexity, a decrease in effective dimensionality, and/or an entropy reduction (to be defined more precisely shortly) (Mandell, 1981). Evidence for this and similar points of view has come, and continues to come, from studies of a variety of neurobiological systems.

In this context, we have recently reviewed neurobiological evidence that the clinical phenomena of neurological disorders with dopaminergic neostriatal involvement reflect losses in complexity of another kind. Here, the global simplicity takes the form of frozen posturing and inertia, and involuntary behavior such as chorea, ballismus, myoclonus, and athetosis (Kelleher and Mandell, 1990). We have suggested that competing influences in the nigrostriatal system, analogous to physical spin-coupling of magnets tending to order, and thermal randomizing

Copyright © 1992 by Academic Press, Inc.
All rights of reproduction in any form reserved.
ISBN 0-12-484815-X

effects tending to disorder, are resolved in favor of the former leading to long-range temporal coherence. This decrease in complexity results in coarsening, simplification, and polarization of the characteristic times of motor behavior, ranging from thrashing to a frozen, catatonic posture.

The postulated locus for the *nearest neighbor coupling* is the *Spiny I* neuron network of the caudate. This system is dominated by GABAergic, inhibitory neostriatal efferents and (like the cerebellum and visual systems) serves to filter and regulate neocortical motor systems via a lateral inhibitory network (Ward, 1968; Kemp and Powell, 1971; Potegal, 1972; Groves, 1983). The *thermal bath field influence* on the caudate lateral inhibitory network is postulated to involve the diffusely projecting, multiply-bifurcating midbrain dopamine neurons (Schwyn and Fox, 1974; Swanson, 1982). As these axons divide, their diameter decreases along with their conduction velocity, thus *decorrelating* the arrival times and *mixing* the signal, hampering the tendency for the caudate *Spiny I* neuron network to couple and organize.

The dynamical interaction between the nigral dopaminergic, randomizing field and the caudate–putamen lateral inhibitory network involves mutual feedback loops (Carpenter, 1976), which both require and modulate the *timing of impulses in the system* (Buchwald *et al.*, 1979). It is the temporal pattern of behavior of these *thermal field* systems (e.g., the brain stem norepinephrine and dopamine neurons) that is of interest here. If temporal complexity below some critical level is pathophysiological, how does sufficient order for the transport of information using sequential inter-event times take place? How can it be described quantitatively?

II.   Temporal Complexity as a Characteristic of Normal Neuronal Behavior

Neurobiological experiments and theory directed toward an understanding of information encoding and transport in the nervous system have generally involved:

Issues of place: where, in the nervous system, changes in activity evoked by sensory pertubation can be located, and

Issues of time: correlations in electrophysiological fluctuations between several locations presumed to support (unknown) dynamical representations of the information being processed.

Recently, studies in the visual and olfactory systems have used temporal and spatial high cross-correlations—in particular, frequency bands—as evidence that multicentered, cooperative processes are *involved phenomena*. What that may mean, either mechanistically or with respect to abstract information processes, has generally not been addressed.

The idea of periodic coherence, e.g., 40 Hz *mode resonances,* in electrophysiological signals as evidence of information transport between structures is less than attractive for at least two reasons:

(1) There is growing evidence in neuronal, global, electrophysiological, neuroendocrine, and cardiac systems that periodic coherence is associated with pathophysiological conditions marked by communication *failure* seen in the loss of adaptation to regulatory information. For example, induced epileptic foci are characterized by rhythmic neuronal firing patterns (Tepper and Mandell, 1987) and measures of complexity are characteristically reduced in the global brain wave patterns of epileptic patients (Babloyantz and Dextexhe, 1986).

(2) Shannon's second theorem (stated intuitively) suggests that reliable information transport requires greater (dynamical) complexity in the receiving channel than that being generated by the sender. Isometries such as frequency resonances between receiver and sender have zero entropy.

Thus, on both empirical and theoretical grounds, it is unlikely that the usual physical *resonances* (with sharp maxima along an energy or energy-equivalent scale) are relevant to information encoding and transport in brain.

Further, singly periodic frequency-locking is an unlikely candidate as a mechanism for information transport because of the intrinsic multiplicity of time scales found in neuronal dynamics. For example, Fig. 1 depicts such a range of times for presynaptic neurotransmitter regulation in a representative brain stem biogenic amine neuron. Since autoreceptor control of firing patterns requires the presence of these mechanisms for neurotransmitter production, it is clear that even with nonlinear coupling, no single time can serve as a communication channel.

What are the alternatives to mode resonances in the dynamical brain processes of recognition required for information transport, given the condition that this system must operate in the context of the generation of (positive) entropy at some rate per unit time?

A remarkable recent series of studies has demonstrated measure-theoretic entropic near-isomorphisms in the human brain's electroencephalographic electrovoltage fluctuations in relationship to quite subtly different objectifiable states of arousal and sleep (Babloyantz *et al.,* 1985; Rapp *et al.,* 1985; Mayer-Kress and Layne, 1987; Graf and Elbert, 1989; Roschke and Bazar, 1989). A good example is the study by Roschke and Aldenhoff (1990). Using the Grassberger and Procaccia correlation integral, *fractal dimension, D_2,* they successfully

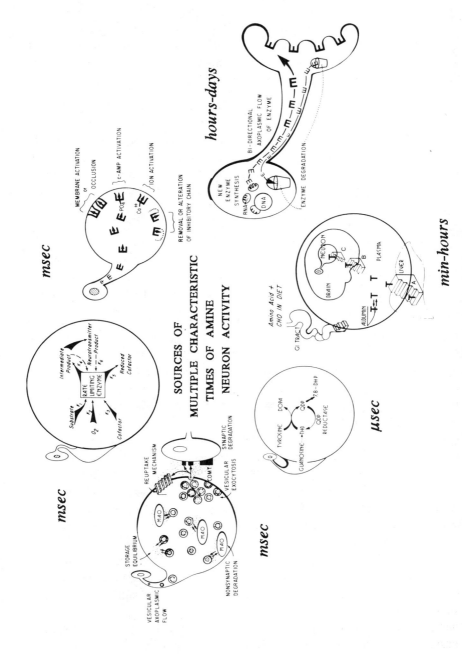

SOURCES OF
MULTIPLE CHARACTERISTIC
TIMES OF AMINE
NEURON ACTIVITY

discriminated the Rechtschaffen and Kales sleep stages in 10 normal subjects: 12–14 Hz, *descending stage* I, $D_2 = 6.51 \pm 0.43$; 8–10 Hz, stage II, $D_2 = 5.67 \pm 0.45$; 4–6 Hz, stage III, $D_2 = 4.69 \pm 0.41$; 1–3 Hz, stage IV, $D_2 = 4.25 \pm 0.23$; and 10–12 Hz, *emerging stage* I, *dreaming*, $D_2 = 6.08 \pm 0.46$.

The work of Joseph Kamiya and Barbara Brown in the late 1960s and others has shown that these kinds of global electrophysiological states can serve as signals for behavior in man. For example, the same broad power spectral bands shown to be equivalent to EEG states (defined by their correlation dimension), when paired to colored lights, resulted in experimental subjects learning to selectively increase their measure within specific frequency ranges during a standardized period of observation. These studies encourage the use of the same kinds of equivalence relations, for neurobiological information transport, as those used in categorical approaches to abstract dynamical systems (i.e., topological, statistical, and generally entropic).

One, *topological conjugacy,* involves relations that can exploit near-equivalence of the *topological entropy* h_T, which (roughly stated) measures the orbital complexity of the system.

For $\varphi^t{:}X \rightarrow X$ as a continuous self-map on a compact metric space and $\varepsilon > 0$ as a small positive real, two t orbits, $\varphi^t x, \varphi^t y$, are ε-distinguishable if the distance $d(\varphi^t x, \varphi^t y) > \varepsilon$. If $n(t,\varepsilon,\varphi)$ is the largest number of ε-distinguishable t orbits, n grows at most exponentially. This approach requires the embedding of the dynamics into a phase space of suitable dimension, the *partition of phase space*, and *symbolic dynamics on that partition* act by sacrificing orbital details but preserving the topology of the behavior. Its form is the result of the action of a map, $\psi{:}X \rightarrow \Sigma$, which assigns a sequence of points in phase space to string Σ of symbols ω_i in sequence space $\Sigma\omega_i$, from which estimates can be made on asymptotic rate of the appearance of new subsequences (*words of length N*) as a function of symbol sequence length $\Sigma\omega_i$. For an expanding dynamical system, it is obvious that the smaller the subdivisions of the partition, the faster the rate of appearance of new subsequences, $\partial N/\partial\Sigma$, with the so-called *generating partition* maximizing the rate $= \log \partial N/\partial\Sigma = h_\text{T}$.

The measure-theoretic entropy h_M reflects the distribution of the measure μ

FIGURE 1.   Each biogenic amine neuron, in addition to perturbation by a variety of arrival times of axons from sensory and other integrative neurons, regulates its own activities via autoreceptors sensitive to the patterns of release of its own neurotransmitters. This sketch summarizes the multiplicity of neurochemical process times intrinsic to neurotransmitter biosynthesis. These processes influence autoreceptor control of intrinsic firing patterns.

on the asymptotic space of subsequences, $p(N) = \mu(N)$, and the Shannon entropy, $h_M = -\Sigma\mu(N)\log\mu(N)$. h_M is maximal when $\mu = 1/N$. It is also the case that the system's entropy is maximal when $h_T - h_M = 0$.

If we partition $\Sigma\omega_i$ as $[+, -]$, $N = 2$ (states), we can invoke another entropy, which we can call $h(R)$: $\log_{(1/p)}[\partial R_{1/(+)}/\partial_i]$, where p represents the bias of the binary process analogous to the *fairness* or *unfairness* of a coin required to generate the observed growth rate of the longest run length as a random process, $\partial R_{(+)}/\partial_i$. Maximal $h(R)$, as in the case of $1/N$, $N = 2$, $h(R)_M$ can be expressed as $\log_{(1/0.5 = 2)}$.

An estimate of the upper bounds on h_T can be obtained by a computation of the leading Lyapounov exponent, $\Lambda_{max} = \lim t \to i, 1/t \log|D(\tau_t)|$ (Grassberger and Procaccia, 1983; Wolf *et al.*, 1985), which also quantitates the divergence rate of nearby initial conditions and *mixing,* a property that relates to the convergence rate of measures such as the (auto)correlation function (Parry, 1981). The autocorrelation function $C(\tau) = [\langle\tau(t + i)\tau(t)\rangle - \tau(t) > ^2] \to 0$ as $t \to \infty$ at a rate defining the rate of mixing. Its Fourier transformation yields the frequency (power) spectrum, $S(\omega) = \int C(\tau)\exp[-i\omega t]d\tau$.

Estimation of the global characteristic time of variation of the inter-event times of a sequence of neuronal discharges presents some difficulties. As will be seen shortly, the decay of $C(\tau)$ is nonexponential and its Fourier transformation yields multiple broadband peaks. It appears that the global characteristic time is composed of a multiplicity of local times, which when averaged (as seen shortly) yield density distribution functions, $\rho(\tau)$, with nonconvergent higher moments. In fact, intermittent bursting of interspike intervals generates distributions of times with variances that *diverge* with n.

We have tried to obtain a *metric-free* estimate of relative times, a more topological description of patterns of sequences that might scale across absolute time. We derived our estimate from the ratio of two exponential (inverse) rates. The dividend is an *extrinsic measure* of the mixing influence of neuronal input, the largest Lyapounov exponent, Λ_{max}, the exponential separation rate of nearby initial conditions. The divisor represents an *intrinsic measure* reflecting the membrane repolarization mechanism (say, some bound on the extra-membrane ion diffusivity process after depolarization) such that fast repolarization rates lead to the tendency to burst. This process is represented by *probability as a time;* that is, as deviation from the minimum ($p = 0.5$) exponential growth rate of the longest run in the binarily reduced *symbolic dynamical system.* (Following the partition of the interspike interval between short and long, the orbit is coded in 0's and 1's.)

With $G(\tau)$ representing the global characteristic time and $\mathbf{T} = |h(R)_{max} - h(R)| \neq 0$:

$$G(\tau) = \int k \exp(-\Lambda_{max} \cdot \tau/\mathbf{T})d\tau. \tag{1}$$

Our general goal is the application of approaches similar to those of Kolmogorov and Sinai (1958, 1959), Anosov (1967), Ornstein (1970), and Parry and Tuncel's (1982) techniques involving isomorphisms of abstract dynamical systems to equivalence relations in neurobiological states (Mandell, 1987; Paulus and Mandell, 1990; Mandell and Shlesinger, 1990; Mandell and Kelso, 1991; Mandell and Selz, 1991).

A neurobiological way of looking at the role of the entropy of elemental processes, such as the intervals between neural discharges (reflecting their latencies to excitation and refractoriness), might involve the length and *shape* of the envelope representing the cooperative electrical events around regions of neuronal communication. These events involve the time constants of synaptic events and dendritic fields, which extend far longer than those of single-neuron depolarization events. Electroencephalographic patterns that correlate with states of arousal and cognition can be seen as the electromagnetic field representation of the global state of these regions in cooperative interaction. We view the *thermal field neurons* of the brain stem (biogenic amine and reticular systems) as *mixing systems* that control the rate of decay of correlation functions as characteristic times in downstream systems such as the neostriatum.

III.   Elemental Mechanics of Single-Neuron Activation

A walk through the history of understanding the physical mechanisms underlying membrane action potentials, from the discovery by Arvanitaki (1939) of post-tetanic, subthreshold *spontaneous* oscillations in millivolts per 10 msec in *Sepia* axons, through the decades of studies of voltage-clamped membrane ion charge carrier conductances (Cole, 1972), to the present era of instabilities and fluctuations in transmembrane polypeptide helical channels as protein dynamics at a water–lipid interface subject to the electrostatic influences of charge separation, would exemplify Feynman's remarks about the persistent value of the correct equations, though their apparent physics may evolve and change over time.

The *equivalent LRC circuit* (L ≡ inductance, R ≡ resistance, C ≡ capacitance) for the relaxation oscillations of the membrane potential were first thought to be sustained by a negative conductance element, $f(V)$. The Hopf bifurcation theorem (proven in 1942) showed how the stable fixed point of nonlinear differential equations could be altered by a critical change in the nonlinear parameter, $r \geq r_c$, to a repeller, and the dynamics changed to self-sustaining stable and unstable limit cycles. Now we see that it is the neurobiological information that enters through $f(V)$.

The time-dependent dynamics of this membrane equivalent circuit can be represented generally as:

$$\frac{dV}{dt} = \frac{i + f(V)}{C}, \tag{2}$$

$$\frac{di}{dt} = \frac{V - Ri}{L}, \tag{3}$$

and the i/V relation, assuming the first-order partial derivatives with respect to i and V of $f(i,V)$ exist and are continuous, is the quotient of Eqs. (1) and (2):

$$\frac{di}{dV} = \frac{V - Ri}{i + f(V)}. \tag{4}$$

Van der Pol activated a passive *RLC* circuit like Eqs. (1), (2), and (3) by a vacuum tube, $f(V)$, generating periodic signals. The resistor element R, *defining the functional relationship between current and voltage,* $V_R = f(i_R)$, had an Ohm's law equivalent to a cubic characteristic such that there were three values of V for a given small range of i. This N characteristic in the form of negative slopes and *wrong way* currents was observed in the squid axon membrane ionic current-potential characteristic during action potentials by Cole and Curtis in 1939 (Cole, 1972). Arvanitaki's subthreshold oscillations and this cubic i/V characteristic are sufficient to generate nonuniformities in phase progression (fast–slow manifolds) in the driven, nonlinear system.

The values of i between R, L, and C are equal except for the sign. Kirchhoff's law of currents and his law for voltage says that $\Sigma V_{R,L,C} = 0$. Combining these constraints with Faraday's law for the inductance, $L(di/dt) = V_L^i$, and the general equation for the capacitance, $C(dV/dt) = ic^i$, we can reduce the six-dimensional vector space of the *autonomous* system, $i^{-i}(L,R,C) \cdot V^i(L,R,C)$, to a state space in $\mathbb{R}^2$, since $i^{-i}(L)$ and $V^i(C)$ can be shown to constrain the other four coordinates. If we normalize L and C to 1 and regard current i as parametrized by a rate, *amount* flowing past a point per unit time, a *velocity,* and V its potential (*energy*) for doing so, we can differentiate by parts such that the phase space dynamics described in Eq. (3) become more generally a plot of the derivative of the potential, a velocity v versus the derivative of the velocity, v', which demonstrates wide variations.

More literal modeling using the Hodgkin–Huxley equations exploit as observables the ion conductances in voltage-clamped nerve membrane preparations. We remind ourselves that the original reduction of the H and H system to a pair of nonlinear differential equations with two parameters (excitability and refractoriness) was inspired by the mathematical nerve modeling of Kenneth Cole, who first suggested that the forced van der Pol equation could serve as a reduced H and H equation after his discussions with Solomon Lefschetz (who is said to have brought what is now considered modern dynamical systems theory to America from Russia in the 1930s). The topological reduction can be sketched as:

From $\mathbb{R}^4 \to \mathbb{R}^2$ (5)

$$\partial(Vt) \equiv \text{perturbation in resting membrane potential;}$$

fast:

$$m'(Na^+) \equiv \text{rate of sodium influx}$$

$$n'(K^+) \equiv \text{rate of potassium efflux}$$

slow:

$$h'(Na^+) \equiv \text{rate of sodium efflux}$$

Such that $x' \equiv (\delta, m')$, and $y' \equiv (n', h')$ and

$$x' = -(x + y)$$ (6)

for the fast (excitation) and slow (recovery) processes
$$y' = x$$

Inserting the cubic discontinuous excitability threshold characteristic, indicating the nonlinear *forced = disappative term r*, parametrically scaling x in the y' equation, and using z as the input perturbation, we have the classical Fitzhugh–Nagumo nerve membrane equations in $\mathbb{R}^2$

$$x' = y - r(x - x^3/3) + z,$$ (7)

$$y' = -\frac{x}{r} + z.$$

In the context of a perturbation-requiring theory, a neural membrane does not speak unless spoken to. The critical nullclines (x' and $y' = 0$) intersect at what was thought to be a stable fixed point, and in this context, a sufficient perturbation of z was required to break out of it and to achieve a membrane activation–recovery cycle. However, Kenneth Cole and (about the same time) Russian membrane physiologists found continuous subthreshold oscillations in membrane potentials such that z became $f(t)$. With trigonometric functions serving in addition as a convenient metric for the phase space, Z became $b \cos \lambda t$, and the equation with $r > r_c$ and $b > b_c$ (above some critical value), opened the possibility for high nonlinearity and amplitude, and the Cartwright–Littlewood–Levinson first strange attractor:

$$x' = y - r(x - x^3/3), \tag{8}$$

$$y' = -\frac{x}{r} + b \cos\lambda t.$$

As r and b grow, the relaxation oscillator, slow around the $y \equiv 0$ saddle sinks and fast between them, goes through a homoclinic tangency, $W^S \perp W^U$, and then an intersection such that (unproven) the saddle sinks can be reached in positive and negative time in a homoclinic orbit. This transition is portrayed in Fig. 2 as the result of a small increase in the nonlinear parameter r. Since we are studying the variations in the inverse velocity = time as a function of time, $\partial(\partial x/\partial t)^{-1}/\partial t = \partial\tau/\partial t$, the relevant phase portrait of Eq. (8) is not a plot of x against $\partial x/\partial t$, but a plot of the velocity (as an inverse time) against its derivative (i.e., $\partial\tau/\partial t$). Figure 3 is an analog computer simulation (EAI-680) of Eq. (8) as described across systematic changes in the nonlinear parameter r and the frequency λ.

We see that parametric control and measurement of the characteristic *dwell* times, how many times the orbit goes around the central saddle sink, and *return* times, how long it takes to get back, are not trivial problems. We remind ourselves that we have defined a completed recurrent orbit, small or large as an activation–refractory cycle, with its size inverse to its rate—staying in the saddle sink implies bursting with a characteristic time τ_i, and leaving and returning to the saddle sink bursting regime is another time, τ_j. Whereas τ_i is under local control via the eigenvalue λ_i of the unstable fixed points, τ_j involves the global character of the system's dynamics.

The issues surrounding parametric control of the tangency to intersection transition have become more complex. Recent results of Palais and Takens (1987) demonstrate that if the eigenvalue of the unstable manifold of the saddle sink fixed point is much greater than that of the stable one, $\lambda 1(W^u) \gg \lambda 2(W^S)$, then there are *premature horseshoes*—homoclinic intersection and chaos (i.e., sensitivity to initial conditions, undecomposability, and dense unstable periodic orbits) *before tangency* and, therefore, bursting. This suggests that there may be random neuronal interspike intervals *on both sides of bursting behavior* along a parametric path. Homoclinic orbits in the region of saddle nodes and saddle node bifurcations are generic (i.e., typical). The Palais–Takens condition, $dH(W^S \perp W^U) > 0$, the homoclinic tangency that generates bursting in its time series, can also be found in association with saddle periodic orbits, complex sets with a range of phase space structures, and a dissipative torus in dimensions 3 and 4.

Figure 4 represents a frequency (power) spectral transformation of an $n = 500$ time series of normalized values (0,1) sampled every 100 milliseconds from

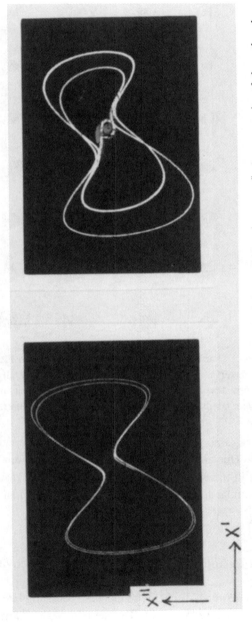

$\mathrm{X}^{\prime} \longrightarrow$

$\mathrm{X}^{\prime\prime} \longleftarrow$

FIGURE 2. Oscilloscopic photographs of the normalized, $(0,1) \times (0,1)$, x' versus x'' phase portraits of the reduced Hodgkin–Huxley-like membrane equation, Eq. (8), portraying a tangent bifurcation around $x' = x'' = 0$ as τ is changed from Fig. 2a to b. The nonuniformity of the single relaxation oscillation orbital transit times (bright regions indicating lower velocities and longer dwell times) bifurcate into a pattern of intermittent *shorts* and *longs*. These studies were conducted using an EAI-680 analog computer.

535

λ   0.1   0.3   0.5   0.8   1.0   1.5   2.0

r[0,1]

0.15 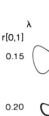

0.20

0.25

0.30

0.40

0.50

0.60

FIGURE 3.   A parameter space chart of oscilloscopic phase portraits as in Fig. 2 across changing nonlinearity parameter τ and the forcing frequency, λ. Generally, *but not perfectly,* the number of recurrent, aperiodic orbits increases with nonlinearity *and* driving frequency. These studies were conducted using an EAI-680 analog computer.

the analog computer at the same parameter values for *r* and λ represented in the phase portraits of Fig. 3. This amounts to a sample sequence of numbers (0,1) representing the changes in times with time. Note the parameter-sensitive multiplicity of broadbands similar to that which will be seen in the real neuronal records next.

IV.   Time Scaling and Entropies in Intermittent Neuronal Activities

A characteristic pattern of neuronal discharge sequences is bursting intermittency (Plant and Kim, 1975; Rinzel, 1985, 1987). The neurobiological *advantage* of bursting patterns of neural discharge has yet to be established. It is of interest to note, however, that the clustering of firefly-sending communication intervals,

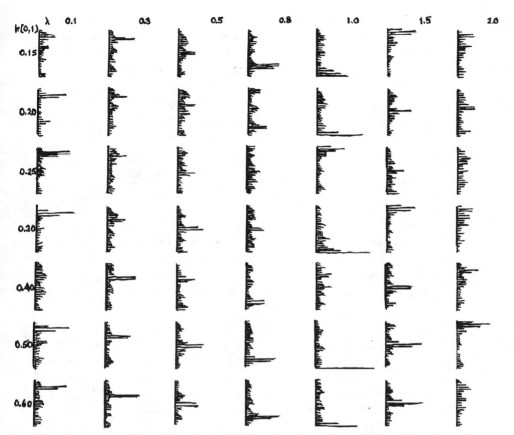

FIGURE 4.   Normalized, (amplitude, 0,1) frequency (power) spectral transformations of the discrete time series $\Sigma x'(t)$, $t = 1 - 500$, sampled every 100 msec. A parameter-sensitive multiplicity of broadband modes as seen in transformations of neuronal interspike intervals is in evidence. $x'(t)^{-1}$ is a time roughly modeling the observable, inter-event interval from the same attractors across parameters r and λ, seen in Fig. 3. The EAI-680 analog computer coupled through a PC to a SUN 3-110 workstation was used in these studies.

which occur in *clumps,* has been decoded with respect to the complexity of the behavior that they support. For example, courting behaviors have shorter intervals, finer resolution resulting in a *language* of large coding distances, and high selectivity. The mimicry associated with hunting is supported by a coarser-grained partition with reduced selectivity (Soucek and Carlson, 1987). This suggests how thresholds, partitions, and bursting may represent components in a *meaningful* code.

The generalized Hodgkin–Huxley equations representing the dynamics of neural membrane permeability with respect to sodium, potassium, anions, and other cations, such as calcium, have been studied from the point of view of the geometric theory of differential equations (Carpenter, 1979), demonstrating a variety of parameter-dependent burst patterns. Studies in topological dynamics indicate that this deterministic intermittency is generic in nonlinear dynamical systems in parameter regions engendering inverse saddle node (Manneville, 1980), saddle node (van Damme and Valkering, 1987), tangent in one-dimensional maps (Pomeau and Manneville, 1980), and pitchfork bifurcations; homoclinic tangencies in dimension 2 (Newhouse *et al.*, 1983), behavior characteristic of orbits at *fractal basin boundaries* of complex attractors (Ditto *et al.*, 1989); or around *chaotic band mergings* in low-dimensional discrete dynamical systems (Post *et al.*, 1988).

In neurophysiological terms, the *bursts* are of nearly regular short interspike intervals, bounded from below by the limit on membrane polarization and/or measurement resolution, interspersed with irregularly longer interspike intervals. Such neuronal patterns have been well-documented in a variety of neuronal systems beginning with those of the molluscan nervous system (Arvanitaki and Chalazonitis, 1961; Strumwasser, 1965; Frazier *et al.*, 1967).

Of interest to this theoretical development are the bursting patterns reported form single-neuron recordings of the biogenic amine cells of the brain stem in mammals whose axons are diffusely distributed throughout the brain. These cell groups release substances such as norepinephrine, dopamine, and serotonin, which are thought to regulate short- and long-lasting *states,* including sleeping, wakefulness, mood, rage, attention, and pathological phenomena such as depression and mania. The actions of most psychotropic drugs, from hallucinogens like LSD to tranquilizers and antidepressants, are rationalized by invoking the actions of these systems.

Bursting activity has been reported in midbrain dopamine neurons by several groups (Bunney *et al.*, 1973; Grace and Bunney, 1983, 1984; Wang, 1981; Chiodo *et al.*, 1984; Grenhoff *et al.*, 1986; Shepard and German, 1988; Gariano *et al.*, 1989; Carlson and Foote, 1989). Those cells that were more ventral in the midbrain (called *A-10* dopamine cells by neuroanatomists) were reported as more *bursty* than those that were more dorsal-lateral (called *A-9*). In contrast, the interspike interval patterns of the neurons in the locus coeruleus norepinephrine cells are not dominated by the bursting *runs of shorts* (Foote *et al.*, 1983).

Recent studies by Robert Roth and his group (Department of Pharmacology, Yale University, 1990, personal communication with A. M.) have shown that *lumping* in time of a fixed number of electrical stimuli to cells of the dopamine system into critically sized *burst packets* increased markedly the release of its neurotransmitter. This suggests that there is some organization of the interspike interval sequence $\Sigma\tau_i$, which can be seen as deviations from maximal h_M, a decrease in Λ_{max}, an increase in $|h(R)_{max} - h(R)|$, and, therefore, an increase in $G(\tau)$. Generally, an increase in coherence beyond the threshold for an action is reflected in some $G(\tau) > G(\tau)_c$.

Neuronal communication via interspike interval sequences has (at least) another major regulatory parameter: the partitioning of the interspike intervals into *shorts* and *longs*. This time scaling by the receiver, here the neural membrane receptor, has direct ramifications for the definitions of the four entropies. Whereas a *generating partition* that maximizes h_T and h_M can be invoked in an abstract dynamical system, it is the message receiver that co-determines the partition in a neurophysiological system, and, therefore, directs the system's entropies. In some sense, the receptive neurobiological mechanism is the viewer $\psi: X \rightarrow \Sigma$, which either *spreads* or *lumps* the sequence of interspike intervals, $\Sigma\tau_i$, partitioning τ_i into shorts or longs. The translation, dilation, and contraction of responsive receptor frequency bands (called *tuning* has been studied in many systems. (See, for example, Kiang and Moxon, 1974.) Such a mechanism accounting for variable impulse thresholds in neural receptive systems has not been previously entertained (Fohlmeister, 1982).

From the standpoint of the abstract algebra of the system, the partition of a set A of interspike intervals into mutually disjoint, nonempty subsets P, such that the union of $P = A$, is the same as establishing an equivalence relation on A. In this way, by parsing the interspike interval train $\Sigma\omega_i$, the receptive mechanism sets up the machinery for similarity and differences between informational sets. The entropy of these systems has biological meaning, since the run lengths of shorts associated with an increase in $h_T - h_M$ determines the action (i.e., the release of neurotransmitter). It is in the *scaling of time* that regulation of the system's effective entropy can be realized.

The intermittency pattern of neuronal firing is uniquely suited to such global receptive viewer influence, since a fair coin, $h_T - h_M = 0$, Bernoulli process would be the same process at any partition of time. In contrast, the unfair coin, $h_T - h_M > 0$, intermittency pattern would be sensitive to the influence of the global rescalings of time. Parry (1986) has developed a model of the *synchronization* of the measure of maximal entropy and the Sinai–Ruelle–Bowen measure (a density distribution describing the *amount of time* spent in the various subsets of partition space), characterized by a unique function that changes the *velocity* of the process (i.e., the phase and sequence space distance covered per unit time). This same synchronization (or, more particularly, *desynchronization*) of the two measures can result from a rescaling of time via a receptor-mediated change in the viewer's partition. The condition, *isomorphism up to a rescaling of time,* is prominent in Ornstein's development. (See his recent informal explication, 1989.)

V.   Databases and Numerical Computations

The time series of interspike intervals of ventral tegmental and substantia nigral dopamine cells, as well as those of locus coeruleus norepinephrine cells, were taken from the work of Carlson and Foote. (See preprint reference, 1989; Foote and Morrison, 1987; Carlson, 1989, and personal communication with A. M..) The time series of *in vitro* pituitary cell perifusion studies were from the laboratory of Roger Guillemin. (See Guillemin *et al.,* 1983, and personal communication.) The bursting release of hormones from endocrine organs is well-established in several systems and was reported first *in vivo* in man during all-night monitoring of urinary hormones in sleep (Mandell *et al.,* 1966a, b).

The inter-discharge intervals of the microwave popcorn popped under standard conditions were recorded on a Sony TCS-470 cassette recorder and then played through a sound transducer into a Grass Model 70D polygraph with 7P511 electroencephalographic amplifiers, which transformed the sounds of popcorn discharges onto the rolling paper via an event recorder with a minimal interpop interval resolution of 0.05 seconds. The popping epoch was divided into the phase of increasing rate of popping up to a transiently stable state; the phase of semi-stable intermittency; and the phase of exhaustion. It was the sequence of interpop intervals during the middle phase of intermittency that was compared with the neuronal and neurohormonal observables.

The topological entropy h_T is defined for the inter-event interval generating function on the space of sequences $f:\Sigma \to \Sigma$ as a real number that indicates the number of different orbits f is generating per unit time. This orbit is traced via a binary symbol sequence, S_i, a one-sided Markov subshift of finite type on $[1,0]$ in which short $\equiv 1$ and $0 \equiv$ long. *As noted before, the receptive partition serves as the partition-equivalence relation determining the meaning of short and long inter-event intervals.* We have called this a *meaningful partition* (Paulus *et al.,* 1990). For heuristic reasons, we have here set it at 10% of the range of the inter-event intervals.

The Frobenius–Perron theorem (Gantmacher, 1959) guarantees the existence of a largest eigenvalue χ of this nonnegative $[1,0]$ incidence matrix M, such that the asymptotic growth rate of its trace, $\text{Tr}(M)^t$, as $t \to n$ (where $n \equiv$ sample length), serves as an easily implemented computational method, since $\log(\chi) \leq h_T$ (Parry and Tuncel, 1981). This computation works in part because expanding dynamical systems (in which unstable periodic points are dense) collect periodic orbits (unstable fixed points), which are equidistributed with respect to the measure of all new orbits (Bowen, 1971). These are *counted* as the growth rate of the trace.

The *metric entropy* hM reflects that statistical structure of the time-averaged distribution of *most points* with respect to the orbital complexity hT. It is a real number computed as $-\Sigma p_i \log(p_i)$ on the asymptotic composition of the Markoff (stochastic—each row adds up to probability one) matrix $\mathcal{M}^t$ (transformed from the transition matrix generated by $\Sigma \omega_i$), which converges in finite time as $t \rightarrow n$ (Katok *et al.*, 1990). This computation of hM is possible because it is known that irreducible stochastic Markov matrices have a unique stationary distribution, $\mathcal{B}$, and the ergodic property: Powers of matrix $\mathcal{M}^k$ converge geometrically fast to $\mathcal{B}$.

The *run entropy* hR, is a measure made on $\Sigma \omega_i$, [1,0], which is derived from the Erdos–Renyi theorem (1970), which considered the length $\mathcal{L}_n$ of the longest run of heads in the first n tosses of a fair coin. They proved that $\mathcal{L}_n$ grows as $\log(\text{base}_{1/p})n$, where p represents the fairness of the coin; with $\mathcal{L}_n$ for a fair coin growing like $\log_2 n$. $h(R)$ can be expressed as the base of the logarithm (as $1/p$) of the coin that would generate the observed growth rate of $\mathcal{L}_n$.

The choice of the partition of $\Sigma \tau_i$ into shorts $\equiv 1$ and longs $\equiv 0$ appears arbitrary. The abstract dynamical system's approach with respect to maximizing hT involves a *generating partition* with the usual criteria that it distinguishes between any two different points of the underlying dynamical system if a *sufficient* time of observation is invoked. Time in a neurophysiological system is finite, in fact, relatively short. Drift and other evidence of nonstationarity are characteristic of the observables of brain function. In addition, as indicated previously, it is likely that the receptive mechanism adjusts this partition as part of the processes of sensitization and desensitization, and it is in this way that we justify the arbitrary choice of partition but maintain it across preparations so that it can serve in the comparative process.

VI. Interspike Interval Patterns

Figure 5a represents a characteristic series of 500 interspike intervals of 50 to 1,200 milliseconds duration normalized to [0,1] as recorded from a locus coeruleus norepinephrine cell (Fig. 5b), a small section magnified so as to note the detail, particularly the relative absence of runs of shorts. This record is consistent with the relative lack of bursting (compared with dopamine cells) reported for this cell group (Carlson and Foote, 1989; Foote *et al.*, 1983). Figure 5c is a 500-event series on a *substantia nigra* dopamine cell (A-9), ranging from 50 to

a.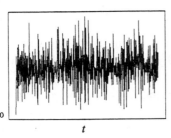

b.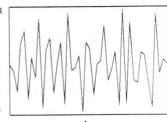

c.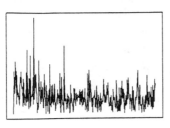

d.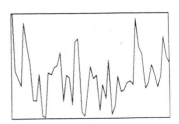

e.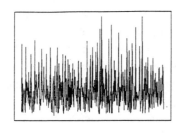

f.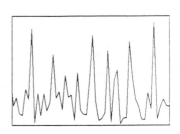

g.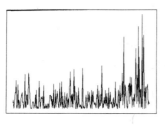

h.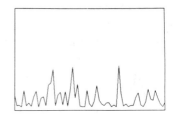

i.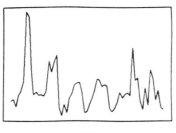

542

1,500 milliseconds, normalized to [0,1], which demonstrates a bit more bursting activity more clearly seen in the magnified sample (Fig. 5d). Figure 5e represents the interspike interval sequence in a ventral tegmental dopamine cell (A-10), intervals ranging from 50 to 1,500 milliseconds normalized to [0,1], in which runs of shorts are more in evidence, and seen more clearly in Fig 5f. This finding from the data of Carlson and Foote (1989) was of the sort first reported by Bunney *et al.* (1973) and confirmed by a number of groups since. (See the preceding.) Figure 5g represents a sequence of 299 interpop intervals from a typical microwave popcorn preparation recorded in the intermittent regime, with fine structure seen in Fig. 5h. It is clear that both the popcorn and the A-10 dopamine cell pattern are the most dominated by runs of shorts. Figure 5i demonstrates a slower intermittency pattern, ranging from 78 to 320 ng/min (normalized to [0,1]) over 61 consecutive samples, as seen in the recording of the release of growth hormone by pituitary cells in a perifusion preparation. This pattern is consistent with that reported by Guillemin *et al.* (1983).

   Figure 6 shows the scaled phase portraits of the consecutive interspike intervals, τ, as τ versus $\tau + 1$ for the event sequences portrayed in Figure 1: norepinephrine neuron, A-9 dopamine neuron, A-10 dopamine neuron, pituitary cell, and microwave popcorn. The increasing amount of short-interval bursting patterns from Figs. 6a to e, and the qualitative resemblance between the patterns of activity of the ventral tegmental dopamine neuron and microwave popcorn discharges, are apparent.

   Figure 7a, from top to bottom, represents the probability density distribution, $p(\tau)$, the autocorrelation graph, $C(\tau) \cong [\langle \tau(t + i)\tau(t) \rangle - \langle \tau(t) \rangle)^2$, and the power spectrum $S(w) = C(t)\exp(iwt)$ of 500 consecutive interspike intervals from a typical locus coeruleus norepinephrine cell demonstrating the characteristic symmetric $p(\tau)$, a waxing and waning (slow near periodicity) in the decay of $C(\tau)$, and an $S(\omega)$ composed of (from left to right) a $1/f$-like piece, a spectral *gap* (Mandell, 1986), and then multiple unstable *resonances* (Ruelle, 1989). The latter can be interpreted as broadband complex singularities of the power spectrum of the interspike interval sequence, poles at complex values of the frequencies,

---

FIGURE 5.   Graphs of time series of (normalized, 0,1) 500 interspike intervals from typical brain stem amine neurons, microwave popcorn, and *in vitro* pituitary cell hormone release. The neuronal data is that of Carlson and Foote (1988, 1989); pituitary data is that of Guillemin (1984–1986): (a), and magnified (b), locus coeruleus norepinephrine; (c), and magnified (d), A-9 dopamine; (e), and magnified (f), A-10 dopamine; (g), and magnified (h), interpop intervals of microwave popcorn; (i) hormone release intervals from the pituitary preparation. See text.

$\tau + 1$

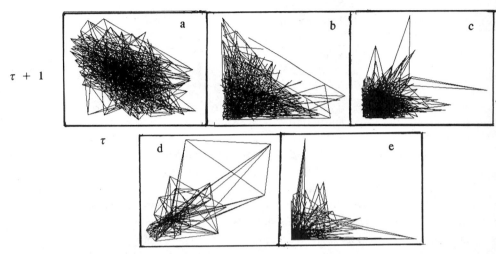

τ

FIGURE 6.   Normalized (0,1) phase portraits, $\tau/\tau + 1$, from the time series of interspike intervals in Fig. 5: (a) locus coeruleus norepinephrine; (b) A-9 dopamine; (c) A-10 dopamine; (d) pituitary cell; (e) microwave popcorn. Increasing bursting is seen as dense regions in the region of short interspike intervals along with rarer long ones.

ω_i. The $1/f$-like piece is consistent with a hierarchy of time scales, including long-range time correlations. The gaps, intermediate frequency regions without resonances, change systematically in size across the three neurons and the popcorn examples an will be discussed within the context of the *autophase spectra* represented in Fig. 8.

The meaning of power spectra of time series with multiple broadband resonances with respect to either the underlying topological dynamics or various measures of complexity, including the Lyapounov characteristic exponents and the lower and upper bounds on the entropies, has constituted a difficult problem in the analysis of *chaotic dynamics* for several years (Crutchfield *et al.,* 1980; Blacher and Perdang, 1981).

A multiplicity of broadband resonances in $S(\omega)$ have been observed in the transformation of time series from one-dimensional quadratic maps, the Rossler and Lorenz systems, and the driven van der Pol and Duffing equations, as well as the Henon and other 2D maps. Ruelle (1989) has discussed resonances in the context of their influence on the decay properties of $C(\tau)$'s and, therefore, the mixing properties of the time evolution (here of the sequential inter-event times). For example, the presence of more broadband resonances in the $S(\omega)$ of Figs. 7c and d is consistent with the slower oscillating decay of their $C(\tau)$'s and decreased mixing of the A-10 neuronal and popcorn *bursty* systems.

The observer of a hierarchy of characteristic times, standing on one of them,

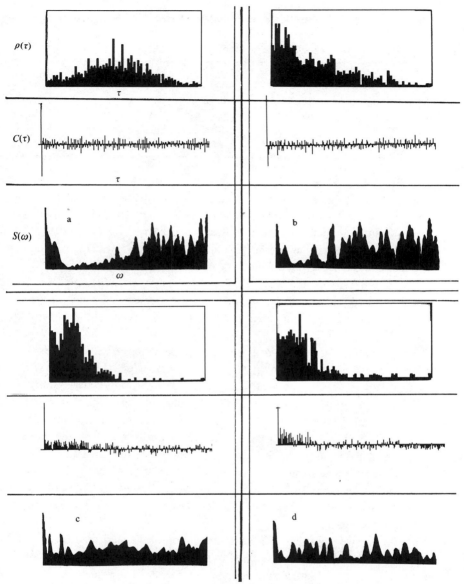

$\rho(\tau)$

τ

$C(\tau)$

τ

$S(\omega)$

a

b

ω

c

d

FIGURE 7. Four sets of three transformations each of neuronal interspike intervals: (a) locus coeruleus norepinephrine; (b) A-9 dopamine; (c) A-10 dopamine; (d) microwave popcorn. The top graph of each set is the probability density, $p(\tau)$; the middle graph is the autocorrelation function, $C(\tau)$; and the bottom, $S(\omega)$, the frequency (power) spectrum. There are differences in the probability density distribution and the power spectra with greater bursting densities in the A-10 and microwave popcorn time series. $p(\tau)$ reflects the differences in the moments seen in Table I. The decreased rate of decay of $C(\tau)$, and more broadbands in the power spectrum with more bursting, is evident.

545

TABLE I.   Statistical Moments on Neurons and Popcorn.

	SD/Mean	Skew	Kurt	Lyapoun
Norepi	0.41	0.063	2.605	0.66
DA A-9	0.802	0.82	2.841	0.79
DA A-10	0.73	1.725	9.577	0.26
Popcorn	1.132	2.807	14.317	0.15

will see the next longer one as a slow wave and the next shorter one as bursting (i.e., *rates* are a function of the observational time scale). Recent kinetic analyses of ion channel dynamics, which focused on channel lifetimes (*openings* and *closings*) and their densities, suggest that these primary neuronal events are *fractals,* that is, represent a hierarchy of characteristic times (Liebovitch and Sullivan, 1987; Liebovitch *et al.,* 1987; French and Stockbridge, 1988). Bursting can be observed at several time scales in neural membranes, much in the same way that neuronal bursting frequencies occur across scale in the $S(\omega)$ of Fig. 7c.

Table I summarizes the normalized standard deviation, σ^2/τ, skewness, asymmetry of the density distribution, $\cong \hat{\tau}^3/\sigma^3$, kurtosis, $\hat{\tau}^4/\sigma^4$, (Lumley, 1970), and the leading Lyapounov exponent, $\Lambda_{max} = \lim, t \rightarrow_i, 1/t \log\|D(\tau_i)\|$ (Grassberger and Procaccia, 1983; Wolf *et al.,* 1985) for the four $\Sigma\tau_i$, which quantitatively confirm the impressions of Fig. 7 made by the distributions of interspike intervals, $p(\tau)$, as well as suggesting a *mixing* hypothesis with respect to the *statistical dynamics* that discriminate between the behaviors of the three kinds of brain stem biogenic amine neurons. Next, we will suggest that these differences may have both mechanistic and functional significance. The lower values of Λ_{max} are observed in the neuronal systems with the larger high moments (Table I) and the more skewed $p(\tau)$.

Mixing, as we use it here, suggests the notion of the statistical independence of the $\Sigma\tau_i$ and the associated properties of convergence to a mean and a distribution, $p(\tau)$, with finite higher moments, $\hat{\tau}^n/\sigma^n$. It also reflects the degree of interleaving of the sets of short, A_0, and long, B_1, interspike intervals such that the (volume) measure μ of their intersection equals the intersection of their measures, $\mu(A_0 \cap B_1) = \mu(A_0) \cap \mu(B_1)$. Mixing also determines the rate at which the (auto)correlation function decays, $C(\tau) = [\langle\tau(t + i)\tau(t)\rangle - \langle\tau(t)\rangle^2]$ $\rightarrow 0$ as $t \rightarrow \infty$. For independent random processes, $C(\tau) \cong \exp - (t/T)$, in which T represents the characteristic time. We can regard Λ_{max} as an *average inverse multiplier* such that

$$C(\tau) \cong \exp[-\Lambda_{max}\, t/\tau] \tag{9}$$

relates the decreased rate of decay of $C(\tau)$ in the A-10 dopamine and microwave popcorn transformations (Figs. 7c and d) to both the increased higher moments and decreased values for the Lyapounov exponent seen in Table I.

The power spectra of Fig. 7 are the results of standard algorithms for the Fourier transformations of $C(\tau)$ (the *fast Fourier transformation*) and, in combination with the results of Table I, suggest the *coexistence of bursting patterns across multiple temporal domains*. Intermittency spreads across larger times scales in the A-10 and microwave popcorn discharge patterns filling up the spectral gap seen in the intermediate inverse times, $\mathcal{P}(\omega)$, of the locus coeruleus pattern (Fig. 7a). The delayed decay of $C(\tau)$, *more amplitude toward the right,* is associated with an increase in densities *toward the left* in $\mathcal{P}(\omega)$, both, of course, representing longer-range correlations.

We note that Λ_{max} represents the *expansion rate,* an exponential term describing the velocity of separation of nearby initial conditions, *mixing* related to the rate of convergence of invariant measures such as $p(\tau)$, and the *decorrelation rate,* the decay of $C(\tau)$. Longer-range correlations associated with larger T are associated with a *decrease* in the decay rate of $C(\tau)$, as indicated in Eq. (9). An increase in the expansion rate (and the upper bound on the topological entropy, h_T), Λ_{max}, *increases* the decay rate of $C(\tau)$.

Computations of characteristic times in intermittent systems present both theoretical and computational difficulties, since correlation functions, $C(\tau)$, tend to decay algebraically (if at all) within physiologically realistic sequence lengths.

In place of $C(\tau)$, we have derived a metric-free estimate using Eq. (1), which describes the ratio of two competing exponential *distortions* of the regularity in the sequence of inter-event times, $\Sigma\tau_i$: an expansive process reflected in the separation rate of nearby initial conditions, Λ_{max}, and a *sticky* process reflected in the growth rate of the longest run as a function of t. The ratio of these two *derivatives* yields a dimensionless number, *useful in comparisons between processes,* which we call $G(\tau)$. When τ is set equal to 1 and Eq. (1) is used to compute $G(\tau)$ using the data for $\Sigma\tau_i$, the results are those seen in Table II. Increased density of bursting (as in A-10 and the popcorn sequences) is associated with larger values for $G(\tau)$ and smaller values for the scaled ($h_T - h_M$) metric entropy, while less bursting is coincident with a diminished $G(\tau)$ value and a larger scaled metric entropy.

The frequency (power) spectral transformations (Fig. 7) suggest that *bursting as a form of phase coherence* may be viewed as a characteristic of phase dynamics occurring across frequencies (Priestly, 1981). For example, an autocovariance approach to a frequency-domain description of a time series, a sample cross-spectrum, is a complex quantity that can be written as a product of a cross-amplitude spectrum and a *phase spectrum*. The cross-amplitude spectrum shows

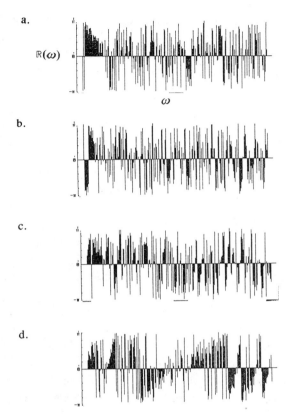

FIGURE 8. Autophase spectral transformations of interspike interval sequences from the data in Figs. 5, 6, and 7: (a) locus coeruleus norepinephrine; (b) A-9 dopamine; (c) A-10 dopamine; and (d) microwave popcorn. Bursting intermittency seen as *phase aggregation* across frequencies is more prominent in c and d.

whether frequency components in one region of the sequence are associated with large or small amplitudes at the same frequency in another region. Similarly, the *phase spectrum*, $\varphi(\omega_i)$, $(-\pi/2, +\pi/2)$, shows whether frequency components in one neighborhood lead or lag the components at the same frequency in another neighborhood in the series. A graph of this latter distribution across frequencies indicates the amount of *phase bunching* or *phase diffusion*. It can serve as an indicator of coherence (Jenkins, 1965) that transcends frequency, a partial ordering process that does not require the (single) frequency locking associated with nervous system pathophysiology. If the system is generated by an independent random process, with coherence (not defined here) about zero, $\varphi(\omega_i)$ is uniformly distributed over $(-\pi/2, +\pi/2)$.

Figure 8 is a graph of $\varphi(\omega_i)$ across the *autophase spectra* of (top to bottom) the locus coeruleus norepinephrine, A-9 dopamine, A-10 dopamine, and microwave popcorn time series. Following the initial bunching at the slower frequencies in the locus and A-9 records, the *phase bunching* across frequencies is less marked than that seen in the A-10 transformation or in the microwave popcorn record. Here, we see the spread of burst bunching across a large range of frequency domains. We recall from the intuitive description before that in a system with hierarchical time scales, any given (inverse) frequency borders on a faster scale with an apparently bursting process.

VII.   Discussion

Intermittency, bursting patterns of discharge of an irregular or regular sort, is a characteristic feature of single neuron behavior and can be predicted from nonlinear global neural membrane models when studied in the appropriate parameter space (e.g., Rinzel, 1985, 1987). Figure 2 represents a reduction of a Hodgkin–Huxley-like system, Eqs. (2)–(8), in a region demonstrating a transition between near periodic membrane oscillations (and resulting neuronal discharges given a physiological threshold) and fluctuations between long and short circuit times. Under increasing values for nonlinear parameter r, the orbits achieve nonuniform velocities, *kink* in region of slowing, become tangent, and finally wrap around the $f''(x) = f(x) = 0$ singular point at the origin, as a function of r and λ (Fig. 3).

In Figs. 2 and 3, and Eq. (8), the $\cos(\lambda t)$ driving term can be regarded generally as the aggregate influence of the input, idealized as periodic. We recall that these brain stem amine (and reticular formation) cells belong to the systems that Cahal and Ramon (1909) saw as second- or third-order sensory and second- or third-order motor in function, involved with the regulation and integration of sensory inflows and extra-pyramidal outflows.

The term controlling the nonlinearity in the systems, r, can be viewed as a global function of neural membrane mechanisms and reflected in a *Reynold's number*-like ratio of latency to excitation to the refractory period (Mandell, 1987). A physiological theory of the ionic mechanism involved in neuronal bursting (Kandel, 1976) involves voltage-sensitive, slow Na^+ and K^+ channels, which change membrane potential (thresholds) quasi-rhythmically at a time scale much longer than that of single activation events. This results in the bunching of spikes and the appearance of long intervals between them.

Another suggested ionic mechanism (with a quadratic maximum-like nonlinear dependence) implicates the intracellular accumulation of more slowly diffusing Ca^{++} (Gorman and Thomas, 1978). A *slow manifold* of excitability fluctuations is created by the relatively slow efflux of the ion. Interspike intervals first shorten, then lengthen as Ca^{++}-activated K^+ channels create post-firing hyperpolarized states. Hyperpolarization at first facilitates multiple firings, and, at higher resting membrane potentials, inhibits it. The membrane returns to its original state as Ca^{++} ions leave the cell and the spiking begins again. Chay (1985) has used a Hodgkin–Huxley-like model using intracellular calcium ions as a critical parameter (connected to the regulation of voltage-insensitive K^+ channels) in the numerical generation of bursting intermittency.

In vivo dopamine neurons manifest two currents consistent with our *r*-dependent, multiple-characteristic time model (Bunney *et al.*, 1973; Grace and Bunney, 1984). A slow depolarization (*latency to excitation*) and an after-hyperpolarization (*refractory period*) mediated by a calcium-activated potassium conductance generate burst firing in which hyperpolarization amplitude is proportional to the length of the preceding burst train. A smaller burst density is characteristic of the A-9 dopamine neurons when compared with those of A-10 (Freeman and Bunney, 1987).

Locus coeruleus norepinephrine neurons are not characterized by burst firing (Foote *et al.*, 1983). Although varying widely in average rate as a function of sleep state and arousal, a computable mean and finite variance of interspike intervals contrasts with cases in which the variance grows with sample size, as in the case of the A-10 neurons.

Listening to the patterns of sound made by popping microwave popcorn and using the interpop intervals as observables, we were struck by the similarities to neuronal systems in the use of a variety of patterns of intermittency to *report out* the current state of the system. The microwave popcorn *epoch* manifests three distinct phases:

1. *Kindling*—the increasing rate of popping up to a transiently stable state.
2. The transiently stable state of *intermittency*.
3. *Exhaustion*—the decrease in the finite number of available kernels, including a residue composed of those with the resistance to popping at even high internal heat.

Physical measures made on heat and microwave processes in popping corn demonstrate conditions such as oil content, salt concentrations, breed, and harvest pattern to alter the pattern of discontinuities in the temperature curve, expansion volume, and residue number (Lyerly, 1942; Hoseney *et al.*, 1983; Lin and Anantheswaran, 1988).

Some aspects of the cooperative physics of the microwave popcorn system and the neuron embedded in a network can be analogized. The variety of short-range connectivities created by axosomatic, axodendridic, and autorecurrent axons are simulated by the variety of *packing arrangements* among the prepopped kernels, which release latent heat locally in the form of steam at the time of their popping, leading naturally to intermittent behavior. The diffusive heat gradient *field* of the microwave oven is analogous to the *nonspecific* activating systems of the brain (the reticular formation and the biogenic amine neurons), which serve as the field of long-range connectivities, altering the pattern of excitability of neurons globally.

On theoretical and experimental grounds, using studies of the dynamics of heart, neuroendocrine systems, single neurons, electroencephalic recordings, and drug-perturbed animal behavior, we have concluded that the complexity required in biological information transport as *entropy generated per unit (discrete) time with respect to the partition ξ and the system generator of times, $f'(\tau)$*, was characteristically above zero and less than one in healthy systems [h_o,h_{max}] (Mandell, 1987; Paulus et al., 1989, 1990). A fixed point or limit cycle, H_o, carries no potential for the entropy reduction of encoding, a system at h_{max}; though with maximal potential for information transport, it can be said to be without sufficient *in place* code in the form of reduced entropy to receive messages about *readiness* for reception and transmission, called *arousal* and *attention* in neurobiology. Informally, between h_{max} and h_{min}, there is room for adaptive regulation (Mandell and Kelso, 1991) via changes in nonuniformity in the expansion by biological systems in the direction of both increases and decreases in h.

As an abstract ergodic system, the actions of a transformation are those that preserve the structure of the probability space. Its measure, μ, is invariant over time evolution. However, without the uniform expansion condition of Anosov, Axiom-A, horseshoe-like processes, the ergodic, measure-theoretic sensitivity to initial conditions becomes more difficult to characterize. Hadamard (1898), Hedlund (1939), and others showed that *geodesic* (minimal length) flows on surfaces of constant (and non-constant) negative curvature (analogous to expansive flows in real space) were ergodic. Uniform and equal maximal expansion in the tangent space of the unstable manifold and contraction in that of the stable manifold (so-called resonance) generates a point set in the Poincaré section with h_{max}.

Though bounds on the entropies of nonuniform hyperbolic systems as geodesic flows on manifolds of nonpositive curvature have been proven (Manning, 1981), little in the way of knowledge about the generic internal structure of such systems is available. In differential systems, theorems of this sort involve conditions on what Dennis Sullivan calls *bounded variation* of the logarithmic derivative of the first derivative.

With respect to manifolds of negative curvature, early theorems of Kuiper (personal communication) demonstrate that they can be extended but not continuously deformed, suggesting some universal structural features. Modern dynamical systems theorists have referred to these topological constraints on metric spaces as *rigidity*. The recent (unpublished) work by Gromov (1989–90) is also moving in the direction of universalities in the hyperbolic *parameters* of path metric spaces in what he calls near- or δ-hyperbolic systems.

We here treat our nonuniformly hyperbolic systems as measurable on experimental partitions of the one-dimensional space of the actions of $f'(\tau)$. Measures of maximal and minimal entropy and their difference (nonuniformity) serve as group theoretic quasi-isomorphisms with reflexive, symmetric, and transitive properties (Ornstein, 1989). Quasi-isomorphic equivalence invokes these properties on equal entropy spaces *except for sets of measure zero*. It is here that Ornstein's 1969 result plays an important role in spaces of dynamical systems: Two coin-flipping processes (with variously unfair coins) of equal entropies are spatially isomorphic. It is this equivalence relation of entropic quasi-isomorphism that serves as the coding scheme for the reporting out of internal conditions by neuron types and microwave popcorn.

h_{max} and h_{min} and other invariants of the dynamical system, $f':A \to A$, are characteristically computed as the supremum taken over all measurable partitions, $\xi(A)$, or are dependent on the condition that ξ is *generating*. Crudely, that ξ is such that no more than one point is contained in a component of the partition. In the entropy formalism, it is this condition that is required for h_{max} and h_{min} to be invariants of both ξ and f'. In effect, this defines bounds on the arbitrariness with which ξ can be manipulated to generate unique entropies.

In applied work, ξ can be derived from intrinsic properties of the real system (such as measurement resolution and/or characteristic behavior) and used in an empirical way. We have called this a meaningful partition (Paulus *et al.*, 1990); for example, held constant in comparisons among treatment conditions. In this way, ξ can be nonarbitrarily *fixed* without being generating. h_{max} and h_{min} in this context will be used with an implicit reference to the *experimentally determined* ξ rather than with respect to some limiting value of the partition as in abstract ergodic theory.

With respect to information transport, if f'_1 and f'_2 have finitely determined (not generating, but experimentally determined and held constant) partitions and $h(f'_1) = h(f'_2)$, then f'_1 and f'_2 are metrically isomorphic (modified from Ornstein). Information transport in brain functions in global behaviors such as *avoid* and *approach* (experimentally discriminable from a measure of complexity on the electrophysiological signals from electrodes in the hippocampus and other limbic sites in cats) are conjectured to be encoded in such metric equivalencies of ergodic redundancies in the neural activities of local neural networks. Bowen has commented (Bowen, 1977) that since these dynamical (topological and metric) entropies were *crude invariants,* it was likely that they could serve some applied purposes.

VIII. $G(\tau)$ as a Global Characteristic Time

In Eqs. (7) and (8), we see that increasing r slows down vertical orbital motion as $y \to 0$ due to the smallness of $-x/r$ and the $y = 0$ scale of the horizontal (*fast eigenvalue*) and vertical (*slow eigenvalue*) manifolds separate as a function of r. In the x'/x'' phase plane (Figs. 2 and 3), the fast large loops and slow smaller orbits around the $x'' = x' = 0 = p$ fixed point represent this temporal division more directly. Slow motion of the *temporal variation in time* in the vicinity of p allows the periodic forcing term, $\cos(\lambda t)$, a chance to capture the orbit in small-scale recurrent motion. The obvious analogy is with the slower temporal dynamics of Ca^{++} and the faster membrane motions of Na^+ and K^+ supplying the critical ratios of characteristic times associated with transitions between bursting and more random distributions of interspike intervals. Zeeman (1972, 1976), using both the unforced van der Pol and Duffing differential equations to model neuronal dynamics, emphasized the role of the disjunction between slow and fast manifolds in bifurcation behavior of the systems.

The temporal structure of intermittent systems has been handled generally as a statistical scaling problem (Schlesinger, 1988). Conventional descriptions such as (normalized) probability distributions, mean values, and variances fail. (See Table I.) Moments diverge with n, and, generally, the relative weight of large fluctuations in inter-event times grows continually for long statistical samples.

In the place of statistical indicators, we have used a ratio of invariant measures (Erdos and Renyi, 1970; Ruelle, 1989), an invariant measure, $G(\tau)$. Equation (1) is the negative exponential ratio of the (positive) Lyapounov characteristic exponent to the deviation from 0.5 of the p of an unfair coin, which, if flipped randomly, would display the growth rate of the longest run that was observed in the system.

Steve Bunney and co-workers (as seen earlier) have shown that the Parkinsonism-mimetic drug family, the phenothiazines, increase burst density in midbrain neurons. A dominant characteristic of Parkinsonian behavior is the long relaxation times (*rigidity*) in both mental and motor behavior (Mandell *et al.,*

TABLE II. Normalized Metric Entropy and Global
Characteristic Times in Neurons and Popcorn.

System	$h(T) - h(M)$	$G(T)$
Locus	0.5777	0.026
A-9-DOPA	0.0804	0.137
A-10-DOPA	0.0764	0.539
Popcorn	0.0351	0.704

1962). $G(\tau)$ captures this quality in neuronal firing patterns (Table II) and converts the observable into a global index of critical coherence and slow relaxation.

Recent evidence from several laboratories has indicated that bursts release up to sixfold more neurotransmitter, dopamine, than a pattern of random firing with the same average frequency. In addition, the release of peptide co-transmitters appears to be burst and not random-discharge-related. (Personal communication with A. M., Steven Bunney, 1991). In this context, $G(\tau)$ can be interpreted as a crude, continuous variable code containing messages about amount of neurotransmission.

These findings appear consistent with our theory that neurons may code using the nonuniformity of neurobiological time (Mandell, 1983; Mandell and Kelso, 1991).

Acknowledgment

This work was supported by the Office of Naval Research (Biophysics).

References

ANOSOV, D. V. (1967). "Geodesic Flows on Closed Riemann Manifolds with Negative Curvature," *Tr. Mat. Inst. Steklov* **90**, 210–217.

ARVANITAKI, A. (1939). "Recherches sur la Reponse Oscillatoire Locale de l'Axone Geant Isole de 'Sepia,'" *Arch. Intl. Physiol.* **49**, 209–256.

ARVANITAKI, A., and CHALAZONITIS, N. (1961). "Variations Lentes et Periodiques du Potentiel de Membrane Associées a des groupes de Pointes," *C. R. Acad. Sci.* **225**, 1523–1525.

BABLOYANTZ, A., and DESTEXHE, A. (1986). "Low-Dimensional Chaos in an Instance of Epilepsy," *Proc. Nat. Acad. Sci.* **83**, 3513–3517.

BABLOYANTZ, A., NICOLIS, C., and SALAZAR, M. (1985). "Evidence for Chaotic Dynamics of Brain Activity during the Sleep Cycle," *Phys. Lett. A,*111, 152–156.

BLACHER, S., and PERDANG, J. (1981). "The Power of Chaos," *Physica* **30**, 512–529.

BOWEN, R. (1971). "Periodic Points and Measures for Axiom A Diffeomorphisms," *Trans. Am. Math. Soc.* **154,** 377–397.

BOWEN, R. (1977). "A Model for Conette Flow," *Lect. Notes Math.* **615,** 117–134.

BUCHWALD, N. A., HULL, C. D., and LEVINE, M. S. (1979). "Neuronal Activity of the Basal Ganglia Related to the Development of 'behavioral sets,'" in M. A. B. Brazier (ed.), *Brain Mechanisms in Memory and Learning* (pp. 93–104). Raven, New York.

BUNNEY, B. S., WALTERS, J. R., ROTH, R. H., and OGHAJANIAN, G. K. (1973). "Dopaminergic Neurons: Effect of Antipsychotic Drugs and Amphetamine on Single Cell Activity," *JPET* **185,** 560–571.

CAHAL, S., and RAMON, Y. (1909). *"Histologie du Système Nerveux de l'Homme et des Vértébres, Vol. I.* A. Malone, Paris.

CARLSON, J. H. (1989). "Characterization of Discharge Patterns of Dopamine Neurons," *NARSAD Symposium,* Washington, D.C.

CARLSON, J. H., and FOOTE, S. L. (1988). "Effects of Local Infusion of Pharmacological Agents on Tonic Activity of *substantia nigra* Dopamine Neurons," *Soc. Neuroscience* **14,** 407.

CARLSON, J. H., and FOOTE, S. (1989). "Dopamine Discharge Patterns, *Time Series.* Dept. of Psychiatry, U.C.S.D., La Jolla, California.

CARPENTER, G. A. (1979). "Bursting Phenomena in Excitable Membranes," *SIAM J. Appl. Math.* **36,** 334–352.

CARPENTER, M. M. (1976). "Anatomical Organization of the Corpus Striatum and Related Nuclei," in M. D. Yahr (ed.), *The Basal Ganglia* (pp. 1–36). Raven, New York.

CHAY, T. R. (1985). "Chaos in a Three-Variable Model of an Excitable Cell," *Physica* **16D,** 233–242.

CHIODO, L. A., BANNAN, M. J., GRACE, A. A., ROTH, R. H., and BUNNEY, B. S. (1984). "Evidence for the Absence of Impulse Regulating Somadendritic and Synthesis Modulating Autoreceptors on Subpopulations of Mesocortical Dopamine Neurons," *Neuroscience* **12,** 1–16.

COLE, K. S. (1972). *Membranes, Ions, and Impulses.* University of California Press, Berkeley, California.

CRUTCHFIELD, J., FARMER, D., PACKARD, N., SHAW, R., JONES, G., and DONNELLY, R. J. (1980). *Phys. Lett.* **76A,** 1–4.

DITTO, W. L., RAUSEO, S., CAWLEY, R., GREBOGI, C., HSU, G. H., KOSTELICH, E., OTT, E., SAVAGE, H. T., SEGNAM, R., SPANO, M. L., and YORKE, J. A. (1989). "Experimental Observation of Crisis-Induced Intermittency and Its Critical Exponent," *Phys. Rev. Lett.* **63,** 923–926.

ECKMANN, J.-P., and RUELLE, D. (1985). "Ergodic Theory of Chaos and Strange Attractors," *Rev. Mod. Physics* **57,** 617–655.

ERDOS, R., and RENYI, A. (1970). "On a New Law of Large Numbers," *J. Analyse Math.* **22,** 103–111.

FOHLMEISTER, J. F. (1982). "The Role of a Variable Impulse Threshold in the Impulse Frequency Response of a Tonic Sensory Neuron," *J. Theoret. Neurobiol.* **1,** 251–267.

FOOTE, S. L., BLOOM, F. E., and ASTON-JONES, G. (1983). "The Nucleus Locus Coeruleus: New Evidence of Anatomical and Physiological Specificity," *Physiol. Rev.* **63**, 844–914.

FOOTE, S. L., and MORRISON, J. H. (1987). "Extrathalamic Modulation of Cortical Function," *Ann. Rev. Neurosci.* **10**, 67–95.

FRAZIER, W. R., KANDEL, E. R., KUPFERMANN, I., WAZIRI, R., and COGGESHALL, R. E. (1967). "Identifiable Cells in the Abdominal Ganglion of *Aplysia californica,*" *J. Neurophysiol.* **30**, 1288–1351.

FREEMAN, A. S., and BUNNEY, B. S. (1987). "Activity of A9 and A10 Dopaminergic Neurons in Unrestrained Rats: Further Characterization and Effects of Apomorphine and Cholecystokinin," *Brain Res.* **405**, 46–55.

FRENCH, A. S., and STOCKBRIDGE, L. L. (1988). "Fractal and Markov Behavior in Ion Channel Kinetics," *Can. J. Physiol. Pharmacol.* **66**, 967–970.

GANTMACHER, F. R. (1959). *The Theory of Matrices, Vol. I.* Chelsea, New York.

GARRIANO, R. F., SAWYER, S. F., TEPPER, J. M., YOUNG, S. N., and GROVES, P. M. (1989). "Mesocortical Dopaminergic Neurons," *Brain Res. Bull.* **22**, 517–523.

GORMAN, A. L. F., and THOMAS, M. V. (1978). "Changes in the Intracellular Concentration of Free Calcium Ions in Pacemaker Neurons Measured with the Metallochromic Indicator Dye Arsenazo III," *J. Physiol.* **275**, 357–376.

GRACE, A. A., and BUNNEY, B. S. (1983). "Intracellular and Extracellular Electrophysiology of Nigral Dopaminergic Neurons: II. Action Potential Generating Mechanisms and Morphological Correlates," *Neuroscience* **10**, 317–331.

GRACE, A. A., and BUNNEY, B. S. (1984). "The Control of Firing Pattern in Nigral Dopamine Neurons: Single Spike Firing," *J. Neurosci.* **4**, 2866–2876.

GRAF, K. E., and ELBERT, T. (1989). "Dimensional Analysis of the Waking EEG," in E. Basar and T. H. Bullock (eds.), *Brain Dynamics, Progress and Perspectives* (pp. 174–191). Springer-Verlag, New York.

GRASSBERGER, P., and PROCACCIA, I. (1983). "Measuring the Strangeness of Strange Attractors," *Physica* **9D**, 183–208.

GROVES, P. M. (1983). "A Theory of the Functional Organization of the Neostriatum and the Neostriatal Control of Voluntary Movement," *Brain Res. Reviews* **5**, 109–151.

GUILLEMIN, R. C., BRAZEAU, P., BRISKIN, A., and MANDELL, A. J. (1983). "Evidence for Synergetic Dynamics in a Mammalian Pituitary Cell Perifusion System," in E. Basar, H. Flohr, H. Haken, and A. J. Mandell (eds.), *Synergetics of Brain* (pp. 365–376). Springer-Verlag, New York.

HADAMARD, J. (1898). "Les Surfaces a Courbures Opposées et Leurs Lignes Geodesiques," *J. Math. Pures. et Appl* **4**, 27–73.

HEDLUND, G. (1939). "The Dynamics of Geodesic Flows," *Bull. Am. Math. Soc.* **45**, 241–246.

HOSENEY, R. C., ZELEZNAK, K., and AHDEBRAHMAN, A. (1983). "Mechanism of Popcorn Popping," *J. Cereal Sci* **1**, 43–54.

JENKINS, G. M. (1965). "A Survey of Spectral Analysis," *App. Stat.* **14**, 2–32.

KANDEL, E. P. (1976). *The Cellular Basis of Behavior* (pp. 261–268). Freeman, San Francisco.

KATOK, A., KNEIPER, G., POLLICOTT, M., and WEISS, H. (1990). "Differentiability of Entropy for Anosov and Geodesic Flows," *Bull. Am. Math. Soc.* **22,** 285–294.

KELLEHER, J. F., and MANDELL, A. J. (1990). "Dystonia Musculorum Deformans: A 'Critical Phenomenon' Model Involving Nigral Dopaminergic and Caudate Pathways," *Medical Hypoth.* **31,** 55–58.

KEMP, J. M., and POWELL, T. P. S. (1971). "The Structure of the Caudate Nucleus of the Cat: Light and Electron Microscopy," *Phil. Trans.* **B262,** 383–402.

KIANG, N. Y. S., and MOXON, E. C. (1974). "Tails of Tuning Curves of Auditory Nerve Fibers," *J. Acoust. Soc. Am.* **55,** 620–630.

KOLMOGOROV, A. N. (1959). "On the Entropy per Time Unit as a Metric Invariant of Automorphisms," *Dokl. Akad. Nauk.* **124,** 754–755.

LIEBOVITCH, L. L., FISCHBARG, J., KONIAREK, J. P., TODOSOVA, I., and WANG, M. (1987). "Fractal Model of Ion Channel Kinetics," *Biochim. Biophys. Acta* **896,** 173–180.

LIEBOVITCH, L. S., and SULLIVAN, J. M. (1987). "Fractal Analysis of a Voltage-Dependent Potassium Channel from Cultured Mouse Hippocampal Neurons," *Biophys. J.* **52,** 979–988.

LIN, Y. E., and ANANTHESWARAM, R. C. (1988). "Studies of Popping of Popcorn in a Microwave Oven," *J. Food Sci.* **53,** 1746–1749.

LYERLY, P. J. (1942). "Some Genetic and Morphological Characters Affecting the Popping Expansion of Popcorn," *J. Am. Soc. Agron.* **34,** 986–997.

MANDELL, A. J. (1981). "Statistical Stability in Random Brain Systems," *Adv. Subst. Abuse* **2,** 299–341.

MANDELL, A. J. (1986). "Toward a Neuropsychopharmacology of Habituation; a Vertical Integration," *Math. Model.* **7,** 809–888.

MANDELL, A. J. (1987). "Dynamical Complexity and Pathological Order in the Cardiac Monitoring Problem," *Physica* **27D,** 235–242.

MANDELL, A. J., CHAFFEY, B., BEILL, P., MANDELL, M. P., RODNICK, J., RUBIN, R. T., and SHEFF, R. (1966). "Dreaming Sleep in Man: Changes in Urine Volume and Osmolarity," *Science* **151,** 1558–1560.

MANDELL, A. J., and KELSO, J. A. S. (1991). "Neurobiological Coding in Nonuniform Times," in J. A. Ellison and H. Uberall (eds.), *Essays on Classical and Quantum Dynamics.* Gordon and Beach, New York.

MANDELL, A. J., MARKHAM, C. H., TALLMAN, F. F., and MANDELL, M. P. (1962). "Motivation and Ability to Move," *Am. J. Psychiat.* **119,** 544–549.

MANDELL, A. J., and SELZ, K. A. (1990). "Heterochrony as a Generalizable Principle in Biological Dynamics," in E. Stanley and N. Ostrowsky (eds.), *Correlations and Connectivity; Geometric Aspects of Physics, Chemistry and Biology* (pp. 281–296).

MANDELL, A. J., and SELZ, K. A. (1991). "A Nonthermodynamic Formalism for Biological Information Systems: Hierarchical Lacunarity in Partition Size of Intermittency," in A. Babloyantz (ed.), *Self-Organization, Emerging Properties and Learning*. Plenum, New York (in press).

MANDELL, A. J., and SHLESINGER, M. F. (1990). "Lost Choices: Parallelism and Topological Entropy Decrements in Neurobiological Aging," in S. Krasner (ed.), *The Ubiquity of Chaos* (pp. 35–46). Am. Assoc. Adv. Sci., Washington, D.C.

MANDELL, M. P., MANDELL, A. J., RUBIN, R. T., RODNICK, J., SHEFF, R., and CHAFFEY, B. (1966). "Activation of the Pituitary–Adrenal Axis during Rapid Eye Movement Sleep in Man," *Life Sciences* **5**, 583–587.

MANNEVILLE, P. (1980). "Intermittency in Dissipative Dynamical Systems," *Phys. Lett.* **79A**, 33–37.

MANNING, A. (1981). "Curvature Bounds for the Entropy of the Geodesic Flow on a Surface," *J. London Math. Soc.* **24**, 351–357.

MAYER-KRESS, G., and LAYNE, S. P. (1987). "Dimensionality of the Human Electroencephalogram," *Ann. N. Y. Acad. Sci.* **504**, 62–87.

NEWHOUSE, S., PALIS, J., and TAKENS, F. (1983). "Bifurcation and Stability of Families of Diffeomorphisms," *Pub. Math. I. H. E. S.* **57**, 5–72.

ORNSTEIN, D. S. (1970). "Bernoulli Shifts with the Same Entropy Are Isomorphic," *Adv. in Math.* **4**, 337–352.

ORNSTEIN, D. S. (1989). "Ergodic Theory, Randomness, and Chaos," *Science* **243**, 182–186.

PALIS, J., and TAKENS, F. (1987). *Ann. Math.* **125**, 337–374.

PARRY, W. (1981). *Topics in Ergodic Theory*. Cambridge University Press, Cambridge, U.K.

PARRY, W. (1986). "Synchronization of Canonical Measures for Hyperbolic Attractors," *Commun. Math. Phys.* **106**, 267–275.

PARRY, W., and TUNCEL, S. (1981). "On the Stochastic and Topological Structure of Markov Chains," *Bull. London Math. Soc.* **14**, 16–27.

PAULUS, M. P., GASS, S. F., and MANDELL, A. J. (1989). "A Realistic, Minimal 'Middle Layer' for Neural Networks," *Physica* **40D**, 135–155.

PAULUS, M. P., GEYER, M. A., GOLD, L. H., and MANDELL, A. J. (1990). "Ergodic Measures of Complexity in Rat Exploratory Behavior," *Proc. Nat. Acad. Sci.* **87**, 723–727.

PLANT, R. E., and KIM, M. (1975). "On the Mechanism Underlying Bursting in the Aplysia Abdominal Ganglion R-15 Cell," *Math. Bioscience* **26**, 357–375.

POMEAU, Y., and MANNEVILLE, P. (1980). "Intermittent Transitions to Turbulence in Dissipative Dynamical Systems," *Commun. Math. Phys.* **74**, 189–196.

POST, T., CAPEL, H. W., and VAN DER WEELE (1988). "Short-Phase Anomalies in Intermittent Band Switching," *Phys. Lett. A,* **133**, 373–377.

POTEGAL, M. (1972). "The Caudate Nucleus Egocentric Localization System," *Acta Neurobiol. Exp.* **32**, 479–491.

PRIESTLY, M. B. (1981). *Spectral Analysis and Time Series, Vol. II* (pp. 702–706). Academic Press, New York.

RAPP, P. E., ZIMMERMANN, I. D., ALBANO, A. M., DEGUZMAN, G. C., GREENBAUM, N. N., and BASHORE, T. R. (1985). "Experimental Studies of Chaotic Neural Behavior; Cellular Activity and Electroencephalic Signals," in H. G. Othman (ed.), *Nonlinear Oscillations in Biology and Chemistry* (pp. 175–205). Springer-Verlag, New York.

RINZEL, J. (1985). "Bursting Oscillations in an Excitable Membrane Model," *Lect. Notes Math., Vol. 1151.*

RINZEL, J. (1987). "A Formal Classification of Bursting Mechanisms in Excitable Systems," *Lect. Notes Biomath.* **71**, 267–281.

ROSCHKE, J., and ALDENHOFF, J. (1990). "The Dimensionality of Human's Electroencephalogram during Sleep," *Biol. Cyber.* (in press).

ROSCHKE, J., and BASAR, E. (1989). "Correlation Dimensions in Various Parts of Cat and Human Brain in Different States," in E. Basar and T. H. Bullock (eds.), *Brain Dynamics; Progress and Perspectives* (pp. 131–148). Springer-Verlag, New York.

RUELLE, D. (1989). *Chaotic Evolution and Strange Attractors.* Cambridge University Press, Cambridge, U.K.

SCHWYN, R. C., and FOX, C. A. (1974). "The Primate *substantia nigra;* a Golgi and Electron Microscopic Study," *J. Hirnforsch* **16**, 95–126.

SHEPARD, P. D., and GERMAN, D. D. (1988). "A Subpopulation of Mesocortical Dopamine Neurons Possess Autoreceptors," *Eur. J. Pharmacol* **114**, 401–402.

SHLESINGER, M. F. (1988). "Fractal Time in Condensed Matter," *Am. Rev. Phys. Chem.* **38**, 269–290.

SINAI, J. (1964). "On Weak Isomorphisms of Transformations with Invariant Measures," *Math USSR Sb.* **63**, 23–42.

SOUCEK, B., and CARLSON, A. D. (1987). "Brain Windows Language of Fireflies," *J. Theor. Biol.* **21**, 93–103.

STRUMWASSER, F. (1965). "The Demonstration and Manipulation of a circadian Rhythm in a Single Neuron," in J. Ashoff (ed.), *Circadian Clocks* (pp. 442–462). North-Holland, Amsterdam.

SWANSON, L. W. (1982). "The Projections of the Ventral Tegmental Area and Adjacent Regions: A Combined Fluorescent Retrograde Tracer and Immunofluorescence Study in the Rat," *Brain Res. Bull.* **9**, 321–353.

TEPPER, J. Z., and MANDELL, A. J. (1987). "Time and Frequency Domain Properties of a Cortical Functional Unit: The Discrete Penicillin Focus in the Rat," *Am. N.Y. Acad. Sci.* **504**, 301–304.

VAN DAMME, R., and VALKERING, T. P. (1987). "Intermittency," *J. Phys. A; Math. Gen.* **20**, 4161–4171.

WANG, R. Y. (1981). "Dopaminergic Neurons in the Rat Ventral Tegmental Area: I. Identification and Characterization," *Brain Res. Rev.* **3**, 123–140.

WARD, A. A. (1968). "The Function of the Basal Ganglia," in P. J. Vinken and G. W. Bruyn (eds.), *Handbook of Clinical Neurology, Vol. 6* (pp. 90–96). Elsevier–North-Holland, New York.

WOLF, A., SWIFT, J. B., SWINNEY, H. L., and VASTANO, J. A. (1985). "Determining Lyapounov Spectra from a Time Series," *Physica* **16D,** 285–317.

ZEEMAN, C. (1972). "Differential Equations for the Heartbeat and Nerve Impulse," in C. H. Waddington (ed.), *Towards a Theoretical Biology, Vol. 4* (pp. 8–67). Edinburgh University Press, Edinburgh, Scotland.

ZEEMAN, C. (1976). "Duffing's Equation in Brain Modeling," *Bull. Inst. Math. Appl.* **12,** 207–214.

Chapter 21 — A Heuristic Approach to Stochastic Models of Single Neurons

CHARLES E. SMITH

Department of Statistics
Biomathematics Program
North Carolina State University
Raleigh, North Carolina

I. Introduction

A. *Biological and Mathematical Framework*

This chapter concentrates on some heuristic approximation methods for the firing times obtained from several stochastic one-compartment or lumped models of a neuron. The approach is that of a pictorial guide rather than emphasis on the mathematical formalism. The uncertainty or stochasticity is in the timing of the generation of action potentials in a nerve cell or neuron, and this is assumed to be due to noisy synaptic input to the cell, which, in turn, makes the membrane potential a random process. The models considered are of the voltage threshold type; namely, when the voltage across the cell membrane at a particular spatial location, called the *trigger zone,* exceeds a voltage threshold, then a brief electrical pulse known as an *action potential* or *spike* is generated. The noisy input is due to the timing and quantal nature of the input synapse release process. If many inputs are present, central limit theory-type arguments produce a diffusion approximation to the voltage process, i.e., the membrane potential is relatively smooth between action potentials. On the other hand, when one or a few synapses

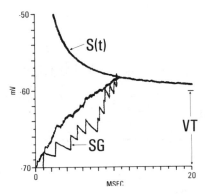

FIGURE 1. Simulated sample paths with small and with large EPSP sizes of the one-compartment neural model.

dominate the behavior of the membrane potential near the trigger site, e.g., a calyx type, the membrane potential appears to make jumps at the time of synaptic inputs and a filtered Poisson model may be more appropriate. Figure 1 illustrates the preceding scenario for the two types of membrane potential sample paths and with the voltage threshold, $S(t)$, being a decreasing function to mimic a relative refractory or recovery period.

Said another way, the membrane potential, $x(t)$, will be modeled as a continuous time random process with either a continuous state space or first-order discontinuities. These processes are often generated by a stochastic differential equation (SDE), that is,

$$dx = a(x,t) \, dt + b(x) \, dW(t) \tag{1a}$$

or

$$dx = a(x,t) \, dt + b(x) \, dP(\lambda,t), \tag{1b}$$

where $W(t)$ is a standard Wiener process (a Gaussian random process with independent increments that are distributed as a normal random variable with zero mean and a variance equal to the width of the increment), and $P(\lambda,t)$ is a homogeneous Poisson process with rate parameter λ; i.e., the times between synaptic inputs are independent and from the density $\lambda e^{\lambda t}$. Intuitively, you can think of $x(t)$ as the voltage trajectory in an RC circuit (with, in general, nonlinear and time-varying elements) attached to a noisy current source. The term $a(x,t)$ reflects the voltage-current relationship of the membrane and may be nonlinear and/or time-varying due to nonlinear and time-varying conductances. The term $b(x)$ reflects the dependence of the synaptic input on the membrane potential, e.g., due to a synaptic reversal potential. $a(x)$ is sometimes called the infini-

tesimal drift and $b^2(x)$ the infinitesimal variance. The initial condition for Eq.
(1) could either be fixed, i.e., same reset value following a spike, or could vary
due to, say, cumulative after-hyperpolarization.

The time for $x(t)$ to reach the voltage threshold $S(t)$ for the first time, given
that it started at x_0, is called a *first-passage time, $t(S|x_0)$*, and is a random variable,
since it varies from one spike to the next. Mathematically, it is often very difficult
to specify completely; hence, our emphasis here on heuristic approximations.
The first-passage time is an example of a more general level-crossing problem.
An extensive bibliography of level-crossing problems with emphasis on engineering applications can be found in Blake and Lindsey (1973). An introduction
to the case where $x(t)$ is a diffusion process (roughly, Markov with continuous
sample paths) can be found in Chapter 3 of Ricciardi (1977) and a review for
neural diffusion models in Lansky *et al.* (1990). More general reviews of stochastic problems in neurobiology can be found, for example, in Holden (1976),
Yang and Chen (1978), and Tuckwell (1988, 1989). Here, the noisiness is due
to noisy synaptic input; if one is dealing with very thin afferent processes or
dendrites, other forms of noise may dominate the behavior. (See, e.g., Holden,
1976, Chapter 1.) The vast literature on the stochastic behavior of single-ion
channels is also not considered in this chapter. (See, e.g., Tuckwell, 1989,
Chapter 8.) The emphasis here will be on stationary spike trains due to resting
discharge or repetitive discharge in response to constant stimulus conditions.

B.   Role of Noise in Neural Systems

Noise is usually considered to be undesirable in a system and to be eliminated
by some appropriate averaging or filtering. It is worth noting, however, that
noise can play other roles as well:

(1) Noise can help to linearize a firing rate versus stimulus relationship; that
    is, the response more smoothly goes to zero, rather than abruptly as in
    current injection (e.g., Jack *et al.,* 1983, pp. 368–369). This represents
    a noisy but temporal alternative to the population averaging described in
    Chapter 12 by Koch and Poggio.
(2) Noise can increase the bandwidth for transmission and minimize distortion
    (*loc. cit.*).
(3) An irregularly spaced spike train can provide an effective way to use a
    facilitating synapse, provided response time is not the main objective.
(4) Noise can change the domain of attraction of a dynamic system. Said
    another way, the system seems to switch states, *cf.* Chapter 19 by Bulsara
    et al. It is worth noting that the changing of states is due to the effects

of the noise rather than systematically changing some parameter such as injected current.

Deterministic nonlinear dynamics and bifurcation theory have also provided useful models for neural activity, particularly oscillatory behavior (e.g., Rinzel and Ermentrout, 1989; Winslow, 1989). Stochastic bifurcation theory is also being pursued with some success in biological applications (e.g., Kliemann, 1983; Cobb and Zacks, 1985; Sherman *et al.,* 1988).

The effect of noise on dynamical systems seems less well-known than the other three aforementioned roles for noise. The next two figures illustrate the role of noise in terms of the context of Eq. (1), with the terms $a(x)$ and $b(x)$ describing the membrane characteristics and the state dependence of the synaptic input process. In Fig. 2, the membrane I-V relationship is cubic and the noise term does not depend on the membrane potential. The trajectory shown alternatively oscillates about the two stable roots of the cubic (± 1) switching between

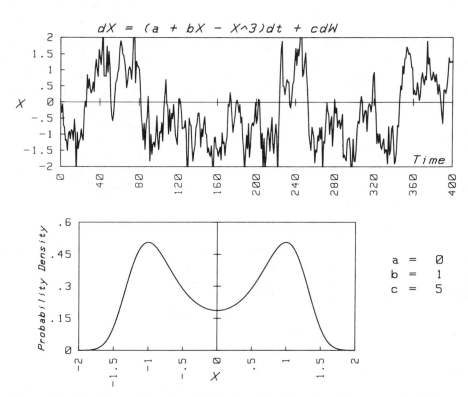

FIGURE 2.   Simulation of sample path for Eq. (1) with cubic drift term and additive Gaussian noise.

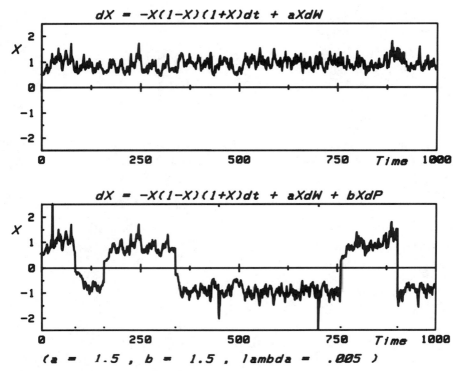

FIGURE 3.   Simulation of two sample paths for Eq. (1) with cubic drift term and state-dependent noise. Upper graph has additive Gaussian noise and lower graph has both Poisson and Gaussian noise.

them due to excursions that pass the unstable root at 0. The lower portion of the figure gives the distribution of the membrane potential collected over a long time record. This bimodal density can be thought of as the time a particle would spend in a bistable potential well formed by flipping over the density. The main point is that the trajectory randomly switches between the two wells of attraction. If a threshold is added at, say, 2.5, the conditional time to firing, given that we are in the upper state, is on average much shorter than that from the lower state.

Figure 3 introduces a state dependence on the noisy input process. In the upper graph, the Wiener process is used, while in the lower graph, a Poisson process is added to the same Wiener process used in the upper graph. The differences are striking. The upper graph is restricted to positive values, while the lower graph shows the trajectory visiting both domains of attraction. One other point is the asymmetry of the noise due to the state dependence. At larger positive or negative values, the fluctuations are larger. This idealized picture is intended to caricature the kicking in of a few additional excitatory and inhibitory synapses near the trigger site.

II.   First-Passage Times as Neural Firing Times

A.   *Types of Models*

Here, some approximation methods will be illustrated for the first-passage times through four neural models. $x(t)$ will be the membrane voltage process referenced to the resting level, which is set to zero. The models are:

1. The Ornstein–Uhlenbeck process (e.g., Sato, 1978; Ricciardi and Sacerdote, 1979; Ricciardi *et al.*, 1983, 1984; Nobile *et al.*, 1985),

$$dx = \frac{-x}{\tau} \, dt + \sigma \, dW(t), \tag{2}$$

   with τ and σ positive constants.
2. Stein's model (Stein, 1967),

$$dx = \frac{-x}{\tau} \, dt + a_e \, dP(\lambda_e, t) + a_i \, dP(\lambda_i, t), \tag{3}$$

   with τ, λ_e, λ_i, a_e, and $(-a_i)$ all positive constants.
3. Stein's model with a reversal potential (e.g., Tuckwell, 1979; Smith and Smith, 1984),

$$dx = \frac{-x}{\tau} \, dt + a_e \left(1 - \frac{x}{V_e} \right) dP(\lambda_e, t) + a_i \left(1 - \frac{x}{V_i} \right) dP(\lambda_i, t), \tag{4}$$

   with the parameters as in model 2 and the reversal potentials $V_e > 0$ and $V_i < 0$.
4. A stochastic after-hyperpolarization model used to model the firing of vestibular nerve fibers (Smith and Goldberg, 1986),

$$x(t) = \frac{\{g_S V_S + g_K(t) V_K + V_P\}}{1 + g_S + g_K(t)}, \tag{5}$$

   where V_S and V_K are positive (synaptic) and negative (potassium) equilibrium potentials, respectively; see Fig. 4 for a circuit model. The g's are normalized membrane conductances (normalized by the leakage conduct-

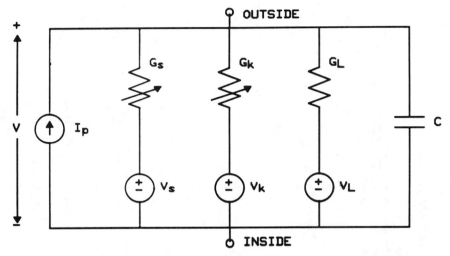

FIGURE 4.   After-hyperpolarization model. See text for symbols. The G_K conductance varies as a function of postspike time. The mean G_S conductance is constant for any steady-state condition, but varies with natural stimulation.

ance), with $g_K(t)$ being a decaying exponential and g_S being a shot noise process produced by passing a Poisson impulse train through a causal, rectangular finite impulse response (FIR) filter of duration 0.5 msec. V_P is the postsynaptic current source I_P normalized by the leakage conductance.

Models 1, 2, and 3 fit into the SDE framework of Eq. (1), while model 4 is a time-varying, nonlinear transformation of a shot noise process. Model 4 arises from a more complicated lumped-circuit model of the trigger site with three parallel conductances: a leakage conductance, a synaptic conductance, and a potassium channel conductance. The voltage across the membrane satisfies a differential equation, which is approximated by Eq. (5) when the membrane time constant is short compared to the time constant of $g_K(t)$. (See Smith and Goldberg, 1986, for details, parameter values, and biological justification for the approximation.) The threshold function $S(t)$ must also be specified. A constant threshold of 10–15 mV is often used. A decaying threshold function is in some ways trying to mimic the time-varying conductances as depicted explicitly in model 4.

At this point, it is worth examining the purpose of the preceding models. One situation is that the values for the neuron's parameters are fixed and the input

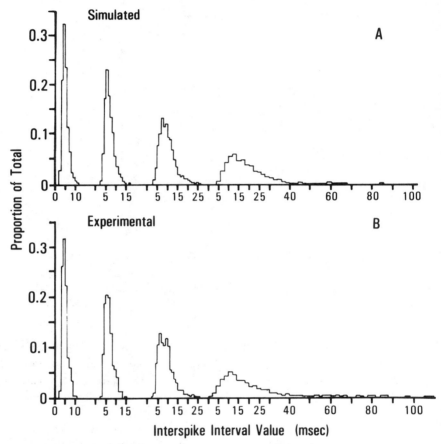

FIGURE 5.   Simulated and experimental interspike interval histograms at four firing rates.
(a) Simulated histograms corresponding to $TS = 25$ msec and $\lambda = 2.7, 2.1, 1.6$, and
1.0/msec from left to right. The corresponding firing rates in sp/sec (and cv's) are 197.23
(.279), 151.98 (.314), 101.02 (.370), and 52.32 (.528). Each histogram represents interspike values for 1,000 simulated intervals. (b) Experimental histograms from horizontal
canal unit 33-1. The stimulus was a velocity trapezoid as described in Goldberg and
Fernandez (1971). The firing rates in spikes/sec (and cv's) from left to right are 201.33
(.282), 155.88 (.324), 106.95 (.406), and 48.19 (.799). The corresponding number of
intervals are 199, 311, 535, and 1,680.

process changes to reflect different stimulus levels. Figure 5 shows this situation
for model 3 with excitatory synapses only. The comparison for the four histograms in the upper and lower graphs is quite good. The only parameter changing
to produce the four histograms in the model was the release rate λ of the Poisson
process having values of 2.7, 2.1, 1.6, and 1.0/msec, τ was 5 msec; resting

EPSP sizes, a_e, were 2 mV; reversal potential V_e was 70 mV; and the threshold function in mV was $S(t) = 10 + 1/(e^{t/25} - 1)$, with time in msec. While the model matched these histograms well, it was not possible to produce the range of variability seen in vestibular units by changing the time constant of the threshold process and at the same time produce good agreement with other stimuli such as short shocks. Model 4 was then used instead, using the insight gained from initial work with model 3.

B.   First-Passage Time Approximations

The approximation methods for first passage times will be classified according to whether or not the mean voltage $\mu(t)$ of the process $x(t)$ crosses the threshold $S(t)$. The term *deterministic crossing* will be used when the mean voltage crosses $S(t)$, and *nondeterministic crossing* when the mean does not cross $S(t)$. For the deterministic crossing, we further distinguish two cases, long correlation time and short correlation time relative to the standard deviation of first-passage time, $t(S|x_0)$. Figure 6 shows a firing rate versus Poisson release rate λ for model 3, with the time constant of the threshold function $S(t)$ changing to produce the different solid curves. The faster threshold functions produce the steeper firing rate curves. The 25 msec value used in Fig. 5 is the fifth most sensitive curve. All other parameters are the same as in Fig. 5. The dashed vertical line in Fig. 6 indicates the boundary between nondeterministic (to the left with lower firing rates) and deterministic crossings (to the right with higher rates for each curve). The fact that it is a vertical line reiterates that the neuron's parameters are fixed with the exception of the threshold function to produce different curves.

For small fluctuations about the mean voltage, the long correlation time case becomes a transformation of the voltage process at the time of deterministic crossing (Stein, 1967), while the short correlation case is approximated by a Wiener process with drift producing locally an inverse Gaussian distribution for the first-passage time (Lerche, 1986). For nondeterministic crossings, the times of crossing become rare events as the threshold level becomes very large. In this case, the first-passage times have a limiting exponential distribution, which is characteristic of a Poisson process. When $S(t)$ is not a constant, the threshold can be subtracted from the mean voltage to give an *effective* mean voltage, which can be thought of as a mean postspike recovery process. The problem now becomes: When does the effective voltage process reach zero? The three approximation cases are examined in more detail shortly and illustrated in Fig. 7 for model 4. Units 1, 3, 5 of Smith and Goldberg (1986) are used to show the mean (middle curve) and range of voltage variability (mean ± standard deviation

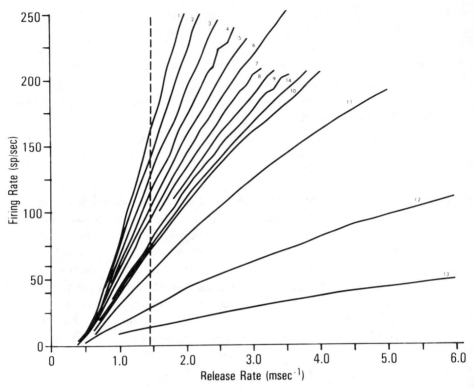

FIGURE 6.   Firing rate versus release rate (λ) for time-varying $S(t)$ boundary described in text. Firing rate is in spikes/sec and release rate in 1/msec. Curve 1 is for a horizontal boundary. Curves 2–14 are for boundaries with TS = 5, 10, 15, 20, 25, 30, 35, 40, 50, 75, 200, 500, and 45 msec, respectively. All points represent 1,000 simulated intervals. Other parameters are: $a_e = SG = 2$ mV; $VT = 10$ mV; $V_e = 70$ mV; and $\tau_m = 5$ msec. The release rate above which the mean membrane potential intersects the threshold boundary is denoted by the dashed vertical line, i.e., $\lambda = 1.49$ msec.

as bordering curves) for a constant output criterion; namely, each has a mean interval of 7 msec. Curves a, b, and c depict the three cases outlined before. a and b have deterministic crossings, while c is nondeterministic, reaching the threshold of 10 mV only through positive fluctuations from the mean.

Finally, in Figs. 8 and 9, we depict five consecutive interspike intervals for model 4 units 1 and 5, the most regular and irregular units, respectively. Voltage trajectories are shown in a, while b shows the corresponding shot noise of the synaptic conductance. The lower set of curves, c, shows the corresponding values of the time-varying K conductance. The cumulative after-hyperpolarization is

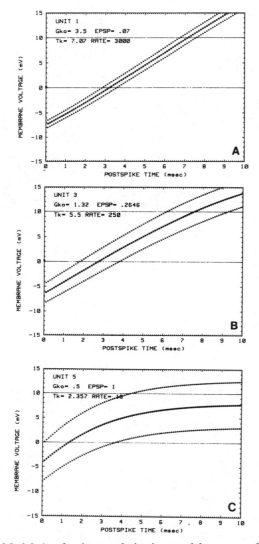

FIGURE 7. For Model 4, after-hyperpolarization model, mean voltage trajectory ± standard deviation for three units that all produce the same mean interspike interval of 7 msec. a is unit 1, b is unit 3, and c is unit 5 in Smith and Goldberg (1986), and correspond to a regular, intermediate, and irregular unit, respectively. a, b, c also illustrate the approximation cases 1, 2, 3, respectively. a and b represent deterministic crossings; c represents nondeterministic crossings.

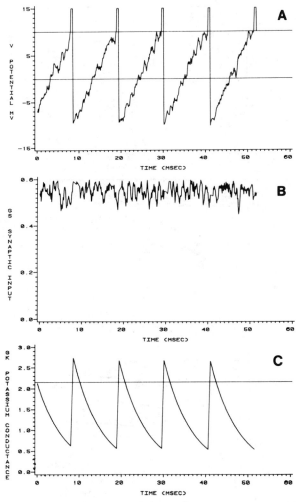

FIGURE 8. Illustration of the model with shot noise synaptic input resulting from Poisson transmitter release. a, b, c represent unit 2 with $p = 1$, i.e., cumulative after-hyperpolarization. Five consecutive spikes following an isolated spike are shown in a for model unit 2 with $p = 1$. The voltage trajectories in a are related to the synaptic input g_S (b), and the potassium conductance g_K (c), via Eq. (5). Spikes in a are shown as 0.5 msec pulses. The lower horizontal reference line is the resting potential; the upper line is the threshold for firing and has a value of 10 mV. The mean of g_S in b is 0.5444, corresponding to the mean synaptic input due to a release rate λ of 560/msec in model unit 2. The horizontal references line in c is the initial value of g_{KO} following an isolated spike.

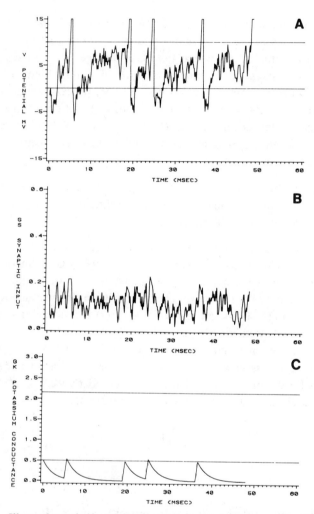

FIGURE 9.   Illustration of the model for unit 5. As in the previous figure, a, b, c corresponds to $p = 1$. The release rate λ is 14.6/msec. The other parameters for unit 5 are in text. The upper reference line for the potassium conductance has the same value as in Fig. 8; the lower reference line is the initial value of g_{KO} for an isolated spike in unit 5.

quite apparent in Fig. 8, as the residual conductance is added to the starting value for the next interspike interval. Both units have a mean interval of 10 msec but their coefficient of variation (standard deviation divided by the mean) differs by more than a factor of 10. Case 1, described next, is well-illustrated by Fig. 8 and case 3 by Fig. 9.

1.   Case 1: Deterministic Crossing, Long Correlation Time.   In this scenario the fluctuations in the first-passage time (FPT) are basically small perturbations about the time t^*, at which the mean voltage $\mu(t)$ crosses $S(t)$. Our description is a simple generalization of the method used by Stein (1967, p. 53) for model 2. Let $r(t) = S(t) - \mu(t)$ be the recovery process mentioned earlier, σ_t the standard deviation of the first-passage time distribution, $g(S,t|x_0)$, and σ^* our approximated standard deviation. The FPT p.d.f. $g(S,t|x_0)$ can be approximated, at least locally, by a transformation of the marginal distribution of $x(t)$ evaluated at t^* when:

1. The voltage distribution does not change its shape drastically near t^*.
2. σ_t is considerably less than the correlation time of $x(t)$ around t^*.
3. $r(t)$ is invertible and sufficiently smooth.

Let h be the inverse function of $r(t)$, that is, $h[r(t)] = t$. Then $g(S,t|x_0)$ is approximately $f(x)/|dh(x)/dx|$ evaluated at t^*, where $f(x)$ is the marginal distribution of $x(t)$. This is the usual Jacobian transformation of random variables; i.e., we are treating the membrane voltage process like a singular random process. For example, if $f(x)$ is Gaussian or normal and $r(t)$ is a decaying exponential, then $g(S,t|x_0)$ is approximately lognormal.

In many cases, we may only be interested in the first few moments of the FPT. The function h is now expanded in a Taylor series about $r(t^*)$. The approximations for the mean and variance are given next with $y = r(t^*)$, and μ_n is the nth central moment of the random variable $x(t^*)$:

$$E(t) \approx t^* + h''(y)\,\frac{\mu_2}{2} + h'''(y)\,\frac{\mu_3}{6} + \dots, \tag{6}$$

$$\mathrm{Var}(t) \approx (h')^2\,\mu_2 + h'h''\,\mu_3 - \left(h''\,\frac{\mu_2}{2} + h'''\,\frac{\mu_3}{6}\right)^2 + \dots, \tag{7}$$

where prime denotes differentiation with respect to voltage. Recall that the first derivative of an inverse function can be expressed as the reciprocal of the derivative of the original function, so that

$$h' = \frac{1}{dr(t)/dt} = \frac{1}{dS(t)/dt - d\mu(t)/dt}, \tag{8}$$

which is simply the derivative of the recovery process at t^*.

Stein's original approximation method was the first term of Eqs. (6) and (7) and can be illustrated graphically using Fig. 7. Note that increasing h'' decreases the approximate mean interval; hence, Stein required for condition (3) that $r(t)$ be nearly linear in the neighborhood of t^*.

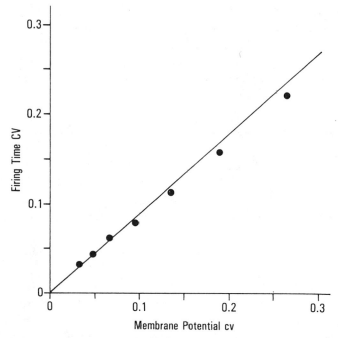

FIGURE 10.   Relation between cv of interspike intervals and cv of membrane potential for a slow threshold function. Points were obtained from simulation of one compartment model without a reversal potential and time-varying boundary with $TS = 200$ msec. The product λa_e was held constant at 8 mV/msec. EPSP sizes of 4, 2, 1, .5, .25, .125, and .0625 mV were used. All points had a firing rate of around 116 sp/sec. The solid line was obtained using the approximation method of Stein; see text for details.

While we might expect such a procedure to work for diffusion processes, Fig. 10 (model 2, excitatory inputs only) and Table 1 of Smith and Smith (1984) (model 3, excitatory inputs only) show that this approximation method also works for Poisson driven systems. In Fig. 10, the coefficient of variation, cv, (standard deviation divided by the mean) of the FPT is plotted against the cv of the voltage process for the case where the mean voltage trajectory is fixed by keeping the product $a_e \lambda_e$ fixed at a constant value of 8 mV/msec. The solid line is that predicted from the preceding equations using only the first term. It has a slope of 0.92 and is a good approximation over this range. The correlation time of the process is, in the worst case, more than twice the resultant σ_t.

This method also gives us a way to qualitatively think about the variability of firing times by looking at the first term for the variance. The standard deviation of the firing time is approximately the standard deviation of the voltage process (due to, say, synaptic noise) divided by the slope of the recovery process. So, large synaptic noise produces larger firing time variability, while fast recovery processes, if acting around t^*, compress the firing time variability. What happens if the three preceding conditions are not met; in particular, what if the correlation time is quite short compared to σ_t? This brings us to our next approximation method.

2.   Case 2: Deterministic Crossing, Short Correlation Time.   In the framework we are considering, we want the correlation time to be quite short and for the variance of $x(t)$ to be increasing linearly locally around t^*, the deterministic crossing time. Note this implies the standard deviation will be increasing as a square root with time. Then $x(t)$ will be approximated as a Wiener process with a linear drift. The linear drift is simply the slope of the recovery process at t^* and the slope of the variance of the membrane potential gives the intensity or scale parameter of the Wiener process. Said another way, around t^*, we have:

$$x(t) \approx a\,t\,+\,b\,W(t), \tag{9}$$

where $W(t)$ is a standard Wiener process, i.e., zero mean and variance t. Fig. 7b is an example of this situation. The first passage time to a constant threshold for a Wiener process with linear drift has an analytic solution that gives the well-known inverse Gaussian distribution for $g(S,t|x_0)$, the density of the first passage time. (For references on the inverse Gaussian distribution, see Chhikara and Folks, 1989.) Generalizations of this idea and a rigorous development of the so-called tangent method can be found in the recent monograph by Lerche (1986). Parameter estimation in the context of neural models for this situation has been examined by Lansky (1983).

What clues can we get from an empirical first-passage time distribution that this approximation might be appropriate? The inverse Gaussian distribution is skewed to the right and the skew increases with increasing cv; more specifically, the skew is three times the cv when skew is measured as the square root of Pearson's beta-1, i.e.,

$$\text{skew} \,=\, \sqrt{\beta_1} \,=\, E\left[\frac{(t - E(t))^3)}{\sigma_t^3}\right].$$

In Fig. 11, this relationship between skew and cv is shown for a simulation series in model 4 with a constant threshold. Each point represents 2,000 simulated intervals and different points starting with the leftmost correspond to a decreasing set of values of the rate parameter that drives the shot noise process. All other parameters were fixed except one, which controlled the serial correlation between interspike intervals, namely, the presence or absence of cumulative after-hy-

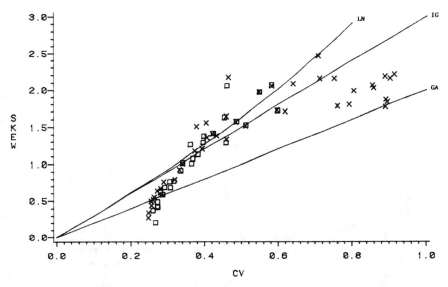

FIGURE 11. The relation between skew and cv for simulations of model 4. Skew is measured as Pearson's square root of beta-1. Each symbol represents 2,000 simulated intervals. Different points correspond to different values of the Poisson release rate that drives the shot noise process. The higher rates produce smaller cv values. The three reference curves are the corresponding relations for the following two parameter distributions: lognormal (LN), inverse gaussian (IG), and gamma (GA). Note that an exponential distribution has a cv of 1 and a skew of 2.

perpolarization. The two values of this parameter are denoted by x and boxes, and do not have a large effect on the plot. (See Smith and Chen, 1986, for more details.) Other two parameter distributions that are also positively skewed are shown for comparison, the gamma (GA) and lognormal (LN), which produce curves of 2 cv and 3 cv + cv^3, respectively. For the middle range of values of cv, which correspond to deterministic crossings and short correlation times, the simulated values bounce around the inverse Gaussian curve (IG). For larger values of cv, we no longer have deterministic crossings, which will be considered shortly as our final approximation method. Figure 12 shows a different measure, minimum Anderson–Darling distance (Boos, 1981), of how well the data of Fig. 11 fits a lognormal distribution as a function of the release rate of the shot noise process. Even with this more sensitive method, the presence or absence of cumulative after-hyperpolarization does not alter the goodness-of-fit measure dramatically.

Another illustration of the use of empirical moments is seen in Fig. 13, with pigeon vestibular data from Dr. Correia's lab (Anastasio *et al.*, 1985). Each

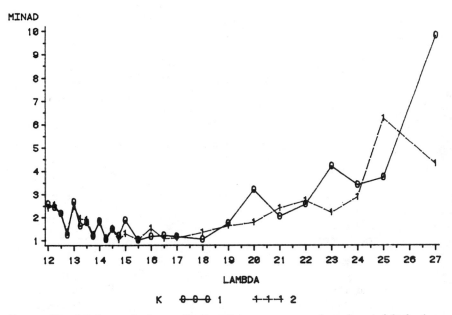

FIGURE 12. Minimum Anderson–Darling distance measure of goodness-of-fit for lognormal distribution fit to data in Fig. 11. Solid (dashed) curve corresponds to presence (absence) of cumulative after-hyperpolarization. λ values correspond to release rate of shot noise synaptic process with large values corresponding to small cv's in Fig. 11.

symbol corresponds to an interspike interval histogram of 1,024 intervals. Here, in what is known as a Pearson plot, normalized fourth and third central moments are used to produce the axes of excess versus skewness measured as β_1, the square of the measure used before. The four curves beginning with the upper correspond to the moment relationships for a lognormal, inverse Gaussian, gamma, and first-passage time of an Ornstein–Uhlenbeck process. The error bars are centered at the exponential distribution, a special case of the gamma, and represent standard errors due to sampling expected for a data size of 1,024 interspike intervals—the point being that larger data records are needed to distinguish between candidate curves on this type plot. An alternative is to use a plot as in Fig. 11 with smaller standard errors, or a goodness-of-fit measure as in Fig. 12.

3. Case 3: Nondeterministic Crossings, Large Threshold. Intuitively, what happens in this case is that the asymptotic value of the mean of the process is below the threshold, and the threshold when measured in units of the standard deviation of the steady-state voltage process is far away from the steady-state mean voltage. Crossings of the process will be rare and roughly a Poisson process.

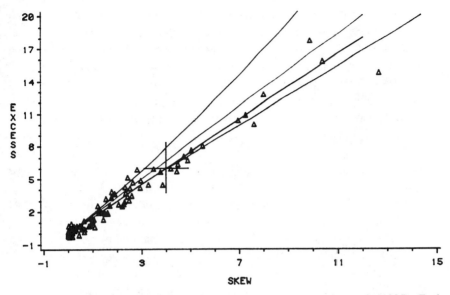

FIGURE 13.   Pearson plot for pigeon vestibular data of Anastasio *et al.* (1985). Each symbol represents 1,024 interspike intervals. Data courtesy of Prof. Correia. Four curves represent relationships for lognormal, inverse Gaussian, gamma, and first-passage time of Ornstein–Uhlenbeck process. Error bars correspond to standard errors for an exponential distribution with 1,024 intervals.

The time between arrivals for a Poisson process is exponentially distributed, so the first-passage time will have an exponential distribution with the rate parameter being the reciprocal of the mean first-passage time. This idea has been around for quite a while, with Newell showing the result for some diffusion processes in 1962, and Keilson gave a weak convergence result in 1966. More recently, for several types of diffusion processes, Ricciardi and colleagues (1985, 1986) have established this result and asymptotic series for the rate parameter of the exponential distribution. Ricciardi and Sato (1983, 1984, 1986) have asymptotic results for nonconstant boundaries.

Using the Ornstein–Uhlenbeck process (model 1), starting from the resting level of zero and having a constant threshold as an example, we will present a sampling of the type of asymptotic results available and note some other approaches for obtaining them. The results are more general in that the behavior of other models (3 and 4) are seen to be well-approximated by the OU process in the limit of large thresholds. The scenario for the next figure is as follows: The mean first-passage time for the OU process is tabulated and, for thresholds more than 3 σ_v away, are well-approximated by the first time of the asymptotic expansion of the OU first-passage time starting from 0. Models 3 and 4 are not stationary processes, so we add a dead time equal to the time at which the mean and variance of $X(t)$ reaches 95% of its steady-state level, time is measured in units of the steady state process's correlation time, and voltages are normalized by the steady-state standard deviation.

We can now compare these processes to the normalized OU process. Only the mean FPT is considered, since it is the only parameter needed to characterize the limiting exponential distribution. In Fig. 14, the shot noise model is seen to be well-approximated by the OU process if the scaling and dead time are included. A similar agreement is found with model 3. Note neither of these processes have continuous sample paths nor Gaussian marginal distributions. What is similar is the correlation structure for small time differences. Model 3 is similar to the OU process in that its linear drift term dominates at large threshold values.

Since the OU process is a diffusion process, the method of Seigert outlined in Ricciardi (1977, Chapter 3) can be used. The starting value will be taken to be 0. The stationary distribution is normal with zero mean and a variance of $\sigma^2\tau/2$, cf. Eq. (2), so we normalize the voltages (including the threshold) by

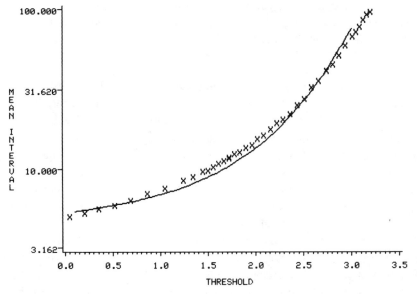

FIGURE 14. Asymptotic Ornstein–Uhlenbeck approximation to model 4. Abscissa is normalized threshold and ordinate is mean interval. A dead time of 2.25 times the time constant is used, and voltages are normalized by the standard deviation of the steady-state voltage process. The solid curve is the relationship for the OU process and the x's are from simulations of model 4 (see Sato, 1977; Favella *et al.*, 1982).

dividing by $\sigma\sqrt{\tau/2}$. The stationary distribution is now Gaussian, with zero mean and unit variance. From Keilson and Ross (1975), the leading term in the asymptotic series, as $S \to \infty$ for the mean FPT, is

$$E(t(S|0)) \sim \sqrt{2\pi} \left(\frac{1}{S^*}\right) \exp\left(\frac{(S^*)^2}{2}\right), \tag{10}$$

where S^* is the normalized threshold. The rest of the asymptotic series in Eq. (10) can be obtained by multiplying the right-hand side by the generalized hypergeometric function, $_2F_0$ $(1,1/(S^*\sqrt{2}), 1/(2S^{*2}))$. The leading term agrees with the actual result within 10% for $S^* > 3$. The leading term can be obtained in another way, using the method of stochastic perturbation of dynamic systems (Williams, 1982; Freidlin and Wentzello, 1984; Hanson and Tier, 1981). Consider the stochastic differential equation,

$$dx = a(x)\, dt + \sqrt{\varepsilon}\, \sigma(x)\, dW(t), \tag{11}$$

and examine what happens as $\varepsilon \to 0$. For the OU process, $\sigma(x) = \sigma$ and $a(x) = -x/\tau$. The exit time out of a domain D, which in our case is $(-\infty, S)$, is shown using generating function and boundary layer theory to be exponential, and an expression for the rate parameter is given (Williams, 1982, p. 151). When we substitute for the parameters of the OU process, the leading term in the asymptotic series (10) is obtained, since letting ε go to zero means the normalized threshold goes to infinity. However, the results from stochastic perturbation theory apply to a wider class of problems, provided we scale the threshold appropriately. The restrictions on $a(x)$ and $\sigma(x)$ are continuity and smoothness (e.g., Lipschitz) and that $x = 0$ be the only attractor of the deterministic ($\sigma(x) = 0$) system. The problem can also be viewed as a particle in a potential well. Kipnis and Newman (1985) have extended the exit time results to bi- and multi-stable potential wells, and Day (1987) has extended earlier results to include nonsmooth quasipotentials. This is just a sampling of recent work in these active areas. Said another way, most commonly reported membrane I-V relationships could be used.

Finally, we mention a third way to obtain the leading term in Eq. (10), using the Poisson nature of the rare crossings, namely, extreme value theory of statistics. Using quite different arguments than the previous two methods, Leadbetter et al. (1983, p. 236) obtain the leading term in Eq. (10) as the rate parameter for the Poisson process of the so-called ε upcrossings for the OU process.

Returning to the neural interpretation, the effect of the recovery process for case 3 can be incorporated by adding a dead time as mentioned earlier. The effect on the cv versus threshold relationship is depicted in Fig. 15. The OU process starting from zero has a cv > 1 for small thresholds. This seems at odds

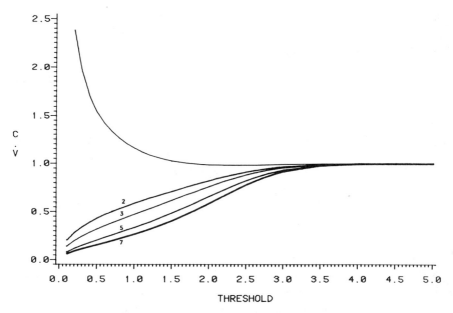

FIGURE 15.   Coefficient of variation versus threshold level for first-passage time of Ornstein–Uhlenbeck process. Asymptotic level of 1 corresponds to an exponential distribution. Lower curves represent effect of various dead times, mimicking an initial recovery process, on cv relationship.

with most interspike interval results. The lower curves illustrate that the cv is dramatically lowered by a recovery process and in better agreement with measured cv's.

In summary of this section, we use Fig. 16. The cv versus mean interval is shown for a number of simulations of model IV, *cf.* Fig. 2 in Smith and Goldberg, 1986, for three model units. Each point represents 500 or 2,000 simulated interspike intervals. The solid lines correspond to the presence of cumulative after-hyperpolarization (CAHP) as in *loc. cit.*, and the open symbols to the absence of CAHP. Very little difference is seen between the solid curves and open symbols. This result can be seen also from our approximation methods. The lower left triangular region can be approximated by case 1, the diagonal region by case 2, and the upper right region by case 3 with a dead time correction.

III.   Discussion

We have illustrated, in terms of four relatively simple neural models, the three ranges of behavior of first-passage times where heuristic approximation methods

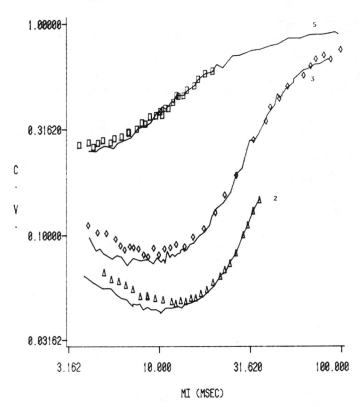

FIGURE 16.   Relations between coefficient of variation (cv) and mean interval (MI) for
$p = 0$ (points) and $p = 1$ (connected lines) versions of model units 2($\triangle$), 3($\blacklozenge$) and
5($\square$). The unit number is displayed. Note the scale is log–log. For units 2 and 5, the
number of simulated interspike intervals, N, represented by each plotted point is 2,000;
for unit 3, $N = 500$.

can be useful. While the approximation methods are no substitute for careful
mathematical analysis and for simulations of models too complicated to be fully
analyzed mathematically, they can be used as a guide for preliminary work,
insight, and in ascertaining the robustness of the model's structure and sensitivity
to its parameter values. Even with current computers, simulations of some first-
passage time problems require a prohibitive amount of cpu time. The preliminary
use of approximations here seems almost mandatory. If the heuristics prove
interesting, collaboration on a more realistic and detailed model with a local
statistician, probabilist, or applied mathematician could evolve more efficiently.

Various interspike interval moment plots, e.g., the plot of skew versus coefficient of variation, were seen to be a diagnostic for locating regions of inverse
Gaussian and of exponential behavior of the p.d.f. of the interspike intervals.

This plot, with standard errors included, can also be used as a guide to how long a data record is needed to distinguish among different models, and can serve as a preliminary guide for statistical goodness-of-fit tests. These plots can also be used in sensitivity analysis on the model parameters. Lansky and Smith (1989), and Lansky and Musila (1991), have examined the effect of a random initial condition in several integrate-and-fire models.

For the asymptotic behavior of the first-passage time in the nondeterministic crossing case, it was noted that the rate parameter of the limiting exponential distribution has been obtained in three ways for the Ornstein–Uhlenbeck process. This interplay between the three approaches (asymptotic series in Seigert's approach, perturbation of dynamic systems via the Wentzell–Freidlin method, and Poisson nature of upcrossings from extreme value theory) appears to be a rich area for gaining insight. Further study of the relation between recent results in extreme value theory (e.g., Aldous, 1989; Leadbetter and Rootzen, 1988) and in exit times for one-dimensional diffusion processes, along the lines of Newell's work (1962), seems warranted.

As mentioned in the introduction, this chapter is somewhat limited in scope to be able to illustrate some basic ways of thinking about stochastic neural

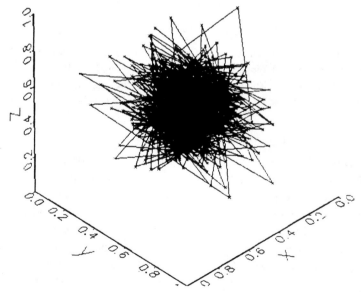

FIGURE 17.   Lack of evidence for chaos (low-dimensional attractor) in simulations of after-hyperpolarization stochastic model; 500 interspike intervals; $X = i$th interval, $Y = (i + 1)$st interval, $Z = (i + 2)$nd interval. Axes are scaled to maximum interval.

problems. Multiple neuron recording also has a strong statistical and stochastic modeling component (e.g., Aertsen *et al.*, 1989). Serial dependence in a spike train was only briefly illustrated with the after-hyperpolarization model. Johnson *et al.* (1986) characterized this dependence in spontaneous activity and used it to predict responses to time-varying stimuli in the cat LSO. In this volume, Chapter 22 by Teich will show a form of long-term dependence in auditory nerve fibers as well.

Finally, there was no mention of chaos in this chapter before now. Figure 17 illustrates why. The plot of three consecutive interspike intervals for 500 intervals from the AHP model produces a pincushion even though there was a statistically significant first-order negative serial correlation. Rotating the graph fails to graphically reveal any low-dimension manifolds. In summary, if your corresponding plot does show more interesting structure, the author would like to suggest the methods of Ellner *et al.* (1991). Their methods seem to require less data for reliable Lyapunov exponent estimation than some other competitors.

Acknowledgments

Support for this work was furnished by the Office of Naval Research Contract N00014-90-J-1646-1. Thanks to Profs. J.M. Goldberg and M. Correia for use of data from their labs.

References

AERTSEN, A. M. H. J., GERSTEIN, G. L., HABIB, M. K., and PALM, G. (1989). "Dynamics of Neuronal Firing Correlation: Modulation of 'Effective Connectivity,'" *J. Neurophysiol.* **61**, 900–917.

ALDOUS, D. (1989). *Probability Approximations via the Poisson Clumping Heuristic.* Springer-Verlag, New York.

ANASTASIO, T. J., CORREIA, M. J., and PERACHIO, A. A. (1985). "Spontaneous and Driven Responses of Semicircular Canal Primary Afferents in the Unanesthetized Pigeon," *J. Neurophysiol.* **54**, 335–347.

BLAKE, I. F., and LINDSEY, W. C. (1973). "Level-Crossing Problems for Random Processes," *IEEE Trans. Info. Theor.* **19**, 295–315.

BOOS, D. D. (1981). "Minimum Distance Estimators for Location and Goodness of Fit," *J. Amer. Stat. Assoc.* **76**, 663–670.

CHHIKARA, R. S., and FOLKS, J. L. (1989). *The Inverse Gaussian Distribution: Theory, Methodology and Applications.* Marcel Dekker, New York.

COBB, L., and ZACKS, S. (1985). "Applications of Catastrophe Theory for Statistical Modeling in the Biosciences," *J. Amer. Stat. Assoc.* **80**, 793–802.

DAY, M. V. (1987). "Recent Progress on the Small Parameter Exit Problem," *Stochastics* **20,** 121–150.

ELLNER, S., GALLANT, A. R., McCAFFERY, D., and NYCHKA, D. (1991). "Convergence Rates and Data Requirements for Jacobian-Based Estimates of Lyapunov Exponents from Data," *Physics Letters A,* **153,** 357–363.

FAVELLA, L., REINERI, M. T., RICCIARDI, L. M., and SACERDOTE, L. (1982). "First Passage Time Problems and Related Computational Methods," *Cybernetics and Systems* **13,** 95–128.

FREIDLIN, M. I., and WENTZELL, A. D. (1984). *Random Perturbations of Dynamical Systems.* Springer-Verlag, New York.

GIORNO, V., NOBILE, A. G., RICCIARDI, L. M., and SACERDOTE, L. (1986). "Some Remarks on the Rayleigh Process," *J. Appl. Prob.* **23,** 398–408.

GOLDBERG, J. M., and FERNANDEZ, C. (1971). "Physiology of Peripheral Neurons Innervating Semicircular Canals of the Squirrel Monkey: III. Variation among Units in Their Discharge Properties," *J. Neurophysiol.* **34,** 676–684.

HANSON, F. B., and TIER, C. (1981). "An Asymptotic Solution of the First Passage Problem for Singular Diffusion in Population Biology," *SIAM J. Appl. Math* **40,** 113–132.

HOLDEN, A. V. (1976). *Models of the Stochastic Activity of Neurons, Lecture Notes in Biomathematics, Vol. 12.* Springer-Verlag, Berlin.

JACK, J. J. B., NOBLE, D., and TSIEN, R. W. (1983). *Electric Current Flow in Excitable Cells.* Clarendon Press, Oxford, U.K.

JOHNSON, D. H., TSUCHITANI, C., LINEBARGER, D. A., and JOHNSON, M. J. (1986). "Application of a Point Process Model to Responses of Cat Superior Olive Units to Ipsilateral Tones," *Hearing Res.* **21,** 135–159.

KEILSON, J. (1966). "A Limit Theorem for Passage Times in Ergodic Regenerative Processes," *Ann. Math. Stat* **37,** 866–870.

KEILSON, J., and ROSS, H. F. (1975). "Passage Time Distributions for Gaussian Markov (Ornstein–Uhlenbeck) Statistical Processes," *Selected Tables in Math. Stat. III,* 233–327.

KIPNIS, C., and NEWMAN, C. M. (1985). "The Metastable Behavior of Infrequently Observed, Weakly Random, One-Dimensional Diffusion Processes," *SIAM J. Appl. Math.* **45**(6), 972.

KLIEMANN, W. (1983). "Qualitative Theory of Stochastic Dynamical Systems—Applications to Life Sciences," *Bulletin Math. Biol.* **45,** 483–506.

LANSKY, P. (1983). "Inference for the Diffusion Models of Neuronal Activity," *Math. Biosci.* **67,** 247–260.

LANSKY, P., and MUSILA, M. (1991). "Variable Initial Depolarization in Stein's Neuronal Model with Synaptic Reversal Potentials," *Biol. Cybernet.* **64,** 285–291.

LANSKY, P., and SMITH, C. E. (1989). "The Effect of a Random Initial Value in Neural First-Passage-Time Models," *Math. Biosci.* **93,** 191–215.

LANSKY, P., SMITH, C. E., and RICCIARDI, L. M. (1990). "One-Dimensional Stochastic Diffusion Models of Neuronal Activity and Related First Passage Time Problems," *Trends in Biological Cybernetics* **1,** 153–162.

LEADBETTER, M. R., LINDGREN, G., and ROOTZEN, H. (1983). *Extremes and Related Properties of Random Sequences and Processes*. Springer-Verlag, New York.

LEADBETTER, M. R., and ROOTZEN, H. (1988). "External Theory for Stochastic Processes," *Annals Prob.* **16**, 431–478.

LERCHE, H. R. (1986). *Boundary Crossing of Brownian Motion, Lecture Notes in Statistics, Vol. 40.* Springer-Verlag, Berlin.

NEWELL, G. F. (1962). "Asymptotic Extreme Value Distribution for One-Dimensional Diffusion Process," *J. Mathematics and Mechanics* **11**, 481–496.

NOBILE, A. G., RICCIARDI, L. M., and SACERDOTE, L. (1985). "Exponential Trends of Ornstein–Uhlenbeck First-Passage-Time Densities," *J. Appl. Prob.* **22**, 360–369.

RICCIARDI, L. M. (1977). *Diffusion Processes and Related Topics in Biology, Lecture Notes in Biomathematics,* (Notes by C. E. Smith) *Vol. 14.* Springer-Verlag, Berlin.

RICCIARDI, L. M., and SACERDOTE, L. (1979). "The Ornstein–Uhlenbeck Process as a Model for Neuronal Activity: I. Mean and Variance of the Firing Time," *Biol. Cybernetics* **35**, 1–9.

RICCIARDI, L. M., SACERDOTE, L., and SATO, S. (1983). "Diffusion Approximation and First Passage Time Problem for a Model Neuron II: Outline of a Computational Method," *Math. Biosci.* **64**, 29–44.

RICCIARDI, L. M., SACERDOTE, L., and SATO, S. (1984). "On an Integral Equation for First Passage Time Probability Densities," *J. Appl. Prob.* **21**, 302–314.

RICCIARDI, L. M., and SATO, S. (1983). "A Note on the First Passage Time Problems for Gaussian Processes and Varying Boundaries," *IEEE Trans. Info. Theory* **29**, 454–457.

RICCIARDI, L. M., and SATO, S. (1984). "A Note on the Evaluation of First Passage Time Probability Densities," *J. Appl. Prob.* **20**, 197–201.

RICCIARDI, L. M., and SATO, S. (1986). "On the Evaluation of First Passage Time Densities for Gaussian Processes," *Signal Processing* **11**, 339–357.

RINZEL, J., and ERMENTROUT, G. B. (1989). "Analysis of Neural Excitability and Oscillations," in C. Koch and I. Segev (eds.), *Methods in Neuronal Modeling* (pp. 135–169). MIT Press, Cambridge, Massachusetts.

SATO, S. (1977). "Evaluation of the First-Passage Time Probability to a Square Root Boundary for the Wiener Process," *J. Appl. Prob.* **14**, 850–856.

SATO, S. (1978). "On the Moments of the Firing Interval of the Diffusion Approximated Model Neuron," *Math. Biosci.* **39**, 53–70.

SHERMAN, A., RINZEL, J., and KEIZER, J. (1988). "Emergence of Organized Bursting in Clusters of Pancreatic β-Cells by Channel Sharing," *Biophysical J.* **54**, 411–425.

SMITH, C. E., and CHEN, C. L. (1986). "Serial Dependency in Neural Point Processes due to Cumulative Afterhyperpolarization," *Mimeo Series 1691,* Institute of Statistics, North Carolina State University, Raleigh, North Carolina.

SMITH, C. E., and GOLDBERG, J. M. (1986). "A Stochastic Afterhyperpolarization Model of Repetitive Activity in Vestibular Afferents," *Biol. Cybernetics* **54**, 41–51.

SMITH, C. E., and SMITH, M. V. (1984). "Moments of Voltage Trajectories for Stein's Model with Synaptic Reversal Potentials," *J. Theor. Neurobiol.* **3**, 67–77.

STEIN, R. B. (1967). "Some Models of Neuronal Variability," *Biophys. J.* **7**, 36–68.

TUCKWELL, H. C. (1979). "Synaptic Transmission in a Model for Stochastic Neural Activity," *J. Theor. Biol.* **77,** 65–81.

TUCKWELL, H. C. (1988). *Introduction to Theoretical Neurobiology, Vol. 2: Nonlinear and Stochastic Theories.* Cambridge University Press, Cambridge, U.K.

TUCKWELL, H. C. (1989). *Stochastic Processes in the Neurosciences, SIAM-CBMS Regional Conference in Mathematics.* SIAM, Philadelphia.

WILLAMS, M. (1982). "Asymptotic Exit Time Distributions," *SIAM J. Appl. Math.* **42,** 149–154.

WINSLOW, R. L. (1989). "Bifurcation Analysis of Nonlinear Retinal Horizontal Cell Models: I. Properties of Isolated Cells," *J. Neurophysiol.* **62,** 738–749.

YANG, G. L., and CHEN, T. C. (1978). "On Statistical Methods in Neuronal Spike Train Analysis," *Math. Biosci.* **38,** 1–34.

Chapter 22   Fractal Neuronal Firing Patterns

MALVIN C. TEICH

Department of Electrical Engineering
Columbia University
New York, New York

I. Introduction

The sequence of action potentials produced by a neuron is best characterized in terms of a stochastic point process (Teich, 1989; Teich and Khanna, 1985). This is because the information is encoded in the occurrence times, rather than in the magnitudes, of the unitary neural events, and these occurrence times are random. The mathematical process that has been traditionally used in auditory, and other branches, of sensory neurophysiology has been the *dead-time-modified Poisson point process,* denoted DTMP (Gaumond *et al.,* 1982; Gray, 1967; Kuffler *et al.,* 1957; Mueller, 1954; Prucnal and Teich, 1983; Teich *et al.,* 1978; Young and Barta, 1986). Theoretical results for the DTMP process are widely available in the literature (Cox, 1962; Müller, 1974; Prucnal and Teich, 1983; Ricciardi and Esposito, 1966; Teich, 1985). This model of neuronal firing achieved its principal successes in describing *interspike-interval histograms* (or pulse-interval distributions, PIDs) and *post-stimulus-time histograms* (PSTs), measures that reset at relatively short times and are therefore insensitive to long-time correlations in spike occurrences. The predictions of the DTMP model turn out to be at odds with many other observed statistical measures of neurophysiological data. In particular, it is now quite clear that the DTMP fails to provide a proper characterization of the sequence of action potentials in the auditory neurons of a number of species (Teich *et al.,* 1990a).

Auditory signals transmitted from the hair-cell receptor in the cochlea to the cortex (and beyond) pass through many way stations located along the auditory pathways. In recent years, the responses of receptor cells, as well as the primary auditory-nerve-fiber neurons that attach to them, have been studied in great detail in many species, both in the presence and in the absence of various kinds of acoustic stimulation. The *patterns* of neuronal firing have been examined in single mammalian neurons at various locations along this pathway, including the auditory nerve (AN), the cochlear nucleus (CN), which is the first way station, and the lateral superior olivary complex (LSO).

The sequences of action potentials observed from single neurons at the AN, CN, and LSO turn out to exhibit long-term correlations that are not captured in the interspike-interval and post-stimulus-time histograms (Teich, 1989; Teich and Khanna, 1985). The properties of these spike trains are unusual. They manifest highly irregular spike rates, even when the integration time is very long; broad pulse-number distributions (which are histograms of the relative frequency of observing a given number of spikes versus the spike number); fractional power-law growth of the variance-to-mean ratio with the counting time T (Fano-factor time curve) with an exponent that depends on the level of stimulation; and $1/f$-type behavior in the spectrum (Teich, 1989; Teich et al., 1990a, b; Woo et al., 1992a).

The firing patterns of these neurons are characterized as *fractal* because the spike-rate fluctuations are self-similar over a large range of integration times. This self-similarity is also evident in the spectrum. These properties bespeak correlation that decays in power-law fashion, and therefore long-term memory, at the periphery of the auditory system.

Fractal patterns exhibit order within apparent randomness, and reveal the presence of multiple scales of time and/or space. All primary auditory neurons examined to date, both in the presence and in the absence of an acoustic stimulus, exhibit this behavior for sufficiently large observation times, as has now been confirmed in cat and chinchilla in a number of laboratories (Powers, 1991; Powers et al., 1991; Teich et al., 1990b; Woo, 1991; Woo et al., 1992). Fractal behavior also appears to be present at the CN (Shofner and Dye, 1989) and may be present at the LSO (Turcott et al., 1991).

In contrast to this behavior, the patterns associated with primary vestibular neurons have not been found to exhibit long-term correlations and appear to be non-fractal (Teich, 1989). It may be that fractal neural firings in the auditory

FIGURE 1.   (a) Spontaneous firing rate of a primary auditory neuron (unit A, CF = 10.2 kHz). Two different time windows were used to compute the rate: $T = 0.5$ s (solid curve) and $T = 5.0$ s (dashed curve). (b) The firing rate of a simulated dead-time-modified Poisson (DTMP) point process with the same time windows. From Teich et al. (1990b).

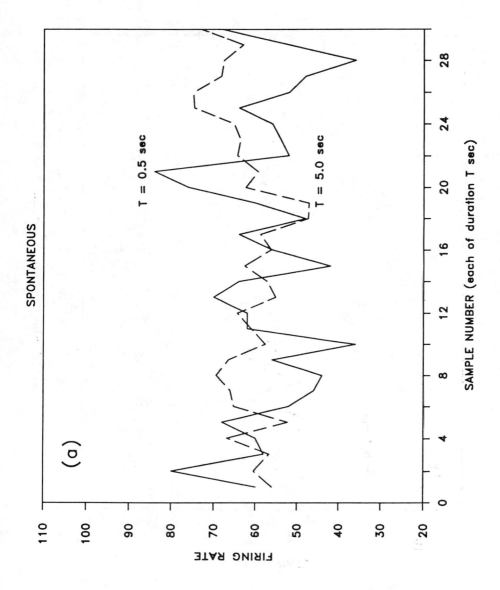

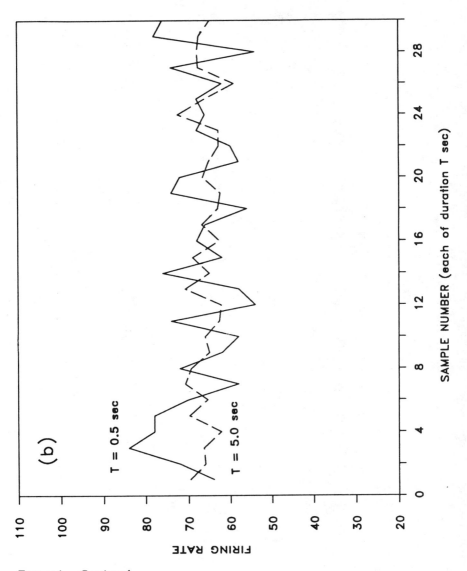

FIGURE 1. *Continued*

system serve to provide efficient sampling of natural fractal sounds. Several biophysical mechanisms present themselves as possible origins of this behavior, as discussed later in this chapter.

In seeking to identify the point process that properly models auditory neural firings, we have constructed a doubly stochastic Poisson point process (DSPP) driven by fractal shot noise (FSN), abbreviated FDSPP, or FSNDP (Lowen and Teich, 1990, 1991). With the incorporation of dead-time effects (absolute refractoriness) and/or sick-time effects (relative refractoriness), this process appears to work remarkably well in describing the fractal firing patterns of primary auditory-nerve neurons, both in the presence of a pure-tone stimulus and in its absence (spontaneous firings) (Teich *et al.*, 1990a).

II.   Self-Similarity of Neuronal Firing Rates

Perhaps the simplest measure of a sequence of action potentials is its rate, i.e., the number of spikes registered per unit time. In primary auditory neurons, even this straightforward measure has unusual properties; the magnitude of the fluctuations of the rate does not decrease appreciably, even when a very long averaging period (time window) is used to compute the rate. This property reflects fractal behavior and is in direct opposition to the predictions of the DTMP process.

In Fig. 1a, we illustrate the firing rate of a spontaneously active adult-cat auditory neuron (unit A) with a characteristic frequency (CF) = 10.2 kHz. Two different time windows were used to compute the rate: $T = 0.5$ s (solid curve) and $T = 5.0$ s (dashed curve). The total time duration of the solid curve is 15 s (30 consecutive time windows, each of 0.5 s), whereas the total time duration of the dashed curve is 150 s (30 consecutive time windows, each of 5.0 s). Evidently, increasing the averaging time by a factor of 10 does not appreciably reduce the magnitude of the fluctuations.

The firing rate of a simulated DTMP point process is illustrated in Fig. 1b. The rate and time windows were chosen to be the same as those for the auditory data shown in Fig. 1a, and the (fixed, nonparalyzable) dead time was taken to be $\tau_d = 2.95$ ms (for reasons that will become apparent in Section III). The $T = 5.0$ s computer data (dashed curve) exhibits noticeably smaller fluctuations than does the $T = 0.5$ s computer data (solid curve). This smoothing with increased averaging time does not occur with the neural data.

The contrast is even more dramatic in the case of driven activity, as illustrated in Fig. 2a. Firing-rate data are shown for the same neuron as illustrated in Fig. 1a, but now with continuous-tone stimulation at the CF. The rate is generally higher than that in Fig. 1a because the neuron is driven. Three different time windows are used to compute the rate: $T = 0.5$ s (solid curve), $T = 5.0$ s (dashed curve), and $T = 50$ s (dotted curve). The total time duration of the solid curve is 15 s (30 consecutive time windows, each of 0.5 s), the total time duration of the dashed curve is 150 s (30 consecutive time windows, each of 5.0 s), and the total time duration of the dotted curve is 550 s (11 consecutive time windows, each of 50 s). To minimize the effects of nonstationarity arising from adaptation, the data presented in Fig. 2a begins 250 s after the onset of the stimulus. The fluctuations of the rate do not appear to be smoothed, even though the change of time scale is a factor of 100 (from $T = 0.5$ s to $T = 50$ s); the process may be said to be self-similar (Mandelbrot, 1983; Teich, 1989).

The firing rate of a simulated DTMP process with the same rate, but now with a dead time $\tau_d = 2.48$ ms, is illustrated in Fig. 2b for comparison. The time windows are the same as those used in Fig. 2a. The substantial smoothing of the rate fluctuations with increasing counting time is in dramatic contrast to the self-similar behavior apparent in the auditory data. This clearly shows that the DTMP is not a satisfactory model for the auditory data.

III.   Power-Law Growth of the Spike-Number Variance-to-Mean Ratio

The *pulse-number distribution* (PND) is a commonly used characteristic of a point process. It is an estimate of the probability $p(n,T)$ of observing n spikes in the observation time T versus the number of spikes n. A useful statistic of the PND is provided by the spike-number (count) variance-to-mean ratio

$$F = \frac{\mathrm{Var}(n)}{\langle n \rangle}. \tag{1}$$

FIGURE 2.   (a) Firing rate of an auditory neuron (unit A, same cell as displayed in Fig. 1) driven by a continuous tone at the characteristic frequency. Three time windows were used to compute the rate: $T = 0.5$ s (solid curve), $T = 5.0$ s (dashed curve), and $T = 50$ s (dotted curve). Increasing the averaging time by a factor of 100 does not appreciably decrease the magnitude of the fluctuations; the continuous-tone-driven process is self-similar. (b) The firing rate of a simulated dead-time-modified Poisson point process with the same time windows as in a. The evident smoothing of the rate fluctuations with increasing counting time is in dramatic contrast to the self-similar behavior observed in the auditory data. From Teich *et al.* (1990b).

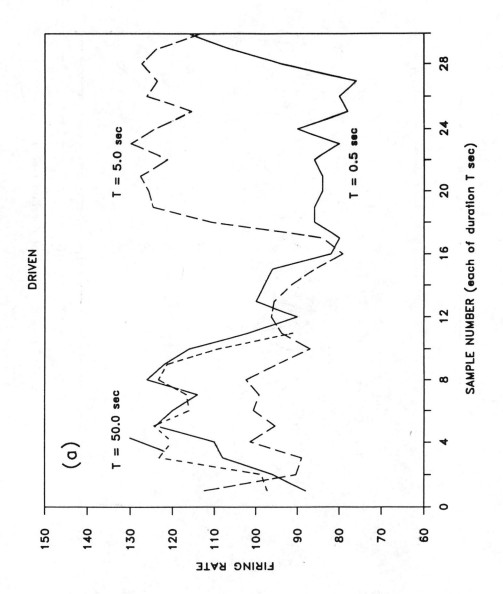

DRIVEN

(a)

T = 50.0 sec

T = 5.0 sec

T = 0.5 sec

FIRING RATE

SAMPLE NUMBER (each of duration T sec)

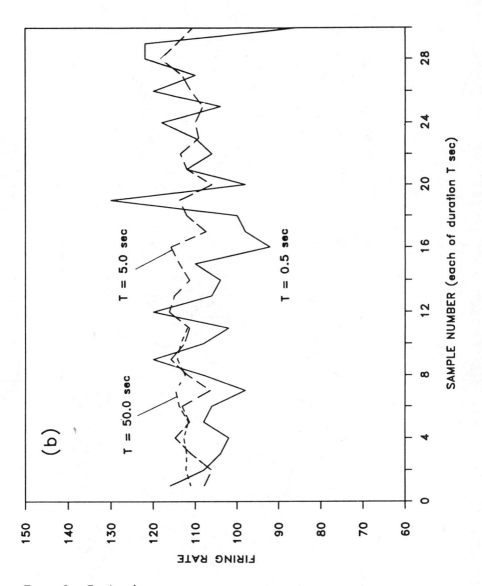

FIGURE 2. *Continued*

This quantity is often referred to as the *Fano factor,* since it was first used by Fano (1947) as a measure of the statistical fluctuations of the number of ions generated by individual fast charged particles.

The *Fano-factor time curve* (FFC) is the ratio of the count variance to the count mean for different counting times T. It is designated $F(T)$, and is shown in Fig. 3 for unit A under conditions of spontaneous firing (dashed curve). This plot is obtained using the same spike train that provided the firing rates shown in Figs. 1a and 2a. (The spontaneous firing rate was about 60 s^{-1} and the overall duration of the experiment was $L = 400$ s.) The FFC is seen to assume a value of unity at short counting times, dip below unity for counting times above about 1 ms (where refractoriness comes into play), and finally to increase with T in power-law fashion (with an exponent $\alpha \approx 0.68$ for this particular neuron) when the counting time exceeds 400 ms. All primary auditory neurons examined exhibited this characteristic FFC shape. For spontaneous activity in the absence of a stimulus, F typically assumes a minimum value $F_{\min}$ between about 0.6 and 1.0 at counting times in the tens of ms. The onset of pure power-law behavior (with exponents in the range between about 0.3 and 0.9) occurs at counting times between about 0.1 and 1.0 s. The fractal nature of the process manifests itself in the power-law regime.

The FFC for a simulated dead-time-modified Poisson process (DTMP) is shown for comparison (solid curve). It assumes a value of unity when $T \ll \tau_d$ (as expected for a Bernoulli process with low probability of success), dips below unity when the counting time T approaches τ_d, and remains approximately constant, at a value below unity, for all values of $T \gg \tau_d$. The asymptotic value assumed by the DTMP Fano factor F_d for counting times large in comparison with the dead time is (Teich, 1985):

$$F_d \approx (1 - \lambda\tau_d)^2, \tag{2}$$

where λ represents the post-dead-time firing rate. The minimum value of the Fano factor observed for unit A, $F_d \approx 0.68$, requires $\tau_d = 2.95$ ms, since $\lambda = 60$ s^{-1}. The dead-time simulations exhibited in Figs. 3, 1b, and 2b all make use of values of τ_d that satisfy Eq. (2) when the observed values of λ and $F_{\min}$ are used.

Poisson processes modified by stochastic dead time or by sick time, which are physiologically more realistic, lead to Fano-factor time curves similar to those for fixed dead time (Teich and Diament, 1980; Teich *et al.,* 1978; Young and Barta, 1986). The fixed dead-time approximation is usually adequate when considering count (as opposed to interval) measures of the spike train.

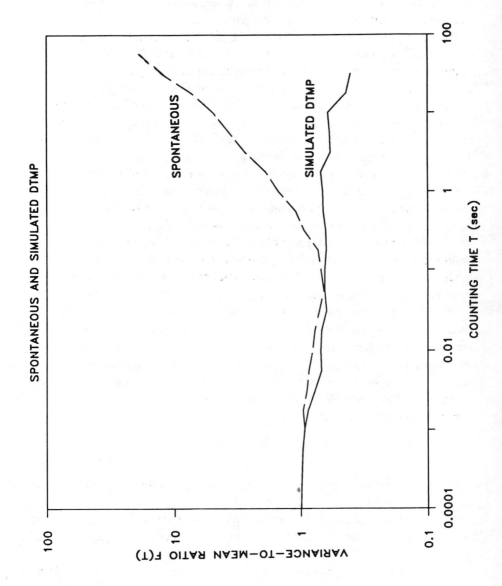

SPONTANEOUS AND SIMULATED DTMP

SPONTANEOUS

SIMULATED DTMP

COUNTING TIME T (sec)

VARIANCE-TO-MEAN RATIO F(T)

IV.   Fractal Dimension of the Firing Pattern

The fractal dimension of the firing pattern is a measure of the degree of spike correlations that is preserved over different time scales. It falls between the topological dimension $D_T = 0$ and the Euclidian dimension $E = 1$ of a one-dimensional point process (or *dust*) (Mandelbrot, 1983). For the auditory neural spike train, the fractal dimension is appropriately defined as the exponent α in the Fano-factor time relation $F(T) \propto T^\alpha$ in the domain where it follows this power-law behavior. If the Fano factor is measured at two *sufficiently large* counting times, T_1 and T_2, then α may be estimated from the relation $F(T) \propto T^\alpha$ as:

$$\alpha \approx \frac{\log[F(T_2)/F(T_1)]}{\log (T_2/T_1)}. \tag{3}$$

Using this formula to calculate α for primary auditory neurons leads to values lying between 0.3 and 1.0. (For the data illustrated in Fig. 3, $\alpha \approx 0.68$.) It is important to note that a spike train of sufficiently long duration, typically several hundred seconds, is useful for obtaining a reliable estimate for α; the estimated value of α usually increases, and the variance of the estimate apparently decreases, with increasing L (Woo, 1991, Table 3.6). Furthermore, the value of α generally depends on which portion of a data set is examined.

V.   Alteration of the Firing Pattern Engendered by Stimulation

The FFC (variance-to-mean ratio versus counting time T) for this same auditory neuron (unit A) is shown in Fig. 4 for driven activity collected for $L = 800$ s when the stimulus is a continuous tone at the CF (solid curve). The spontaneous data shown in Fig. 3 are repeated for comparison (dashed curve). Although the

---

FIGURE 3.   Fano-factor time curve (FFC) for the spontaneous firing (no stimulus) of a primary auditory neuron (unit A) in an experiment of duration $L = 400$ s (dashed curve). The spontaneous rate fluctuations for this neuron are displayed in Fig. 1a. The Fano-factor time curve assumes a value of unity at short counting times, dips below unity for counting times $\gtrsim 1$ ms, and finally increases in pure power-law fashion when the counting time exceeds about 400 ms. In contrast, the FFC for a simulated dead-time-modified Poisson process (solid curve) remains approximately constant at a value below unity for all values of T larger than the dead time. The Fano factor for a Poisson point process in the absence of dead time is always precisely unity, whatever the value of T. From Teich *et al.* (1990b).

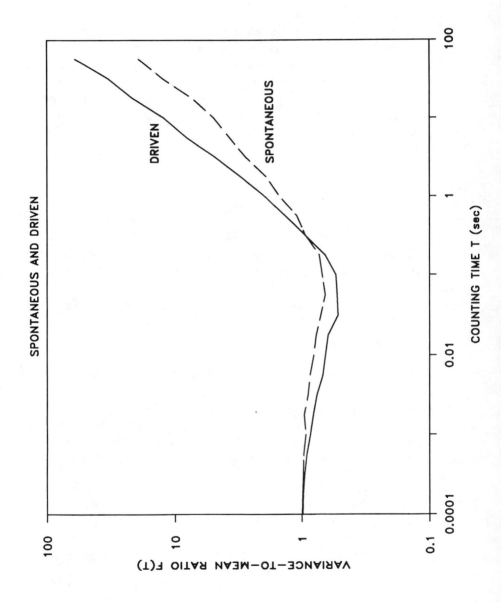

SPONTANEOUS AND DRIVEN

DRIVEN

SPONTANEOUS

VARIANCE—TO—MEAN RATIO F(T)

COUNTING TIME T (sec)

shapes of the FFCs are similar, the minimum Fano factor F_{min} is lower under stimulation (because dead time has a greater effect on F when the rate is higher), and the power-law exponent is greater. (For this particular neuron, it increases from 0.68 to 0.85 when the tone is applied.) This represents an increase in the fractal dimension, indicating that an acoustic stimulus serves not only to alter the rate of action-potential firing but the pattern of firing as well. The presence of the stimulus results in larger rate fluctuations.

For driven firing, F typically assumes a minimum value F_{min} between about 0.5 and 0.9 at counting times in the tens of ms. The onset of pure power-law behavior (with exponents in the range between about 0.7 and 1.0) occurs at counting times between about 0.1 and 0.5 s. Figure 5 illustrates the change in the power-law exponent for several primary auditory neurons when continuous-tone stimulation is applied. For primary auditory neurons, this exponent, representing the fractal dimension, generally increases in the presence of such stimulation.

Because the power-law exponent associated with driven activity is generally greater than that associated with spontaneous activity, as shown in Fig. 5, estimates of the mean firing rate for driven activity will converge more slowly than estimates for spontaneous activity for integration times longer than several hundred milliseconds.

VI.   Comparison of Auditory and Vestibular Firing Patterns

As a counterpoint to the behavior of primary auditory neurons, which have Fano factors substantially larger than zero even when the counting time is relatively short, spontaneously firing low-skew vestibular neurons have narrow PNDs and very small Fano factors. It is well-known that such neurons fire in a far more regular pattern than do auditory neurons (Walsh *et al.*, 1972). Indeed, the firing pattern of a low-skew vestibular neuron is similar to that of a mammalian retinal ganglion cell at high luminance levels (Barlow and Levick, 1969). In Figs. 6a and 6b, we present short-counting-time (T = 51.2 ms, 2,000 samples) and

FIGURE 4.   Fano-factor time curve for continuous-tone-driven firing of unit A in an experiment of duration L = 800 s (solid curve). The Fano-factor time curve for the spontaneous firing of this neuron, shown in Fig. 3, is repeated here for purposes of comparison (dashed curve). The driven rate fluctuations for this cell are displayed in Fig. 2a. The shapes of the Fano-factor time curves are similar for the spontaneous and driven firing; however, the power-law exponent increases from 0.68 for spontaneous firing to 0.85 for driven firing. From Teich *et al.* (1990b).

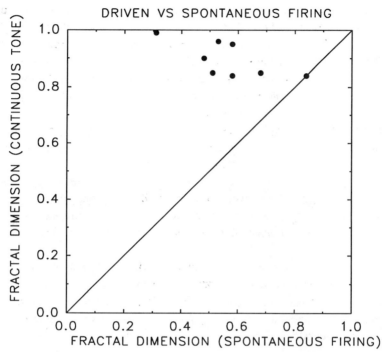

FIGURE 5.   Relationship of the power-law exponents for several auditory neurons, under conditions of continuous-tone-driven and spontaneous firing. The exponent, which represents the fractal dimension of the process, generally increases under continuous-tone stimulation.

slightly longer counting-time (T = 204.8 ms, 500 samples) spontaneous vestibular PNDs (denoted VES), respectively, for one such low-skew cell. Both of these PNDs were constructed from the same neural spike train. They exhibit count means of 1.95 and 7.81, and Fano factors that are very low, $F(T = 51.2$ ms) = 0.04 and $F(T = 204.8$ ms) = 0.03, respectively. The small values of $F(T)$ indicate that these vestibular firings tick along with the near regularity of a clock, at least for counting times $\leqslant$204.8 ms. It will be of interest to measure vestibular PNDs using longer counting times.

The vestibular PNDs are compared with PNDs from auditory neurons (AUD), and from simulated-Poisson data (POI), all with the same approximate spike rate ($\approx$40 s^{-1}) for both counting times. Like the vestibular PNDs, the auditory PNDs were constructed from the same underlying sequence of neural events. It is clear from Fig. 6 that the vestibular PNDs are the narrowest of the three. For these particular counting times, the auditory PNDs are narrower than the simulated-Poisson PNDs, but for longer counting times this reverses. The scalloping

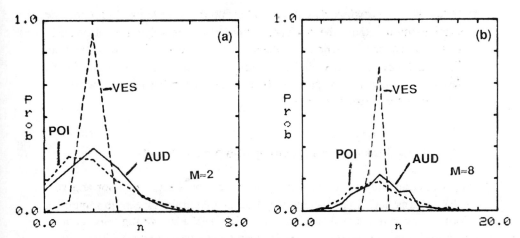

FIGURE 6. PNDs for cat vestibular (VES) and auditory (AUD) primary neurons, and simulated-Poisson (POI) data, with the same approximate spike rate ($\approx$ 40 s^{-1}). T = 51.2-ms PNDs are shown in a; T = 204.8-ms PNDs are shown in b. The vestibular PNDs have count means of 1.95 and 7.81, and Fano factors of 0.04 and 0.03 in a and b, respectively. The auditory data are drawn from Teich and Khanna (1985, Fig. 8) with stimulation at a frequency of 1,445 Hz and a level of 6 dB SPL. The auditory PNDs have count means of 2.01 and 8.03, and Fano factors of 0.70 and 0.81 in a and b, respectively. This particular data set was chosen because its spike rate is quite close to that of the vestibular neuron. The simulated-Poisson PNDs have count means of 1.92 and 7.85, and Fano factors of 1.02 and 1.08 in a and b, respectively. The number of samples is 2,000 and 500 for the short and longer counting-time vestibular PNDs, respectively, whereas it is 1,000 and 250 for the short and longer counting-time auditory and simulated-Poisson PNDs, respectively. From Teich (1989), © IEEE.

evident in the auditory PNDs (Teich and Turcott, 1988; Teich, 1989) appears to diminish as the number of samples increases.

VII. Fractal Firing Patterns at Higher Auditory Centers

Recent experiments carried out by Shofner and Dye (1989) in the gerbil (using a counting time of T = 400 ms) reveal similar behavior in the spike train at the cochlear nucleus, which derives its information from primary auditory nerve-fiber inputs.

We have carried out a preliminary analysis of single-neuron firing at the LSO, and fractal firing patterns may also be present at that locus (Turcott *et al.*, 1991). Because LSO neurons exhibit markedly nonstationary firing rates (in many cells the rate falls in approximately exponential fashion), we have had to remove the nonstationarity to expose the stochastic nature of the underlying point process.

Thus, the sequence of action potentials at three distinct loci in the mammalian auditory pathway (AN, CN, and LSO) all appear to share a similarity in the underlying stochastic point process: fractal neuronal firing patterns when the observation time is sufficiently large.

VIII.   Neural Information Processing with Fractal Events

Why would auditory neuronal firing patterns exhibit fractal behavior and vestibular patterns not? Since the auditory neural-spike train appears to sample an information-carrying signal (Khanna and Teich, 1989a,b), we suggest that these unusual patterns may serve to sample fractal signals and natural fractal noises in an efficient correlated manner (Teich, 1989), Indeed, the instantaneous audio power of music and speech, and the instantaneous frequency (rate of zero crossings) of music, exhibit fractal ($1/f$-type) properties over a substantial range of low frequencies (Voss and Clarke, 1978). The potential benefits to be gained from such sampling, such as bandwidth compression, need to be established from an information-theoretic point of view.

An analogous argument for the visual system would suggest that fractal firing patterns may be present at loci where fractal image information is sampled, e.g., at the striate cortex. Indeed, spike bursts and recurrences of bursts do appear to occur at that locus (Legéndy and Salcman, 1985).

The vestibular system is designed to estimate angular acceleration with high accuracy. The information is slowly varying so that fractal behavior, if present, would be evident only for extremely long counting times. It will be of use to discover where fractal neuronal firing patterns do, and do not, occur as a prelude to understanding why they occur.

If fractal firings are useful for the sampling and decoding of fractal information-bearing signals, they are, by virtue of their noisiness, a liability for the detection of weak acoustic signals. Psychophysical tasks involving the detection of weak signals, such as intensity discrimination and loudness estimation, can be understood in terms of the relationship of the count variance to the count mean of an underlying point process representing neural activity (McGill and Teich, 1991a,b). Since the Fano-factor time curve for primary auditory fibers typically achieves its minimum value for counting times in the range of 50 to 500 ms, it would appear that the neural count at the periphery of the auditory system is least noisy over this range of integration times. Information processing tasks that rely on low variance for good performance would, it seems, be best served by using integration times in this range, and these psychophysical tasks do, indeed, exhibit behavior that appears to accord with this. Perhaps the system has its cake—and eats it, too—by making use of short integration times (to maximize signal-to-noise ratio) for tasks involving the detection, discrimination, and estimation of weak signals, and long integration times (to maximize memory) for tasks involving the extraction of information from strong fractal signals.

IX.   Biophysical Origins of the Fractal Behavior

The mathematical point process used to describe the sequence of action potentials in the peripheral auditory system should be consistent with the underlying physiological behavior of the system. There are a number of possible origins of the observed fractal behavior, three of which appear to merit further consideration (Teich *et al.*, 1991):

1. Slow decay of intracellular calcium in the receptor hair cell.
2. Fractal ion-channel statistics (Liebovitch and Tóth, 1990; Teich, 1989).
3. Self-organized criticality in ion-channel behavior.

The first two of these models may be cast in the form of a *sick-time-modified* FDSPP.

X.   Identifying the Mathematical Point Process

Three mathematical models are provided for describing the point process underlying auditory neural firings. The first model is applicable for an arbitrary stationary point process with constant rate; it therefore admits both correlation and dead (or sick) time. However, its range of prediction is limited to second-order statistics, e.g., quantities such as the FFC; it cannot be used to calculate measures such as the PND or PID. The second model is a *fractal doubly stochastic Poisson point process* (FDSPP) that we developed on the basis of plausible physiological arguments (Teich *et al.*, 1990a). The use of a specific model such as this has the advantage that its range of prediction is unlimited; any measure obtained from the neural events can, in principle, be calculated for the process. Although the inclusion of dead (or sick) time destroys the DSPP character of this process, thereby making it difficult to obtain analytical results, we have managed to incorporate the effects of refractoriness in a simulated version of the FDSPP. This process appears to behave very much like the neural events.

The third model is a generalization of the first that is suitable for a time-varying rate or stimulus.

A.   *General Stationary Point Process with Constant Rate*

For an arbitrary stationary point process with constant rate, there is a unique relation between the Fano-factor time curve $F(T)$ and the joint probability of event-pair coincidences $\lambda^2 g(\tau)$ (Cox and Lewis, 1966, pp. 72–75; Teich, 1989; Teich and Saleh, 1988, Eq. (2.16)):

$$F(T) = 1 + 2\lambda \int_0^T \left(1 - \frac{\tau}{T}\right) [g(\tau) - 1] \, d\tau. \qquad (4)$$

Here, λ is the mean rate of the point process, and τ is the delay time between the events. The normalized coincidence rate $g(\tau)$ plays the role of the correlation function for continuous processes.

A simple coincidence rate may be constructed by including idealized models of absolute refractoriness and correlation:

(1) For delay times less than the average refractory period τ_d, the coincidence rate is taken to be zero, since action potentials cannot follow one another within this time.

(2) At the termination of the refractoriness period, the normalized coincidence rate rises abruptly to unity, since the event occurrences are then presumed to be uncorrelated.

(3) Finally, for delay times longer than the fractal onset time τ_f, the normalized coincidence rate increases above unity and falls in power-law fashion toward unity, representing a slowly decreasing correlation of the spike occurrences with increasing delay time.

The idealized normalized coincidence rate is then:

$$g(\tau) = \begin{cases} 0, & |\tau| < \tau_d, \\[2mm] 1, & \tau_d \leq |\tau| \leq \tau_f, \\[2mm] 1 + \dfrac{\delta}{\lambda} \left(\dfrac{|\tau|}{\tau_f}\right)^{\alpha - 1}, & |\tau| > \tau_f, \end{cases} \qquad (5)$$

where δ is a constant (units of s^{-1}), and $\alpha - 1$ $(0 < \alpha < 1)$ is the power-law exponent of the normalized delay time. The correlation between a pair of spikes is typically rather small; values of δ/λ typically range from 0.02 to 0.04, and seldom stretch above 0.1. Equation (5) is a generalized form of Eq. (4) in Teich (1989); the power-law exponent is $\alpha - 1$ rather than $-\frac{1}{2}$, and δ/λ is used in place of δ. A more realistic coincidence rate would rise gradually, rather than abruptly, to account for relative refractoriness (Teich, 1989).

Substituting this coincidence rate into Eq. (4) leads to a Fano-factor time function given by:

$$F(T) = \begin{cases} 1 - \lambda T, & T < \tau_d \\[2mm] 1 - \lambda\tau_d[2 - \tau_d/T], & \tau_d \leqslant T \leqslant \tau_f \\[2mm] 1 - \lambda\tau_d[2 - \tau_d/T] + \dfrac{2}{\alpha(\alpha + 1)}\,\delta\tau_f[(T/\tau_f)^\alpha + \alpha(\tau_f/T) \\ \qquad\qquad\qquad\qquad\qquad - (\alpha + 1)], & T > \tau_f. \end{cases} \tag{6}$$

Equation (6) is a generalization of Eq. (5) in Teich (1989); the power-law exponent is α rather than $\frac{1}{2}$, and the power-law term has a coefficient proportional to $\delta\tau_f$ rather than $\delta\lambda\tau_f$. (This accommodates the empirical independence of δ, as defined in Eq. (6), on λ, and results from the use of the coefficient δ/λ in Eq. (5).) Each of the five panels in Fig. 7 illustrates the dependence of $F(T)$, as given in Eq. (6), on one of the essential parameters: λ, τ_d, δ, τ_f, and α. For purposes of illustration, we use parameters with physiologically reasonable values, *viz.*, $\lambda = 100$ s^{-1}, $\tau_d = 1.5$ ms, $\delta = 2$ s^{-1}, $\tau_f = 0.1$ s, and $\alpha = 0.5$. As expected, the theoretical curves always assume a value of unity as $T \to 0$, and dip below unity as T increases and dead time comes into play. The power-law growth of the Fano factor for large T, $F(T) \propto (T/\tau_f)^\alpha$, follows from the delay-time dependence of the coincidence rate for large τ, $g(\tau) - 1 \propto (|\tau|/\tau_f)^{\alpha - 1}$. Although the correlation between a single pair of spikes is typically rather small, the Fano factor can become quite large since it integrates the many correlations from different pairs of spikes within the time window T. (See Eq. (4).)

We have used Eq. (6) (with the value of λ set at the experimental spike rate), in conjunction with the curve-fitting routine in Jandel Scientific's Sigma-Plot™ software package (version 4.0), to fit the FFCs of eight primary auditory nerve-fiber spike trains, both in the presence and in the absence of a pure-tone stimulus. The software routine makes use of the Marquardt–Levenberg algorithm, which finds parameters that minimize the square difference of the theory and data. We chose to minimize the square difference of the logarithms of the theory and data. The results shown in Fig. 8 for unit A are typical; the theoretical variance-to-mean ratio $F(T)$ given in Eq. (6) nicely describes both the spontaneous and driven experimental data (which is the same data as shown in Fig. 4) with reasonable physiological parameters. Evidently, the idealized coincidence rate postulated in Eq. (5) captures the essential elements inherent in auditory neuronal firing patterns (refractoriness and decaying power-law correlation with Poisson underlying events), at least to second order. The theory is, of course, least satisfactory in the region where relative refractoriness plays its principal role, *viz.*, from about 1 to 40 ms. The best-fitting parameters for all eight units, using Eq. (6), are compared under conditions of continuous-tone stimulation at the CF, and spontaneous firing, in Fig. 9. Aside from causing λ to increase, the presence of the tone generally causes α to increase; however, it appears to have little effect on τ_d, τ_f, and δ. The durations of these experiments range from 50 to nearly 2,000 s.

The power spectral density $S(f)$ of a random process, when it exists, is

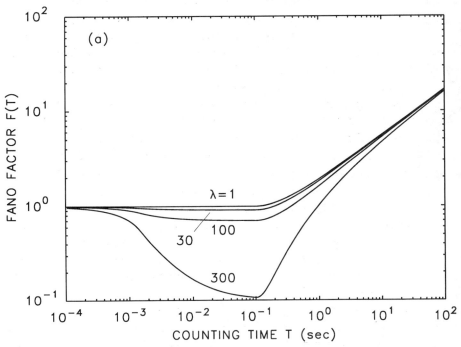

FIGURE 7.   Variance-to-mean ratio $F(T)$ versus counting time T plotted in accordance with Eq. (6), using the default parameters $\lambda = 100$ s^{-1}, $\tau_d = 1.5$ ms, $\delta = 2$ s^{-1}, $\tau_f = 0.1$ s, and $\alpha = 0.5$. (a) Behavior of $F(T)$ for different values of λ. (b) Behavior of $F(T)$ for different values of τ_d. (c) Behavior of $F(T)$ for different values of δ. (d) Behavior of $F(T)$ for different values of τ_f. (e) Behavior of $F(T)$ for different values of α.

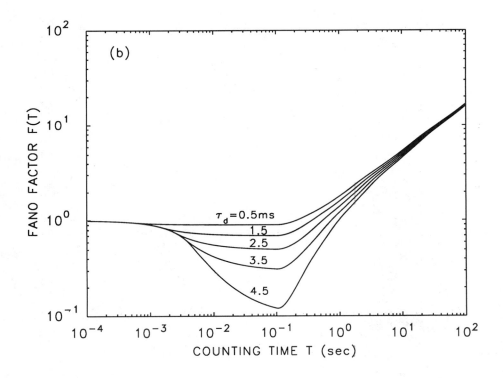

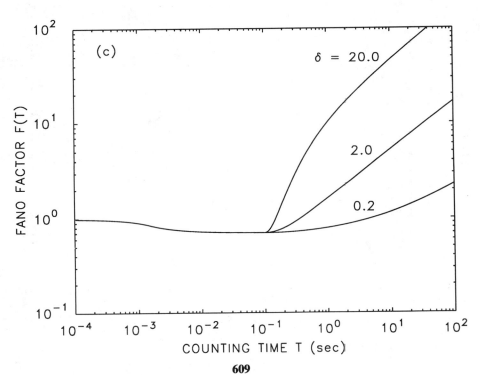

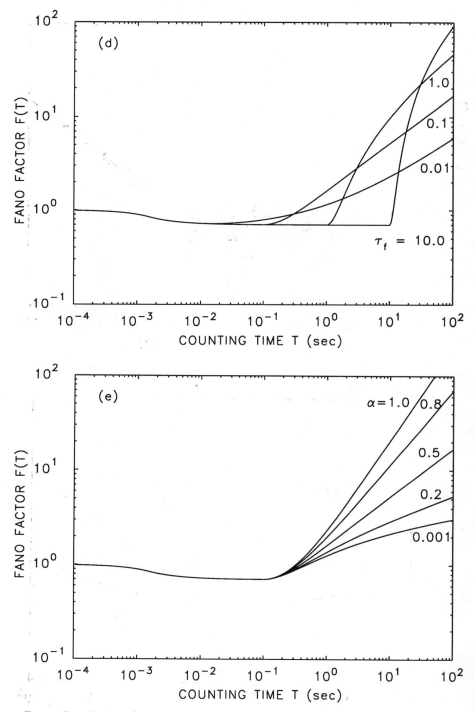

FIGURE 7. *Continued*

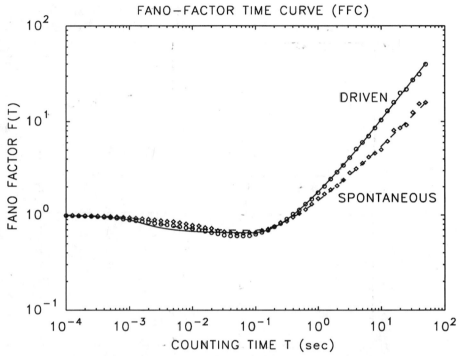

FIGURE 8.   Fit of the theoretical Fano factor given in Eq. (6) (dashed and solid curves) to spontaneous and driven experimental data points for unit A. For the spontaneous fits, the parameters take on the physiologically plausible values $\lambda = 65$ s^{-1}, $\tau_d = 2.4$ ms, $\delta = 1.34$ s^{-1}, $\tau_f = 87$ ms, and $\alpha = 0.68$. For the driven fits, the parameter values are $\lambda = 113$ s^{-1}, $\tau_d = 1.6$ ms, $\delta = 1.63$ s^{-1}, $\tau_f = 88$ ms, and $\alpha = 0.85$.

determined from the coincidence rate by means of the Wiener–Khinchin theorem. When $g(\tau)$ takes the form indicated in Eq. (5), $S(f)$ behaves as

$$S(f) \propto (1/\lambda)f^{-\alpha}, \quad f \lesssim f_f, \tag{7}$$

in the low-frequency (large delay-time) regime. The quantity f_f is the fractal cutoff frequency; it is inversely proportional to τ_f. Since $0 < \alpha < 1$, this represents $1/f$-type noise (Woo et al., 1992).

With this model in hand, we may now quantitatively consider the self-similarity of the rate fluctuations illustrated in Figs. 1a and 2a. Our considerations are restricted to counting times sufficiently long so that we are in the power-law (fractal) regime. The degree to which a process with firing rate λ_T may be considered to be self-similar is established by determining the dependence of its standard deviation σ_λ on the time window T. Since $\lambda_T = n/T$, where n is the random number of neural spikes in the time T, the variance of the firing rate is

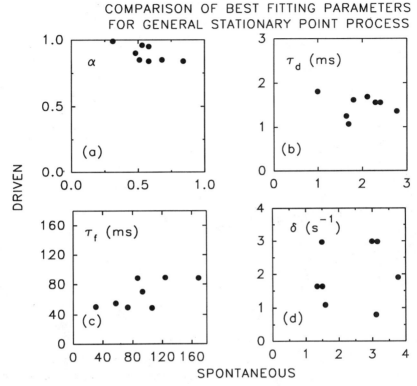

FIGURE 9.   Comparison of best-fitting parameters for eight units, under driven and spontaneously active firing conditions, using Eq. (6) for the variance-to-mean ratio.

$Var(\lambda_T) = (1/T^2)Var(n)$. By definition, however, $Var(n) = \langle n \rangle F(T)$, where $\langle n \rangle$ is the mean number of events in the time T, and F is the Fano factor, so that $Var(\lambda_T) = (1/T^2) \langle n \rangle F(T)$. Since $\langle n \rangle \propto T$ and $F(T) \propto T^\alpha$ for sufficiently large T, we obtain:

$$Var(\lambda_T) \propto \frac{1}{T^{1-\alpha}}. \qquad (8)$$

The standard deviation of the rate therefore is given by

$$\sigma_\lambda \propto \frac{1}{T^{(1-\alpha)/2}}. \qquad (9)$$

Non-fractal processes have a fractal dimension $\alpha = 0$, so that the standard deviation of the rate is proportional to $1/T^{1/2}$; thus, the rate converges relatively quickly with increased averaging. Processes with $\alpha \to 1$ have a standard deviation that is independent of T so that there is no convergence with time averaging, and the rate process is fully self-similar. Estimates of the mean rate, therefore, converge more slowly with increasing T for fractal processes than for non-fractal processes, with the rate of convergence depending on the fractal dimension α. In the region where refractoriness is operative (below several hundred milliseconds), the process is non-fractal and the standard deviation of the rate behaves as $1/T^{1/2}$. The firing rate, therefore, is most accurately estimated by using counting times in this range.

B. Fractal Doubly Stochastic Poisson Point Process

The approach presented in the preceding, though valid for an arbitrary stationary point process, is based on a phenomenological construct for the coincidence rate and is limited to providing second-order, and some first-order, statistics. We have developed a particular type of fractal point process, the *dead-time-modified fractal doubly stochastic Poisson point process* (FDSPP), that exhibits behavior consistent with all of the experimental statistics of spontaneous and pure-tone-driven VIIIth-nerve action potentials examined to date, including the pulse-number and interspike-interval distributions.

Particular attention is devoted to two specific examples of this process: the *fractal-shot-noise-driven* (FSND) DSPP and one of its special cases, the *fractal-Gaussian-noise-driven* (FGND) DSPP (Lowen and Teich, 1991). Fractal behavior in the FSND DSPP is assured by choosing an impulse-response function that decays in a power-law manner with a certain range of exponents. This particular process is physiologically plausible for certain nerve-spike generation models, as mentioned in Section IX. Analytical results have been derived for many features of this process, including the coincidence rate, Fano factor, and spectrum. For the range of parameters of interest to the problem at hand, and for sufficiently long delay and counting times, these latter quantities turn out to be [Lowen and Teich, 1991, Eqs. (37), (48), and (52)]:

$$g(\tau) = 1 + (k_1/\lambda)|\tau|^{\alpha - 1}, \tag{10}$$

$$F(T) = 1 + k_2 T^{\alpha}, \tag{11}$$

and

$$S(f) = (k_3/\lambda)f^{-\alpha}, \tag{12}$$

where k_1, k_2, and k_3 are constants. These equations assume the same form as those set forth in Eqs. (5)–(7) in Section X.A. This is expected, since we are limiting our attention to long delay and counting times where the effects of dead time are negligible, and both approaches incorporate a power-law decreasing coincidence rate. Since the phenomenological Fano factor developed in Section X.A is in accord with the FFC data, obviously Eq. (11) will be too.

In the case of the FDSPP, however, we have knowledge of the full point process. This allows us to calculate other measures (e.g., the PND and PID, as well as the serial count correlation coefficient (SCC) and rescaled range analysis (R/S), both of which will be explained shortly), and to compare them with the neural data. We can, therefore, perform more stringent tests for identifying the mathematical point process that characterizes auditory neuronal firings.

Indeed, the collection of data in the form of the PID, PND, FFC, SCC, and R/S provides a rather comprehensive picture of a neural spike train and enables us to make reasonable conjectures about the underlying mathematical point process (Teich *et al.*, 1990a). The PID, PND, and FFC are, by now, well-known statistics that have been described in detail elsewhere (Teich, 1989; Teich and Khanna, 1985; Teich and Turcott, 1988).

The SCC and R/S provide estimates of the degree of serial correlation in the data set. The SCC gives the correlation between the numbers of neural spikes in adjacent counting periods, and is generally a function of the counting time

---

FIGURE 10.　(a) Pulse-number distribution (PND) constructed from the spontaneous spike train of auditory unit A (as in Figs. 1a, 3, and 8), using a counting time $T = 1$ sec (solid curve). PNDs obtained from simulations of three theoretical models are also shown. The model parameters were chosen to give the same mean count as the data. The PND obtained from the FDSPP resembles the data. On the other hand, the PND obtained from the dead-time-modified Poisson point process (DTMP, denoted PP here) is narrower than the data, while that obtained from the renewal fractal process (RFP) is far broader than the data. (b) The Fano-factor time curve (FFC) is constructed from the PND. For auditory-nerve data, $F(T)$ typically grows in power-law fashion as T^α ($0 < \alpha < 1$) for sufficiently large counting times T, implying a power-law-decreasing normalized coincidence rate and a power-law form for the power spectral density at low frequencies. Again the FFC obtained from the FDSPP resembles the data quite closely, whereas the FFCs of the PP and the RFP deviate substantially from it, even though the latter does exhibit power-law behavior. (c) For the PP, $\alpha = 0$, so that $C(T) = 0$ for the serial count correlation coefficient (SCC), as is evident in the figure for sufficiently large counting times. (The dip in the curve in the vicinity of 2 msec, which arises from dead time, would be moderated were sick time used instead.) Once again the SSC obtained from the FDSPP closely resembles the data, while the SSCs associated with the PP and the RFP deviate substantially from it. (d) In rescaled range (R/S) analysis, the renewal nature of the PP and RFP cause $R(k)$ to behave as $k^{\frac{1}{2}}$, where k is the number of interspike intervals; the data and the results from the FDSPP rise more steeply, indicating positive correlation for collections of large numbers of interspike intervals. From Teich *et al.* (1990a).

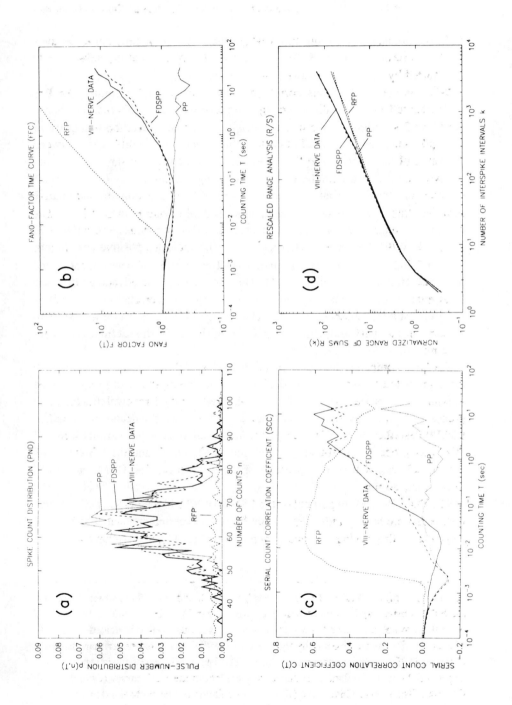

T. In the domain of counting times where the Fano-factor time curve behaves as T^{α}, the serial count correlation coefficient $C(T)$ plateaus at the value $2^{\alpha} - 1$, as required by the relation between these measures.

Whereas the SCC reflects correlations between successive counts, the R/S parameter reflects correlations among interspike intervals. This measure is obtained by first estimating the interval mean and standard deviation in a block of interspike intervals of size k. For each of the k intervals, the difference between the value of the interval and the mean value is obtained and the result is successively added to a cumulative sum. The range is defined as the difference between the maximum and minimum values achieved within the cumulative sum, and this is normalized by the sample standard deviation to give $R(k)$. The normalized range of sums $R(k)$ is estimated for increasing values of k and plotted against k. With $R(k)$ proportional to k^h, $h > 0.5$ indicates positive correlation, $h < 0.5$ indicates negative correlation, and $h = 0.5$ indicates uncorrelated intervals (Hurst, 1951; Feller, 1951). This measure has the advantage of being valid even when the data exhibit extremely long-term correlations, as well as large (or infinite) variance; these are characteristics that can cause a process to appear nonstationary and, consequently, seriously impair the usefulness of standard measures (Mandelbrot, 1983).

We have compared the experimental behavior of these various statistics with those predicted by several theoretical models, including the dead-time-modified Poisson point process (DTMP), the dead-time-modified renewal fractal point process (RFP), as described by Teich *et al.* (1990a), and the dead-time-modified FDSPP described here. We have performed simulations using various forms of this latter process, and found that both the FSND DSPP and the FGND DSPP exhibit behavior that accords with all of these statistics. The DTMP and RFP, in contrast, do not.

The FDSPP—and, in particular, the FSND DSPP and the FGND DSPP—give results that are largely indistinguishable from the experimental statistics both for spontaneous firings (Fig. 10) and driven firings (Fig. 11), and we have identified

---

FIGURE 11.   (a) PND ($T = 1$ sec), (b) FFC, (c) SCC, and (d) R/S constructed from the same 800-s driven spike train of unit A as that used in Figs. 2a, 4, and 8. The renewal processes are not represented, since their behavior is similar to that indicated in Fig. 10. The fractal dimension revealed by the slope of the FFC curve is greater than the value observed in Fig. 10b for spontaneous firing. The larger FFC exponent is reflected in a larger serial count correlation estimate, as is evident in the SCC curve. (Compare with Fig. 10c.) The increase in α and $C(T)$ under stimulation has been observed in all primary auditory-nerve cells examined to date. From Teich *et al.* (1990a).

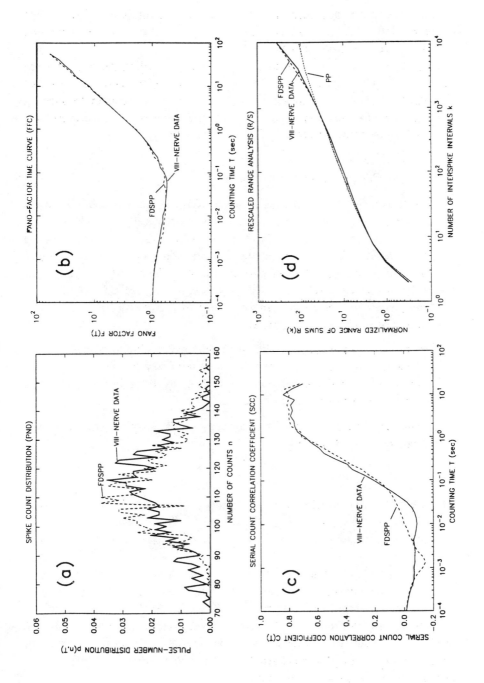

the FDSPP as the point process that characterizes the auditory neural spike train (Teich *et al.*, 1990a). The essence of its behavior arises from Poisson underlying events with decaying power-law correlations.

These correlations can be removed by randomly shuffling the interspike intervals, for both spontaneous and driven data, as shown by Teich *et al.* (1990a). The shuffling serves to destroy the long-term correlations inherent in the ordering of the individual events. Indeed, shuffling alters the experimental measures, and their FDSPP theoretical counterparts, in precisely the same way.

C. *General Point Process with Variable Rate*

The formulas provided in Sections X.A and X.B are not applicable when the mean rate is varying (either deterministically or stochastically). This is the case, for example, when the stimulus is Gaussian noise or an information-bearing signal that varies with time, rather than a pure tone, or when adaptation is present. The stimulus itself then introduces another degree of variability, and the results described previously must be generalized to account for this. The resulting process will be a *triply stochastic Poisson point process* (TSPP), with three forms of stochasticity arising from:

1. Rate variations associated with adaptation and/or stimulus variability.
2. A biophysical mechanism involving long-term correlations.
3. An action-potential generation mechanism involving intrinsic auditory-neuron fluctuations and refractoriness.

Consider an arbitrary counting distribution $p(n,T|W)$ conditioned on an integrated rate (energy) W given by:

$$W = \int_{t_0}^{t_0+T} \lambda_t \, dt, \qquad (13)$$

where λ_t is the time-varying rate, t_0 is the beginning of the counting interval, and T is the counting time (Cox and Lewis, 1966; Prucnal and Teich, 1979). If the rate λ_t is constant (homogeneous) and fixed at λ_0, the integrated rate W is simply $\lambda_0 T$.

On the other hand, if the rate λ_t exhibits a stochastic or deterministic time dependence, with characteristic time τ_c, the integrated rate W may be described in terms of a probability density function $P(W)$. When $T \gg \tau_c$, depending on the nature of the fluctuations of λ_t, W may become constant, in which case the results become identical to those for a homogeneous rate; in the other limit, when $T \ll \tau_c$, λ_t may be considered to be slowly varying, so that it can be removed from the integral, whereupon

$$W = \lambda_t T, \tag{14}$$

and the statistics of W mimic those of λ_t.

Removing the conditioning from $p(n,T|W)$ provides the unconditional counting distribution $p(n,T)$ and the PND statistics in the presence of the stimulus or rate fluctuations,

$$p(n,T) = \int_W p(n,T|W) \, P(W) \, dW. \tag{15}$$

Equation (15) reveals that the counting distribution of the underlying kernel obtained in the absence of rate variations is smeared (broadened) by these variations. Note that the kernel, although often taken to be Poisson (Cox and Lewis, 1966; Prucnal and Teich, 1979), can assume an arbitrary form, such as that associated with the DTMP distribution (Cantor and Teich, 1975), a fractal counting distribution, or the Neyman Type-A distribution (Teich, 1981).

Using Eq. (15) to calculate the unconditional mean of $p(n,T)$, we find

$$\langle n \rangle = \sum_{n=0}^{\infty} np(n,T) = \int_W dW \, P(W) \left[\sum_{n=0}^{\infty} np(n,T|W) \right]$$

$$= \int_W dW \, P(W) \, \langle n|W \rangle = \int_W dW \, P(W) \, W = \langle W \rangle, \tag{16}$$

provided that

$$\langle n|W \rangle = W. \tag{17}$$

This indicates that the mean integrated rate $\langle W \rangle$ is sufficient for calculating the unconditional mean count $\langle n \rangle$.

The unconditional variance is also readily calculated for processes with known conditional variance. Again, using Eq. (15), the unconditional mean-square count is:

$$\langle n^2 \rangle = \sum_{n=0}^{\infty} n^2 p(n,T) = \int_W dW \, P(W) \, \langle n^2|W \rangle. \tag{18}$$

The conditional mean-square count is, of course, expressible in terms of the conditional variance and conditional mean as:

$$\langle n^2|W \rangle = \mathrm{Var}(n|W) + \langle n|W \rangle^2. \tag{19}$$

For auditory neurons with a constant mean rate, the conditional variance is related to the conditional mean by the conditional Fano factor $F(T|W)$ given in Eq. (6):

$$\mathrm{Var}(n|W) = F(T|W)\langle n|W \rangle. \tag{20}$$

Thus, since $\langle n|W\rangle = W$,

$$\langle n^2|W\rangle = F(T|W)\langle n|W\rangle + \langle n|W\rangle^2 = F(T|W)W + W^2. \tag{21}$$

Using Eq. (21) in Eq. (18) provides:

$$\langle n^2\rangle = \int_W dW\, P(W)\,[F(T|W)W + W^2]. \tag{22}$$

We can now make use of an abbreviated form of Eq. (6), valid when the dead time can be ignored,

$$F(T|W) \approx 1 + c_1, \tag{23}$$

where c_1 is a function of α, δ, τ_f, and T. To facilitate the analysis, c_1 is assumed to be independent of λ and therefore W, as is often the case. Inserting Eq. (23) into Eq. (22), we obtain:

$$\langle n^2\rangle = \langle W\rangle + c_1\langle W\rangle + \langle W^2\rangle, \tag{24}$$

so that

$$\mathrm{Var}(n) = \langle n^2\rangle - \langle n\rangle^2 = \langle W\rangle + c_1\langle W\rangle + (\langle W^2\rangle - \langle W\rangle^2). \tag{25}$$

Finally, then, the Fano factor in the presence of rate variations turns out to be:

$$F(T) = \frac{\mathrm{Var}(n)}{\langle n\rangle} = 1 + c_1 + F_W, \tag{26}$$

where

$$F_W = \frac{\mathrm{Var}(W)}{\langle W\rangle} = \frac{\langle W^2\rangle - \langle W\rangle^2}{\langle W\rangle}. \tag{27}$$

Equation (26) represents the Fano factor associated with a triply stochastic distribution: when the kernel is Poisson ($c_1 = 0$), we obtain the well-known doubly stochastic Poisson result, $F(T) = 1 + F_W$ (Teich and Saleh, 1988). When, in addition, the rate is nonvarying, $F_W = 0$ and we obtain the homogeneous Poisson result, $F(T) = 1$ for all T.

We may examine the dependence of the unconditional Fano factor in Eq. (26) on the counting time by explicitly writing c_1 in terms of its dependence on T:

$$F(T) \approx 1 + \frac{2}{\alpha(\alpha + 1)}\,\delta\tau_f\left(\frac{T}{\tau_f}\right)^\alpha + F_W. \tag{28}$$

For linearly and exponentially varying rates (Prucnal and Teich, 1979), and for the intensity fluctuations of Gaussian noise (Teich and Saleh, 1988, Eq. (2.19)),

$F_W = \langle\lambda\rangle T/M$, where M is a degrees-of-freedom parameter that is an increasing function of T, so that Eq. (28) becomes:

$$F(T) \approx 1 + \frac{2}{\alpha(\alpha + 1)} \delta\tau_f \left(\frac{T}{\tau_f}\right)^\alpha + \frac{\langle\lambda\rangle T}{M}. \tag{29}$$

Finally, then, the dependence of the unconditional Fano factor on the mean rate $\langle\lambda\rangle$ and on the counting time T is contained in the succinct expression:

$$F(T) \approx 1 + c_2 T^\alpha + \frac{\langle\lambda\rangle T}{M}, \tag{30}$$

where $c_2 = 2\delta\tau_f^{1-\alpha}/\alpha(\alpha + 1)$.

We conclude that the dependence of the Fano-factor time curve on the counting time T comprises two components in the long-counting time limit: The first increases as T^α, as in the case of pure-tone stimulation; the second increases with T/M (where M itself depends on T) and arises from mean rate fluctuations. This result is based on certain specific characteristics for the rate fluctuations (e.g., exponential or linear variation, or integrated intensity fluctuations associated with Gaussian noise) and is predicated on the counting time being large. A more detailed analysis of this general case can be obtained on the basis of the moment-generating functional for cascades of filtered Poisson point processes (Matsuo et al., 1982).

We now explicitly consider a nonstationary rate arising from adaptation rather than from a stochastic stimulus. In auditory theory, two useful examples emerge in this context (Harrison, 1985; Lütkenhöner and Smith, 1986). The first is provided by an instantaneous rate λ_t that linearly decreases from a maximum value λ_{max} to a minimum value λ_{min}. For a deterministic nonstationarity such as this, the duration of the experiment L plays the role of τ_c because the rate continues to change continuously over L. In auditory experiments, $T \ll L$ so that the linearly advancing starting time t_0 of successive counting intervals serves to uniformly and exhaustively sample the rate over its entire range. In this case, the probability density function $P(\lambda)$ is easily shown to be uniformly distributed on the interval $[\lambda_{min},\lambda_{max}]$, with mean (Prucnal and Teich, 1979; Teich and Diament, 1969)

$$\langle\lambda\rangle = \frac{\lambda_{max} + \lambda_{min}}{2} \tag{31}$$

and variance

$$\text{Var}(\lambda) = \frac{(\lambda_{max} - \lambda_{min})^2}{12}. \tag{32}$$

Alternatively, λ_{min} can be expressed in terms of λ_{max}, L, and the slope of the decline.

The second example is embodied by a rate that decreases from a maximum value λ_{max} to a minimum value λ_{min} in accordance with a simple time-decaying exponential function of time constant τ,

$$\lambda(t) = \lambda_{max}\exp(-t/\tau). \tag{33}$$

The probability density function $P(\lambda)$ is then distributed as $1/\lambda$ over an appropriate range of λ (Prucnal and Teich, 1983; Teich and Card, 1979), and, for an experiment of duration L, the mean and variance turn out to be:

$$\langle\lambda\rangle = \lambda_{max}(\tau/L)[1 - \exp(-L/\tau)] \tag{34}$$

and

$$\text{Var}(\lambda) = \lambda_{max}^2\left(\frac{\tau}{L}\right)\left[\left(\frac{1}{2} - \frac{\tau}{L}\right) + 2\left(\frac{\tau}{L}\right)\exp\left(-\frac{L}{\tau}\right)\right. \tag{35}$$
$$\left. - \left(\frac{1}{2} + \frac{\tau}{L}\right)\exp\left(-\frac{2L}{\tau}\right)\right].$$

Since $T \ll L$, Eq. (15) can be written as:

$$p(n,T) = \int_\lambda p(n,T|\lambda)\, P(\lambda)\, d\lambda, \tag{36}$$

so that all of the equations obtained thereafter can be considered as being conditioned on λ rather than on W. The unconditional Fano factor can then be obtained from Eq. (28) by using $F_{\lambda T}$ in place of F_W. A calculation of this type has been carried out earlier (Teich, 1989).

Acknowledgments

I am grateful to S. B. Lowen, R. G. Turcott, T. W. Woo, and N. Powers for useful discussions. This work was supported by the Joint Services Electronics Program and by the Office of Naval Research.

References

BARLOW, H. B., and LEVICK, W. R. (1969). "Changes in the Maintained Discharge with Adaptation Level in the Cat Retina," *Journal of Physiology (London)* **202**, 699–718.

CANTOR, B. I., and TEICH, M. C. (1975). "Dead-Time-Corrected Photocounting Distributions for Laser Radiation," *Journal of the Optical Society of America* **65**, 786–791.

Cox, D. R. (1962). *Renewal Theory*. Methuen, London.

Cox, D. R., and Lewis, P. A. W. (1966). *The Statistical Analysis of Series of Events*. Methuen, London.

Fano, U. (1947). "Ionization Yield of Radiations. II. The Fluctuations of the Number of Ions," *Physical Review* **72**, 26–29.

Feller, W. (1951). "The Asymptotic Distribution of the Range of Sums of Independent Random Variables," *Annals of Mathematical Statistics* **22**, 427–432.

Gaumond, R. P., Molnar, C. E., and Kim, D. O. (1982). "Stimulus and Recovery Dependence of Cat Cochlear Nerve Fiber Spike Discharge Probability," *Journal of Neurophysiology* **48**, 856–873.

Gray, P. F. (1967). "Conditional Probability Analyses of the Spike Activity of Single Neurons," *Biophysical Journal* **7**, 759–777.

Harrison, R. V. (1985). "Long-Term Adaptation in Afferent Neurones from the Normal and Pathological Cochlea," *Journal of the Acoustical Society of America Supplement 1* **77**, S94.

Hurst, H. E. (1951). "Long-Term Storage Capacity of Reservoirs," *Transactions of the American Society of Civil Engineers* **116**, 770–808.

Khanna, S. M., and Teich, M. C. (1989a). "Spectral Characteristics of the Responses of Primary Auditory-Nerve Fibers to Amplitude-Modulated Signals," *Hearing Research* **39**, 143–158.

Khanna, S. M., and Teich, M. C. (1989b). "Spectral Characteristics of the Responses of Primary Auditory-Nerve Fibers to Frequency-Modulated Signals," *Hearing Research* **39**, 159–176.

Kuffler, S. W., FitzHugh, R., and Barlow, H. B. (1957). "Maintained Activity in the Cat's Retina in Light and Darkness," *Journal of General Physiology* **40**, 683–702.

Legéndy, C. R., and Salcman, M. (1985). "Bursts and Recurrences of Bursts in the Spike Trains of Spontaneously Active Striate Cortex Neurons," *Journal of Neurophysiology* **53**, 926–939.

Liebovitch, L. S., and Tóth, T. I. (1990). "Using Fractals to Understand the Opening and Closing of Ion Channels," *Annals of Biomedical Engineering* **18**, 177–194.

Lowen, S. B., and Teich, M. C. (1990). "Power-Law Shot Noise," *IEEE Transactions on Information Theory* **36**, 1302–1318.

Lowen, S. B., and Teich, M. C. (1991). "Doubly Stochastic Poisson Point Process Driven by Fractal Shot Noise," *Physical Review A* **43**, 4192–4215.

Lütkenhöner, B., and Smith, R. L. (1986). "Rapid Adaptation of Auditory-Nerve Fibers: Fine Structure at High Stimulus Intensities," *Hearing Research* **24**, 289–294.

Mandelbrot, B. B. (1983). *The Fractal Geometry of Nature*. Freeman, New York.

Matsuo, K., Saleh, B. E. A., and Teich, M. C. (1982). "Cascaded Poisson Processes," *Journal of Mathematical Physics* **23**, 2353–2364.

McGill, W. J., and Teich, M. C. (1991a). "Auditory Signal Detection and Amplification in a Neural Transmission Network," in M. L. Commons, J. A. Nevin, and M. C. Davison (eds.), *Signal Detection* (pp. 1–37). Lawrence Erlbaum, Hillsdale, New Jersey.

McGILL, W. J., and TEICH, M. C. (1991b). *Simple Models of Sensory Transmission* (Report no. CHIP 132). Center for Human Information Processing, University of California (San Diego), La Jolla, California.

MUELLER, C. G. (1954). "A Quantitative Theory of Visual Excitation for the Single Photoreceptor," *Proceedings of the National Academy of Sciences* **40**, 853–863.

MÜLLER, J. W. (1974). "Some Formulae for a Dead-Time-Distorted Poisson Process," *Nuclear Instrumentation and Methods* **117**, 401–404.

POWERS, N. (1991). *Discharge Rate Fluctuations in the Auditory Nerve of the Chinchilla.* Unpublished Doctoral Dissertation. State University of New York at Buffalo.

POWERS, N. L., SALVI, R. J., and SAUNDERS, S. S. (1991). "Discharge Rate Fluctuations in the Auditory Nerve of the Chinchilla (Abstract no. 411)," in D. J. Lim (ed.), *Abstracts of the Fourteenth Midwinter Research Meeting of the Association for Research in Otolaryngology,* p. 129. Association for Research in Otolaryngology, Des Moines, Iowa.

PRUCNAL, P. R., and TEICH, M. C. (1979). "Statistical Properties of Counting Distributions for Intensity-Modulated Sources," *Journal of the Optical Society of America* **69**, 539–544.

PRUCNAL, P. R., and TEICH, M. C. (1983). "Refractory Effects in Neural Counting Processes with Exponentially Decaying Rates," *IEEE Transactions on Systems, Man and Cybernetics* **SMC-13**, 1028–1033.

RICCIARDI, L. M., and ESPOSITO, F. (1966). "On Some Distribution Functions for Non-Linear Switching Elements with Finite Dead Time," *Kybernetik (Biological Cybernetics)* **3**, 148–152.

SHOFNER, W. P., and DYE, R. H., JR. (1989). "Statistical and Receiver Operating Characteristic Analysis of Empirical Spike-Count Distributions: Quantifying the Ability of Cochlear Nucleus Units to Signal Intensity Changes," *Journal of the Acoustical Society of America* **86**, 2172–2184.

TEICH, M. C. (1981). "Role of the Doubly Stochastic Neyman Type-A and Thomas Counting Distributions in Photon Detection," *Applied Optics* **20**, 2457–2467.

TEICH, M. C. (1985). "Normalizing Transformations for Dead-Time-Modified Poisson Counting Distributions," *Biological Cybernetics* **53**, 121–124.

TEICH, M. C. (1989). "Fractal Character of the Auditory Neural Spike Train, *IEEE Transactions on Biomedical Engineering* **36**, 150–160.

TEICH, M. C., and CARD, H. C. (1979). "Photocounting Distributions for Exponentially Decaying Sources," *Optics Letters* **4**, 146–148.

TEICH, M. C., and DIAMENT, P. (1969). "Flat Counting Distribution for Triangularly Modulated Poisson Process," *Physics Letters* **30A**, 93–94.

TEICH, M. C., and DIAMENT, P. (1980). "Relative Refractoriness in Visual Information Processing," *Biological Cybernetics* **38**, 187–191.

TEICH, M. C., and KHANNA, S. M. (1985). "Pulse-Number Distribution for the Neural Spike Train in the Cat's Auditory Nerve," *Journal of the Acoustical Society of America* **77**, 1110–1128.

TEICH, M. C., and SALEH, B. E. A. (1988). "Photon Bunching and Antibunching," in E. Wolf (ed.), *Progress in Optics, Vol. 26* (pp. 1–104). Elsevier, Amsterdam.

TEICH, M. C., and TURCOTT, R. G. (1988). "Multinomial Pulse-Number Distributions for Neural Spikes in Primary Auditory Fibers: Theory," *Biological Cybernetics* **59**, 91–102.

TEICH, M. C., LOWEN, S. B., and TURCOTT, R. G. (1991). "On Possible Peripheral Origins of the Fractal Auditory Neural Spike Train (Abstract no. 154)," in D. J. Lim (ed.), *Abstracts of the Fourteenth Midwinter Research Meeting of the Association for Research in Otolaryngology*, p. 50. Association for Research in Otolaryngology, Des Moines, Iowa.

TEICH, M. C., MATIN, L., and CANTOR, B. I. (1978). "Refractoriness in the Maintained Discharge of the Cat's Retinal Ganglion Cell," *Journal of the Optical Society of America* **68**, 386–402.

TEICH, M. C., TURCOTT, R. G., and LOWEN, S. B. (1990a). "The Fractal Doubly Stochastic Poisson Point Process as a Model for the Cochlear Neural Spike Train," in P. Dallos, C. D. Geisler, J. W. Matthews, M. A. Ruggero, and C. R. Steele (eds.), *The Mechanics and Biophysics of Hearing (Lecture Notes in Biomathematics, Vol. 87)* (pp. 354–361). Springer-Verlag, New York.

TEICH, M. C., JOHNSON, D. H., KUMAR, A. R., and TURCOTT, R. G. (1990b). "Rate Fluctuations and Fractional Power-Law Noise Recorded from Cells in the Lower Auditory Pathway of the Cat," *Hearing Research* **46**, 41–52.

TURCOTT, R. G., LOWEN, S. B., TEICH, M. C., JOHNSON, D. H., and TSUCHITANI, C. (1991). Personal communication.

VOSS, R. F., and CLARKE, J. (1978). " '1/f Noise' in Music: Music from 1/f Noise," *Journal of the Acoustical Society of America* **63**, 258–263.

WALSH, B. T., MILLER, J. B., GACEK, R. R., and KIANG, N. Y-S. (1972). "Spontaneous Activity in the Eighth Cranial Nerve of the Cat," *International Journal of Neuroscience* **3**, 221–236.

WOO, T. W. (1991). *Fractals in Auditory-Nerve Spike Trains.* Unpublished Master's Dissertation. Johns Hopkins University, Baltimore.

WOO, T. W., SACHS, M. B., and TEICH, M. C. (1992). "1/f-like Spectra in Cochlear Neural Spike Trains," in *Abstracts of the Fifteenth Midwinter Research Meeting of the Association for Research in Otolaryngology.* Association for Research in Otolaryngology, Des Moines, Iowa.

YOUNG, E. D., and BARTA, P. E. (1986). "Rate Responses of Auditory Nerve Fibers to Tones in Noise near Masked Threshold," *Journal of the Acoustical Society of America* **79**, 426–442.

Index

A

Absence seizures, 286
Absolute refractoriness, 486–487, 593, 606
Acetylcholine
  afferent and association fiber strengths and, 456
  depolarization, 267–269, 287
  directional selectivity and, 328, 330
  innervation, 248
Action potentials
  analog in time, 421
  antidromic spread, 186–188
  auditory neuron, 589–590
  cardiac-like, 253
  compartmental model, 180, 186–189
  digital amplitude pulse, 416, 421
  generation in thalamic neurons, 269–270
  Hodgkin–Huxley mechanism, 89–92, 174, 190

  pulse height and width, 483–484
  single or repetitive, 236
  stochasticity, 561, 589
  threshold mechanism, 316–317, 482
A-current, 235
Adaptation, 621
Adaptive filtering, 504
Additive noise, 504–506, 515–518
Address-event data representation, 416–418, 432
Adenylyl cyclase, 281
Adiabatic approximation, 514–515, 518–519
Adiabatic elimination, 505, 508
Adiabatic regime, 507
Adrenergic agonist, 245–247
Afterdepolarization, 248
Afterdischarges, 465–466
Afterhyperpolarization
  calcium-dependent, 244–247
  cumulative, 576–577, 582

Afterhyperpolarization *continued*
  hyperexcitability after, 491–492
  long-lasting, 241–243, 248
  pharmacologic sensitivity, 245–246
  spike-activated, 249
  stochastic model, 566–567
Algorithms
  finite difference approximation, 69–72
  learning, 423, 432–433
  Marquardt–Levenberg, 607
  Newton–Raphson, 13
  polynomial, 317, 324
  synthesis from multipliers, 324–325
Alpha function, 84–85, 292, 294–295, 299,
    362
4-Aminopyridine, 252–254, 278
AMPA, 108, 291
Amphibian, retinal output, 349
Amplitude modulation, 308
Analog processing, 173–194
Analog VLSI neurons, 413–433
Anderson–Darling distance, 577–578
AND gate, 316
AND-like gate, 329, 331
AND-NOT gate, 328, 402
Anesthesia, responses during, 200–202,
    446
A-9 dopamine cells, 538, 541–550
Anomalous rectification
  neostriatal neurons, 149–161, 166
  Purkinje cells, 182
  synaptic integration and, 162–165, 168
Antibodies, 175
Apamin, 245–247
Apical nexus, 201, 221–223
Apical tufts, 207, 215–225
Approach, 552
APV, 207, 331–332, 470
Arousal, 551
Artificial neural networks
  analog–digital, 416
  architectures, 318–319
  delay-line architecture, 488, 494
  noise, 503–504
  real neurons, differences from, 81–93, 113
  scaling, 382
  silicon, 423, 432–433
Aspartate, 330

Association fiber system, 439, 442, 448,
    453–456
Associative learning, 340, 453–455
Associative memory
  Hopfield neurons, 504
  models, 453–455
  olfactory object recognition and, 452–453
Associative storage, 504
ASTAP, 32–33
Asymmetric coupling, 429–430, 432
Asymmetric inhibition, 349–350
Asymmetry index, 217–218
A-10 dopamine cells, 538, 543–550
Athetosis, 525
Attention, 551
Auditory neurons
  action potentials, 589–590
  firing patterns, 601–604
  long-time dependence, 585
  self-similarity of firing rates, 593–594
Auditory pathways, 590
Auditory signals, 589–590
Auto-association matrix memory, 441
Autophase spectra, 544, 548
Avoid, 552
Axoaxonic cell, 392
Axon
  address-event representation, 418
  communications network, 414
  models, 180–181, 189–190
  propagation, 173–194
  timing device, 191
AXONTREE, 181, 193

B

Babcock–Westervelt model, 504–506, 513,
    517, 520
Back-propagation learning technique, 325
Backward projections, 200–202, 224
Baclofen, 395
Ballismus, 525
Bandpass filter, 165, 286, 495, 497
Bandwidth compression, 604
Bandwidth processing, 496–497
BAPTA, 247
Barium, 249
Barlow–Levick model, 379, 402–403

Basin boundaries, 504, 538
Basket cell, 392–393, 395
Behavior
  characteristic times, 526
  courting, 537
  firing patterns and, 49
  involuntary, 525
  neuronal activity and, 285
Bernoulli process, 539, 597
Betz cells
  input-output relations, 235–256
  two classes, 250–251, 254, 256
Bicuculline, 397, 401
Bifurcations, 506, 518, 538
Bifurcation theory, 564
Binomial probability function, 84
Biocytin, 144, 175, 207
Biogenic amine cells, 529, 531, 538, 546, 549
Biophysical parameters, estimating, 208
Bistable systems, 519–520
Bi-threshold neurons, 485–493, 497–498
Bi-threshold phenomenon
  continuous models, 485
  discrete models, 485–487, 498
BK channels, 252
Boltzmann machine learning algorithm, 423
Boolean functions, 317–319
Brain
  electrical activity patterns, 259–260
  information processing, 291, 316, 340
  module theory, 322–324
  slice preparation, 238
Brain stem neurons
  discharge patterns, 525–554
  regulating respiration, 463
Broca's band, 440
Bursting, 536, 547, 549–550
Burst packets, 538

C

CA1, 202, 306
CABLE, 32–33
Cable equations, 69, 117–120
  finite difference discretization, 176–177
Cable models
  amacrine cell, 352–355, 360, 362–367, 371
  branching systems, 30

dendritic neurons, 176–177
dendritic signal processing, 117–123,
    125–126, 128–136
  limitations, 119–121
  physiological models compared to, 293,
    295, 308–309
  Purkinje cell, 184
  squid giant axon, 120
  two-port electrical networks and, 68–69
Calcium
  calmodulin binding, 338–339
  dendritic spine concentration, 21–24
  intracellular accumulation, 550
  role in long-term potentiation, 21, 23–24
Calcium buffer, dendritic spine concentration,
    21–24
Calcium–calmodulin-dependent protein kinase,
    339
Calcium currents
  high-threshold, 263
  high-voltage-activated, 468
  low-threshold, 262–265, 273–279, 283,
    285, 288, 464
  model, 22–24
  oscillations and, 464
  transient, 262, 269–271, 275–278, 283
Calcium spike generation, 263, 265, 268,
    270–271, 275–287
Calmodulin, 338–339
Camera lucida, 62
CaMKII, 339
CANON, 391
Canonical circuits, 37, 51
Canonical microcircuit, 379, 394–402, 405,
    409
Canonical neurons, 27–54
Cartwright–Littlewood–Levinson strange
    attractor, 533–534
Catatonic posture, 526
Cation channel, NMDA-activated, 470
Cation current
  calcium-dependent, 248
  hyperpolarization-activated, 263, 265, 269–
    271, 274, 278–282, 285, 288
  slow inward, 249–251
Caudate, spiny I neuron network, 526
C-current, 468
Cells, temporal behavior, 292

Central nervous system
  mammalian, structure/function relations, 438
  oscillatory behaviors, 463
Central pattern generator, 463
Cerebellar granule cells, number of, 38
Cerebellar Purkinje cells
  action potentials, 236
  activity patterns, 49
  delay-line architecture, 494
  dendritic branches, 29
  dendritic tree model, 174–189, 194
  membrane properties, 181–184
  morphologies, 260
Cerebral cortex, 142
  network organization, 438–439, 457
Cesium, 182–184, 249
Chandelier cell, 392–393
Chaos, 534, 584–585
  high-dimensional noise and, 520
  LCR model, 504
Chaotic band mergings, 538
Chaotic dynamics, 544
Chemical signal, 481
Chips
  communication between, 415–416, 421
  integration on, 419, 423, 428–432
  learning on, 423
  macro-cell, 54
  modification based on past history, 414
Chloride current, calcium-dependent, 248
Chloride-mediated inhibition, 121–122, 133
Chorea, 525
Circuit
  basic, 36
  canonical, 37
  cortical, 381–409
  directional selectivity, 347–372
  equivalent, 87
  hippocampal slice model, 464
  integrated, 54
  LCR, 504, 513
  local, 292, 305–306
  LRC, 531–532
  lumped, 294, 297–299, 304, 309, 567
  RC, 87–88, 297–299, 504, 562
  RLC, 93
  two-port, 68–69, 103–104

winner-take-all, 305–306
Circuit analysis programs, 32–33
Climate models, stochastic resonance, 514
CMOS neuron model, 419–420
CMOS VLSI, 413–414, 423, 432
Cochlear nucleus
  delay-line architecture, 494
  fractal firing patterns, 603–604
Cognition, 54, 285
Coincidence neuron, 332–333
Coincidence rate, 606–608, 611, 613–614
Compartmental modeling programs, 32–34
Compartmental models, 31
  axonal trees 180–181
  biological detail, 86
  dendritic neurons, 176–177
  electrotonic structure from, 12–14
  finite difference approximations, 70
  hippocampal pyramidal neurons, 61–62
  layer 5 pyramidal cell, 389, 391
  olfactory granule cells, 42–44
  olfactory receptors, 38–39
  physiological models compared to, 293, 308–310
  piriform cortex, 443–445
  Purkinje cell, 184
  pyramidal neurons, 49–53
  thalamic neuronal activity, 269–285
Complex attractors, 538
Complexity
  decreased, 525–527
  required in biological information transport, 527, 551–553
  temporal, 526–531
Computation, single neuron, 7–24
Computer-assisted reconstruction, 62
Computer-microscope systems, 62, 64
Conditional probability density, 505, 510
Conditioning, classical, 489–491, 498
Conductances, nonuniform distribution, 17–20
Conduction delay, 418, 488
Connectionist models, 193
Cooperative binding, 338–339
Correlation models, 351
Cortex
  learning in, 322–324
  mammalian, 62–63
  quasi-crystalline, 394

Cortex *continued*
three-layered and six-layered, 45
Cortical microcircuits, 381–409
Cortical neurons
electrotonic models, 67
role, 381–382
Cortical pyramidal neurons
abstraction of, 383–391
cartoon, 14
ionic channels, 17
model, 15–16
olfactory, 45–54
time constants, 11–12
types, 46–47, 395
Cortical slice preparation, 203–204
Cortico-cortical connections, 200, 224
Corticomotoneuronal neurons, 236
firing patterns, 255–256
Corticostriatal neurons, 166
Co-transmission model, 352
Courting behaviors, 537
Crank–Nicolson scheme, 71
Cross-amplitude spectrum, 547
Cross-correlation function, 490–491, 498
Current source density analysis, 451–452
Curve fitting, PSP measurements, 295–296,
299, 309
Cyclic AMP, 269

D

Data compression, 414
Dead-time effects, 593, 597, 605, 613–614
Dead-time-modified Poisson point process
(DTMP), 589–597
Decaying exponentials, 295
Decaying power-law correlations, 618
Decorrelation rate, 547
Degrees of freedom
compartmental models, 13
reduction, 8–9, 24
Déja vu, 380, 471
Delay-differential systems, 520
Delayed excitability, 487–488, 498
Delay-line architecture, 488, 494–495
Delay loops, 488
Delta functions, 295
Demultiplexing, 493–498

Dendrites
CA1, 202
calcium spikes, 468, 471
cerebellar Purkinje cell, 202
computational models, 31
contribution, 30
electrotonic models, 7–24
equivalent, 14
Hebbian computations, 81–113
hippocampal pyramidal cell, 202, 468–469,
471
length, 63–64
logic operations in, 53–54, 92–93, 97
membrane, 179–180
neocortical pyramidal cell, 202
piriform pyramidal cell, 439, 452, 456–457
spinal motoneuron, 202
subunits, 42
surface area, 145–149, 177–179
synaptic interaction on, 128–129
synaptic self–organization and, 74, 112
very distal, functions of, 199–225
voltage gradients, 93–102
Dendritic branches
tracing, 62
types, 28–29, 35
Dendritic integration, 173–194
Dendritic spines
biophysical model, 108–109
calcium concentration, 21–24
cerebellar Purkinje cell model, 177–189
counts of actual densities, 212–213
current attenuation, 166–167
dimensions, 20–21
explicit simulation, 210–211
interactions, 51–53
logic operations, 117
membrane, 179–180
mitral cell, 304
model, 14–16
numbers of, 177
olfactory granule cell, 42–44
shapes and calcium concentration, 22–24
shortening in development, 136
simulation by area insertion, 210–214
surface area, 145–149, 177–179
synaptic efficacy and, 186–187
synaptic integration, 117–136

Dendritic spines *continued*
  synaptic modification, 20–24
  synaptic strength and, 158–162
  voltage gradients, 93–102
Dendritic tree
  address-event representation, 418
  cerebellar Purkinje cells, 174–189
  compartmental model, 87
  delay line, 190
  equivalent cylinder, 67–68
  lengths and diameters, 8–11
  logical units, 193
  morphology in equivalent cylinder model, 10
  multiplicative interaction locus, 332
  neostriatal spiny neurons, 144–146
  RC properties, 87
  variety of channels, 193
Dentate gyrus, 20–22, 63–64
Depression, 538
Desynchronization, 539
Deterministic nonlinear dynamics, 564
Development, dendritic spine shortening in, 136
Dialysis, 66–67
Difference-of-exponentials functions, 295, 308, 310
Differential equations
  geometric theory, 538
  N-neuron, 505
  partial, 70, 180–181
  stochastic, 562, 581
Digital delay lines, 418
Digital processing, 173–194
Digital weight storage, 423
Dimensionality, decrease, 525
Dirac delta functions, 482–483, 509, 515
Direct fitting method, 68
Directional difference, 348
Directional index, 365–368
Directional selectivity
  asymmetry required for, 348–356, 370
  Barlow–Levick model, 379, 402–403
  canonical mechanism, 370–371
  classical explanation, 402
  development, 370
  model circuit, 347–372, 402–408
  motion detectors with, 324
  multiplication interaction, 320
  retinal, 328–329
  versus difference, 348
Direction-selective cells, squaring operation, 334–336
Discrete functions, 483
Discretization, 70, 483
Disinhibition, 128
DNQX, 204
Dopamine, 538
Doubly stochastic Poisson point process (DSPP), 593
Duffing equation, 553
Dwell times, 534
Dyes, voltage- and ion-dependent, 175
Dynamical disorders, 525
Dynamic analog weight storage, 423

E

EEG-synchronized sleep, 285
EEPROM process, 422
Electrical distance, 189–190
Electrical signal, 481
Electrodiffusion
  model, 117–118, 121–134, 136
  synaptic integration by, 117–136
Electroencephalogram (EEG), 259
  α and θ rhythms, 492
  fluctuations, 527–529
  oscillations, 464
  patterns, 531
  piriform cortex, 446, 449–450, 464
Electronic structure, spatial representation, 102–106
Electrotonic distance, 102–103, 105
Electrotonic length
  spiny neuron, 148–150, 156–157, 166–167
  time constant and, 156–157
Electrotonic models, 7–24
Electrotonic structure
  compartment models, 12–14
  complex neurons, 103–104
  equivalent cylinder models, 9–12
  estimation, 8–16
  hippocampal neurons, 64–67
Embedded signals, extracting, 497–498
EMG activity, 236
Entorhinal cortex, 439–440
Entorhinal cortical neurons, oscillation, 464

Entropy
  bounds, 551
  effective, 539
  elementary processes, 531
  intermittent neuronal activities, 536–540
  maximal, 539
  measure-theoretic, 527, 529
  metric, 541, 547, 553–554
  per unit time, 551
  reduction, 525
  run, 541
  Shannon, 530
  time scaling and, 539
  topological, 529, 540, 553
  zero, 527
Epilepsy, brain wave patterns, 527
Epileptiform bursts, 253
Epileptogenesis, 492, 498
Equivalent cylinder models, 30–31, 67–68,
    178–179
  limitations, 9–12
Equivalent electrical circuits, 87
Erdos–Renyi theorem, 541
Escape times, 518–519
Ethosuximide, 286
Euclidian dimension, 599
Euler approximations, 71–72
Event-pair coincidences, 606
Examples, 322–324
Expansion rate, 547
Exponential nonlinearities, 337
Extreme value theory, 581, 584

F

Facilitation
  mechanisms, 350, 358
  silent disinhibition and, 329–330
Fano factors, 599–601, 611, 613, 619–622
Fano-factor time curve (FFC), 590, 597–599,
    604–607, 614, 621
Faraday's law, 532
Feedback inhibition, 426
Feedback loops, 526
Feedforward inhibition, 426
F factor, 125–130
F-H model, 322
Finite difference approximation algorithms,
    69–72

Finite difference discretization, 176–177
Finite impulse response filter, 567
Firefly sending intervals, 536–537
Firing
  rates, product of, 332–333
  repetitive, 235–237
Firing-rate function, 300–301
First-passage times, 563, 566–584
  approximations, 569–582
Fitzhugh–Nagumo equations, 533
Fly
  directionally selective retinal output, 349
  directional selectivity reversal, 360
  motion detection, 351
  visual system, 330, 337
F model, 322
Fokker–Planck equation, 505–506, 510–511,
    513, 517–518
Forward computation, 12
Fourier series, 325
Fourier transformation, fast, 547
Fractal doubly stochastic Poisson point process
    (FDSPP), 593, 605, 613–618
Fractal-Gaussian-noise-driven DSPP, 613
Fractals
  basin boundaries, 504, 538
  counting distribution, 619
  cutoff frequency, 611
  dimension, 527, 599–601, 612
  firing patterns, 589–622
  ion channel dynamics, 546
  shot noise, 593
  statistics, 605
Fractal-shot-noise-driven DSPP, 613
Free energy minimum, 517
Frequency doubling, 320, 329, 339
Frequency locking, 527
  pathophysiology and, 548
Frequency-to-spatial transformation, 301–304,
    310–311
Frobenius–Perron theorem, 540
Frozen posturing, 525

G

GABA, 390, 392–393
$GABA_A$
  directional selectivity and, 328, 330
  receptors, 134, 326

GABAergic pathways, 350, 358
GABA$_B$ receptors, 129, 134
Gain control, 337
Gamma distribution, 577–578
Gamma polynomials, 325
Ganglion cells
   directional selectivity, 328–330, 339, 350,
      352, 358–359
   DS recordings, 367–370
Gaussian
   density function, 84
   distribution, 510
   inverse, 576, 578, 583
   noise, 620–621
Generating partition, 541
GENESIS, 32–33, 310, 445
Geniculo-cortical input, 442
Geodesic flows, 551
Gigaseal patch electrode technique, 213
Globus pallidus, 143
Glutamate, 330, 464
   quisqualate-type effect, 467
γ-D-Glutamylglycine, 466
Golgi staining, 28, 35–36, 62–64, 144
Goodness-of-fit tests, 584
G proteins, 130
Gradient descent, 324
Granule cells, morphometry, 20–22, 63–64

H

Hair-cell receptor, 589, 605
Hard potential model, 519
Hebb–anti-Hebb correlation, 370
Hebbian computations, 81–113
Hebbian learning rule
   delayed, 489
   olfactory network, 453
   silicon synapse implementing, 423, 425
   synaptic modification, 110
Hebbian long-term potentiation, 82, 106–108,
      331, 425
Hebbian synapse
   biophysical model, 108–109
   delayed modification, 488–489, 498
Hebb-modifiable inhibitory connections, 426
Heuristic approximation methods, 561–585
Hidden layer, 317–319, 325

Hierarchical clustering, 292
Hierarchical organization, 39–40
High-voltage electron microscopy, 145–148
Hippocampal dendrites
   Hebbian computations, 81–113
   spines, 51, 117
Hippocampal neurons
   bi-threshold phenomenon, 485
   current attenuation, 167
   epileptic kindling, 492
Hippocampal pyramidal neurons
   action potentials, 236
   compartmental models, 49–50, 87–88
   computational models, 61–76
   conductance of synapses, 126–128
   dendritic spines, 51
   electrotonic compactness, 96, 103
   electrotonic structure, 64–67
   morphometry, 63–64
   potassium channel, 358
   slow potassium currents, 245–247
   synapses on spines, 117
   voltage-dependent conductances, 92–94
Hippocampal slice, 464
Hippocampus
   memory storage, 439
   synchronized multiple bursts, 463–471
Histamine, 267–269, 281
Hodgkin–Huxley equations, 90–92, 117,
      188–189, 300
   reduced, 532–533, 535, 537–538
Hodgkin–Huxley membrane kinetics,
      186–188, 193
Hodgkin–Huxley model, 30, 485
Hopf bifurcation theorem, 531
Hopfield model, 504–505, 507, 513, 517
Horizontal layer I (HLI) preparation, 203–207
Hormones, bursting release, 540
Horseradish peroxidase, 63–64, 175,
      181–182, 383
Hyperpolarization, embedded signal in, 498
Hyperpolarizing inhibition, 121, 123, 134
Hysteresis effects, 506

I

Ice ages, periodicity, 506
Implicit method, 72

Inertial system, 516–517
Inferior colliculus, delay-line architecture, 494
Inferior olivary neurons
    bi-threshold phenomenon, 485
    oscillation, 464
Infinitesimal drift, 562–563
Infinitesimal variance, 563
Information processing
    fractal events, 604–605
    nervous system, 173
    population rhythms and, 463
    temporal, 291–311
Information transfer
    complexity required, 527, 551–553
    noise enhanced, 506–507, 515–516,
        518–519
Inhibition
    cortical, 391–394, 408–409
    neocortical, 391–392
    on-path, 127, 356–357
    posthyperpolarization, 254, 256
    potassium-mediated, 121–122, 129–130,
        134
    symmetric, 350
    see also Shunting inhibition, Silent
        inhibition
Insects
    directionally selective retinal output, 349
    optomotor response, 320–321, 339
Integrate-and-fire method, 300
Integrate-and-fire models, 332–333, 584
Integrated circuits, canonical neuron and, 54
Intensity discrimination, 604
Interchip communication, 415–416, 421
Interface, 414–415
Interspike-interval histograms (PIDs), 589,
    614
Interspike intervals, 484, 493–495, 497–498
    higher-order, 495–496
    intermittent bursting, 538, 539
    locus coeruleus norepinephrine cells, 540–
        549
    partitioning into shorts and longs, 539
    probability density functions, 507, 519
    substantia nigra dopamine cells, 540–549
    time series, 540–549
    ventral tegmental dopamine cells, 540–549
Inverse computation, 12–14

Ion channels
    distribution, 17
    fractal statistics, 605
    kinetic analyses, 546
    self-organized criticality, 605
    stochastic behavior, 563
    variable membrane conductances and, 92
    voltage-dependent, antibodies to, 175
Ionic currents, 235–256
    modeling, 271–273
    see also Calcium currents, Cation currents,
        Chloride currents, Potassium currents,
        Sodium currents
Ionic signaling, 481
Isocortex, 394
Isopotentiality, 97–102, 113, 193
Iterative techniques, 324

J

Jitter, 333

K

Kainate, 330
Katz–Miledi model, 174
Kinetics
    Hodgkin–Huxley, 186–188, 193
    intracellular calcium regulation, 464
Kirchhoff's law, 532
Kramers formula, 514
Kramers rate, 514, 516

L

Labeled-line approach, 418
Labeled-time-slot approach, 418
Labeling, intracellular, 63
Langevin noise, 504–505, 509, 511, 512–
    513
Lateral olfactory tract, simulation, 441–448,
    451
Lateral superior olivary complex, 590
    fractal firing patterns, 603–604
Layer I of neocortex, 199–225
    synaptic inputs, 221–224
    significance, 199–202

Layer V cells, 202
  computational model, 207–215
LCR circuit, 504, 513
Leakage conductance, 420–421, 566–567
Leaky integrate-and-fire neuron, 332–333
Learning
  associative, 340, 453–455
  back-propagation technique, 325
  dendritic spine shortening in, 136
  hippocampus and, 61
  noise and, 503–504
  on-chip, 423
  silicon neuron, 428–432
  theories, 322–324, 340
  see also Hebbian learning rule
Learning algorithms
  Boltzmann machine, 423
  silicon neural network, 423, 432–433
Learning rate constant, 488
Level-crossing problem, 563
Ligand-gated receptor channels, 481
Limit cycles, 532
Local circuits, 292, 305–306
Locus coeruleus norepinephrine cells, 538,
  550
Log–exp transform, 337
Logic operations, dendritic, 53–54, 92–93,
  97, 117
Lognormal distribution, 577–578
Long-term potentiation
  AMPA channels and, 291
  biochemical cascade underlying, 339
  biophysical mechanism, 108–109
  calcium and, 21
  differential expression, 311
  Hebbian form, 82, 106–110, 331, 425
  induction rule, 306–307
  NMDA channels and, 224, 291
  silicon Hebbian synapse and, 425
  synaptic gradients and, 62
  synaptic plasticity, 106
  synchronized multiple bursts and, 465
  theta-burst stimulus, 292
Loudness estimation, 604
LRC circuit, 531–532
Lumped circuits, 294, 297–299, 304, 309,
  567
Lyapunov exponents, 530, 544–547, 553, 585

M

Magnesium block, 330
Mania, 538
MANUEL, 32
Mapping
  memories, 340
  nonlinear, 324
  topographical, 497
  visual cortex, 394
Markov matrices, 541
Markov random process, 92
Marquardt–Levenberg algorithm, 607
McCulloch–Pitts model, 29, 40, 52–54, 316
Meaningful partition, 540, 552
Membrane
  area insertion, 209–214
  conductances, 89–92
  dendrite, 179–180
  excitability, 180
  kinetics, 186–188, 193
  nonlinear dynamics, 88–93
  patches, potassium channels in, 241–243
  postsynaptic, 420, 422
  potential, 151–153, 297, 326–327, 531,
    561–562
  Purkinje cell, 181–184
  resting potential, 484
  time constants, 87, 151–153, 180
  see also Resistance, membrane; Resistivity,
    membrane
Memory
  associative, 452–455, 504
  hippocampus and, 61, 76
  long-term, 590
  mapping, 340
  storage in piriform cortex model, 439, 455–
    457
Midbrain dopamine neurons, bursting activity,
  538
Mimicry, 537
Mitral cells, 39–41, 45
Mitral/tufted cells, 301, 304, 310
Mixing hypothesis, 546–547
Models
  faithful, 176
  reduced, 14
  requirements of, 7–8

Mode resonances, 527
Molecular families, 36
Molluscan nervous system, 538
Morphoelectrotonic transform, 104–106
Morphometric techniques, 174–175
  history, 62–64
  see also Golgi staining
MOSIS, 420, 432
Motion
  detection, 348
  direction, 348
  perception, 315, 320–322, 339
Motoneurons
  compartmental models, 31
  dendritic branches, 29
  firing patterns, 49
  morphology representations, 14
  time constants, 11–12
  vagal, 209, 217–218
Multimodal probability density function,
    505–506
Multiplexing, 416, 493–498
Multiplication
  mechanisms, 325–339
  squaring operation implementation,
    333–337, 339
  synaptic and neuronal, 315–340
Multiplicative neurons, 315, 340
Multiplicative noise, 504–506, 513, 516, 518
Multistability, 505
Multistate neuronal activity, 285–288
Multi-threshold neurons, signal processing,
    481–498
Muscarine, 246, 248
Muscarinic agonists, 245–247
Myoclonus, 525

N

Nearest neighbor coupling, 526
Neocortex
  behavioral states and neuronal activity, 285
  inhibition, 391–392
  input to layers I and II, 200–202
  motor systems, 526
  number of layers, 45
  relative uniformity, 394
Neocortical neurons, 46–48

action potentials, 236
general classification, 48
Neocortical pyramidal cells
  axial resistivity, 208, 221
  dendritic spines, 53–54
  layer I synapses, 199–225
  steady-state and transient responses,
    215–221
Neostriatal neurons, functional properties,
    141–168
Neostriatum
  function, 165–168
  pathways, 162
  transformation of input, 142
Nernst–Planck equations, 118, 120, 122, 131,
    133
Nernst potential, 121, 125, 128–129, 132,
    135–136
Neural modeling programs, 32–34
Neural networks
  analog-digital hybrid, 416
  architectures, 318–319
  dynamics simulation, 445–450
  standard, nonlinear synaptic interaction and,
    327–328
  temporal processing, 292
Neurological disorders, 525
NEURON, 32–33, 72, 74, 88, 207–208
Neuron doctrine, 28, 40
Neurons
  bipolar, 29
  chaos, 520
  computational representation, development
    of, 30–34
  dynamic polarization, 28
  dynamic properties of single and networks,
    437–458
  dynamic range of computations, 16–20, 24
  electrical properties, 260
  firing threshold, 484–487
  first intracellular recording, 30
  history as biological entities, 3, 28–30
  input–output properties, 190, 481–483, 493
  network behavior, 260
  noise-free, 503
  noisy, 506, 517
  reduced representations, 29
  resting state, 17–19

Neurons *continued*
  standard model, 316
  summing nodes, 29–30
  time delay, 520
NEUROS, 32
Neurotransmitters
  biosynthesis process times, 527–528
  quantal release, 83–85
Newton–Raphson algorithm, 13
Neyman Type-A distribution, 619
Nigrostriatal system, 525
NMDA channels
  long-term potentiation and, 224
  time course, 291
NMDA conductance, 467–468, 470
NMDA currents, 330–331
NMDA receptors, 330–332
  LTP induction and, 108–109
N-neuron problem, 505, 511, 517
Noise
  additive, 504–506, 515–518
  Gaussian, 620–621
  high-dimensional, 520
  information transfer and, 506–507,
      515–516, 518–519
  learning and, 503–504
  multiplicative, 504–506, 513, 516, 518
  neural, 503–504, 506, 517, 563–566
  shot, 570, 572, 593
Noisy data sets, 503–504
Noisy input, 561–563, 565
Nonlinear optical systems, 506
Norepinephrine, 267–269, 281, 287, 538
Nucleus reticularis, 264
Nudibranch molluscs, nonbursting neurons,
    235
Numerical methods, 177, 180
Nystatin patch technique, 66

O

Olfactory bulb, 440
  frequency-to-spatial transformation,
      301–304, 310–311
  glomerular structure, 292
  information processed, 439–440
  model, 441
Olfactory cortex

connections, 292
  dynamic properties, 437–458
  local-circuit architecture, 305
  modeling, 440–452
  network models, 306
  simplicity, 45
Olfactory granule cell, 41–44
Olfactory nerve, 301, 304, 311
Olfactory neurons
  broad tuning, 453
  compartmental models, 31
  dendritic spines, 51
Olfactory object recognition, 452–453
Olfactory receptors, 37–39
  compartmental models, 38–39
  number of, 37–38
Olfactory stimulus space, 453
Olfactory system
  canonical neurons, 27, 37–44
  interactions in, 292, 310
  lack of processing stages, 440
  temporal and spatial cross-correlations, 526–
      527
On-chip learning, 423
$1/f$-type behavior, 590, 604, 611
On-path inhibition, 127, 356–357
Optical flow, 322
Ornstein–Uhlenbeck process, 566, 578–581,
    584
Oscillations
  hippocampal neuronal population, 463–471
  intrinsic, 467
  LCR model, 504
  membrane potential, 531
  models, 564
  periodic bi-stable, 492
  piriform cortex, 445–452, 456–457

P

Pacemaking neurons, 492
Palais–Takens condition, 534
Parkinsonian tremor, 286, 492
Parkinson-mimetic drugs, 553
Partial differential equations
  fast integration methods, 181
  membrane excitability, 180
  solution methods, 70

Particle in a well, 504, 581
Passive models, 175–177
Patch clamp, whole-cell, 11, 66–67, 205
Patch pipette, 66
Pathophysiology
  frequency locking and, 548
  loss of complexity in, 525–527
Pattern association, 109–112
Pattern recognition, Hopfield neurons, 504
Pearson plot, 578–579
PEDRO, 32
Peeling exponentials, 11–12, 68, 157, 183
Penicillin, 466
Perceptron, 318–319
Perforated patch technique, 66–67
Periodic coherence, 527
Peripheral sensory information, 439
Phase bunching, 548–549
Phase coherence, 547
Phase diffusion, 548
Phase invariance, 320, 329, 339
Phase spectrum, 547–548
Phenothiazines, 553
Photodetectors, retina-like array, 415
Photoreceptors, 322, 324
Physiological models, 293–311
Picrotoxin, 465, 467–468
Pigeon vestibular data, 577–579
Piriform cortex
  architectural features, 292
  dynamic properties, 437–458
  EEG oscillations, 464
  periodic activity patterns, 445–452
  pyramidal cells, 442–445
Pituitary cell perifusion, time series, 540, 543
Point neuron, 494–495
Point processes
  mathematical, 605–622
  Poisson, 599
Poisson probability function, 84
Poisson process, 562, 565, 568–569, 572, 578–579
Polynomial algorithms, 317, 324
Polynomial equations, 300
Polynomial networks, 315, 319
Polynomial neurons, 315–316
Polynomial units, 327–328

Popcorn, microwave popped, inter-discharge intervals, 540, 543–551, 554
Posthyperpolarization excitation, 251, 254, 256
Posthyperpolarization inhibition, 254, 256
Post-inhibitory rebound, 491–492, 498
Post-stimulus-time histograms, 589
Postsynaptic membrane, leaky integrator, 420, 422
Postsynaptic potential functions, 295–299, 307–308
Postsynaptic potentials
  nonlinearity, 191
  physiological models, 293–301
  signal processing, 482
Potassium channels
  time- and voltage-dependent, 358
  voltage-insensitive, 550
  voltage-sensitive, 549
Potassium current
  calcium-activated, 263
  calcium-dependent, 236, 241, 243–248
  delayed rectifier, 269–271, 278
  $GABA_B$-elicited, 466
  leak, 263, 269–271, 279, 284–285
  rapidly and slowly inactivating, 264
  slow, 238–239, 241
  sodium-dependent, 241–243
  transient, 235
  voltage-activated, 269–271
  voltage-gated, 251–254
Potassium-mediated inhibition, 121–122, 129–130, 134
Potential
  membrane, 151–153, 297, 326–327, 484, 531, 561–562
  postsynaptic, 191, 293–301, 482
  reversal, 118, 566
  steady-state, symmetry, 215–217
  thermodynamic, 516, 518
  see also Action potential
Potential well
  bi-0 and multi-stable, 581
  double, 504, 517
  one-dimensional, 514
  residence time, 507
Power spectral density, random process, 608–611
Preferred/null waveforms, 348, 359–360

Premature horseshoes, 534
Primary vestibular neurons, 590
Processing elements, 81–83, 86, 113
Pseudoinverse techniques, 324
Psychotropic drugs, 538
Pulse-number distribution, 594, 601–603, 614
Purkinje cell
    anomalous rectification, 182
    dendrites, 174–189, 202
    dendritic spines, 177–189
    models, 184
    synchronized multiple bursts in, 465, 468
    see also Cerebellar Purkinje cells
Pyramidal neurons
    canonical form, 47, 49–53
    compartmental models, 49–53, 61–62, 389,
        391
    computational models, 49–54
    firing patterns, 47–49
    layer V, 235–236
    olfactory, 51
    piriform cortex, 442–445
    rectification, 205
    repetitive firing, 235
    time constants, 208–209, 456
    see also Cortical pyramidal neurons,
        Hippocampal pyramidal neurons,
        Neocortical pyramidal cells

Q

Quantal content, 84
Quantal size, 83
Quantum, 83–84
Quisqualate, 330, 467, 470–471

R

Radial basis functions, 325
Rall model, 30–32
Random fluctuations, 83–86
Raster-scanning mechanisms, 416
RC circuit
    CA1 pyramidal cell, 87–88
    Hopfield neuron model, 504
    PSP transient from, 297–299
    voltage trajectory, 562
Rectification

anomalous, 149–166, 168, 182
    fast anomalous, 162–165
    fast inward, 153
    pyramidal neuron, 205
Refractory periods, 300–301, 487
    see also Absolute refractoriness, Relative
        refractoriness
Refractory/reset mechanism, 519
Regenerative depolarization, 240
Reichardt model, elaborated, 322, 339
Reinforcement signal, 488–490
Relative refractoriness, 300–301, 487, 593,
    607–608
Reptile, directionally selective retinal output,
    349
Rescaled range, 614–616
Residence times theory, 518–519
Resistance
    dendritic spines, 42
    diffusional, 21
    input, 148–156
    spine neck, 20–21
Resistance, membrane
    dendritic spines and, 149–150
    effective, 180
    shunting inhibition and, 119
    spiny neuron, 153, 167
    true value, 96–97
Resistivity
    assumptions about, 14, 213
    axial, 208, 221
    membrane, 9–12, 14
Resonances, multiple unstable, 543
Respiration, neurons regulating, 463
Resting membrane potential, 484
Retina
    adaptive, 414
    biological and silicon, 414
    directional selectivity, 328–329, 347–
        372
    ganglion cells, firing patterns, 601
    rods, number of, 38
    silicon, 414, 416, 423, 432
    structure, 348
    transmission from, 267
    turtle, DS recordings, 367–370
Return times, 534
Reversal potential, 118, 566
rf SQUID, 506

Rhythmic oscillation, 261, 273, 286
  ionic basis, 264–265
Rhythms, generation, 492, 498
Rigidity
  Parkinsonian, 553
  topological constraints as, 552
Ring laser cavity, 506
RLC circuit, 93
Runs of shorts, 538–539, 541–543

S

SAAM, 32
SABER, 32–33, 310
Saddle node, 538
Saturation, 360
Schaffer collateral/commissural synapses,
    106–108
Schmitt trigger, 519
Second messengers, 65, 481
Seizure fragment, 465, 471
Self-connection, 505, 508, 512
Self-coupling term, 505, 513
Self-diffusion, one-neuron, 512–513
Self-organization, 74, 109–113
Sensitivity analysis, 584
Sensorimotor cortex, layer V pyramidal
    neurons, 235–236
Sensorimotor feedback systems, 414
Sensorimotor processing, real-time, 413
Sensory processing, fast high-density, 413
Serial count correlation coefficient, 614–616
Serotonin, 269, 281, 538
Shannon's second theorem, 527
Shape modulation, 308
Sharp electrodes
  resistances measured by, 67, 96–97, 213
  tip diameter, 65
Shot noise, 570, 572, 593
Shrinkage, tissue, 8–9
  compensation for, 209–212, 215, 219–223
Shunting inhibition, 326
  AND–NOT-like operation, 117
  anomalous rectifier as, 157
  directional selectivity and, 329–330, 408
  effective, 119, 121, 125, 133
  functions, 134
  membrane resistance and, 119
  modeling, 123–124

multiplication-like operation, 339
  reversal potential for, 118
Sick-time effects, 593, 597, 605
Sick-time-modified FDSPP, 605
Sigma–pi networks, 322
Sigma–pi units, 112, 316–317
Sigma–Plot^{TM}, 607
Sigmoidal network, 325
Sigmoid function, 301, 319–320, 513
Signal processing, multi-threshold neurons,
    481–498
Signal switching, 492–493, 498
Signal-to-noise ratio, 506, 515–519
Silent disinhibition, 329–330
Silent inhibition
  directional selectivity and, 329
  membrane potential and, 297
  modeling, 294
  multiplication-like operation, 339
  nonlinear synaptic interaction via, 325–329
Silicon
  canonical neuron and, 54
  high-bandwidth communication, 416
Silicon neurons
  Hebbian synapses, 423, 425, 428–429, 432
  integration on a chip, 419, 428–432
  models, 419–422, 432–433
  sequence-learning, 428–432
Silicon retina, 414, 416, 423, 432
Sinai–Ruelle–Bowen measure, 539
Single-cell models, 293
Single-compartment models, 68
Single-microelectrode clamp, 65–66, 238, 240
SK channels, 245
Slaving principle, 517
Sleep stages, 529
Slice technique, 175
Slow-wave sleep, 259
Smooth neurons
  types, 392
  visual cortical, 394, 399–401, 407–408
Sniffing rate, cortical EEGs correlated with,
    450, 456
Sodium channels, voltage-sensitive, 549
Sodium current
  fast transient, 263
  persistent, 239–241, 263
  transient, 269–271
Sodium–potassium pump, 135, 241

Soma
  compartmental model, 87
  shunt, 11
Soma-centric hypothesis, 3, 30
Somatic leak conductance, 212–215, 219–223
Somatic recording, 175
Somatosensory cortex, axon model, 174,
      189–190, 192–193
Space constant λ, 102–103
Spatio-temporal energy model, 322, 336
Spatio-temporal parameter, 494–495
SPICE, 32–33, 185, 306–307, 310
Spike codes, 483–484, 493–494
Spike frequency adaptation, 237
Spike train
  bandpass filtering, 495
  Dirac delta function representation, 482–483
  pulse duration, 483
Spike-number variance-to-mean ratio,
      594–598, 607–608
Spikes
  baseline, 485
  low-threshold, 485
Spinal motoneurons, 236, 250
Spin-coupling, 525
Spindle waves, 264, 286, 464
Spin-glass analogies, 508
Spiny neurons
  bandpass filter, 165
  firing pattern, 143
  model, 147–149
  neostriatal, 142
  relative silence, 143, 165–166
  synaptic inputs, 143–146
  synaptic integration, 157–158
  visual cortical, 407–408
Spiny I neuron network, 526
Spiny stellate cells, 47–48
Squaring operation, multiplication
      implemented by, 333–337
Squid giant axon
  axial resistivity, 208
  bi-threshold phenomenon, 485
  cable model, 120
  intracellular recordings, 260
  wrong way currents, 532
Starburst amacrine cells
  asymmetric input, 328, 352

cable model, 352–355, 360, 362–367, 371
cholinergic release, 355
dendritic subunits, 42
rabbit, 361
simulations, 360–367
Starburst dendrites, 352–360, 371
Starfish egg, fast inward rectification, 153
States, regulation of, 538
Statistical dynamics, 546
Stein's model, 566, 574–575
Step functions, 295
Stochastic differential equation, 562, 581
Stochastic models of neurons, 503–520,
      561–585
Stochastic resonance, 506–507, 513–518, 520
Stone–Weierstrass theorem, 320
Strange attractors, 520, 533–534
Stratonovich equation, 511
Stratonovich Fokker–Planck equation, 509
Stratum lacunosum-moleculare, 64
Stratum oriens, 64
Striate cortex, fractal firing patterns, 604
Substantia nigra, 143
Sum and fire operation, 193
Summation nodes, 29–30, 81–83, 86, 223
Superposition theorem, 297, 299–300
Switch states, 563
Symmetric inhibition, 350
Synapses
  clusters, 111–112
  cooperativity among, 167–168
  fundamental units, 40–42
  linear interaction, 325
  modification, 8, 20–24
  nonlinear interaction, 325–329
  nonlinearity, 158–159, 167
  plasticity, 62, 101–102, 106–109
  self-organization, 74, 109–113
  silicon model, 422–428
  strength, 158–162
Synaptic doctrine, 41
Synaptic efficacy, 217, 219–220, 224
  dendritic spines and, 186–187
Synaptic integration, 117–136
  fast anomalous rectification and,
      162–165
  spiny neurons, 157–158
Synaptic potentiation, 306–307

Synaptic transfer function, 337, 339
Synaptic transmission, Katz–Miledi model, 174
Synaptic weight, 20, 422–428, 433
  delayed change, 488–491
  stabilization, 431–432
  storage in silicon, 422–423
Synchronized firing, 491–492, 498
Synchronized multiple bursts, 463–471
  biological significance, 471
  mechanisms, 466–471
  number of cells needed for, 466

T

Tangential interneurons, 329–330
Taylor series expansion, 326–327
T-current, 262–263
Temporal complexity, 526–531
Temporal integration, 301–307
Tetraethylammonium, 247, 251–253
Tetrodotoxin, 248, 251–252, 268, 278
  sodium blocker, 239–241
Thalamic neurons
  action potentials, 236
  bi-threshold phenomenon, 485
  signal switching, 493
  tuning, 381
Thalamic relay neurons
  computational model, 269–285
  electrophysiological properties, 261–266
  morphologies, 260
  neuromodulation, 267–269
  two-state firing, 279–285
Thalamocortical neurons
  oscillation, 464
  phasing of spontaneous firing, 492
Thalamostriatal neurons, firing, 166
Thalamus
  behavioral states and neuronal activity, 285
  information from, 439
  projections to layer I, 200
  state-dependent processing, 259–288
Thermal bath field influence, 526
Thermal field neurons, 531
Thermal randomization, 525–526
Thermodynamic potential, 516, 518

Theta-burst stimulus, 292
Theta rhythm, olfactory-hippocampal, 305
3/2 power law, 68, 123
Threshold function
  boolean linear, 317
  boolean polynomial, 319
Threshold mechanism, 316
Time
  characteristic, 525–554
  global characteristic, 553–554
  rescaling, 540
  scaling, 539
Time constants
  cell membrane, 87
  compartmental models, 12–13
  dendrite membrane, 180
  dendritic spine membrane, 180
  equivalent cylinder model, 10–12
  length constant and, 156–157
  membrane potential and, 151–153
  neocortical pyramidal neuron, 208–209
  olfactory pyramidal cells, 456
  peeled, 11–12, 183
Time series, autocovariance approach, 547–548
Topographical mapping, 497
Topological conjugacy, 529
Topological dimension, 599
Topological dynamics, 538
Topological functional units, 39
Tracing technique, 62
Transconductance amplifier, 419–420
Transduction, 235
Transients, time constants peeled from, 11–12
Trigger zone, 561–562
Triply stochastic Poisson point process (TSPP), 618, 620
Tufted cells, 39–41, 45
Tuning, 450, 539
Two-port electrical circuits, 68–69, 103–104

U

Ultraviolet programmable floating gate, 423–424

V

van der Pol equation, 533, 553
Vestibular nerve fibers, 566
Vestibular neurons, 601–604
Visual cortex
  mapping connections, 394
  neuron types, 395
  NMDA currents, 331
Visual system
  fractal firing patterns, 604
  temporal and spatial cross-correlations,
    526–527
Voltage attenuation, 104–105
Voltage clamp

isopotentiality and, 97
single-microelectrode clamp, 65–66
somatic, 97–101
two electrodes, 66
Voltage gradients, dendrites and spines,
  93–102
Volterra series, 322

W

Weierstrass approximation theorem, 317, 319
Whole cell recordings, 96–97
Wiener–Khinchin theorem, 515, 611
Wiener process, 562, 565, 576
Winner-take-all circuits, 305–306